I0063433

Wound Repair and Regeneration

Special Issue Editor
Allison Cowin

MDPI • Basel • Beijing • Wuhan • Barcelona • Belgrade

MDPI

Special Issue Editor
Allison Cowin
University of South Australia
Australia

Editorial Office
MDPI AG
St. Alban-Anlage 66
Basel, Switzerland

This edition is a reprint of the Special Issue published online in the open access journal *International Journal of Molecular Sciences* (ISSN 1422-0067) from 2016–2017 (available at: http://www.mdpi.com/journal/ijms/special_issues/wound_repair).

For citation purposes, cite each article independently as indicated on the article page online and as indicated below:

Lastname, F.M.; Lastname, F.M. Article title. *Journal Name*. **Year**. *Article number, page range.*

First Edition 2018

ISBN 978-3-03842-767-4 (Pbk)
ISBN 978-3-03842-768-1 (PDF)

Articles in this volume are Open Access and distributed under the Creative Commons Attribution license (CC BY), which allows users to download, copy and build upon published articles even for commercial purposes, as long as the author and publisher are properly credited, which ensures maximum dissemination and a wider impact of our publications. The book taken as a whole is © 2018 MDPI, Basel, Switzerland, distributed under the terms and conditions of the Creative Commons license CC BY-NC-ND (http://creativecommons.org/licenses/by-nc-nd/4.0/).

Table of Contents

About the Special Issue Editor

Allison Cowin is Professor of Regenerative Medicine and NHMRC Senior Research Fellow at the University of South Australia who is internationally recognised as a leader in the field of wound healing and scar formation. Her research on the cytoskeletal protein Flightless I, has resulted in over 100 research articles, editorials and book chapters in leading international journals. She was awarded fellowship of Wounds Australia in 2014, was an SA finalist in the Telstra Women's Business awards in 2015 and her research "Novel drug delivery of therapeutic antibodies to wounds" made the finals of the "Australian Innovation Challenge" (2015). She was the founding President of the Australasian Wound & Tissue Repair Society, is the Editor of Wound Practice and Research, Associate Editor for US Journal Wound Repair and Regeneration, and is a Board member of the Australian Society for Dermatology Research.

Preface to "Wound Repair and Regeneration"

Wounds have been around since time immemorial. Ancient civilisations first described their approaches to wound management including herbal remedies, lint, animal grease and honey in 1500 BC. Since the 19th century, scientists worldwide have been endeavouring to understand the processes involved in wound healing and there have been many important discoveries that have helped to pave the way to the wound management that we see today. In the last 20 years in particular, our understanding of the molecular and cellular processes that underpin the complexity of wound healing has grown exponentially, yet, despite all our best endeavours, there are still limited wound care products on the market which can actively stimulate the healing process, and none that lead to the perfect regeneration of the skin after injury. Unfortunately, we cannot just sit back and wait another 20 years for new advances to occur. Due to our aging populations, the increase in chronic diseases such as diabetes, and the prevalence of obesity in our society, we are seeing a spiralling epidemic of wounds affecting the lives and well-being of millions of people worldwide.

The time is therefore right for this book to showcase the "state of the art" research that is being undertaken by leading clinicians and academics from all around the world. Our participant authors are at the forefront of their studies and are furthering the understanding of molecular processes that are involved in wound repair. Using the latest knowledge, they are developing new approaches that are leading to improvements in healing in preclinical animal models with the aim that future studies will be performed in people that will lead to exciting and innovate wound management developments in the future.

This book brings together a selection of articles that highlight new developments in therapeutic approaches for wound repair, including the use of biomaterials to deliver cells and/or drugs to promote healing. Cellular responses that underpin angiogenesis, inflammation, proliferation and remodeling, as well as advances in cytoskeletal interactions in keratinocytes and fibroblast cell functions, are included. Wound remodelling and scar formation including the roles of growth factors, cytokines and stem cells are also overviewed. I would very much like to thank all the authors who have generously contributed their articles to this book. By disseminating and sharing our research, we ensure that advances in the field can be made. It is by connecting the dots, engaging in that lightbulb moment or by having that conversation that the next innovation in wound repair will arise. If this book can help get this process started, then we will have achieved something worthwhile.

Allison Cowin
Professor of Regenerative Medicine
Future Industries Institute, Mawson Lakes, University of South Australia,
South Australia 5095, Australia.
Special Issue Editor

International Journal of
Molecular Sciences

MDPI

Article

DNA Damage-Inducible Transcript 4 Is an Innate Surveillant of Hair Follicular Stress in Vitamin D Receptor Knockout Mice and a Regulator of Wound Re-Epithelialization

Hengguang Zhao [1,*], Sandra Rieger [2], Koichiro Abe [3], Martin Hewison [4] and Thomas S. Lisse [2,5,*]

[1] Department of Dermatology, The First Affiliated Hospital of Chongqing Medical University, Chongqing 400016, China

[2] Kathryn W. Davis Center for Regenerative Biology and Medicine, Mount Desert Island Biological Laboratory, 159 Old Bar Harbor Road, Salisbury Cove, ME 04672, USA; srieger@mdibl.org

[3] Department of Molecular Life Science, Tokai University School of Medicine, Isehara, Kanagawa 259-1193, Japan; abeko@is.icc.u-tokai.ac.jp

[4] Institute of Metabolism and Systems Research, College of Medical and Dental Sciences, The University of Birmingham, Birmingham B15 2TH, UK; m.hewison@bham.ac.uk

[5] The Jackson Laboratory, Bar Harbor, ME 04609, USA

* Correspondences: zhgdoc@hotmail.com (H.Z.); thomas.lisse@jax.org (T.S.L.);
Tel: +86-23-8901-2820 (H.Z.); +1-207-288-3605 (T.S.L.); Fax: +86-23-8901-2939 (H.Z.); +1-207-288-2130 (T.S.L.)

Academic Editor: Allison Cowin

Received: 19 October 2016; Accepted: 22 November 2016; Published: 26 November 2016

Abstract: Mice and human patients with impaired vitamin D receptor (VDR) signaling have normal developmental hair growth but display aberrant post-morphogenic hair cycle progression associated with alopecia. In addition, VDR$^{-/-}$ mice exhibit impaired cutaneous wound healing. We undertook experiments to determine whether the stress-inducible regulator of energy homeostasis, DNA damage-inducible transcript 4 (Ddit4), is involved in these processes. By analyzing hair cycle activation in vivo, we show that VDR$^{-/-}$ mice at day 14 exhibit increased Ddit4 expression within follicular stress compartments. At day 29, degenerating VDR$^{-/-}$ follicular keratinocytes, but not bulge stem cells, continue to exhibit an increase in Ddit4 expression. At day 47, when normal follicles and epidermis are quiescent and enriched for Ddit4, VDR$^{-/-}$ skin lacks Ddit4 expression. In a skin wound healing assay, the re-epithelialized epidermis in wildtype (WT) but not VDR$^{-/-}$ animals harbor a population of Ddit4- and Krt10-positive cells. Our study suggests that VDR regulates Ddit4 expression during epidermal homeostasis and the wound healing process, while elevated Ddit4 represents an early growth-arresting stress response within VDR$^{-/-}$ follicles.

Keywords: wound repair; VDR; mTOR; DDIT4; stress; hair follicle; stem cells

---

1. Introduction

Besides its known regulation of mineral homeostasis, the vitamin D receptor (VDR) also plays a functional role within the cutaneous environment. The importance of the cutaneous VDR is demonstrated in certain human patients with hereditary vitamin D-resistant rickets and mice that harbor loss-of-function mutations in the VDR and eventually develop hair loss (alopecia totalis [1]), as well as impaired skin wound repair [2,3]. In utero, the epithelium and underlying mesenchyme interact to form hair follicles during morphogenesis, which ends during the second week of life in mice. During the ensuing post-morphogenic hair cycle, the permanent follicular component is retained, which includes sebaceous glands and the upper outer root sheath that harbors the hair follicle stem cell center called the bulge. In contrast, the lower part of the follicle cycles through episodic periods of growth, regression and quiescence, termed the anagen, catagen and telogen stages respectively. The telogen-to-anagen transition marks a stage in which stem cells become activated to form a new hair bulb with concomitant inner root sheath (IRS) and hair shaft differentiation.

The VDR is thought to play a limited role during the early morphogenic period of follicular development since VDR$^{-/-}$ mice appear grossly normal and generate their first coat of hair [4]. However, VDR$^{-/-}$ mice exhibit compromised post-morphogenic hair cycles and improper anagen initiation and integrity, in which mutant mice do not grow new hair, and instead the follicles become epidermal cysts [5–7]. The signaling defects within VDR-deficient hair follicles are localized to keratinocytes, and not to the mesenchymal component of the follicle [6–8]. Through a series of recovery transgenesis experiments in mice, studies have shown that the zinc finger DNA binding domain of the VDR, but not the ligand binding or activation function 2 domains, was critical for the hair cycle [9–11].

Despite these studies, the biological explanation for hair loss in VDR-deficient mice remains unclear. The VDR was identified as a major molecular signature within hair bulge label-retaining stem cells [12], with subsequent studies suggesting that unliganded VDRs are required for the self-renewal, colony formation and normal lineage specification, as well as function of bulge keratinocyte stem cells (KSCs; [13]). However other follow-up studies observed no functional defects in VDR-deficient KSCs [14]. Hair follicle induction and maintenance are controlled by numerous factors such as fibroblast growth factor, transforming growth factor beta, Hedgehog and Wnt signaling, by controlling the self-renewal and lineage progression of KSCs and activation of progenitor cells. Studies suggest that VDR interactions with the lymphoid enhancer-binding factor 1 (lef1) transactivator may be the mechanism by which the unliganded VDR promotes Wnt signaling [13,15]. Importantly, it was shown that the VDR interaction with the canonical Wnt target gene, *Axin2*, was perturbed in both VDR-null and Lef1-null keratinocytes [16]. Furthermore, the Hedgehog signaling pathway was found to be disrupted in VDR- and Lef1-null keratinocytes, thus contributing to defective follicular signaling [16,17].

Besides its role in hair follicle biology, the VDR is expressed ubiquitously in the basal layer of the interfollicular epidermis [14], and plays a role in epidermal integrity and wound repair. In the epidermis, the biologically active form of vitamin D, 1,25-dihydroxyvitamin D (1,25D$_3$), is known to suppress tumor formation, promote innate antimicrobial immunity, suppress epidermal keratinocyte proliferation and promote differentiation and formation of the permeability barrier (reviewed in [18]). Epidermal abnormalities are observed in VDR$^{-/-}$ mice at as early as one week of age, with age-dependent reduction of differentiation markers and lipid content [4,19,20]. Consequently, the epidermis of VDR$^{-/-}$ mice is generally thicker and more folded, reflecting enhanced epidermal proliferation [21]. In contrast to VDR$^{-/-}$ mice, mice lacking the enzyme that catalyzes synthesis of 1,25D$_3$ (cytochrome p450 27B1; CYP27B1$^{-/-}$) have normal hair, but altered epidermal differentiation signatures, impaired permeability barrier and wound healing responses, suggesting that 1,25D$_3$ is crucial for epidermal integrity [22,23]. The cutaneous wound healing defects in global VDR$^{-/-}$ animals have been characterized by scarcity of infiltrating macrophages in the early inflammatory phase and disrupted vascular invasion of granulation tissue, yet re-epithelialization was reported to be unaffected [2]. In contrast, epithelial-specific deletion of the VDR in mice results in impaired re-epithelialization of wounds [3]. Also, in vitro scratch migration assays using VDR$^{-/-}$ primary keratinocytes showed decreased injury-stimulated migration [14].

Mechanistic/mammalian target of rapamycin (mTOR) is a serine-threonine protein kinase that is considered to be a "master regulator" of cell growth and survival. It is the major checkpoint for sensing of cellular nutrients and oxygen and energy levels to orchestrate the cellular responses to any changes. Upstream inhibitors of mTOR have been characterized, including DNA damage-inducible transcript 4 (Ddit4), also called REDD1, a mitochondrial protein which is involved in the conservation of energy resources. *Ddit4* was first identified as a stress response gene transcriptionally activated upon

UV-induced DNA damage [24], and then during hypoxia, oxidative stress, endoplasmic reticulum (ER) stress, atrophied glucocorticoid treatment and energy/serum deprivation [25–28]. We have recently shown that *Ddit4/DDIT4* is a direct and conserved transcriptional target of the VDR in human and mouse bone-forming osteoblasts to promote differentiation [29,30]. In view of the ubiquitous expression pattern of Ddit4 within the mammalian system, the present study focused on potential cutaneous interactions between VDR and Ddit4 signaling. Accordingly we hypothesize that in mice: (1) *Ddit4* is functionally regulated by the VDR during the hair cycle and epidermal wound repair, and (2) given its known role as a stress-inducible factor, Ddit4 may be utilized as an innate surveyor of VDR-dependent adverse cellular effects. By monitoring the effects on Ddit4 we were able to show that the VDR and Ddit4 function distinctly at the crossroads between hair follicle cycling and the epidermal wound repair process.

2. Results

2.1. DDIT4/Ddit4 Is an Acute Phase Effector of Stress Related to Inflammatory and Other Immune Challenges to Physical or Bacterial Complications

We initially sought a general assessment under which biological stresses regulate *DDIT4/Ddit4* gene expression within keratinocytes and other epithelial cell types. To do this, we appraised the publically available Gene Expression Omnibus (GEO) Profiles repository of curated microarray/next generation sequencing datasets for individual gene expression profiles (available at: http://www.ncbi.nlm.nih.gov/geoprofiles). We identified five (cutaneous) experimental DataSet records whereby *DDIT4/Ddit4* was significantly regulated compared to control samples. In a study to explore the space effects of heavy Fe ion radiation exposure of a rat keratinocyte line, *Ddit4* mRNA expression was decreased, despite the induction of other "DNA repair" genes within the data set [31] and original discovery of *Ddit4* as a major transcript activated by UV irradiation [24] (Figure 1A). In contrast, under the stress of inflammation (i.e., via overexpression of the NF-κB activator IKKβ) [32] or specific types of periodontal pathogens [33], *Ddit4/DDIT4* transcript was increased within mouse skin and human gingival epithelial cells, respectively (Figure 1B,C). In another study, human keratinocytes that were exposed to an enzymatic cell disassociation treatment as a model of epidermal injury [34], resulted in elevated *DDIT4* message (Figure 1D). In a more unperturbed physiological setting, multipotent human hair follicle stem cells were co-cultured with mesenchymal dermal papilla cells to induce keratinocyte progenitor differentiation, whereby *DDIT4* message levels decreased over time [35].

2.2. VDR Positively Regulates Ddit4 mRNA and Protein Expression in Primary Murine Epidermal Keratinocytes

To investigate the VDR-Ddit4 relationship within the mouse cutaneous system, we first determined whether $1,25D_3$-VDR actions could transcriptionally activate *Ddit4* within primary epidermal keratinocytes harvested and cultured from neonatal pups aged 2–3 days derived from either wildtype (WT) or $VDR^{-/-}$ mice on a normalized calcium/phosphorous diet. Since keratinocytes differentiate with calcium, our culture system was established using low-calcium conditions (0.05 mM) to mimic undifferentiated cells in the basal layer that have the highest concentration and effectiveness of the VDR [36]. Treatment of WT, but not $VDR^{-/-}$ cells with $1,25D_3$ (1–100 nM) for 6 h resulted in a dose-dependent induction of both *cyp24a1* (1,25-dihydroxyvitamin D_3 24-hydroxylase), a major transcriptional target and negative feedback regulator of vitamin D, and *Ddit4* (Figure 2A). Expression of *Ddit4* mRNA levels in primary keratinocytes was significantly lower in $VDR^{-/-}$ mice, but transfection of these cells with a murine Vdr transgene resulted in re-expression of *Ddit4*, and enhanced induction after $1,25D_3$ treatment (Figure 2B). These results demonstrate that *Ddit4* is directly controlled by VDR actions.

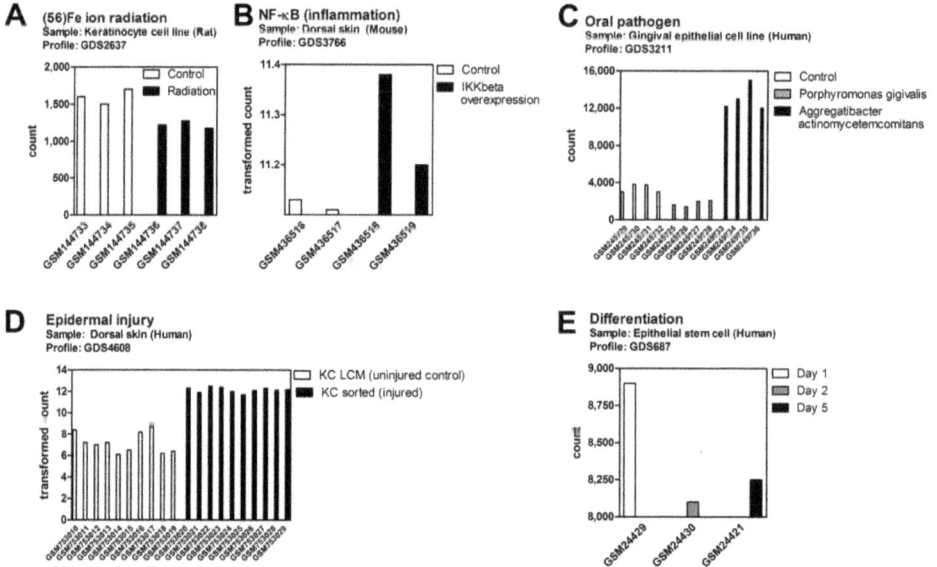

Figure 1. Gene Expression Omnibus (GEO) profile assessment of *Ddit4/DDIT4* responses to biological and stress conditions within epithelial cells. (1.5 column) (**A**) For all DataSets presented in this figure, the graphs represent *Ddit4/DDIT4* transcript levels. DataSet GDS2637 shows that *Ddit4* was not induced after 56FE ion irradiation within a rat keratinocyte cell line; (**B**) DataSet GDS3766 shows that NF-κB activation by IKKβ overexpression in mouse dorsal skin resulted in increased *Ddit4* levels; (**C**) DataSet GDS3211 reports that periodontal pathogen infection of a human gingival epithelial cell line with *Aggregatibacter actinomycetemcomitans* led to *DDIT4* induction; (**D**) DataSet GDS4608 shows that upon injury of human epidermal cells, *DDIT4* transcript levels were increased compared to uninjured samples; (**E**) DataSet GDS687 shows that differentiation of keratinocyte stem cells upon mesenchymal-epithelial interactions resulted in decreased *DDIT4* expression over time. All DataSets can be accessed at: http://www.ncbi.nlm.nih.gov/geoprofiles.

Next we investigated the effects on Ddit4 protein expression within WT and VDR$^{-/-}$ primary keratinocytes using immunofluorescence staining. In WT cells treated with 1,25D$_3$ (10 nM) for 18 h in growth medium supplemented with FBS with reduced-calcium (0.05 mM), we observed an increased number of cells with elevated intracellular accumulation of Ddit4 (Figure 2C, upper middle panel). In addition, these 1,25D$_3$-treated WT keratinocytes appeared to be morphologically distinct (i.e., enlarged and differentiated) compared to their VDR$^{-/-}$ counterparts. The 1,25D$_3$-treated VDR-null cells appear to acquire higher length:width ratios, indicative of an undifferentiated motile state (Figure 2C, lower middle panel). Likewise, after 1,25D$_3$ treatment of VDR-null keratinocytes, there was a comparable decrease in Ddit4 expression to that of WT cells. As a safeguard for energy conservation, Ddit4 expression is known to be enhanced under stress conditions such as growth factor (serum) deprivation [25]. With this in mind, we tested this response in WT keratinocytes and observed increased Ddit4 protein expression after serum starvation (Figure 2C, upper right panel). Interestingly, Ddit4 expression was also enhanced in VDR$^{-/-}$ keratinocytes under serum-deprived conditions (Figure 2C, lower right panel). These results suggest that (1) the Ddit4-mediated energy and growth factor depletion sensing mechanism has the potential to function independent of cellular VDRs; and (2) endogenous Ddit4 activation can be used as an innate surveillant for catabolic stress within the VDR loss-of-function system.

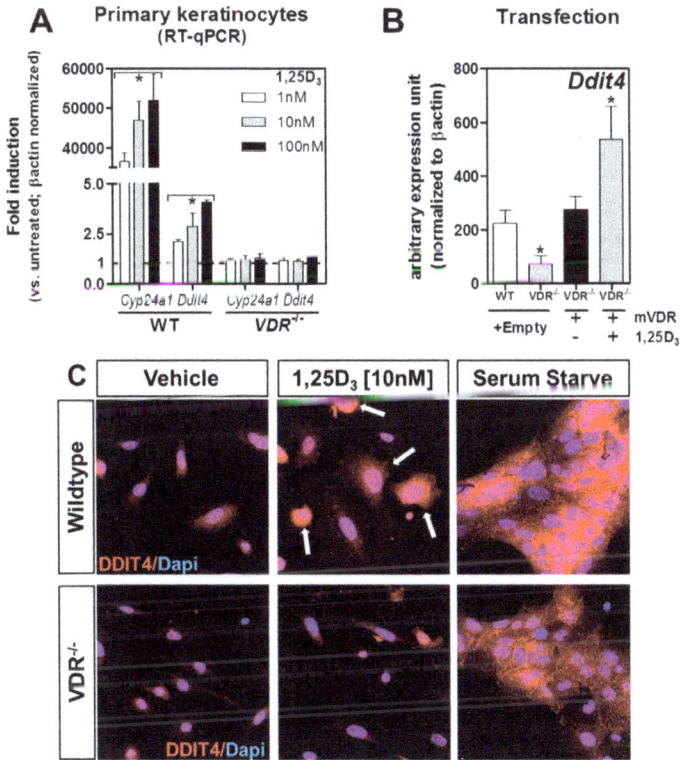

Figure 2. Regulation of the mechanistic/mammalian target of rapamycin (mTOR) inhibitor and stress sensor DNA damage-inducible transcript 4 (Ddit4) by liganded vitamin D receptors (VDRs) within primary keratinocytes (column 1.5). (**A**) Dose-dependent transcriptional induction of *Cyp24a1* and *Ddit4* by 1,25-dihydroxyvitamin D (1,25D$_3$, 6 h) in wildtype (WT), but not VDR$^{-/-}$, primary keratinocytes derived from neonatal pups; (**B**) Transient transfection of VDR$^{-/-}$ primary keratinocytes with a mVDR plasmid restores endogenous *Ddit4* transcript levels. Combined treatment further enhanced *Ddit4* mRNA levels (with mVDR transfection and 1,25D$_3$ of 10 nM for 6 h); (**C**) Immunofluorescence detection of Ddit4 within primary keratinocytes. WT, but not VDR$^{-/-}$, primary keratinocytes exposed to 10 nM 1,25D$_3$ for 18 h resulted in increased intracellular accumulation of Ddit4 (white arrows). Ddit4 upregulation in response to fetal bovine serum deprivation for 24 h in both WT and VDR$^{-/-}$ keratinocytes. One-way ANOVA at an $\alpha = 0.05$ (95% confidence interval) and Tukey's multiple comparison post-tests were utilized. Significance is denoted with asterisks: * $p < 0.05$ ($n = 3$–4 experiments). RT-qPCR: reverse transcription quantitative polymerase chain reaction.

2.3. Ddit4 Is a Direct Transcriptional Effector of the Liganded VDR within Primary Epidermal Keratinocytes

Transcriptional regulation by 1,25D$_3$ involves occupancy of VDREs on effector genes by 1,25D$_3$-bound or unbound VDRs and numerous other co-regulators. Here, we performed chromatin immunoprecipitation (ChIP) qPCR assays to assess potential VDREs and general transcriptional element binding to the *Ddit4* promoter within WT neonatal primary keratinocytes. We tiled the proximal promoter region (i.e., with primers Ddit4-1 and Ddit4-2), which was approximately 500 base pairs upstream of the *Ddit4* transcriptional start site (TSS) on the reverse strand. The distal region was appraised using primer Ddit4-3, which targeted a DNA region 1.7 kilobases upstream of the TSS. All sequence positions were based on the primary assembly GRCm38 released by the Genome Reference Consortium in 2012 based on the *Mus musculus* strain C57BL/6J.

ChIP results show that endogenous VDRs bound strongly to the proximal promoter of *Ddit4* under unstimulated conditions (Figure 3(AI)). Upon 1,25D$_3$ stimulation (10 nM, 15 min) there was increased recruitment of liganded VDRs to the proximal, but not distal, promoter region of *Ddit4* as assessed by ChIP-qPCR (Figure 3B). These results suggest that the promoter region of *Ddit4* in keratinocytes includes functional VDRE(s). Although the unliganded VDR bound to the proximal *Ddit4* promoter, this represented a transcriptionally inactive state as RNA polymerase 2 (RNApol2), a general marker for precursor RNA synthesis, was tethered or poised at the more distal region (Figure 3(AII)). In contrast, there was accompanying RNApol2 binding activity at the proximal *Ddit4* promoter only after 1,25D$_3$ treatment, suggesting direct transcriptional activation of *Ddit4* after ligand stimulation (Figure 3C). For ChIP validation we utilized the murine osteocalcin gene, *Bglap*, which has no VDREs in its promoter region, nor is it transcriptionally influenced by addition of 1,25D$_3$ [37]. We also chose the S1 site within intron 3 and 4 of the murine *Vdr* gene as a potent positive control for both auto-regulatory VDR and RNApol2 binding, as previously shown in MC3T3-E1 murine osteoblasts [38]. Interestingly in primary keratinocytes, VDRs but not RNApol2, bound to the S1 site, suggesting cell type-specific usage of this transcriptional auto-regulatory site. All ChIP-qPCR results were normalized to a validated non-specific (NS) "naked" control target site . Overall, our findings support the notion that the liganded VDR directly targets the *Ddit4* promoter to induce transcription within primary epidermal keratinocytes.

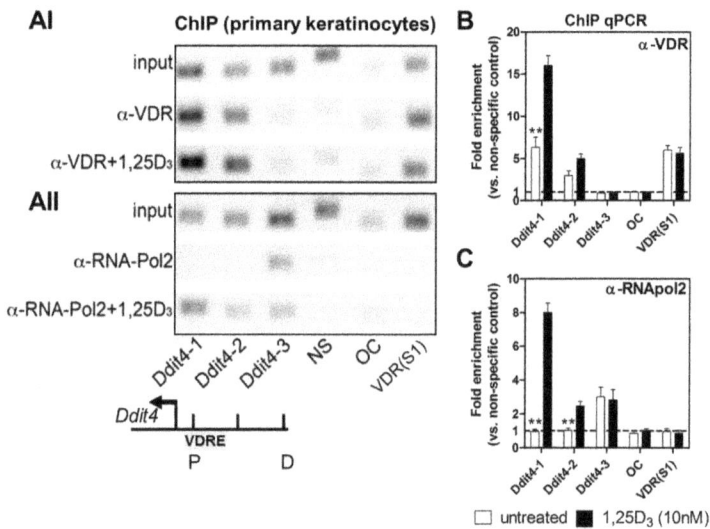

Figure 3. *Ddit4* transcriptional activity is directly regulated by the VDR within primary epidermal keratinocytes (1.5 column). (**A**) Representative chromatin immunoprecipitation (ChIP) gel images of VDR and RNA polymerase 2 (RNApol2) interactions at the *Ddit4* promoter region within WT primary keratinocytes. Figure 3(AI) shows VDR immunoprecipitation, while Figure 3(AII) shows RNApol2 immunoprecipitation results. Cells were treated with 1,25D$_3$ (10 nM) for 15 min, fixed and then the chromatin was purified. NS (non-specific), OC (osteocalcin), and VDR-S1 (VDR intron 3 and 4) genomic sites were used for control purposes. P (proximal), D (distal), VDRE (putative vitamin D response element); (**B**) ChIP-Quantitative PCR (qPCR) analysis of VDR immunoprecipitation; (**C**) ChiP-qPCR analysis of RNApol2 immunoprecipitation. One-way ANOVA at an α = 0.05 (95% confidence interval) and Tukey's multiple comparison post-tests were utilized. Significance is denoted with asterisks: ** $p < 0.01$ ($n = 4$ experiments).

2.4. Formation of Ddit4-Positve Stress Compartments in VDR$^{-/-}$ Morphogenic Follicles and Reduced Ddit4 Epidermal Expression

Having shown a direct transcriptional relationship between the VDR and *Ddit4* within primary keratinocytes, we next sought to decipher the potential in vivo role of Ddit4 during activation and resting stages of the hair follicle cycle and epidermal homeostasis. Hair follicle morphogenesis initiates during embryogenesis and ceases approximately 2–3 weeks postnatally, whereby the ensuing post-morphogenic hair cycle marks a new anagen stage (Figure 4A). Using immunostaining, we monitored Ddit4 protein expression during transitioning hair follicular and epidermal keratinocytes through morphogenic and post-morphogenic stages within WT and VDR$^{-/-}$ mice maintained on a rescue diet. Catagen is a transitional apoptosis-mediated involution stage that signals the end of the active growth of hair. Importantly, the VDR is selectively expressed in hair follicle keratinocytes during late anagen and catagen stages, which reflects the reduced period of proliferation and elevated differentiation of follicular cells [39]. At day 14 during the first postnatal early catagen phase, we observed distinct Ddit4 staining patterns between WT and VDR$^{-/-}$ hair follicles and epidermis derived from littermates (Figure 4B). Both types of mice showed Ddit4 expression in the inner root sheath (IRS) of the hair follicle and moderate levels within the interfollicular epidermal regions (Figure 4B). However, there was slightly higher, atypical Ddit4 expression in VDR$^{-/-}$ follicles. These punctate regions of increased Ddit4 expression within VDR$^{-/-}$ follicles were observed in nearly every follicle within superbasal sections at day 14 (Figure 4(BIII–VI)). Given the stress inducible role of Ddit4, we classified these areas as stress compartments (SC) within VDR-deficient follicles. These results suggest that VDR-deficient hair follicles that are transitioning between catagen-telogen exhibit increased cellular stress.

By day 29, initiation of post-morphogenic anagen in the hair follicle is characterized by increased cell proliferation in the follicular epithelium below the activated bulge. During anagen in WT skin, Ddit4 expression was substantially reduced within the hair follicle and interfollicular regions compared to day 14 (Figure 4C), suggesting a reduction of its growth-inhibitory functions. Despite this reduction, there was more residual Ddit4 within the interfollicular epidermis compared to the follicle. At this stage, VDR-null hair follicles are degenerated and exhibit an abnormal morphology depicted by follicular dystrophy, IRS hyperplasia and massive dilation of the junctional canal consistent with previous work [17]. In contrast to normal hair follicles, VDR-null follicles exhibit increased and variegated expression of Ddit4 throughout the dystrophic follicular epithelium and interfollicular regions. There was a statistically significant increase in normalized average fluorescent intensity (VDR$^{-/-}$: 538 ± 31 a.u.; WT: 74 ± 11 a.u., $n = 8$ follicles, $p < 0.01$) within the bulb region of VDR$^{-/-}$ hair follicles when compared to WT samples at this stage. We also found that the utricules (u), i.e., the epidermal portion of abnormal hair follicles, also expressed Ddit4 in the lining epithelium. By day 48, the hair follicles of normal animals remain in the resting telogen phase and are kept dormant for another five weeks. The epidermis at this stage also expresses major factors of differentiation [4]. In WT animals, the epidermis and follicular epithelium uniformly expressed increased levels of Ddit4 (Figure 4D), presumably to help maintain a growth-inhibiting, dormant state. Furthermore, dermal fibroblasts in WT skin also expressed Ddit4. On the other hand, VDR-null mice at day 48 exhibit epidermal hyperplasia and increased epidermal corneocytes (c), suggesting abnormal shedding of cornified material. Remarkably, Ddit4 expression was absent within VDR-null skin at telogen. Ddit4 expression was suppressed throughout the epidermis, remaining non-exogenic follicular epithelium and within the dermis as well. Furthermore, there was increased dermal cellularity in VDR$^{-/-}$ skin, inversely correlated with the level of Ddit4 (Figure 4D).

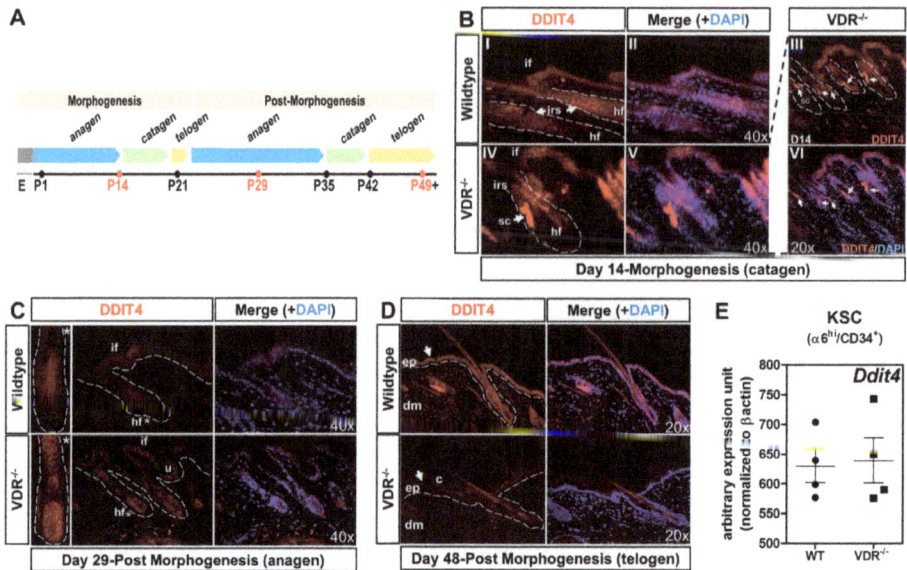

Figure 4. Ddit4α-positive stress compartments present in VDR$^{-/-}$ morphogenic hair follicles (column 2). (**A**) Chart depicting the morphogenic and post-morphogenic stages of the hair cycle up to post-natal day 49 (P49); (**B**) At day 14, hair follicles are within the first postnatal morphogenic early catagen (regression) phase. In WT (**I,II**) and VDR$^{-/-}$ (**IV,V**) animals, Ddit4 is expressed throughout the inner root sheath (irs), but not in the bulb of the hair follicle (hf). In both genotypes, Ddit4 was expressed within the interfollicular (if) epidermis. Hair follicles of VDR$^{-/-}$ animals exhibit Ddit4-positive stress compartments (sc, white arrows) (**IV**). In the right panel (**III,VI**), a lower magnification (20×) is presented where each hair follicle is outlined with a white dotted line. Ddit4 immunostaining is counter-detected with Alexa® 594 and nuclei stained with DAPI; (**C**) At day 29, hair follicles are within the second postnatal anagen (growth) phase. In WT skin, Ddit4 expression was attenuated within the IF and hair follicle compared to day 14. Asterisks reflect the zoom of the respective hair follicle. In VDR$^{-/-}$ hair follicles, there was aberrant and increased Ddit4 expression throughout the length of the follicle (bulb to inner root sheath) compared to WT. There was a similar expression pattern in the utricles (u) and the interfollicular epidermis; (**D**) At day 48, WT skin display telogen hair follicles with quiescent morphology. Ddit4 expression was enhanced and uniform throughout the epidermis (**ep**) (white arrow), dermis (**dm**) and follicular epithelium. In contrast, VDR$^{-/-}$ skin exhibited significant reduction of Ddit4 throughout the skin. c (corneocytes). For immunostaining, representative slides are presented. Dotted white lines outline individual hair follicles or the epidermal-dermal junction; (**E**) Hair follicle bulge stem cells from 29-day-old animals were fluorescence-activated cell sorting (FACS) purified and analyzed with RT-qPCR. There was no significant change in *Ddit4* message level between WT and VDR$^{-/-}$ bulge stem cells (*n* = 4 experiments per genotype).

2.5. The VDR Does Not Regulate Ddit4 within Bulge Keratinocyte Stem Cells

Besides epidermal and follicular keratinocytes, the cutaneous niche consists of many different cell types, including bulge keratinocyte stem cells (KSCs). It was recently shown that hair follicle growth and stem cell exhaustion in mice are linked to mTOR dysregulation as a means to maintain genetic integrity of the stem cell population [40,41]. We therefore hypothesized that the unliganded VDR maintains proper *Ddit4* levels, and hence mTOR function, within KSCs. We compared *Ddit4* transcript levels under unliganded conditions within 29-day-old KSCs, a period in which KSCs are activated (Figure 4C). We isolated living bulge KSC populations from anagen hair follicles of WT and VDR$^{-/-}$ littermates using FACs purification according to the established markers α6^{hi} and CD34^{+} [42].

Based on these preparations, we observed no difference in *Ddit4* message levels between WT and VDR$^{-/-}$ KSCs. These results suggest no relationship between the VDR and mTOR signaling towards possible stem cell exhaustion and no effect on the hair loss phenotype, raising the question of primary defects in progenitor cells instead.

2.6. Impaired Ddit4/Ddit4 and Krt10/Krt10 Expression in the Neo-Epidermis of Wounds from VDR$^{-/-}$ Mice

The cause for delayed onset of cutaneous wound closure in global VDR-null mice is unclear. Based on this, we studied the association between Ddit4 and VDR during the cutaneous wound repair process. Mice aged 47 days mice were subjected to 3.5 mm trunk punch biopsies and tissues were harvested and processed for RT-qPCR, BrdU (5-bromo-2'-deoxyuridine) labeling, and immunocytochemistry six days after injury. First, we monitored and compared *Ddit4* and *Krt10* (keratin 10) message levels between WT and VDR$^{-/-}$ animals. Krt10 is an early marker of differentiating daughter cells in the stratum basale which faces the epidermal surface. We confirmed the lower *Ddit4* message levels within uninjured VDR$^{-/-}$ epidermis, which was accompanied by a significant reduction in *Krt10* mRNA compared to uninjured WT tissue (Figure 5A). Relative to uninjured skin, there was a significant decrease in both *Ddit4* and *Krt10* message levels in WT wounds six days post-injury, representing the early-intermediate phase of the re-epithelialization process. Compared to WT wounds six days after injury, *Ddit4* and *Krt10* mRNA levels were further attenuated in VDR$^{-/-}$ wounds. Histological analysis of wound closure showed larger wound openings (Figure 5B, red lines) in VDR$^{-/-}$ mice after six days. Interestingly, within VDR$^{-/-}$ injured skin, in vivo BrdU labeling revealed an increased number of proliferating epidermal and follicular keratinocytes compared to WT samples at the wound edges, marking a potential reserve of cells impaired in the re-epithelialization process (Figure 5(BII,IV); representative black box). In both genotypes, BrdU also labeled damaged hair follicles. Particular to VDR$^{-/-}$ tissue were BrdU-positive proliferating dermal fibroblasts (red arrows). Next we monitored Ddit4 and Krt10 expression with immunohistochemistry in the newly restored epidermis six days after injury. We observed moderate expression of Ddit4 within the newly re-epithelized wound edge in WT animals (Figure 5(BV,VI); representative red box). In contrast, there was a comparable decrease in Ddit4-positive cells present within the neo-epidermis in VDR$^{-/-}$ animals (Figure 5(BVII,VIII); representative red box). In WT neo-epidermis, there were diffuse cells with high Krt10 immuno-reactivity in contrast to VDR$^{-/-}$ wounds (Figure 5C). Overall, these findings suggest that the loss of Ddit4/Krt10-positive cells in wounded VDR$^{-/-}$ animals contributes to the impaired re-epithelialization process.

2.7. Ddit4-Deficient Mouse Embryonic Fibroblasts Are Resistant to the Pro-Differentiation Actions of Vitamin D

Although the initial data suggests a functional association between VDR and Ddit4, the impact of Ddit4 on vitamin D function is unclear. As Ddit4 is expressed in dermal fibroblasts and dysregulated in VDR$^{-/-}$ skin (Figure 4D), mouse embryonic fibroblasts (MEFs) derived from Ddit4$^{-/-}$ and WT animals (kindly provided by Leif W. Ellisen) were tested for any differences in classic vitamin D anti-proliferative responses. We observed more spindle-shaped cells, characteristic of mitotic cells, in Ddit4-deficient MEFs even following 50 nM 1,25D$_3$ treatment (Figure 6A, right panels). Measurement of cell proliferation in WT MEFs showed that even at the lowest concentration range of 1,25D$_3$, cell proliferation was inhibited compared to the baseline (dotted line) (Figure 6B). At 50 nM 1,25D$_3$, there was an approximate 10% decrease in cell proliferation in WT samples, but no effect in Ddit4$^{-/-}$ MEFs. It was only at the highest level (100 nM) of 1,25D$_3$ that Ddit4$^{-/-}$ MEFs exhibited comparable effects on cell number to WT cells (Figure 6A,B), possibly succumbing to apoptosis [43]. Next, we performed BrdU-incorporation studies using MEFs treated with 50–75 nM 1,25D$_3$ (Figure 6C,D), and observed statistically significant decreases in BrdU incorporation within WT, but not Ddit4$^{-/-}$ MEFs. The fibroblast maturation marker vimentin was also decreased in Ddit4$^{-/-}$ MEFs when compared to WT samples (Figure 6(EI)). Lastly, to gain insight into signaling events, we performed quantitative real-time PCR analysis to monitor *Vdr* expression and induction of Cyp24a1 (Figure 6(EII–IV)). Interestingly, Ddit4$^{-/-}$ MEFs expressed more *Vdr* compared to normal MEFs (Figure 6(EII)). As a result, there was a concomitant increase in induction of *Cyp24a1* in Ddit4$^{-/-}$ MEFs (Figure 6(EIII,IV)). These results suggest that despite the increase in Vdr and its signaling capacity in MEFs lacking Ddit4, vitamin D is unable to differentiate these cells due to specific defects in the DDIT4 signaling cascade. In conclusion, Ddit4-deficient MEFs were resistant to low-to-moderate vitamin D treatments, and Ddit4 is a downstream effector of liganded VDR actions to promote cellular differentiation within fibroblasts.

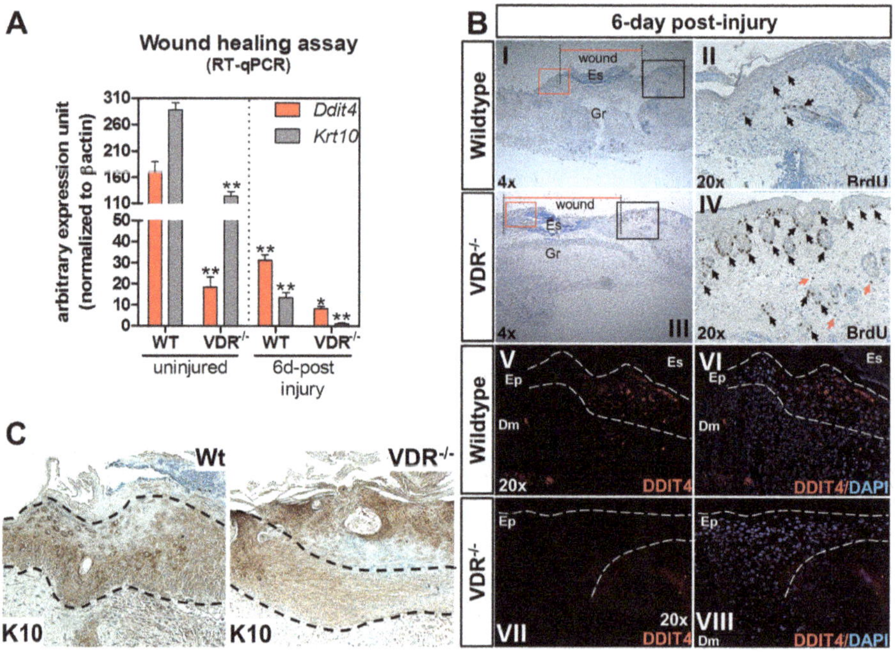

Figure 5. Reduction of Ddit4 in the re-epithelized wounds of VDR$^{-/-}$ animals (column 1.5). (**A**) RT-qPCR analysis of the wound repair process from excised wounds. For uninjured samples, comparisons were made between genotypes. Six days following injury, sample comparisons were made between the respective uninjured genotypes. One-way ANOVA at an $\alpha = 0.05$ (95% confidence interval) and Tukey's multiple comparison post-tests were utilized. Significance is denoted with asterisks: * $p < 0.05$, ** $p < 0.01$ ($n = 4$ samples per genotype and time point); (**B**) Representative slides showing 5-bromo-2′-deoxyuridine (BrdU) immunohistochemical (**I–IV**) and Ddit4 immunofluorescence (**V–VIII**) staining of WT and VDR$^{-/-}$ skins six days following injury. Wound closure is depicted by the horizontal red bars. Black boxed areas in the 4× slides represent the BrdU-labeled magnified regions (**II and IV**). Red-boxed areas represent the Ddit4-labeled magnified regions (**V–VII**). In the BrdU slides, black arrows depict proliferating epidermal and follicular keratinocytes. Red arrows depict proliferating dermal fibroblasts. In the Ddit4 panels, the dotted white line marks the re-epithelialized area. Nuclei are marked in **blue** with 4′,6-diamidino-2-phenylindole (DAPI), while Ddit4 is labeled in **red**. ES: eschar; GR: granulation tissue; EP: epidermis; DM dermis; (**C**) Krt10 immunostaining within the neo-epidermis (demarcated by dashed lines) six days following injury.

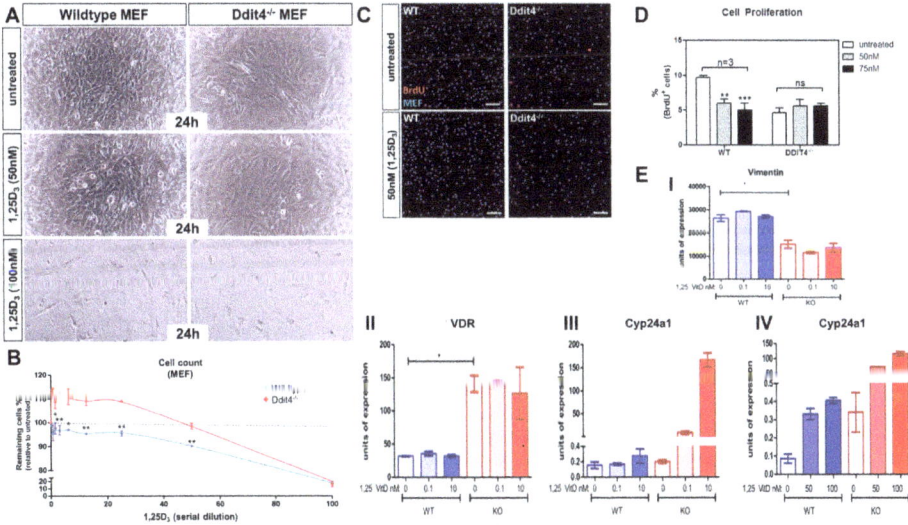

Figure 6. Ddit4$^{-/-}$ mouse embryonic fibroblasts are resistant to the pro-differentiation actions of vitamin D (1 column). (**A**) Representative images of WT and Ddit4$^{-/-}$ mouse embryonic fibroblasts (MEFs). MEFs were untreated or treated with a serial dilution of 1,25D$_3$ (0.39–100 nM) for 24 h; (**B**) Cell counts performed after 1,25D$_3$ treatment to calculate the proliferation % relative to the untreated sample. One-way ANOVA at an α = 0.05 (95% confidence interval) and Tukey's multiple comparison post-tests were utilized. Significance is denoted with asterisks: * $p < 0.05$, ** $p < 0.01$ ($n = 3$ experiments per conditions and genotype); (**C**) Representative BrdU-labeled MEFs shown after 1,25D$_3$ treatment for 12 h. DAPI-positive cells were marked using the Imaris (Bitplane) software to perform quantitative analysis. Bar = 50 μm; (**D**) Quantification of BrdU-labeled MEFs after 1,25D$_3$ treatment. Significance is denoted with asterisks: ** $p < 0.01$, *** $p < 0.001$ ($n = 3$ experiments per conditions and genotype), ns (not significant); (**E**) Quantitative PCR analysis of *Vimentin* in Wt and KO MEFs after 6 h with or without vitamin D, * $p < 0.05$ (**I**). Quantitative PCR analysis of MEFs treated or untreated with low or high concentrations of 1,25D$_3$ for 6 h (**II–IV**). Increased expression of *Vdr* (**II**) and induction of *Cyp24a1* (**III,IV**) within Ddit4$^{-/-}$ MEFs. ($n = 3$ preparations of each cell line).

3. Discussion

3.1. Insights into Ddit4-VDR

Perturbation of VDR expression in humans and mice is associated with alopecia. The results reported here therefore provide novel insights into biological and clinical applications of vitamin D and skin function. Our data show that the mTOR inhibitor, *Ddit4*, is a direct transcriptional target of VDR in epidermal keratinocytes. Regulation of mTOR activity is a major component of "checks and balances" of energy homeostasis within a cell. Ddit4, a mitochondria-resident protein, represses mTOR signaling by activating TSC2 (tuberin), a guanosine triphosphate (GTP) hydrolyzing enzyme (GTPase) activating protein (GAP), which then stimulates the small GTPase Ras homolog enriched in brain (Rheb) in its GDP-bound form to inactivate mTOR [25]. Dysregulation of mTOR signaling can lead to a plethora of diseases including cancer and metabolic disorders. Studies have shown that 1,25D$_3$ or its analogues can suppress tumor cell growth by upregulation of Ddit4 in various cancer cell model systems [44]. Suppression of mTOR via Ddit4 activation results in attenuation in both cell size and growth rate, comparable to that observed after nutrient and growth factor deprivation [45]. Conversely, Ddit4 levels are mitigated under growth-like conditions [27]. The importance of Ddit4 in regulating cytoprotection and survival is highlighted in the Ddit4$^{-/-}$ mouse line that is associated with enhanced mTOR activity [25]. DDIT4$^{-/-}$ mice are known to be resistant to a diverse set of stress conditions such as those caused by oxidative stress in the retina, tobacco smoke-induced emphysema, steroid-induced atrophy in the skin and apoptosis of lung epithelial cells caused by ceramide [46–48]. Mouse *Ddit4* resides in a genomic region on chromosome 10 (10: 59316668-74913026), which forms the only syntenic cluster on human chromosome 10 (10: 53435340-73103214), emphasizing its conservation, genomic and functional importance. In addition, previous studies found that simultaneous loss of Drosophila *scylla* and *charybdis*, which are homologs of the human *DDIT4* and *DDIT4-like* genes, generated flies that showed mild overgrowth [49]. In contrast, enhanced expression of Ddit4 can promote apoptotic cell death or terminal differentiation in certain cell types [50], including VDR-deficient follicular epithelial cells observed in our studies.

3.2. Hair Follicle Defects in VDR-Deficient Animals

Our data support the notion that abnormalities in Ddit4 signaling disrupt follicular energy homeostasis to affect follicular integrity in VDR$^{-/-}$ animals. It was previously reported that the inability of VDR-deficient animals to initiate a new post-morphogenic hair cycle was due to primary defects within KSCs [13]. Additional findings observed no change in differentiation markers of VDR-deficient neonatal keratinocytes in culture, concluding that the mutant follicular keratinocytes during the morphogenic period are normal [5]. In contrast, the findings of this study and others have shown that there exist defects within the follicular epithelium of VDR$^{-/-}$ mice during the catagen-to-telogen morphogenic period [17]. In this regard, it has been shown that the failure of the follicular epidermis to maintain the hair follicle in VDR-deficient animals likely represents the compromised adhesion and motility capacity of surface cells along the follicle at the onset of anagen, and not due to functional defects within label-retaining KSCs [14,19]. These defects can further compromise lineage progression and differentiation status of follicular keratinocytes lacking a functional VDR. Furthermore, we observed no differences in *Ddit4* expression within VDR$^{-/-}$ KSCs, suggesting that Ddit4-related defects reside either in the hair germ or more in differentiated progenitor cells. It is well known that mTOR activation propagates metastasis and matrix-stimulated cell migration, and treatment with mTOR inhibitors such as rapamycin can block cell motility under numerous experimental conditions (reviewed in [51]). There are also several examples how primary defects of the strictly regulated morphogenic catagen-telogen transition can lead to failed initiation of the ensuing anagen hair cycle, much like that observed in VDR$^{-/-}$ animals [52,53]. Overall, our findings may have uncovered a critical cog—mTOR signaling—in the full understanding of how hair follicles become impaired and degenerate in VDR$^{-/-}$ animals.

To date, it is unclear if the stress compartments identified in VDR$^{-/-}$ morphogenic follicles are formed due to systemic and/or local stress stimuli. As a number of sensing cues impinge on the Ddit4-mTOR signaling pathway, it will require additional efforts to identify the specific cues within the VDR$^{-/-}$ follicles. It is commonly known that damaged hair follicles endure premature catagen initiation to inhibit proper hair follicle growth [54], further supporting our findings and suggesting precocious and prolonged catagen-to-telogen transition in VDR-deficient follicles. This speculation is supported by a recent study which performed RT-qPCR analysis on plucked VDR$^{-/-}$ hair follicles during morphogenetic days 13 and 15 [17]. In the study, at day 13 there was a significant increase in pro-apoptotic *CASP3* (caspase 3) transcripts in VDR$^{-/-}$ follicles compared to controls, and at day 15 there was a reported decrease in *Shh* (sonic hedgehog) message levels. Epithelial Shh is a major driver of hair follicle morphogenesis, and its decrease in VDR$^{-/-}$ follicles hints at perturbations in the process. Furthermore, it is unclear if the elevated Ddit4 level in VDR$^{-/-}$ follicles is a harbinger of apoptosis activation [17], as there is a clear link between persistent Ddit4 activation with programmed cell death [55]. Lastly, we are unclear whether the Ddit4-positive stress compartments are associated with clearance of dysfunctional cellular debris, as DDIT4 upregulation and mTOR suppression are positively correlated with autophagocytosis [56]. Overall, we speculate that follicular degeneration in

VDR-null animals does not reflect defects in the VDR-Ddit4 axis within KSCs, rather in the cells that make up the cycling portion of the hair follicle (Figure 7).

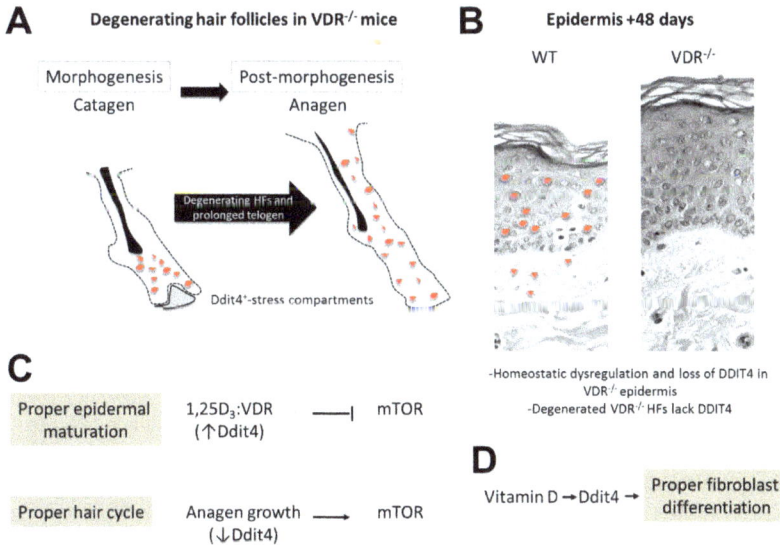

Figure 7. Early stress in VDR$^{-/-}$ hair follicles and epidermal regulation of Ddit4 by the VDR (column 1.5). Schema depicting the major findings. (**A**) VDR$^{-/-}$ hair follicles (HFs) exhibit DDIT4$^+$ stress compartments, marked in red, during morphogenesis which prolongs the telogen phase; (**B**) The VDR regulates Ddit4 expression during epidermal homeostasis and wound healing responses. By day 48, VDR$^{-/-}$ hair follicles have degenerated resulting in loss of Ddit4expression; (**C**) Proposed functional role of Ddit4 and mTOR during epidermal maturation and hair growth. During epidermal maturation, VDR signaling induces *Ddit4* expression to block mTOR cascades. During the initiation of anagen, Ddit4 levels are suppressed in order to stimulate the mTOR growth-promoting cascade; (**D**) Proposed involvement of DDIT4 in regulating the pro-differentiation effects of vitamin D in embryonic fibroblasts.

3.3. Epidermal Wounding Defects

The defect in VDR$^{-/-}$ hair follicles highlights the dichotomy which exists in the epidermis (Figure 7). Our data suggests that within uninjured epidermal keratinocytes, both in vitro and in vivo, the VDR regulates *Ddit4* message levels in a concentration and age-dependent manner, respectively. By day 48 the epidermis normally expresses high levels of differentiation markers, yet animals void of VDR associate with reduced epidermal differentiation [17,57]. This is consistent with reduced levels of Ddit4 within skin of VDR$^{-/-}$ mice. This finding is also consistent with increased Shh in the epidermis of VDR$^{-/-}$ mice [57], which can induce a basal-like phenotype as well as basal cell carcinomas in the skin. Thus, we speculate that the VDR maintains epidermal homeostasis via direct transcriptional control of *Ddit4* over time.

Skin wounding triggers an acute inflammatory response with the innate immune system contributing both to protection against invasive organisms and invasion of inflammatory cells into the wounded area. These cells release a variety of cytokines and growth factors that stimulate the proliferation and migration of dermal and epidermal cells to close the wound. Mice globally lacking the VDR or the enzyme CYP27B1 exhibit decreased lipid content of the lamellar bodies leading to a defective permeability barrier [58], and a defective response of the innate immune system to invading infections acting through dermal TGF-β signaling [2]. We observed delayed wound closure in the

VDR$^{-/-}$ mice, which is consistent with most reports [3,14,59]. This is in contrast to one study that showed no difference in wound closure of VDR$^{-/-}$ animals [2], which may reflect differences in the age of the animals and/or severity and conditions of the wounds. Regardless, studies investigating "epithelial-specific" ablation of the VDR resulted in delayed wound closure attributed to impaired β-catenin signaling within epidermal stem cells [3]. Recently, epithelial-specific ablation of phosphatase and tensin homolog *Pten* and tuberous sclerosis 1 *Tsc1* (both inhibitors of mTOR) has shown that mTOR activation can dramatically increase epithelial cell migration and cutaneous wound closure [40]. Although we observed the opposite phenomena, one could argue for global VDR ablation having systemic effects on the immune system that impedes the subsequent steps of the healing process, such as re-epithelialization.

Fibroblasts also migrate into the wounded area and proliferate to deposit a provisional extracellular matrix consisting of reforming granulation tissue. Keratinocytes migrate across the injured dermis above the provisional matrix and begin to proliferate. By 3–10 days after injury, the wound is filled with granulation tissue, and fibroblasts are recruited to the wound by growth factors from macrophages. Fibroblasts then *trans*-differentiate into myofibroblasts, leading to wound contraction and immature collagen deposition assisting in wound closure. At this stage, the apical wound portion is overlaid with a neo-epidermis associated with fibroblasts. At six days after wounding we observed sporadic Ddit4-postive cells within the re-epithelized wound within the neo-epidermis in WT animals. The identity of these cells is unclear, and may signify the transitioning (myo)fibroblasts during the healing process or keratinocytes. At this stage the levels of both *Ddit4*/Ddit4 and *Krt10*/Krt10 decreased relative to uninjured tissue representing the development-like regenerative steps of healing. In contrast, VDR$^{-/-}$ neo-epidermis did not harbor any Ddit4-positive cells, suggesting loss of or delayed formation and recruitment of this specialized subset of regenerative cells during the healing process. Importantly, we observed primary ligand resistance to VDR signaling within Ddit4$^{-/-}$ MEFs, suggesting potential dermal fibroblastic regulation and defects in VDR$^{-/-}$ animals as well.

4. Materials and Methods

4.1. Animal Maintenance

VDR$^{-/-}$ animals (B6.129S4-Vdr^{tm1Mbd}/J) with targeted ablation of the second zinc finger were purchased from the Jackson Laboratory (JAX: 006133, Bar Harbor, ME, USA). Animal studies were approved by the institutional animal care and use committee (First Affiliated Hospital of Chongqing Medical University; SYXK2012-0001, January 2015). Animals were kept in a clean (virus- and parasite-free) facility under a 12-h light, 12-h dark cycle on a diet enriched with calcium (2%), phosphorus (1.25%) and lactose (20%) to prevent hyperparathyroidism, rickets and osteomalacia, but not alopecia [16].

4.2. Mouse Puncture Assay

Male 48-day-old littermate mice ($n = 4$ mice per time point) were anesthetized and received a 3.5 mm biopsy skin punch on each side of the trunk. Six days after puncture, re-epithelialization was monitored by harvesting tissue in 10% buffered formalin. Paraffin embedded wounds were processed for immunohistochemistry. For reverse transcription quantitative PCR (RT-qPCR) analysis, wounds ($n = 4$ per genotype and time point) were excised, trimmed and then processed using the RNeasy Plus Universal Mini Kit (Qiagen, Gaithersburg, MD, USA).

4.3. Chromatin Immunoprecipitation (ChIP)

We performed the ChIP assays using ChIP-IT$^{®}$ Express Chromatin Immunoprecipitation Kits from Active Motif. Primary keratinocytes were treated for 15 min with 10 nM 1,25D$_3$ (Biomol, Plymouth Meeting, PA, USA) reconstituted in absolute ethanol. Chromatin was cross-linked with 1% formaldehyde, quenched with glycine and processed in 1% sodium dodecyl sulfate

(SDS) cell lysis buffer. Samples were enzyme-treated to yield fragmented chromatin. Chromatin samples were incubated with 10 µg ChIP-grade anti-RNA polymerase II (ab26721; Abcam, Boston, MA, USA) and anti-VDR (sc-1008x; Santa Cruz Biotechnology, Santa Cruz, CA, USA) antibodies. Quantitative PCR (qPCR) was performed with primers flanking putative vitamin D response element (VDRE) consensus sequences in the regulatory region of mouse Ddit4 (based on the GRCm38 mouse assembly; available on: https://www.ncbi.nlm.nih.gov/genome/52). Primers were designed using Primer3 (bioinfo.ut.ee/primer3-0.4.0/primer3). Ddit4 target: Ddit4-1: (forward) 5-ttcccatccttttgcagttc-3, (reverse) 5-ccactgcccaatttcatctt-3; Ddit4-2: (forward) 5-tcagggtcccagtgtcctac-3, (reverse) 5-caattcaatggaacccagga-3; and Ddit4-3: (forward) 5-ggtacctttctcccctgctc-3, (reverse) 5-ctctcccctcgccttagc 3. Control primers: OC (osteocalcin): (forward) 5-caggggcagacactgaaaa-3, (reverse) 5-aggagactgccaggttctga-3; VDR-S1 (VDRE site): (forward) 5-gtagccatccatgtggcttt-3, (reverse) 5-ccagacggaagcctagagaa-3; and non-specific control (calponin) based on [60]. The non-specific control was used to normalize for DNA content and to calculate the relative enrichment of the regulatory regions according to the formula of Livak and Schmittgen [61] and presented as fold enrichment. Samples were run on a 1.5% tris-acetic acid- ethylenediaminetetraacetic acid (TAE) agarose gel for visualization.

4.4. In Vivo/In Vitro 5-Bromo-2′-deoxyuridine (BrdU) Labeling and Immunocytochemistry

For cell proliferation studies, wounded animals were injected intraperitoneally with 5-bromo-2′-deoxyuridine (BrdU) at 0.25 mg/gram (Roche) three hours prior to the time that the animals were sacrificed. Wounds were excised, fixed, and processed for immunocytochemical analysis. Proliferating cells were detected using a BrdU labeling and detection kit (Roche, Indianapolis, IN, USA) according to the manufacturer's instructions. For in vitro studies, sterile filtered 10 µM BrdU was prepared in media. After 12 h of vitamin D treatment, cells were washed with phosphate buffered saline (PBS), and replaced with the BrdU solution for a 2 h labeling period. Cells were fixed and then detected as above. Cell counting analysis was performed using the "spot" function on the Imaris (version 8.4, Bitplane, Concord, MA, USA) software.

4.5. Fluorescence-Activated Cell Sorting of Hair Follicle Stem Cells

Skin from 29-day-old wildtype (WT) and VDR$^{-/-}$ littermates was used to harvest hair follicle stem cells. The purification of bulge stem cells from mice was performed using established expression markers α6^{hi}/Cluster of differentiation 34 (CD34$^+$) [12]. The antibodies used were: Anti-Integrin α6 phycoerythrin (PE) conjugated, ab95703 (Abcam); anti-CD34, ab8158 (Abcam); and Alexa Fluor® 488 (Thermofisher, Waltham, MA, USA) secondary antibody at ~1 µg antibody per 10^6 cells. The double-labeled cell suspensions were placed in FACS tubes on ice for sorting on a BD FACSCalibur™ (BD Biosciences, San Jose, CA, USA). BD Falcon tubes containing 50% chelex-treated FBS in PBS were used to collect double-labeled KSCs. Using four sets (*n* = 4 experiments per genotype) of four 29-day-old postnatal mice, we enriched KSCs for RT-qPCR analysis. KSCs were further collected in the appropriate RNA stabilization reagent supplied in the RNeasy mini kit (Qiagen).

4.6. Immunofluorescent and Immunohistochemical Labeling of Skin

Skin tissues were obtained from the lower dorsal region. Wounds were excised and fixed in 10% formalin/PBS, processed and embedded vertically in paraffin. Tissues sections were mounted on microscope slides and processed for immunofluorescence staining with detergent (0.1% triton X). DDIT4 labeling was detected using an DDIT4 antibody (Abcam, ab106356) at 1:200 dilution followed by secondary staining with Alexa Fluor® 594 (Thermofisher). Slides were mounted in Vectashield® mounting medium with DAPI (Vector Laboratory). Immunohistochemistry was performed using anti-K10 (Abcam, ab9026) at 1:100 dilution with HRP-conjugated secondary antibody and DAB. Slides were imaged using a FV1000 (Olympus, Center Valley, PA, USA) confocal microscope. Control slides were included by negating primary antibodies or substituting with pre-immune sera (data not shown).

For confocal imaging, a series of three dimensional "z-axis" image projections of follicular axial depths were obtained in XYZ scan mode set to 1 µm/slice and a sample speed of 12.5 µs/pixel. All other parameters (e.g., pinhole diameter, gain, laser intensities) were kept constant during imaging. The fluorescence intensity was never saturated (max. 4096 intensity level) during imaging. The series of projected "z-axis" images were used to calculate average fluorescent intensity profiles per 594 nm channel using the freehand analysis tool in the Fluoview software v. 4.1 (Olympus). The proximal bulb region of Ddit4-stained hair follicles were averaged per image, as well as follicles of negative controls averaged per image to generate average background fluorescence intensities. Ddit4 levels in the bulb were relatively compared and background corrected between for eight individual follicles per genotype.

4.7. Ddit4$^{-/-}$ Mouse Embryonic Fibroblasts (MEFs) and Cell Count Measurement

Ddit4$^{-/-}$ and wild type MEFs were generously provided by Leif W. Ellisen as previously described [25], and maintained in Dulbecco's Modified Eagle's medium (DMEM)/10% fetal bovine serum, Pen/Strep. Cells were maintained at 37 °C in a 95% air/5% CO_2 atmosphere. Cells were plated in 24-well tissue culture plates at 2×10^4 cells per cm^2 and then replaced with fresh media and 1,25D3. Cells were trypsinized and counted 24 h later using an automated cell counter (Countess II FL, Thermofisher). Three sample preparations were made per condition in order to calculate the % of cells remaining ($n = 3$).

4.8. mRNA Reverse Transcription Quantitative PCR (RT-qPCR) Analysis

Total RNA was purified using the RNeasy mini kit (Qiagen). cDNA was synthesized from 300 ng total RNA by SuperScript Reverse Transcriptase III (Invitrogen, Carlsbad, CA, USA) using random hexamers. RT-qPCR analysis was performed with a Stratagene MX-3005P instrument utilizing TaqMan system reagents from ABI, and target genes were normalized to ATCB (β-actin) expression. TaqMan® assays used: (1) Control ATCB, β-actin (VIC®-labeled), assay ID 4352341E; (2) 1,25-dihydroxyvitamin D3 24-hydroxylase (Cyp24a1, FAM™-labeled), assay ID 4331182; (3) Ddit4 (FAM™-labeled), assay ID 4331182; and (4) Krt10 (FAM™-labeled), assay ID 4331182. All reactions were performed in triplicate experimental conditions. Data are presented as comparable arbitrary expression units.

4.9. Primary Neonatal Keratinocytes and mVDR Transient Transfection

Primary keratinocytes were harvested from neonatal pups aged 2–3 days using a trypsin floating method. Briefly, the pups were skinned and then floated on 0.05% trypsin (Sigma, St. Louis, MO, USA) at low temperature for 12 h. The epidermal sheets were harvested, minced, and then agitated with a stir bar in a keratinocyte growth medium on ice for 1 h. The medium consisted of: calcium and magnesium-free Eagle's Minimal Essential Medium (Gibco, Carlsbad, CA, USA), FBS with reduced calcium, 2 ng/mL human recombinant epidermal growth factor (EGF) (Novoprotein, Summit, NJ, USA) supplemented with 1× penicillin–streptomycin (Gibco). Calcium in the serum was removed by treating FBS with a chelating resin, Chelex®100 (Bio-Rad Laboratories, Hercules, CA, USA). Calcium concentration was adjusted to 0.05 mM by adding calcium chloride solution. The cell suspension was then filtered using a 40 micron mesh and were seeded (4×10^4 cells/cm^2) in tissue culture plates pre-coated with type I collagen (Gibco, R-011-K). Cultures were incubated in 8% CO_2 and 92% humidified atmosphere at 34 °C, and medium was changed every 2–3 days. The untagged mouse vitamin D receptor expression vector (pCMV6) was purchased from Origene (BC006716). Transfection of primary keratinocytes was conducted using the BioT reagent (Biolands) and protocol.

5. Conclusions

In conclusion, we show that Ddit4 marks early stress compartments within VDR$^{-/-}$ hair follicles which initiate during the morphogenic period. Importantly, we implicate Ddit4 as a functional component of growing anagen follicles during the hair cycle. Ddit4 is also a direct transcriptional target of the VDR within epidermal keratinocytes, highlighting its ligand-dependent genomic role during vitamin D signaling. VDR regulates Ddit4 during epidermal homeostasis and the wound repair process, namely the proper differentiation and stratification of the neo-epidermis post injury.

Acknowledgments: A Small Research Grant awarded to Thomas S. Lisse from the MDI Biological Laboratory and Hengguang Zhao from Natural Science Foundation Project of Chongqing, China, under grant number cstc2012jjA10119 supported the work. Research reported in this publication was also supported by Institutional Development Awards (IDeA) from the National Institute of General Medical Sciences of the National Institutes of Health under grant numbers P20GM0103423 and P20GM104318 (Sandra Rieger).

Author Contributions: Thomas S. Lisse and Hengguang Zhao conceived and designed the experiments; Thomas S. Lisse, Hengguang Zhao and Sandra Rieger carried out the experiments; Martin Hewison, Koichiro Abe, Hengguang Zhao and Sandra Rieger edited and discussed the manuscript; Thomas S. Lisse, Hengguang Zhao, Martin Hewison, Koichiro Abe and Sandra Rieger analyzed the data; Thomas S. Lisse wrote the manuscript.

Conflicts of Interest: The authors declare no conflict of interest.

Abbreviations:

VDR	Vitamin D receptor
Ddit4	DNA damage-inducible transcript 4
VDRE	Vitamin D response element
KSC	Keratinocyte stem cell
mTOR	Mechanistic/mammalian target of rapamycin

References

1. Hughes, M.R.; Malloy, P.J.; Kieback, D.G.; Kesterson, R.A.; Pike, J.W.; Feldman, D.; O'Malley, B.W. Point mutations in the human vitamin D receptor gene associated with hypocalcemic rickets. *Science* **1988**, *242*, 1702–1705. [PubMed]
2. Luderer, H.F.; Nazarian, R.M.; Zhu, E.D.; Demay, M.B. Ligand-dependent actions of the vitamin D receptor are required for activation of TGF-β signaling during the inflammatory response to cutaneous injury. *Endocrinology* **2013**, *154*, 16–24. [PubMed]
3. Oda, Y.; Tu, C.L.; Menendez, A.; Nguyen, T.; Bikle, D.D. Vitamin D and calcium regulation of epidermal wound healing. *J. Steroid Biochem. Mol. Biol.* **2015**, *164*, 379–385. [CrossRef] [PubMed]
4. Xie, Z.; Komuves, L.; Yu, Q.C.; Elalieh, H.; Ng, D.C.; Leary, C.; Chang, S.; Crumrine, D.; Yoshizawa, T.; Kato, S.; et al. Lack of the vitamin D receptor is associated with reduced epidermal differentiation and hair follicle growth. *J. Investig. Dermatol.* **2002**, *118*, 11–16. [CrossRef] [PubMed]
5. Sakai, Y.; Demay, M.B. Evaluation of keratinocyte proliferation and differentiation in vitamin D receptor knockout mice. *Endocrinology* **2000**, *141*, 2043–2049. [CrossRef] [PubMed]
6. Demay, M.B.; MacDonald, P.N.; Skorija, K.; Dowd, D.R.; Cianferotti, L.; Cox, M. Role of the vitamin D receptor in hair follicle biology. *J. Steroid Biochem. Mol. Biol.* **2007**, *103*, 344–346. [CrossRef] [PubMed]
7. Sakai, Y.; Kishimoto, J.; Demay, M.B. Metabolic and cellular analysis of alopecia in vitamin D receptor knockout mice. *J. Clin. Investig.* **2001**, *107*, 961–966. [CrossRef] [PubMed]
8. Chen, C.H.; Sakai, Y.; Demay, M.B. Targeting expression of the human vitamin D receptor to the keratinocytes of vitamin D receptor null mice prevents alopecia. *Endocrinology* **2001**, *142*, 5386–5389. [CrossRef] [PubMed]
9. Li, Y.C.; Pirro, A.E.; Amling, M.; Delling, G.; Baron, R.; Bronson, R.; Demay, M.B. Targeted ablation of the vitamin D receptor: An animal model of vitamin D-dependent rickets type II with alopecia. *Proc. Natl. Acad. Sci. USA* **1997**, *94*, 9831–9835. [CrossRef] [PubMed]
10. Erben, R.G.; Soegiarto, D.W.; Weber, K.; Zeitz, U.; Lieberherr, M.; Gniadecki, R.; Moller, G.; Adamski, J.; Balling, R. Deletion of deoxyribonucleic acid binding domain of the vitamin D receptor abrogates genomic and nongenomic functions of vitamin D. *Mol. Endocrinol.* **2002**, *16*, 1524–1537. [CrossRef] [PubMed]
11. Skorija, K.; Cox, M.; Sisk, J.M.; Dowd, D.R.; MacDonald, P.N.; Thompson, C.C.; Demay, M.B. Ligand-independent actions of the vitamin D receptor maintain hair follicle homeostasis. *Mol. Endocrinol.* **2005**, *19*, 855–862. [CrossRef] [PubMed]

12. Tumbar, T.; Guasch, G.; Greco, V.; Blanpain, C.; Lowry, W.E.; Rendl, M.; Fuchs, E. Defining the epithelial stem cell niche in skin. *Science* **2004**, *303*, 359–363. [CrossRef] [PubMed]

13. Cianferotti, L.; Cox, M.; Skorija, K.; Demay, M.B. Vitamin D receptor is essential for normal keratinocyte stem cell function. *Proc. Natl. Acad. Sci. USA* **2007**, *104*, 9428–9433. [CrossRef] [PubMed]

14. Palmer, H.G.; Martinez, D.; Carmeliet, G.; Watt, F.M. The vitamin D receptor is required for mouse hair cycle progression but not for maintenance of the epidermal stem cell compartment. *J. Investig. Dermatol.* **2008**, *128*, 2113–2117. [CrossRef] [PubMed]

15. Luderer, H.F.; Gori, F.; Demay, M.B. Lymphoid enhancer-binding factor-1 (LEF1) interacts with the DNA-binding domain of the vitamin D receptor. *J. Biol. Chem.* **2011**, *286*, 18444–18451. [CrossRef] [PubMed]

16. Lisse, T.S.; Saini, V.; Zhao, H.; Luderer, H.F.; Gori, F.; Demay, M.B. The vitamin D receptor is required for activation of cWnt and hedgehog signaling in keratinocytes. *Mol. Endocrinol.* **2014**, *28*, 1698–1706. [CrossRef] [PubMed]

17. Teichert, A.; Elalieh, H.; Bikle, D. Disruption of the hedgehog signaling pathway contributes to the hair follicle cycling deficiency in Vdr knockout mice. *J. Cell. Physiol.* **2010**, *225*, 482–489. [CrossRef] [PubMed]

18. Bikle, D.D. Vitamin D and the skin: Physiology and pathophysiology. *Rev. Endoc. Metab. Disord.* **2012**, *13*, 3–19. [CrossRef] [PubMed]

19. Bikle, D.D.; Elalieh, H.; Chang, S.; Xie, Z.; Sundberg, J.P. Development and progression of alopecia in the vitamin D receptor null mouse. *J. Cell. Physiol.* **2006**, *207*, 340–353. [CrossRef] [PubMed]

20. Oda, Y.; Uchida, Y.; Moradian, S.; Crumrine, D.; Elias, P.M.; Bikle, D.D. Vitamin D receptor and coactivators SRC2 and 3 regulate epidermis-specific sphingolipid production and permeability barrier formation. *J. Investig. Dermatol.* **2009**, *129*, 1367–1378. [CrossRef] [PubMed]

21. Zinser, G.M.; Sundberg, J.P.; Welsh, J. Vitamin D3 receptor ablation sensitizes skin to chemically induced tumorigenesis. *Carcinogenesis* **2002**, *23*, 2103–2109. [CrossRef] [PubMed]

22. Bikle, D.D. Vitamin D regulated keratinocyte differentiation. *J. Cell. Biochem.* **2004**, *92*, 436–444. [CrossRef] [PubMed]

23. Schauber, J.; Dorschner, R.A.; Coda, A.B.; Buchau, A.S.; Liu, P.T.; Kiken, D.; Helfrich, Y.R.; Kang, S.; Elalieh, H.Z.; Steinmeyer, A.; et al. Injury enhances TLR2 function and antimicrobial peptide expression through a vitamin D-dependent mechanism. *J. Clin. Investig.* **2007**, *117*, 803–811. [CrossRef] [PubMed]

24. Ellisen, L.W.; Ramsayer, K.D.; Johannessen, C.M.; Yang, A.; Beppu, H.; Minda, K.; Oliner, J.D.; McKeon, F.; Haber, D.A. REDD1, a developmentally regulated transcriptional target of p63 and p53, links p63 to regulation of reactive oxygen species. *Mol. Cell* **2002**, *10*, 995–1005. [CrossRef]

25. Sofer, A.; Lei, K.; Johannessen, C.M.; Ellisen, L.W. Regulation of mTOR and cell growth in response to energy stress by REDD1. *Mol. Cell. Biol.* **2005**, *25*, 5834–5845. [CrossRef] [PubMed]

26. Long, X.; Lin, Y.; Ortiz-Vega, S.; Yonezawa, K.; Avruch, J. Rheb binds and regulates the mTOR kinase. *Curr. Biol.* **2005**, *15*, 702–713. [CrossRef] [PubMed]

27. McGhee, N.K.; Jefferson, L.S.; Kimball, S.R. Elevated corticosterone associated with food deprivation upregulates expression in rat skeletal muscle of the mTORC1 repressor, REDD1. *J. Nutr.* **2009**, *139*, 828–834. [CrossRef] [PubMed]

28. Baida, G.; Bhalla, P.; Kirsanov, K.; Lesovaya, E.; Yakubovskaya, M.; Yuen, K.; Guo, S.; Lavker, R.M.; Readhead, B.; Dudley, J.T.; et al. REDD1 functions at the crossroads between the therapeutic and adverse effects of topical glucocorticoids. *EMBO Mol. Med.* **2015**, *7*, 42–58. [CrossRef] [PubMed]

29. Lisse, T.S.; Liu, T.; Irmler, M.; Beckers, J.; Chen, H.; Adams, J.S.; Hewison, M. Gene targeting by the vitamin D response element binding protein reveals a role for vitamin D in osteoblast mTOR signaling. *FASEB J.* **2011**, *25*, 937–947. [CrossRef] [PubMed]

30. Lisse, T.S.; Vadivel, K.; Bajaj, S.P.; Chun, R.F.; Hewison, M.; Adams, J.S. The heterodimeric structure of heterogeneous nuclear ribonucleoprotein C1/C2 dictates 1,25-dihydroxyvitamin D-directed transcriptional events in osteoblasts. *Bone Res.* **2014**. [CrossRef] [PubMed]

31. Wu, F.; Zhang, R.; Burns, F.J. Gene expression and cell cycle arrest in a rat keratinocyte line exposed to 56Fe ions. *J. Radiat. Res.* **2007**, *48*, 163–170. [CrossRef] [PubMed]

32. Page, A.; Navarro, M.; Garin, M.; Perez, P.; Casanova, M.L.; Moreno, R.; Jorcano, J.L.; Cascallana, J.L.; Bravo, A.; Ramirez, A. IKKβ leads to an inflammatory skin disease resembling interface dermatitis. *J. Investig. Dermatol.* **2010**, *130*, 1598–1610. [CrossRef] [PubMed]

33. Handfield, M.; Mans, J.J.; Zheng, G.; Lopez, M.C.; Mao, S.; Progulske-Fox, A.; Narasimhan, G.; Baker, H.V.; Lamont, R.J. Distinct transcriptional profiles characterize oral epithelium—Microbiota interactions. *Cell. Microbiol.* **2005**, *7*, 811–823. [CrossRef] [PubMed]

34. Kennedy-Crispin, M.; Billick, E.; Mitsui, H.; Gulati, N.; Fujita, H.; Gilleaudeau, P.; Sullivan-Whalen, M.; Johnson-Huang, L.M.; Suarez-Farinas, M.; Krueger, J.G. Human keratinocytes' response to injury upregulates CCL20 and other genes linking innate and adaptive immunity. *J. Investig. Dermatol.* **2012**, *132*, 105–113. [CrossRef] [PubMed]

35. Roh, C.; Tao, Q.; Lyle, S. Dermal papilla-induced hair differentiation of adult epithelial stem cells from human skin. *Physiol. Genom.* **2004**, *19*, 207–217. [CrossRef] [PubMed]

36. Milde, P.; Hauser, U.; Simon, T.; Mall, G.; Ernst, V.; Haussler, M.R.; Frosch, P.; Rauterberg, E.W. Expression of 1,25-dihydroxyvitamin D3 receptors in normal and psoriatic skin. *J. Investig. Dermatol.* **1991**, *97*, 230–239. [CrossRef] [PubMed]

37. Lisse, T.S.; Chun, R.F.; Rieger, S.; Adams, J.S.; Hewison, M. Vitamin D activation of functionally distinct regulatory miRNAs in primary human osteoblasts. *J. Bone Miner. Res.* **2013**, *28*, 1478–1488. [CrossRef] [PubMed]

38. Zella, L.A.; Meyer, M.B.; Nerenz, R.D.; Lee, S.M.; Martowicz, M.L.; Pike, J.W. Multifunctional enhancers regulate mouse and human vitamin D receptor gene transcription. *Mol. Endocrinol.* **2010**, *24*, 128–147. [CrossRef] [PubMed]

39. Reichrath, J.; Schilli, M.; Kerber, A.; Bahmer, F.A.; Czarnetzki, B.M.; Paus, R. Hair follicle expression of 1,25-dihydroxyvitamin D3 receptors during the murine hair cycle. *Br. J. Dermatol.* **1994**, *131*, 477–482. [CrossRef] [PubMed]

40. Squarize, C.H.; Castilho, R.M.; Bugge, T.H.; Gutkind, J.S. Accelerated wound healing by mTOR activation in genetically defined mouse models. *PLoS ONE* **2010**, *5*, e10643. [CrossRef] [PubMed]

41. Castilho, R.M.; Squarize, C.H.; Chodosh, L.A.; Williams, B.O.; Gutkind, J.S. mTOR mediates Wnt-induced epidermal stem cell exhaustion and aging. *Cell Stem Cell* **2009**, *5*, 279–289. [CrossRef] [PubMed]

42. Blanpain, C.; Lowry, W.E.; Geoghegan, A.; Polak, L.; Fuchs, E. Self-renewal, multipotency, and the existence of two cell populations within an epithelial stem cell niche. *Cell* **2004**, *118*, 635–648. [CrossRef] [PubMed]

43. Danielsson, C.; Fehsel, K.; Polly, P.; Carlberg, C. Differential apoptotic response of human melanoma cells to 1α,25-dihydroxyvitamin D3 and its analogues. *Cell Death Differ.* **1998**, *5*, 946–952. [CrossRef] [PubMed]

44. Al-Hendy, A.; Diamond, M.P.; Boyer, T.G.; Halder, S.K. Vitamin D3 inhibits Wnt/β-catenin and mTOR signaling pathways in human uterine fibroid cells. *J. Clin. Endocrinol. Metab.* **2016**, *101*, 1542–1551. [CrossRef] [PubMed]

45. Fingar, D.C.; Salama, S.; Tsou, C.; Harlow, E.; Blenis, J. Mammalian cell size is controlled by mTOR and its downstream targets S6K1 and 4EBP1/eIF4E. *Genes Dev.* **2002**, *16*, 1472–1487. [CrossRef] [PubMed]

46. Brafman, A.; Mett, I.; Shafir, M.; Gottlieb, H.; Damari, G.; Gozlan-Kelner, S.; Vishnevskia-Dai, V.; Skaliter, R.; Einat, P.; Faerman, A.; et al. Inhibition of oxygen-induced retinopathy in RTP801-deficient mice. *Investig. Ophthalmol. Vis. Sci.* **2004**, *45*, 3796–3805. [CrossRef] [PubMed]

47. Yoshida, T.; Mett, I.; Bhunia, A.K.; Bowman, J.; Perez, M.; Zhang, L.; Gandjeva, A.; Zhen, L.; Chukwueke, U.; Mao, T.; et al. Rtp801, a suppressor of mTOR signaling, is an essential mediator of cigarette smoke-induced pulmonary injury and emphysema. *Nat. Med.* **2010**, *16*, 767–773. [CrossRef] [PubMed]

48. Kamocki, K.; van Demark, M.; Fisher, A.; Rush, N.I.; Presson, R.G., Jr.; Hubbard, W.; Berdyshev, E.V.; Adamsky, S.; Feinstein, E.; Gandjeva, A.; et al. RTP801 is required for ceramide-induced cell-specific death in the murine lung. *Am. J. Respir. Cell Mol. Biol.* **2013**, *48*, 87–93. [CrossRef] [PubMed]

49. Reiling, J.H.; Hafen, E. The hypoxia-induced paralogs Scylla and Charybdis inhibit growth by down-regulating S6K activity upstream of TSC in Drosophila. *Genes Dev.* **2004**, *18*, 2879–2892. [CrossRef] [PubMed]

50. Shoshani, T.; Faerman, A.; Mett, I.; Zelin, E.; Tenne, T.; Gorodin, S.; Moshel, Y.; Elbaz, S.; Budanov, A.; Chajut, A.; et al. Identification of a novel hypoxia-inducible factor 1-responsive gene, RTP801, involved in apoptosis. *Mol. Cell. Biol.* **2002**, *22*, 2283–2293. [CrossRef] [PubMed]

51. Liu, L.; Parent, C.A. Review series: TOR kinase complexes and cell migration. *J. Cell Biol.* **2011**, *194*, 815–824. [CrossRef] [PubMed]

52. Bai, X.; Lei, M.; Shi, J.; Yu, Y.; Qiu, W.; Lai, X.; Liu, Y.; Yang, T.; Yang, L.; Widelitz, R.D.; et al. Roles of GasderminA3 in catagen-telogen transition during hair cycling. *J. Investig. Dermatol.* **2015**, *135*, 2162–2172. [CrossRef] [PubMed]
53. Ragone, G.; Bresin, A.; Piermarini, F.; Lazzeri, C.; Picchio, M.C.; Remotti, D.; Kang, S.M.; Cooper, M.D.; Croce, C.M.; Narducci, M.G.; et al. The Tcl1 oncogene defines secondary hair germ cells differentiation at catagen-telogen transition and affects stem-cell marker CD34 expression. *Oncogene* **2009**, *28*, 1329–1338. [CrossRef] [PubMed]
54. Peters, E.M.; Stieglitz, M.G.; Liezman, C.; Overall, R.W.; Nakamura, M.; Hagen, E.; Klapp, B.F.; Arck, P.; Paus, R. p75 Neurotrophin receptor-mediated signaling promotes human hair follicle regression (Catagen). *Am. J. Pathol.* **2006**, *168*, 221–234. [CrossRef] [PubMed]
55. Kim, J.O.; Kim, J.Y.; Kwack, M.H.; Hong, S.H.; Kim, M.K.; Kim, J.C.; Sung, Y.K. Identification of troglitazone responsive genes: Induction of RTP801 during troglitazone-induced apoptosis in Hep 3B cells. *BMB Rep.* **2010**, *43*, 599–603. [CrossRef] [PubMed]
56. Molitoris, J.K.; McColl, K.S.; Swerdlow, S.; Matsuyama, M.; Lam, M.; Finkel, T.H.; Matsuyama, S.; Distelhorst, C.W. Glucocorticoid elevation of dexamethasone-induced gene 2 (Dig2/RTP801/REDD1) protein mediates autophagy in lymphocytes. *J. Biol. Chem.* **2011**, *286*, 30181–30189. [CrossRef] [PubMed]
57. Bikle, D.D. The vitamin D receptor: A tumor suppressor in skin. *Discov. Med.* **2011**, *11*, 7–17. [PubMed]
58. Bikle, D.D.; Chang, S.; Crumrine, D.; Elalieh, H.; Man, M.Q.; Dardenne, O.; Xie, Z.; Arnaud, R.S.; Feingold, K.; Elias, P.M. Mice lacking 25OHD 1α-hydroxylase demonstrate decreased epidermal differentiation and barrier function. *J. Steroid Biochem. Mol. Biol.* **2004**, *89*, 347–353. [CrossRef] [PubMed]
59. Elizondo, R.A.; Yin, Z.; Lu, X.; Watsky, M.A. Effect of vitamin D receptor knockout on cornea epithelium wound healing and tight junctions. *Investig. Ophthalmol. Vis. Sci.* **2014**, *55*, 5245–5251. [CrossRef] [PubMed]
60. Palmer, H.G.; Anjos-Afonso, F.; Carmeliet, G.; Takeda, H.; Watt, F.M. The vitamin D receptor is a Wnt effector that controls hair follicle differentiation and specifies tumor type in adult epidermis. *PLoS ONE* **2008**, *3*, e1483. [CrossRef] [PubMed]
61. Livak, K.J.; Schmittgen, T.D. Analysis of relative gene expression data using real-time quantitative PCR and the $2^{-\Delta\Delta Ct}$ method. *Methods* **2001**, *25*, 402–408. [CrossRef] [PubMed]

© 2016 by the authors. Licensee MDPI, Basel, Switzerland. This article is an open access article distributed under the terms and conditions of the Creative Commons Attribution (CC BY) license (http://creativecommons.org/licenses/by/4.0/).

International Journal of
Molecular Sciences

MDPI

Article

Immune Modulating Topical S100A8/A9 Inhibits Growth of *Pseudomonas aeruginosa* and Mitigates Biofilm Infection in Chronic Wounds

Hannah Trøstrup [1,*], Christian Johann Lerche [1], Lars Christophersen [1], Peter Østrup Jensen [1], Niels Høiby [1,2] and Claus Moser [1]

[1] Department of Clinical Microbiology, Copenhagen University Hospital, Rigshospitalet, 2100 Copenhagen, Denmark; cjl@dadlnet.dk (C.J.L.); lars.christophersen@regionh.dk (L.C.); peter.oestrup.jensen@regionh.dk (P.Ø.J.); hoibyniels@gmail.com (N.H.); moser@dadlnet.dk (C.M.)

[2] Institute for Immunology and Microbiology, University of Copenhagen, 2100 Copenhagen, Denmark

* Correspondence: htroestrup@gmail.com; Tel.: +45-2991391

Received: 29 April 2017; Accepted: 16 June 2017; Published: 26 June 2017

Abstract: *Pseudomonas aeruginosa* biofilm maintains and perturbs local host defense, hindering timely wound healing. Previously, we showed that *P. aeruginosa* suppressed S100A8/A9 of the murine innate host defense. We assessed the potential antimicrobial effect of S100A8/A9 on biofilm-infected wounds in a murine model and *P. aeruginosa* growth in vitro. Seventy-six mice, inflicted with a full-thickness burn wound were challenged subcutaneously (s.c.) by 10^6 colony-forming units (CFUs) of *P. aeruginosa* biofilm. Mice were subsequently randomized into two treatment groups, one group receiving recombinant murine S100A8/A9 and a group of vehicle controls (phosphate-buffered saline, PBS) all treated with s.c. injections daily for up to five days. Wounds were analyzed for quantitative bacteriology and contents of key inflammatory markers. Count of blood polymorphonuclear leukocytes was included. S100A8/A9-treatment ameliorated wound infection, as evaluated by quantitative bacteriology ($p \leq 0.05$). In vitro, growth of *P. aeruginosa* was inhibited dose-dependently by S100A8/A9 in concentrations from 5 to 40 µg/mL, as determined by optical density-measurement (OD-measurement) and quantitative bacteriology. Treatment slightly augmented key inflammatory cytokine Tumor Necrosis Factor-α (TNF-α), but dampened interferon-γ (IFN-γ) levels and blood polymorphonuclear count. In conclusion, topical S100A8/A9 displays remarkable novel immune stimulatory and anti-infective properties in vivo and in vitro. Importantly, treatment by S100A8/A9 provides local infection control. Implications for a role as adjunctive treatment in healing of chronic biofilm-infected wounds are discussed.

Keywords: biofilm infection; chronic wounds; host defense; *Pseudomonas aeruginosa*; S100A8/A9

1. Introduction

Infected chronic wounds are an increasing threat to society and can be fatal for diabetic patients. There is an urgent need for adjunctive therapies taking increasing microbial resistance into consideration. Innate host mechanisms may hold a key to this progress in the understanding of wound chronicity.

A rigorous debridement is necessary for optimal treatment of chronic human wounds. Even though surgical effort has been made in order to optimize the wound bed for healing, failure of split-skin transplantation is often the case, especially when *Pseudomonas aeruginosa* biofilm is present in the wound [1]. *P. aeruginosa* biofilms reside deeply in the wound bed [2] and have an undisputable effect on local host response, as increasingly evidencesd in the last decade [3]. Biofilms are invisible to the naked eye and at present, there is no local marker to guide clinicians. Thus, greater understanding of the clinical impact of biofilms on host response is of importance and may lead to the development of new adjunctive treatment strategies that take sufficient antibiotic therapy as well as the immunomodulatory effect of biofilm infections into consideration [4].

S100A8 and A9 are the most abundant cytoplasmic proteins of neutrophils and monocytes [5] and are released upon activation of these phagocytes [6]. They form the physiologically relevant heterodimeric S100A8/A9, where S100A8 is the active component and S100A9 the regulating subunit [7]. S100A8/A9 has wide antifungal and antimicrobial potential [8]. We previously reported reduced S100A8/A9 levels in colonized human chronic wounds [9,10]. Since *P. aeruginosa* is able to degrade cytokines [11], S100A8/A9 may likewise be degraded in vivo in anatomically close proximity to the *P. aeruginosa* biofilm as part of the evasion of the immune system.

S100A8/A9 may exert its host defense functions locally upon neutrophil death mediated release at sites of tissue infection, thereby controlling growth of *P. aeruginosa* [12]. Animal models may shed light on the complex pathophysiology of chronic wounds and accordingly, possible immunomodulatory and beneficial effects of S100A8/A9 on wound infection were investigated in the present study using a representative animal model. The BALB/c strain of mice is susceptible to *P. aeruginosa* infection and literature describes a potential anti-infective effect of this protein. Thus, we assessed the impact of recombinant S100A8/A9 on the course of infection in BALB/c mice as well as the direct inhibitory effect on S100A8/A9 in vitro.

2. Results

2.1. Quantitative Bacteriology of P. aeruginosa Biofilm-Infected Wounds

In order to determine an anti-infective effect of host response S100A8/A9 on chronic *P. aeruginosa* biofilm-infected wounds in vivo, we subjected BALB/c mice to topical intervention by this recombinant protein.

A significant increase in colony-forming unit (CFU) count from day 1 to day 5 was observed in the control (phosphate-buffered saline, PBS) group ($p < 0.0001$). In contrast, no such increase was seen in the S100A8/A9-treated group. Mean values ($\log_{10}$ CFU/wound) for the treated group compared to controls was 8.266 ± 0.183 vs. 7.804 ± 0.866 at day 1. At day 2, it was 8.628 ± 0.429 to 8.618 ± 0.328; day 3, 8.498 ± 0.500 to 8.981 ± 0.437; and day 5, 8.523 ± 1.221 to 9.317 ± 0.316. At the termination of the experiment on day 5, significantly reduced CFUs were found in the S100A8/A9-treated wounds compared to the control group (Figure 1, $p < 0.05$).

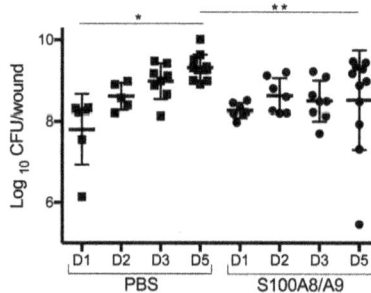

Figure 1. Induction of local wound infection control. $\log_{10}$ colony-forming units (CFUs) of *P. aeruginosa*/wound as expression of time in the intervention group (S100A8/A9) and control group (phosphate-buffered saline, PBS). S100A8/A9-treatment induced infection control. The number of colony-forming units (CFUs) after 5 days of treatment was significantly reduced as compared to the non-treated control group (** $p < 0.05$). Increase in growth over 5 days in the control group only (* $p < 0.0001$). Individual values indicate mean of duplicates with overall mean ± standard deviation. Black squares, PBS-treated; black circles, S100A8/A9-treated.

2.2. Bacterial Growth Inhibition Assay

To verify the anti-infective effect of topical S100A8/A9 on infected wounds, we assessed the direct interaction of S100A8/9 and planktonic PAO1 (a strain of *P. aeruginosa*). A growth inhibition assay was performed to determine whether the reduced bacterial levels in the wounds were due to immune modulation or direct inhibition of *P. aeruginosa* growth. The addition of S100A8/A9 reduced bacterial growth of *P. aeruginosa* time- and dose-dependently, starting at 2.50 µg/mL at 24 h of incubation (Figure 2A, optical density (OD) 0.732 versus 0.999 at blank), 5 µg/mL at 6 h (0.481 versus 0.574) and 10, 20, and 40 µg/mL at 2 h (0.081 versus 0.11). In addition, PAO1 was grown in medium for a maximum of 24 h at 37 °C with or without S100A8/A9. CFU counts the following day confirmed the result with a lower CFU/mL at 1 h of incubation with 20 and 40 µg/mL. At 6 h, bacterial growth was inhibited dose-dependently, from 5 to 40 µg/mL (Figure 2B) and a maximal 0.5 log inhibition with 40 µg/mL. At 24 h, approximately 1 log reduction of CFU/mL was observed.

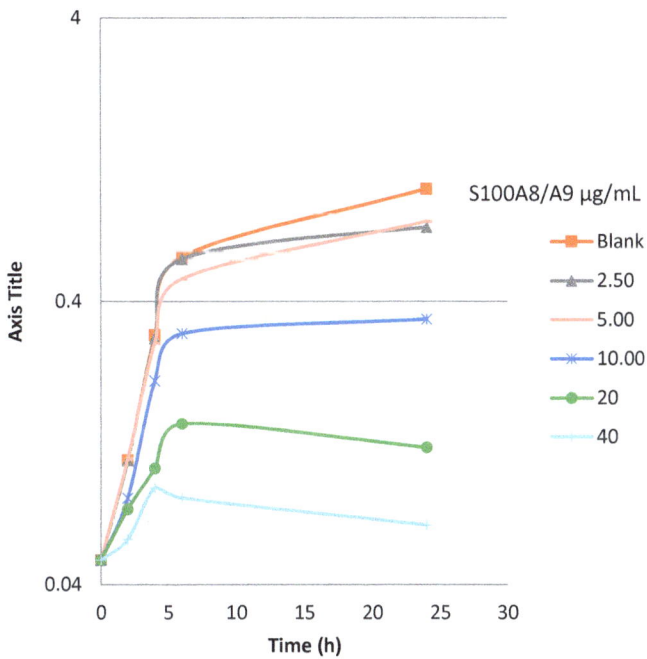

(**A**)

Figure 2. *Cont.*

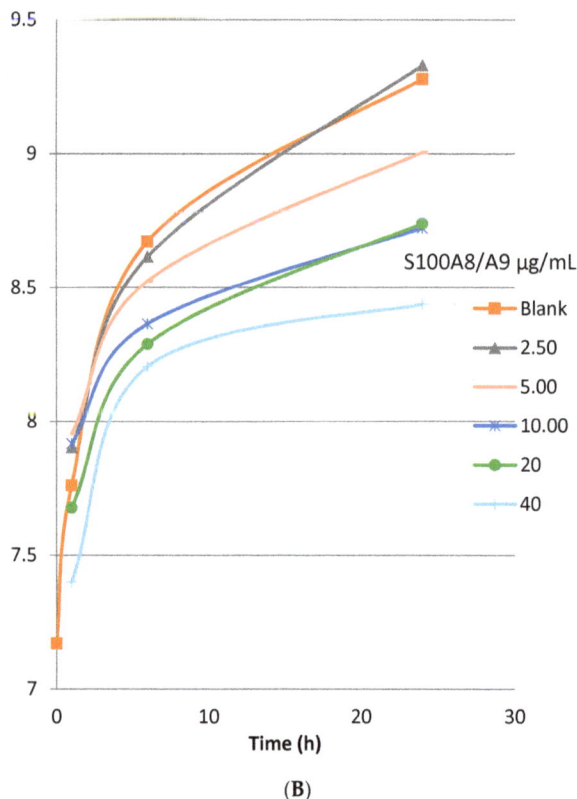

(B)

Figure 2. S100A8/A9 inhibits growth of *Pseudomonas aeruginosa*. (**A**) Growth inhibitory effect of S100A8/A9 on PAO1, shown by OD-curve (read at OD_{600} at time points indicated). (**B**) Growth inhibitory effect of S100A8/A9 on PAO1, shown by bacterial CFU count (1 h, 6 h, 24 h, at concentrations 40, 20, 10, 5, 2.5, or 0 μg/mL). Growth inhibitory effect of S100A8/A9 on PAO1, shown by (**A**) OD-curve (read at OD_{600} at time points indicated) and (**B**) bacterial count as log_{10} CFU (1 h, 6 h, 24 h at concentrations 40, 20, 10, 5, 2.5, or 0 μg/mL).

2.3. The Impact of S100A8/A9-Treatment on Local Wound Inflammatory Markers

We previously described an aggravated Interleukin-1β (IL-1β) response towards *P. aeruginosa* biofilm in the chronic wound model using BALB/c mice [13]. As S100A8/A9 is an endogenous peptide, we speculated that exogenous application would immunomodulate host defense in the wounds.

S100A8/A9 stimulated Tumor Necrosis Factor-α (TNF-α) from day 1 to 5 ($p < 0.0261$) and 3 to 5 ($p < 0.001$) (Figure 3A). Interferon-γ (IFN-γ) levels (Figure 3B) were discretely attenuated with a significant increase from day 1–5, day 2–5, and day 3–5 only in the control group ($p < 0.0023$, $p < 0.0008$, $p < 0.0054$, respectively). In the S100A8/A9 group, no such increase was seen.

There were no significant immunomodulatory effects of S100A8/A9 on the levels of IL-1β (Figure 3C), IL-10 (Figure 3D), Keratinocyte-derived Chemokine (KC) (Figure 3E), Granulocyte-Colony Stimulating Factor (G-CSF) levels (Figure 3F), or IL-17 levels (Figure 3G).

IL-1β, IL-10, KC, and G-CSF levels were all induced by infection (Figures S2–S5). IL-1β levels were induced in 1 day with treatment or PBS, TNF-α in 5 days with treatment or PBS as compared to background mice (Figures S1 and S2).

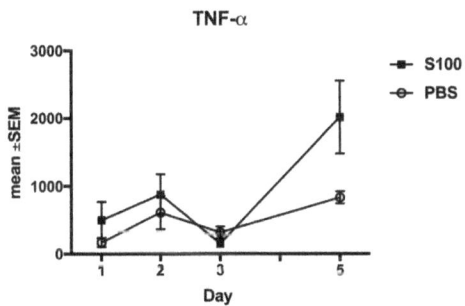

(A)

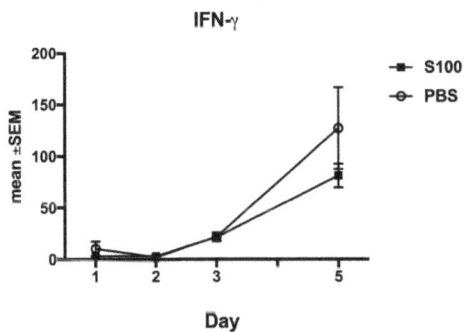

(B)

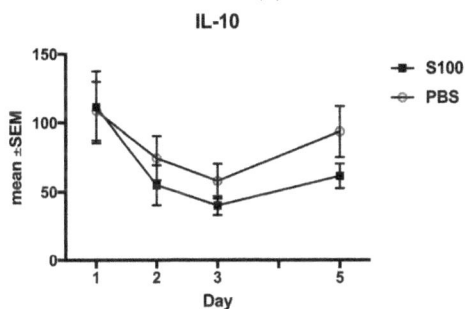

(C)

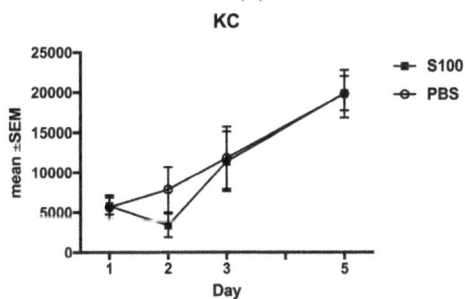

(D)

Figure 3. *Cont.*

(E)

(F)

(G)

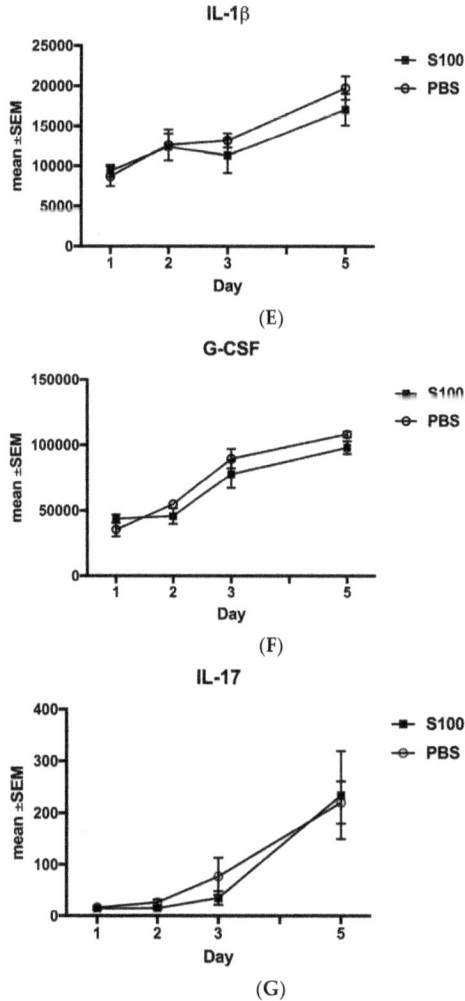

Figure 3. Modulation of proinflammatory cytokines. (**A**) Tumor Necrosis Factor-α (TNF-α) levels were increased in the S100A8/A9 group from day 1 to 5 ($p < 0.0261$) and day 3 to 5 ($p < 0.001$). No such increase was observed in the PBS group; (**B**) Interferon-γ (IFN-γ) levels were increased in the PBS from day 1 to 5 ($p < 0.0023$), 2 to 5 ($p < 0.0008$), and day 3 to 5 ($p < 0.0054$). No increase was observed in the S100A8/A9 group; (**C**) IL-10. No significant changes were observed in either of the two groups, but there was a tendency to an attenuated response in the S100A8/A9 group; (**D**) in the PBS group, Keratinocyte-derived chemokine (KC) increased from day 1 to 5 ($p < 0.0156$). In the S100A8/A9 group, KC increased from day 1 to 5 ($p < 0.0139$) and day 2 to 5 ($p < 0.0017$). (**E**) In both groups, IL-1β increased from day 1 to 5 ($p < 0.0005$ in the PBS group and $p < 0.0482$ in the S100A8/A9 group); (**F**) In both groups, Granulocyte-Colony Stimulating Factor (G-CSF) increased with time for all time points evaluated. (**G**) In both groups, Il-17 increased with time for all time points evaluated. Modulation of inflammatory response by box plots of inflammatory cytokines (**A–E**): Levels of proteins are expressed in pg/mL wound.

2.4. Systemic Polymorphonuclear Count

In the S100A8/A9 group, no significant increase in PMN count was observed from 1 to 5 days of treatment. In contrast, in the control group, an increase in PMN count was seen from 1 to 3 days after infection ($p < 0.03$) and from 1 to 5 days of treatment ($p < 0.003$). Infection induced significantly higher PMN counts in all groups compared to background mice (no wound, no infection, $n = 6$, mean PMNs/mL whole blood $\pm$ SD: $2.7 \times 10^5 \pm 1.4 \times 10^5$). Burn wounding (controls immediately after wound infliction ($n = 3$), and at 2 ($n = 3$) and 4 days ($n = 3$) post wound infliction) and burn wounded mice challenged with infection and sacrificed immediately after infection ($n = 3$) induced higher PMN count than in background mice (Figure 4). Systemic total white blood cell count at 5 days after infection was equal between the groups (Figure S6).

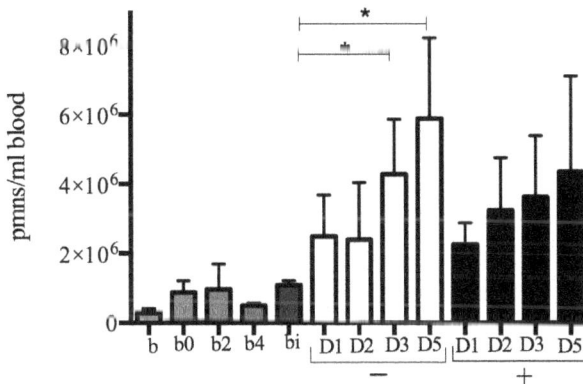

Figure 4. S100A8/A9 dampens systemic polymorphonuclear count. Systemic polymorphonuclear (PMN) count in S100A8/A9 (+) treated group versus PBS (−). Bars are expressed with mean + standard deviation. No significant increase was seen in the S100A8/A9 group. In contrast, in the PBS group, an increase in PMN count was seen from day 1 to 3 ($p < 0.03$) and day 1 to 5 ($p < 0.003$). Abbreviations: b: background ($n = 6$), b0: burned, results obtained immediately after wounding, b2: burned, two days after wounding, b4: burned, 4 days after wounding; bi: burn and infection, results obtained immediately after infection. D1, 2, 3, 5: day 1, 2, 3, 5. *, significant increase.

2.5. Digital Planimetry

To assess a clinical impact on wound closure, we analyzed wound area and necrotic areas by planimetry.

After 5 days of treatment, S100A8/A9 did not affect the total wound area (mean 244.9 mm^2 $\pm$ 74.11 versus 257.1 $\pm$ 60.54 for the control group) or the area of necrosis at day 5 (174.8 mm^2 $\pm$ 52.38 versus 180.1 $\pm$ 34.27 for the control group) (Figures S7 and S8).

2.6. Epidermal Growth Factor Levels

The key growth factor epidermal growth factor (EGF) was quantified as a pseudomarker for wound closure, with results showing no effect of S100A8/A9 on EGF levels in wounds (Figure S9).

3. Discussion

The aim of the present study was to evaluate an antimicrobial effect in vivo and in vitro of S100A8/A9 on *P. aeruginosa*, a clinically important wound pathogen.

Selection of this calcium-binding protein [14] was based on a previous description of reduction of S100A8/A9 in human chronic wounds [9,10].

Degradation of S100A8/A9 in the proteolytic wound environment could explain this reduction. However, the protein displays high protease-resistance, suggesting another route [15]. Regardless of the underlying mechanism behind S100A8/A9 reduction in chronic biofilm-infected wounds, a logical consequence of previous observations was to perform a study of topical S100A8/A9 treatment with dose selection of 1 µg/250 µL. Studies performed by Sroussi et al. showed that a single dose of topically applied ala[42] S100A8 (25 µL of a 1 µg/mL solution intradermal +10 µL directly to excisional murine wounds) reduced bacterial CFU count in mice subject to restraint stress at day 1 and day 5 post wounding. However, the bacteriology was not further described and biofilm presence, in contrast to the present study, was not considered [16].

The present results show, to our knowledge, a novel inhibitory effect of S100A8/A9 on *P. aeruginosa* in a murine model of biofilm-infected wounds as well as in vitro. Growth inhibition by S100A8/A9 has been described by several groups, in multiple species in vitro including *Borrelia burgdorferi*, *Klebsiella pneumoniae*, *Escherichia coli* [12], *Staphylococcus aureus*, *Candida albicans*, and *Salmonella typhimurium* [8,17–20]. The addition of zinc in several of the studies reversed the inhibitory effect and accordingly, S100A8/A9 chelation of zinc and manganese [8] may inhibit growth by nutritional immunity. Another possibility could be the disruption of biofilm alginate polymerization by S100A8/A9 bound calcium and a resultant increase in susceptibility of bacteria superficially situated in the biofilm to phagocytosis by PMNs. Another potential method of action was described by Akerström et al. [21], who found a lysing effect of S100A8/A9 on the anaerobic bacterium *Finegoldia magna* (previously known as *Peptostreptococcus magnus*) by interacting with the cell membrane.

Regarding the modulation of inflammatory mediators in the wounds, treatment augmented TNF-α and decreased IFN-γ levels locally. Vogl and colleagues showed that S100A8/A9 amplifies phagocyte activation during lipopolysaccharide-induced sepsis via activation of TLR-4 upstream of TNF-α response [7]. The reason for a potential decrease in IFN-γ is not clear. We assessed the wounds in the early part of inflammation and speculate that at this point IFN-γ acts as an acute inflammatory marker, related to the wound infliction itself. We have previously shown an improved clinical outcome of lung disease in Th1-reacting C3H/HeN mice (higher IFN-γ levels) as compared to Th2-reacting BALB/c mice [22,23]. The adaptive immune response also emerges only at the termination of the present study. Accordingly, the timing of treatment and improved healing outcome and correlations to IFN-γ levels need further investigation.

Previously, we found an early aggravating effect of *P. aeruginosa* biofilm on local IL-1β levels in BALB/c mice. Treatment by S100A8/A9 in the doses chosen in this study did not counteract this observation.

In regard to the stability and pharmacodynamics of S100A8/A9, we were unable to find reports on the durability of exogenously applied S100A8/A9 in vivo, but further experiments could address this matter. In vitro, S100A8/A9 is extremely protease-resistant [15].

In addition to growth inhibiting properties, S100A8/A9 may also have immunomodulatory leukocyte effects [24]. In a murine model of *P. aeruginosa* keratitis, S100A8/A9 treatment promoted bacterial clearance by increasing production of reactive oxygen species [25]. Interestingly, subunit S100A9 induces neutrophil phagocytosis of *Eschericia coli* [26]. Therefore, an increased phagocytosis caused by appropriate levels of S100A8/A9 might be an additional explanation for the diminished bacterial count in the present study. A role for S100A8/A9 in the oxidative metabolism and functional role of neutrophils should be explored further, as this could explain the reduced CFU count in the S100A8/A9-treated group of mice and may indirectly impact the delayed IL-1β response. Ryckmann et al. showed that S100A8/A9 in the concentration range 10^{-12} to 10^{-10} M acts as a chemotactic factor for neutrophils in vitro. In the same study, equimolar quantities of S100A8 and A9 (0.1–10 µg/mL), dissolved in PBS, attracted neutrophils in an in vivo murine air pouch model [27].

In the present study, treatment reduced blood PMN count although this attenuation was not reflected in the PMN mobilizer G-CSF, nor the two chemo-attractants KC and IL-17. Interestingly, no significant differences in levels of cytokines and chemokines were found between treated and control groups while these inflammatory markers increased with time in both groups. We speculate that the dose of S100A8/A9 chosen was insufficient to cause significant differences. Ongoing studies will substantiate relevant and safe doses of S100A8/A9 in this model.

A reduction of circulating PMNs could theoretically be beneficial, as chronic wounds are trapped in the inflammatory state of healing, showing continuous neutrophil extravasation and excessive host responses regarding TNF-α, but not IL-17, which is part of the adaptive immune system. IL-17 suppression by S100A8/A9 could imply that this protein delays initiation of the adaptive immune response. We hypothesize that S100A8/A9 is a marker of activated, phagocytizing leukocytes. The reduced extracellular levels of this protein, observed in our previous studies, may reflect the inhibition of neutrophil function and phagocytosis caused by rhamnolipid, produced by *P. aeruginosa* biofilms, in the chronically infected wounds, to ensure a fitness advantage and evasion of the immune system [28,29]. However, further insights on the actual impact of topical S100A8/A9 on PMN mobilization and physiology should be addressed in vitro.

The intervention did not have any significant impact on wound healing, as determined by planimetry, or on levels of EGF. This is probably due to the relatively short observation period (up to 10 days after wound infliction or six days post infection).

In conclusion, we have demonstrated that S100A8/A9 mitigates local *P. aeruginosa* biofilm infection by modulating local host response and attenuating systemic reaction in the chronic wound murine model. Accordingly, a clear growth inhibiting effect of recombinant murine S100A8/A9 on *P. aeruginosa* was observed in vitro. Apparently, this innate host defense protein possesses the capacity to combat an established biofilm infection as well as planktonic bacteria. The present study supports a beneficial role of S100A8/A9 as adjunctive treatment in refractory *P. aeruginosa* biofilm-infected wounds.

4. Materials and Methods

4.1. Study Design of Animal Experiments

Mice predisposed for chronic *P. aeruginosa* biofilm infection by infliction of burn wound full-thickness necrosis were treated daily with S100A8/A9 or vehicle (PBS) for a period of 5 days post infection. Wounds were assessed by quantitative bacteriology, quantification of key inflammatory cytokines and growth factors locally, and count of systemically polymorphonuclear (PMN) leukocytes. Digital photoplanimetry described development of wounds within the 5 day treatment period. Assessment of epidermal growth factor (EGF) was chosen as a marker for progression in wound closure.

4.2. Animals

In total, 78 female BALB/c mice (Taconic Europe A/S, Lille Skensved, Denmark) were used for the present study. For comparison of inflammatory reaction to the wound infliction and no consequent treatment, six sham control and burn wounded, but not infected, mice were included. Three mice were evaluated at 2 days and three mice at 4 days post wound infliction. An additional three burn wounded mice were sacrificed immediately after infection for presentation of baseline levels. Two untreated and unburned mice were kept for background levels.

Animals were acclimatized for at least 1 week in the animal facilities before experimentation and were allowed free access to chow and water. Animal care was provided by trained personnel.

The study was approved by the Animal Ethics Committee of Denmark (2015-15-0201-00618, 15 July 2017) and all experiments were performed following National and European Union guidelines.

4.3. Establishment of Chronic Wound Infection

First, structural damage to the skin acting as a predisposing factor for settlement of infection was induced by establishment of third degree burn wounds [13,30]. The area of the thermal lesion was 1.7×1.7 cm (2.9 cm^2) and mice were anesthetized by 0.3 mL hyp/midazolam.

Four days post wound infliction, 100 µL of a 10^7/mL colony forming units (CFU). *P. aeruginosa* biofilm solution (10^6 CFU/wound) was injected subcutaneously (s.c.), centrally, beneath the wound. The biofilm was prepared as previously described [31]. Treatment was initiated 24 h after biofilm injection to allow for the settlement of infection. Mice were sacrificed by intraperitoneal injection of a pentobarbiturate/lidocaine overdose after 1, 2, 3, or 5 days of treatment.

4.4. Immunomodulation of Murine Wounds

Recombinant mouse S100A8/A9 (Cloud Clone, Katy, TX, USA) was reconstituted in sterile-filtered PBS (0.22 µm) to reach a concentration of 4 µg/mL and 250 µL of the solution (1 µg) was given s.c. by injection centrally beneath the wounds. Controls received PBS from the same flask used for reconstitution.

Treatment (+/− S100A8/A9) was given 1, 2, 3, 4, and 5 days after biofilm injection. Fourteen mice were treated for 1 day, 14 for 2 days, and 16 for 3 days. Eleven mice were given PBS for 5 days (one mouse died as a result of the burn wound infliction) and 12 mice were given S100A8/A9 for 5 days.

4.5. Collection of Wounds

To minimize tissue disruption, wounds were carefully retrieved (n = 76) with sterile scalpels. Each wound was placed in 2 mL sterile saline and kept on ice until homogenization for 20 s by 14,000 rpm using a Heidolph Silent Crusher M (Heidolph Instruments, Schwabach, Germany) followed by centrifugation for 15 min at 5000 rpm. Supernatants were sterile-filtered (0.22 µm) and kept at −80 °C until analysis.

4.6. Quantitative Bacteriology

Serial dilutions of wound homogenates were prepared in saline and aliquots of 0.1 mL were spread onto modified Conradi-Drigalski medium (State Serum Institute, Copenhagen, Denmark), selective for Gram-negative rods. CFUs were counted the following day. The actual numbers of CFUs in the samples were calculated by multiplication with the dilution factor.

4.7. Bacterial Growth Inhibition by S100A8/A9

The growth inhibitory effect of recombinant S100A8/A9 (Cloud-Clone Corps, Katy, TX, USA) on *P. aeruginosa* (PAO1, Iglewski) was assessed. Bacteria were grown from an overnight culture to stationary phase in a liquid culture of Luria-Bertani (LB) medium, and diluted ×100 before assays to an OD_{600} of 0.043. One hundred microliters (100 µL) of bacterial suspension was added to 100 µL of recombinant S100A8/A9 heterodimer to reach final concentrations of 40, 20, 10, 5, 2.5, 1.25, and 0 µg/mL and incubated with gentle shaking for 1–24 h at 37 °C. OD_{600} was measured in duplicates at 0, 2, 4, 6, and 24 h. Bacterial growth and inhibition was assessed by plating 10-fold serial dilutions of bacterial suspensions on agar plates at 1, 6, and 24 h, which were then incubated at 37 °C for 24 h. The actual growth was determined by count of CFU/mL. Data are expressed in $\log_{10}$ CFU/mL as a function of time and concentration.

4.8. Quantification of Inflammatory Cytokines and Growth Factors

Inflammatory markers were quantified in sterile-filtered wound homogenates. Levels of TNF-α, IFN-γ, IL-10, KC, IL-1β, G-CSF, and IL-17 (Bio-Rad, Hercules, CA, USA) were analyzed by Luminex Immunoassay (Luminex Corp., Austin, TX, USA). EGF levels were quantified by ELISA (Sigma Aldrich, Saint Louis, MO, USA), as S100A8/A9 also has stimulatory effects on normal, human keratinocytes [32].

4.9. Systemic Polymorphonuclear Count

Count of systemic PMNs in anticoagulated whole blood was performed immediately after sacrifice by flow cytometry, as described by Brochmann et al. [33]. In brief, distribution of PMNs and total white blood cells in anticoagulated blood from all mice was determined by use of a FACSCanto™ flow cytometer (BD Biosciences, San Jose, CA, USA). A 488 nm argon laser and a 530/30 nm band pass emission filter were used for the recording of hydroxyphenyl fluorescence in FL-1. PI fluorescence was collected through a 585/42 nm band pass emission filter, recorded in FL-3. Samples were analyzed at a low flow rate (10 µL/min) and ≥10,000 events were recorded per sample. Cytometer Setup and Tracking Beads (BD Biosciences, San Jose, CA, USA) was used for calibration of the instrument. Flow data were processed and analyzed by Diva (BD Biosciences, San Jose, CA, USA).

4.10. Assessment of Wound Closure

All mice were photographed at a fixed distance by camera (CANON, EOS 650D, Kyushu, Japan) every other day. Mice were anesthetized by halothane during photography. Digital images were analyzed using photoplanimetry (Image J®, version 1.47, Bethesda, MD, USA). Necrotic area and total area of each wound were evaluated.

4.11. Statistics

Statistical calculations on continuous data were performed using the statistical programme GraphPad Prism (version 7.02; GraphPad Software, Inc., San Diego, CA, USA). Quantitative variables (bacteriology) were compared by student's unpaired *t*-test. Cytokine assessment was by two-way ANOVA (Bonferroni). A level of $p \leq 0.05$ was considered statistically significant.

Supplementary Materials: Supplementary materials can be found at www.mdpi.com/1422-0067/18/7/1359/s1.

Author Contributions: Hannah Trøstrup, Christian Johann Lerche, Lars Christophersen, Peter Østrup Jensen, Niels Høiby, and Claus Moser conceived and designed the experiments; Hannah Trøstrup, Christian Johann Lerche, Lars Christophersen, Peter Østrup Jensen, and Claus Moser performed the experiments; Hannah Trøstrup and Claus Moser analyzed the data; Peter Østrup Jensen and Niels Høiby contributed reagents/materials/analysis tools; Hannah Trøstrup wrote the paper.

Conflicts of Interest: The authors declare no conflict of interest.

References

1. Høgsberg, T.; Bjarnsholt, T.; Thomsen, J.S.; Kirketerp-Møller, K. Success rate of split-thickness skin grafting of chronic venous leg ulcers depends on the presence of Pseudomonas aeruginosa: A retrospective study. *PLoS ONE* **2011**, *6*, e20492.

2. Kirketerp-Møller, K.; Jensen, P.O.; Fazli, M.; Madsen, K.G.; Pedersen, J.; Moser, C.; Tolker-Nielsen, T.; Høiby, N.; Givskov, M.; Bjarnsholt, T. Distribution, organization, and ecology of bacteria in chronic wounds. *J. Clin. Microbiol.* **2008**, *46*, 2717–2722. [CrossRef] [PubMed]

3. Malone, M.; Bjarnsholt, T.; McBain, A.J.; James, G.A.; Stoodley, P.; Leaper, D.; Tachi, M.; Schultz, G.; Swanson, T.; Wolcott, R.D. The prevalence of biofilms in chronic wounds: A systematic review and meta-analysis of published data. *J. Wound Care* **2017**, *26*, 20–25. [CrossRef] [PubMed]

4. Trøstrup, H.; Bjarnsholt, T.; Kirketerp-Møller, K.; Høiby, N.; Moser, C. What is new in the understanding of non healing wounds. *Ulcers* **2013**, *2013*, 625934. [CrossRef]

5. Roth, J.; Vogl, T.; Sorg, C.; Sunderkotter, C. Phagocyte-specific S100 proteins: A novel group of proinflammatory molecules. *Trends Immunol.* **2003**, *24*, 155–158. [CrossRef]

6. Rammes, A.; Roth, J.; Goebeler, M.; Klempt, M.; Hartmann, M.; Sorg, C. Myeloid-related protein (MRP) 8 and MRP14, calcium-binding proteins of the S100 family, are secreted by activated monocytes via a novel, tubulin-dependent pathway. *J. Biol. Chem.* **1997**, *272*, 9496–9502. [CrossRef] [PubMed]

7. Vogl, T.; Tenbrock, K.; Ludwig, S.; Leukert, N.; Ehrhardt, C.; van Zoelen, M.A.; Nacken, W.; Foell, D.; van der Poll, T.; Sorg, C.; et al. Mrp8 and Mrp14 are endogenous activators of Toll-like receptor 4, promoting lethal, endotoxin-induced shock. *Nat. Med.* **2007**, *13*, 1042–1049. [CrossRef] [PubMed]

8. Corbin, B.D.; Seeley, E.H.; Raab, A.; Feldmann, J.; Miller, M.R.; Torres, V.J.; Anderson, K.L.; Dattilo, B.M.; Dunman, P.M.; Gerads, R.; et al. Metal chelation and inhibition of bacterial growth in tissue abscesses. *Science* **2008**, *319*, 962–965. [CrossRef] [PubMed]

9. Trøstrup, H.; Lundquist, R.; Christensen, L.H.; Jørgensen, L.N.; Karlsmark, T.; Haab, B.B.; Ågren, M.S. S100A8/A9 deficiency in nonhealing venous leg ulcers uncovered by multiplexed antibody microarray profiling. *Br. J. Dermatol.* **2011**, *165*, 292–301. [CrossRef] [PubMed]
10. Trøstrup, H.; Holstein, P.; Christophersen, L.; Jørgensen, B.; Karlsmark, T.; Høiby, N.; Moser, C.; Ågren, M.S. S100A8/A9 is an important host defence mediator in neuropathic foot ulcers in patients with type 2 diabetes mellitus. *Arch. Dermatol. Res.* **2016**, *308*, 347–355. [CrossRef] [PubMed]
11. Parmely, M.; Gale, A.; Clabaugh, M.; Horvat, R.; Zhou, W.W. Proteolytic inactivation of cytokines by *Pseudomonas aeruginosa*. *Infect. Immun.* **1990**, *58*, 3009–3014. [PubMed]
12. Sohnle, P.G.; Collins-Lech, C.; Wiessner, J.H. Antimicrobial activity of an abundant calcium-binding protein in the cytoplasm of human neutrophils. *J. Infect. Dis.* **1991**, *163*, 187–192. [CrossRef] [PubMed]
13. Trøstrup, H.; Thomsen, K.; Christophersen, L.J.; Hougen, H.P.; Bjarnsholt, T.; Jensen, P.O.; Kirkby, N.; Calum, H.; Høiby, N.; Moser, C. *Pseudomonas aeruginosa* biofilm aggravates skin inflammatory response in BALB/c mice in a novel chronic wound model. *Wound Repair Regen.* **2013**, *21*, 292–299. [CrossRef] [PubMed]
14. Dale, I.; Fagerhol, M.K.; Naesgaard, I. Purification and partial characterization of a highly immunogenic human leukocyte protein, the L1 antigen. *Eur. J. Biochem.* **1983**, *134*, 1–6. [CrossRef] [PubMed]
15. Nacken, W.; Kerkhoff, C. The hetero-oligomeric complex of the S100A8/S100A9 protein is extremely protease resistant. *FEBS Lett.* **2007**, *581*, 5127–5130. [CrossRef] [PubMed]
16. Sroussi, H.Y.; Williams, R.L.; Zhang, Q.L.; Villines, D.; Marucha, P.T. Ala42S100A8 ameliorates psychological-stress impaired cutaneous wound healing. *Brain Behav. Immun.* **2009**, *23*, 755–759. [CrossRef] [PubMed]
17. Lusitani, D.; Malawista, S.E.; Montgomery, R.R. Calprotectin, an abundant cytosolic protein from human polymorphonuclear leukocytes, inhibits the growth of Borrelia burgdorferi. *Infect. Immun.* **2003**, *71*, 4711–4716. [CrossRef] [PubMed]
18. Achouiti, A.; Vogl, T.; Urban, C.F.; Rohm, M.; Hommes, T.J.; van Zoelen, M.A.; Florquin, S.; Roth, J.; van't Veer, C.; de Vos, A.F.; et al. Myeloid-related protein-14 contributes to protective immunity in gram-negative pneumonia derived sepsis. *PLoS Pathog.* **2012**, *8*, e1002987. [CrossRef] [PubMed]
19. Steinbakk, M.; Naess-Andresen, C.F.; Lingaas, E.; Dale, I.; Brandtzaeg, P.; Fagerhol, M.K. Antimicrobial actions of calcium binding leucocyte L1 protein, calprotectin. *Lancet* **1990**, *336*, 763–765. [CrossRef]
20. De Jong, H.K.; Achouiti, A.; Koh, G.C.; Parry, C.M.; Baker, S.; Faiz, M.A.; van Dissel, J.T.; Vollaard, A.M.; van Leeuwen, E.M.; Roelofs, J.J.; et al. Expression and function of S100A8/A9 (calprotectin) in human typhoid fever and the murine Salmonella model. *PLoS Negl. Trop. Dis.* **2015**, *9*, e0003663. [CrossRef] [PubMed]
21. Akerstrom, B.; Bjorck, L. Bacterial surface protein L binds and inactivates neutrophil proteins S100A8/A9. *J. Immunol.* **2009**, *183*, 4583–4592. [CrossRef] [PubMed]
22. Moser, C.; Johansen, H.K.; Song, Z.; Hougen, H.P.; Rygaard, J.; Høiby, N. Chronic *Pseudomonas aeruginosa* lung infection is more severe in Th2 responding BALB/c mice compared to Th1 responding C3H/HeN mice. *APMIS* **1997**, *105*, 838–842. [CrossRef] [PubMed]
23. Moser, C.; Hougen, H.P.; Song, Z.; Rygaard, J.; Kharazmi, A.; Høiby, N. Early immune response in susceptible and resistant mice strains with chronic *Pseudomonas aeruginosa* lung infection determines the type of T-helper cell response. *APMIS* **1999**, *107*, 1093–1100. [CrossRef] [PubMed]
24. Hancock, R.E.; Sahl, H.G. Antimicrobial and host-defense peptides as new anti-infective therapeutic strategies. *Nat. Biotechnol.* **2006**, *24*, 1551–1557. [CrossRef] [PubMed]
25. Deng, Q.; Sun, M.; Yang, K.; Zhu, M.; Chen, K.; Yuan, J.; Wu, M.; Huang, X. MRP8/14 enhances corneal susceptibility to *Pseudomonas aeruginosa* Infection by amplifying inflammatory responses. *Investig. Ophthalmol. Vis. Sci.* **2013**, *54*, 1227–1234. [CrossRef] [PubMed]
26. Simard, J.C.; Simon, M.M.; Tessier, P.A.; Girard, D. Damage-associated molecular pattern S100A9 increases bactericidal activity of human neutrophils by enhancing phagocytosis. *J. Immunol.* **2011**, *186*, 3622–3631. [CrossRef] [PubMed]
27. Ryckman, C.; Vandal, K.; Rouleau, P.; Talbot, M.; Tessier, P.A. Proinflammatory activities of S100: Proteins S100A8, S100A9, and S100A8/A9 induce neutrophil chemotaxis and adhesion. *J. Immunol.* **2003**, *170*, 3233–3242. [CrossRef] [PubMed]

28. Jensen, P.O.; Bjarnsholt, T.; Phipps, R.; Rasmussen, T.B.; Calum, H.; Christoffersen, L.; Moser, C.; Williams, P.; Pressler, T.; Givskov, M.; et al. Rapid necrotic killing of polymorphonuclear leukocytes is caused by quorum-sensing-controlled production of rhamnolipid by *Pseudomonas aeruginosa*. *Microbiology* **2007**, *153*, 1329–1338. [CrossRef] [PubMed]

29. Bjarnsholt, T.; Kirketerp-Møller, K.; Jensen, P.O.; Madsen, K.G.; Phipps, R.; Krogfelt, K.; Høiby, N.; Givskov, M. Why chronic wounds will not heal: A novel hypothesis. *Wound Repair Regen.* **2008**, *16*, 2–10. [CrossRef] [PubMed]

30. Calum, H.; Moser, C.; Jensen, P.O.; Christophersen, L.; Malling, D.S.; van Gennip, M.; Bjarnsholt, T.; Hougen, H.P.; Givskov, M.; Jacobsen, G.K.; et al. Thermal injury induces impaired function in polymorphonuclear neutrophil granulocytes and reduced control of burn wound infection. *Clin. Exp. Immunol.* **2009**, *156*, 102–110. [CrossRef] [PubMed]

31. Christophersen, L.J.; Trøstrup, H.; Malling Damlund, D.S.; Bjarnsholt, T.; Thomsen, K.; Jensen, P.O.; Hougen, H.P.; Høiby, N.; Moser, C. Bead-size directed distribution of *Pseudomonas aeruginosa* results in distinct inflammatory response in a mouse model of chronic lung infection. *Clin. Exp. Immunol.* **2012**, *170*, 222–230. [CrossRef] [PubMed]

32. Nukui, T.; Ehama, R.; Sakaguchi, M.; Sonegawa, H.; Katagiri, C.; Hibino, T.; Huh, N.H. S100A8/A9, a key mediator for positive feedback growth stimulation of normal human keratinocytes. *J. Cell. Biochem.* **2008**, *104*, 453–464. [CrossRef] [PubMed]

33. Brochmann, R.P.; Toft, A.; Ciofu, O.; Briales, A.; Kolpen, M.; Hempel, C.; Bjarnsholt, T.; Høiby, N.; Jensen, P.O. Bactericidal effect of colistin on planktonic *Pseudomonas aeruginosa* is independent of hydroxyl radical formation. *Int. J. Antimicrob. Agents* **2014**, *43*, 140–147. [CrossRef] [PubMed]

© 2017 by the authors. Licensee MDPI, Basel, Switzerland. This article is an open access article distributed under the terms and conditions of the Creative Commons Attribution (CC BY) license (http://creativecommons.org/licenses/by/4.0/).

International Journal of
Molecular Sciences

MDPI

Article

HMGB1 Promotes Intraoral Palatal Wound Healing through RAGE-Dependent Mechanisms

Salunya Tancharoen [1], Satoshi Gando [2], Shrestha Binita [3], Tomoka Nagasato [3],
Kiyoshi Kikuchi [1,4], Yuko Nawa [5], Pornpen Dararat [1], Mika Yamamoto [3], Somphong Narkpinit [6]
and Ikuro Maruyama [3,*]

[1] Department of Pharmacology, Faculty of Dentistry, Mahidol University, Bangkok 10400, Thailand;
    salunya.tan@mahidol.ac.th (S.T.); kikuchi_kiyoshi@kurume-u.ac.jp (K.K.); pornpen.dar@mahidol.ac.th (P.D.)
[2] Department of Emergency and Critical Care, Hokkaido University, Kita-ku, Sapporo 0608648, Japan;
    gando@med.hokudai.ac.jp
[3] Department of Systems Biology in Thromboregulation, Kagoshima University Graduate School of Medical
    and Dental Science, Kagoshima 8908544, Japan; binita@m2.kufm.kagoshima-u.ac.jp (S.B.);
    n-tomoka@m.kufm.kagoshima-u.ac.jp (T.N.); m-yamamo@m2.kufm.kagoshima-u.ac.jp (M.Y.)
[4] Division of Brain Science, Department of Physiology, Kurume University School of Medicine, Asahi-machi,
    Kurume 8300011, Japan
[5] Department of Anesthesiology and Intensive Care, Hokkaido Medical for Child Health and Rehabilitation,
    Sapporo 0060041, Japan; nawa@sapmed.ac.jp
[6] Department of Pathobiology, Faculty of Science, Mahidol University, Bangkok 10400, Thailand;
    somphong.nar@mahidol.ac.th
*   Correspondence: rinken@m3.kufm.kagoshima-u.ac.jp; Tel.: +81-99-275-6474; Fax: +81-99-275-6463

Academic Editor: Allison Cowin
Received: 15 September 2016; Accepted: 15 November 2016; Published: 23 November 2016

Abstract: High mobility group box 1 (HMGB1) is tightly connected to the process of tissue organization upon tissue injury. Here we show that HMGB1 controls epithelium and connective tissue regeneration both in vivo and in vitro during palatal wound healing. Heterozygous HMGB1 ($Hmgb1^{+/-}$) mice and Wild-type (WT) mice were subjected to palatal injury. Maxillary tissues were stained with Mallory Azan or immunostained with anti-HMGB1, anti-proliferating cell nuclear antigen (PCNA), anti-nuclear factor-κB (NF-κB) p50 and anti-vascular endothelial growth factor (VEGF) antibodies. Palatal gingival explants were cultured with recombinant HMGB1 (rHMGB1) co-treated with siRNA targeting receptor for advanced glycation end products (RAGEs) for cell migration and PCNA expression analysis. Measurement of the wound area showed differences between $Hmgb1^{+/-}$ and WT mice on Day 3 after wounding. Mallory Azan staining showed densely packed of collagen fibers in WT mice, whereas in $Hmgb1^{+/-}$ mice weave-like pattern of low density collagen bundles were present. At three and seven days post-surgery, PCNA, NF-κB p50 and VEGF positive keratinocytes of WT mice were greater than that of $Hmgb1^{+/-}$ mice. Knockdown of RAGE prevents the effect of rHMGB1-induced cell migration and PCNA expression in gingival cell cultures. The data suggest that HMGB1/RAGE axis has crucial roles in palatal wound healing.

Keywords: HMGB1; NF-κB; wound healing; palatal mucosa

1. Introduction

In tissues skin and oral mucosa tissues, wound healing encompasses a number of overlapping phases, including inflammation, tissue formation and tissue remodeling [1]. The process of wound repair requires a complex and dynamic interplay of a number of tenant epithelial and mesenchymal cells with hematopoietic cells to accomplish the three stages of wound healing [2]. In the inflammation process, the primary phase is characterized by a local activation of innate immune mechanisms resulting in an initial influx of neutrophils into the injured tissue followed by macrophages accumulation. Subsequently, epithelialization proceeds gradually from the wound edges. Finally, tissue remodeling occurs during maturation of the newly formed tissues, which leads to scar formation [3]. Recently, wound healing analysis in a number of murine knockout models deficient for distinct inflammatory mediators including tumor necrosis factor (TNF)-α, interleukin (IL)-6, monocyte chemotactic protein (MCP)-1, and interferon (IFN)-γ have been investigated [4]. Dermal wound healing is accelerated in TNF-receptor-55 [5] or IFN-γ-deficient mice [6] but is impaired in mice deficient of IL-6 [7]. Likewise, palatal wound healing is accelerated in Smad3 [8] and interleukin (IL)-10 [9] deficient mice.

High-mobility group box 1 (HMGB1) is a ubiquitous 25 kDa nuclear DNA-binding protein that, under typical conditions, is located in the nucleus, where it organizes the basic structural unit of chromatin, DNA replication, and transcription. HMGB1 also acts in the extracellular environment as a primary pro-inflammatory cytokine [10]. Upon *Prevotella intermedia* lipopolysaccharide stimulation, odontoblast-like cells respond by upregulating receptor for advanced glycation end products (RAGEs) and producing HMGB1. In addition, HMGB1, a novel cytokine, is known to contribute to the pathogenesis of various inflammatory diseases [11–13], including periodontal diseases [14]. This suggests that HMGB1 can activate an inflammatory response during infection or injury. On the contrary, a role for HMGB1 in wound repair has been reported by the results of earlier studies. HMGB1 promotes neuronal differentiation of embryonic stem cells and neurite outgrowth in the developing nervous system [15]. Moreover, HMGB1 is a potent chemoattractant and mitogen for blood vessel-associated stem cells [16]. Limana et al. demonstrated that intracardiac HMGB1 injection in a mouse model of myocardial infarction induced new myocyte formation and improved infarcted hearts function [17]. In addition, HMGB1-induced increase of HaCaT keratinocytes proliferation, cell migration, and wound closure via RAGE and extracellular signal-regulated kinase (ERK) pathway [18]. RAGE is a member of the immunoglobulin superfamily and is expressed on gingival epithelial and fibroblast cells [19], mononuclear phagocytes, vascular smooth muscle cells, and neurons [20,21]. RAGE interacts with a range of ligands, including advanced glycation end products (AGEs), HMGB1, and S100/calgranulins [22,23]. Ligand binding results in RAGE-dependent sustained nuclear factor-kappa B (NF-κB) activation [24] as well as in wound healing promotion [25]. These reports indicate that HMGB1 is a multifunctional cytokine involved in inflammatory responses and tissue repair. Despite this, whether and how HMGB1 contributes to protective and/or pathological responses of palatal wound healing in vivo is unclear. In this study, we provide evidence that the loss of *Hmgb1* gene in HMGB1-heterozygous (*Hmgb1*$^{+/-}$) mice results in delayed palatal wound closure. In addition, RAGE silencing reduces closure of an in vitro scratch wound gingival cell migration and proliferating cell nuclear antigen (PCNA) expression. Our results, show for the first time, that HMGB1/RAGE axis has crucial roles in palatal wound healing, by regulating collagen accumulation, cell proliferation and cell migration. Taken together, these results bring scientific support to the possible application of HMGB1 in regenerative medicine.

2. Results

2.1. Identification and Targeted Disruption of the Mouse Gene

DNA extraction from mouse tail and genotyping PCR separated by agarose gel electrophoresis has been performed in *Hmgb1*$^{+/-}$ and Wild-type (WT) mice (Figure 1A). PCR with the HMGB1 primer pair, for the mouse genome identifies WT mice, in which the product is detected at about 500 bp band (Figure 1A, lanes 1–4). On the other hand, PCR with the primer complementary to the gene identifies *Hmgb1*$^{+/-}$ mice, shows a bright band at about 500 bp (amplification control) and another at about 300 bp (Figure 1A, lanes 1–3). The target allele of PCR product is 495 bp for WT mice and 336 bp for *Hmgb1*$^{+/-}$ mice. Our data replicated previous findings that *Hmgb1*$^{+/-}$ mice contained approximately one-half the level of HMGB1 mRNA compared with WT mice [26]. This decrease is confirmed by immunohistochemistry of HMGB1 protein within the unwounded site (Day 0) and experimental wound (Day 7) of hard palate mucosa in mice. By using a specific anti-HMGB1 polyclonal rabbit antibody (Ab) that does not detect HMGB2 and HMGB3 [27], we found that at seven days post-surgery, tissues from the WT mice presented higher expression of HMGB1 in the periphery of the nucleus together with some translocation from the nucleus to the cytoplasm of basal epithelial cells and cells of connective tissue compared with that in unwounded site (Figure 1B). Interestingly, marked differences in the intensity and localization of HMGB1 staining in unwounded and wounded sites were observed between $Hmgb1^{+/-}$ and WT mice. Very faint staining of HMGB1 in $Hmgb1^{+/-}$ mice tissue was shown and the pattern of HMGB1 expression was mainly intranucleus in basal keratinocytes. These results indicate that the positive staining of HMGB1 in hard palate mucosa of WT mice was diminished in $Hmgb1^{+/-}$ mice.

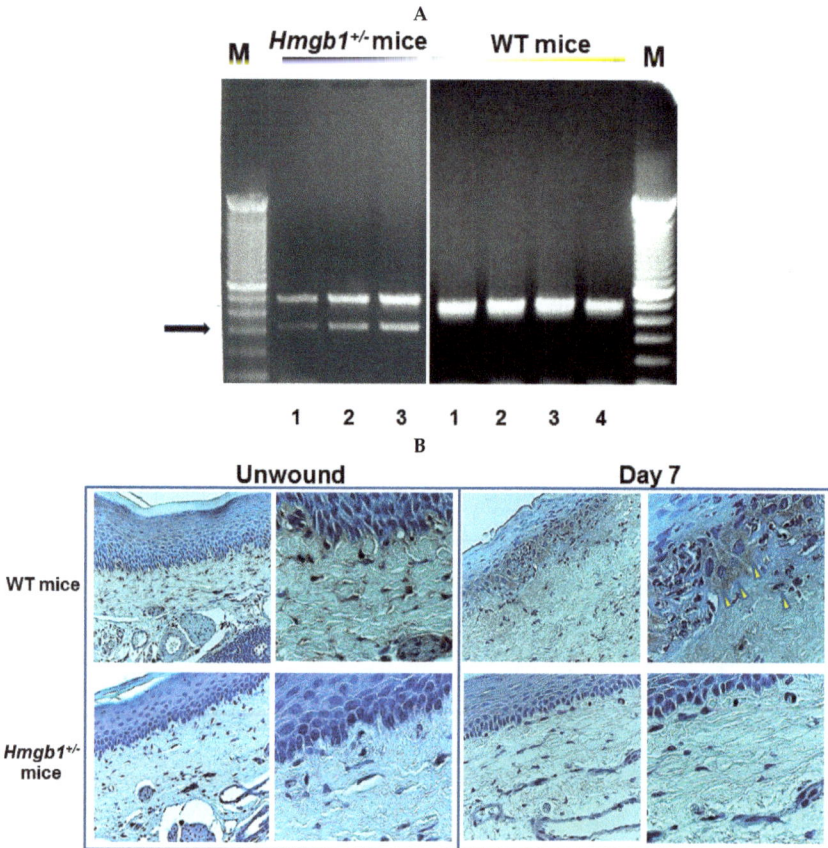

Figure 1. RT-PCR transcripts of mouse tail and immunohistochemistry determination of palatal section for $Hmgb1^{+/-}$ and WT mice. (**A**) PCR products were separated in ethidium-bromide-stained agarose gels with DNA molecular weight marker (M, 100 bp ladder). PCR with the primer complementary to the gene identifies $Hmgb1^{+/-}$ mice, shows two bands, a bright band at 500 bp (**upper**) and another at about 300 bp (**lower**). The predicted size of HMGB1 in WT and $Hmgb1^{+/-}$ mice was 495 and 336 bp, respectively. Arrow indicates target allele of $Hmgb1^{+/-}$ mice and (**B**) Immunohistochemical analysis of HMGB1 in palatal section of unwounded (Day 0) and wounded site (Day 7). Arrowheads indicate HMGB1 positive staining in the cytosol. Original magnifications: left panels 400×, right panels 800×. n = 3–5 for each group.

2.2. Wound Closure Is Attenuated in Hmgb1$^{+/-}$ Mice

Macroscopic wound closure was attenuated in *Hmgb1*$^{+/-}$ mice compared with WT animals. At three days post-surgery, it was observed that mucosal closure was not completed in the *Hmgb1*$^{+/-}$ mice (Figure 2A). The wound of WT mice appeared epithelialized, whereas the mutant wounds showed partial epithelialization. At seven days post-surgery, wound healing was more favorable in incision areas in the WT group than that of *Hmgb1*$^{+/-}$ group. Measurement of the wound area on digital images showed that the differences between *Hmgb1*$^{+/-}$ (55.8% ± 1.48% of Day 0) and WT mice (25.6% ± 0.7% of Day 0) were statistically significant on Day 3 after wounding ($p < 0.001$, Figure 2B). Wound area assessment demonstrated significantly larger wounds in *Hmgb1*$^{+/-}$ mice compared to WT controls at Day 3 (1.2 ± 0.06 mm^2 vs. 0.7 ± 0.04 mm^2; $p < 0.05$) after wounding, whereas there was no statistically significant difference ($p > 0.05$) in the wound area between both groups at Day 7 (Figure 2C). The wound areas on Days 0, 3, and 7 were measured by three examiners. Pearson's correlation coefficient (r) was used to show the correlation between the examiners (Figure S1). In WT mice, we found a statistically significant correlation between examiner 1 and examiner 2 ($r = 0.9992$, $p < 0.001$); examiner 1 and examiner 3 ($r = 0.9992$, $p < 0.001$); and examiner 2 and examiner 3 ($r = 0.9998$, $p < 0.001$). Likewise, we demonstrated a statistically significant correlation between examiner 1 and examiner 2 ($r = 0.9909$, $p < 0.05$); examiner 1 and examiner 3 ($r = 0.9902$, $p < 0.01$); and examiner 2 and examiner 3 ($r = 0.9906$, $p < 0.01$) in *Hmgb1*$^{+/-}$ mice. According to the macroscopic results, we concluded that *Hmgb1*$^{+/-}$ mice showed a significant delay in wound healing as compared to WT mice (seven days vs. three days).

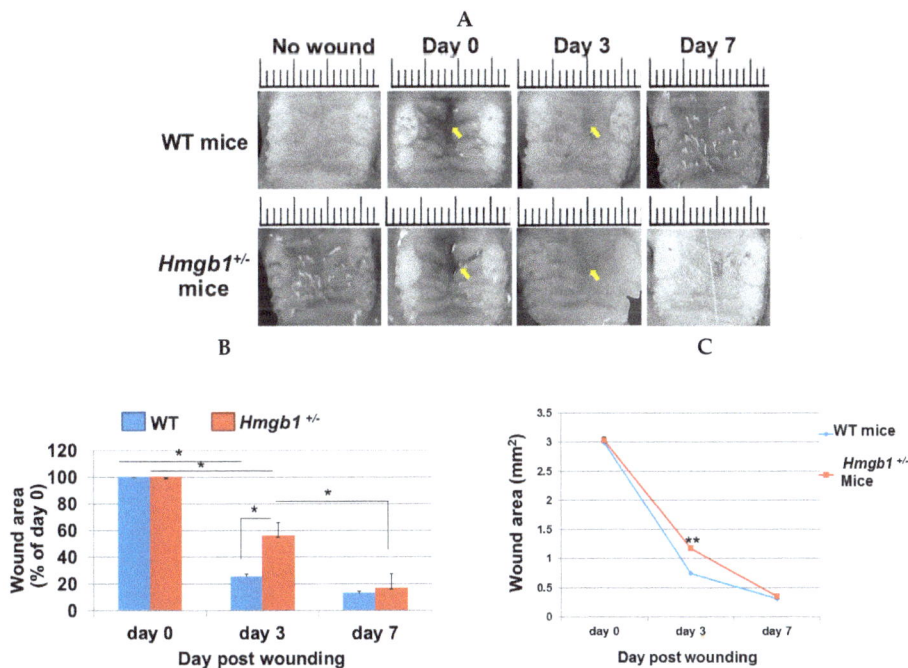

Figure 2. Wound closure is attenuated in *Hmgb1*$^{+/-}$ mice vs. WT mice. (**A**) Macroscopic appearance of wounds of *Hmgb1*$^{+/-}$ and WT mice at indicated time points after injury. Arrows indicate the wound bed or scar margins. Scale bars: 1.0 mm; (**B**) At the time points indicated, the wound area was analyzed and expressed as the percentage of the initial wound areas; (**C**) The reduction in wound area in WT and *Hmgb1*$^{+/-}$ mice in mm^2. Data are expressed as mean ± SD, $n = 5$ wounds for each time point and genotype (* $p < 0.001$, ** $p < 0.05$ vs. the WT group).

2.3. Reduction of Collagen Fibers and Delayed Re-Epithelialization in HMGB1$^{+/-}$ Wound

Collagen components are a major part of oral mucosa development [28]. The macroscopic findings of wound closure were confirmed by histological assessment. At three days post-surgery, delayed wound healing was determined in the *Hmgb1*$^{+/-}$ group compared to the WT group. Mallory Azan staining of Day 3 wound showed well-organized, parallel, densely packed and thick bundles of collagen fibers in WT mice, whereas in *Hmgb1*$^{+/-}$ mice weave-like pattern of low density collagen bundles were present (Figure 3A). At seven days post-surgery, the collagen fibers were prominently mature and re-epithelialization was observed in wound regions of both groups. A corresponding hematoxylin and eosin (H & E)-stained wound section is shown in Figure 3B for comparison. In the *Hmgb1*$^{+/-}$ group, subepithelial healing was evidenced but was not completed and, in connective tissue region, mononuclear cell infiltration was present at three days post-surgery. More prominent infiltration of mononuclear cell infiltration was present in WT mice than in *Hmgb1*$^{+/-}$ mice. At Day 7, the wound healing properties were not different in *Hmgb1*$^{+/-}$ and WT mice. Collectively, these findings show significant delays in wound healing parameters, including epithelialization and decreased collagen formation in *Hmgb1*$^{+/-}$ mice.

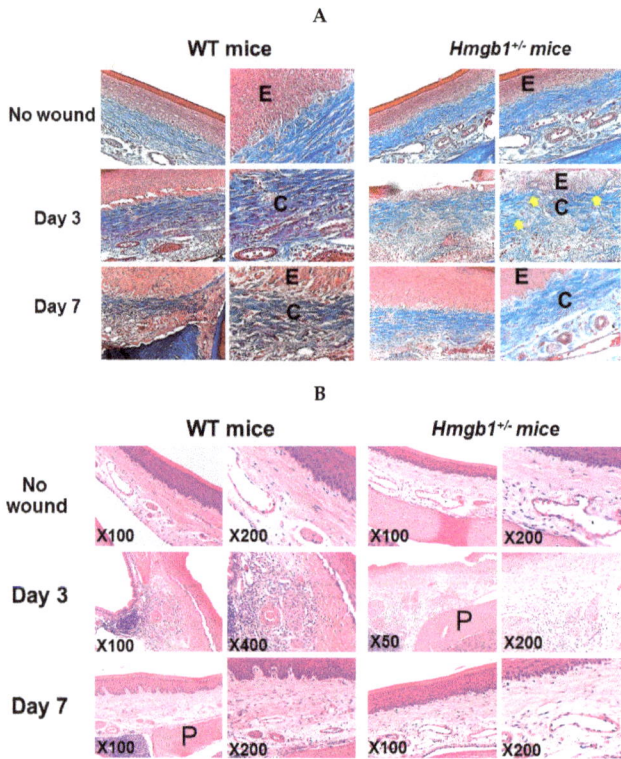

Figure 3. Reduction of collagen fibers and delayed re-epithelialization in *Hmgb1*$^{+/-}$ wound on Day 3 after wound surgery. (**A**) Sections were stained with Mallory's azan. Representative palatal wound sections show the collagen bundles of WT and *Hmgb1*$^{+/-}$ mice. Note that blue indicates collagen bundle stained by Mallory's azan stain. Arrows indicate weave-like pattern of collagen bundles in *Hmgb1*$^{+/-}$ mice. Original magnifications: left panels 200×, right panels 400×; (**B**) Sections were stained with hematoxylin and eosin. E, epithelium; C, collagen bundle; P, palatal bone. *n* = 5 wounds for each time point and genotype.

2.4. Immunohistochemistry Determination of Proliferating Cells at Palatal Wounds in WT and Hmgb1$^{+/-}$ Mice

To identify the mechanism underlying the attenuated palatal wound closure in *Hmgb1$^{+/-}$* mice, we assessed cell proliferation in the repaired tissue by immunohistochemistry, using PCNA Ab. PCNA is expressed in both basal and suprabasal cell layers (Figure 4A). At three days post-surgery, the numbers of PCNA-positive keratinocytes in WT mice wound site (168 ± 18) were significantly greater than *Hmgb1$^{+/-}$* mice (70 ± 8.7) (Figure 4B, $p < 0.001$). At seven days post-surgery, PCNA-positive keratinocytes numbers are reduced in both groups. The values were significantly higher ($p < 0.001$) in WT mice (106.5 ± 10.4) than *Hmgb1$^{+/-}$* mice (55 ± 9.7). From the above results, we conclude that the proliferation marker, PCNA, was significantly lower in the *Hmgb1$^{+/-}$* mice compared with WT mice.

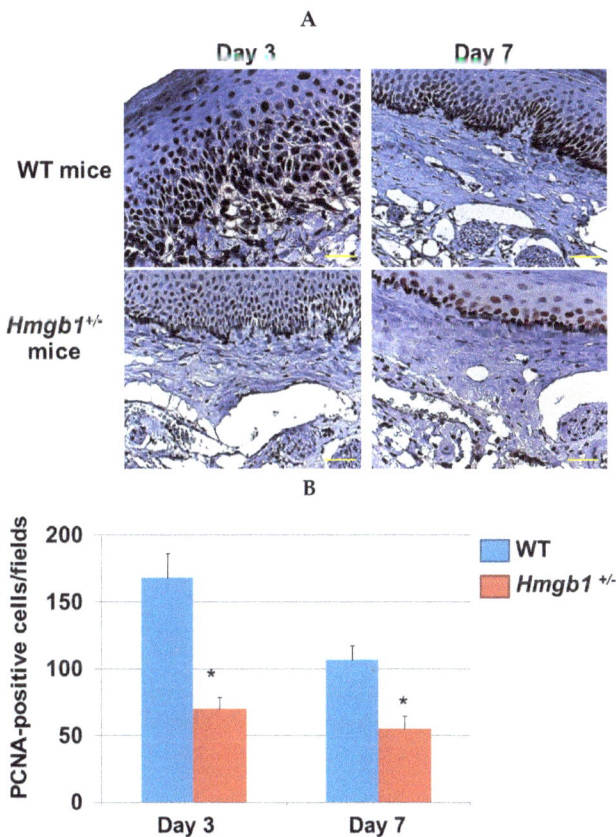

Figure 4. Distribution of proliferating cells at palatal wounds in WT and *Hmgb1$^{+/-}$* mice. (**A**) Analysis of cell proliferation by immunohistochemistry stained with the anti-PCNA Ab in the palatal wounds on Day 3 and Day 7 post surgery. Scale bars = 100 μm; (**B**) The number of PCNA-positive cells per field in palatal wound. Data are expressed as the mean ± SD. (n = 5–8 for each group). * $p < 0.001$ vs. the WT group.

2.5. Localization of NF-κB p50 Isoform at Palatal Wounds in WT and Hmgb1$^{+/-}$ Mice

Blocking HMGB1 can decrease the degree of radiation-induced pulmonary damage, and its mechanism may be related to the promotion of NF-κB p50 activation and its downstream molecular expression [29]. NF-κB p50 antigen in epithelial cells was examined in serial sections of the same tissue block (Figure 5A). At three days post-surgery, NF-κB p50-immunopositive cells in WT mice wound site (75.6 ± 7.8) were significantly greater (*p* < 0.05) than *Hmgb1*$^{+/-}$ mice (35.1 ± 4.9). At seven days post-surgery, NF-κB p50-positive cell numbers are reduced in both groups. The values were significantly higher (*p* < 0.05) in WT mice (45.1 ± 10.5) than *Hmgb1*$^{+/-}$ mice (25.1 ± 2.9) (Figure 5B). These results indicated that pro-inflammatory signaling pathway, NF-κB was significantly lower in the *Hmgb1*$^{+/-}$ mice compared with WT mice. No tissues in samples from WT and *Hmgb1*$^{+/-}$ group were stained with isotype-matched control IgG (Figure S2).

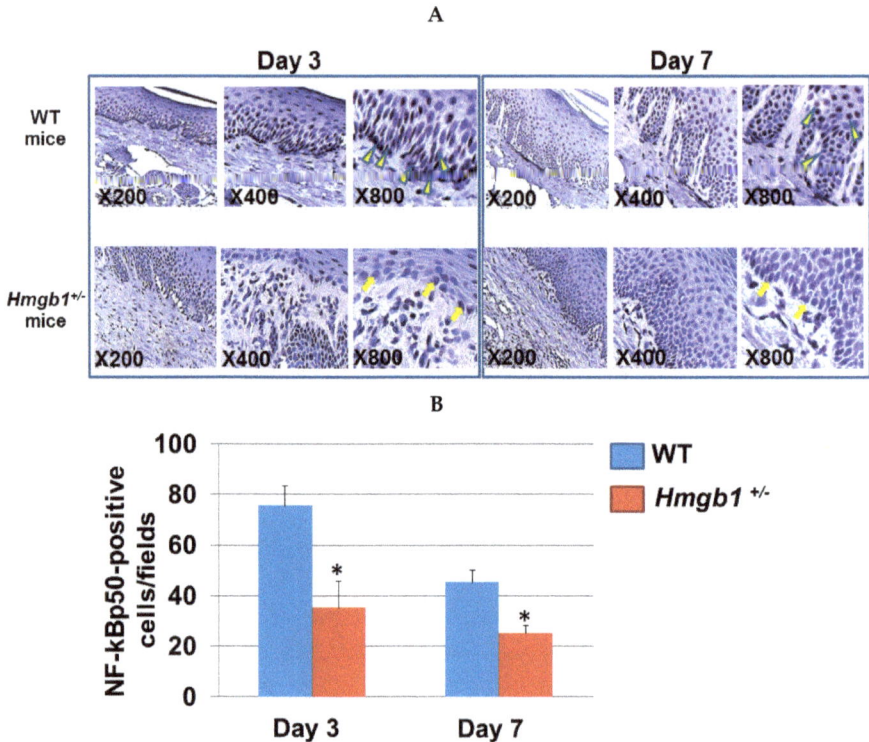

Figure 5. Localization of NF-κB p50 isoform at palatal wounds in WT and *Hmgb1*$^{+/-}$ mice. (**A**) By immunohistochemistry, slides were stained with the anti-NF-κB p50 Ab in the palatal wounds on Day 3 and Day 7 post surgery. Nuclei were counterstained with Mayer's hematoxylin. Arrowheads indicate NF-κB p50-positive stained nuclei of epithelial cells. Arrows indicate faint immunostaining of NF-κB p50 in *Hmgb1*$^{+/-}$ mice; (**B**) The number of NF-κB p50-positive cells per field in palatal wound. Data are expressed as the mean ± SD. Error bars indicate standard deviation (*n* = 5 for each group). * *p* < 0.05 vs. the WT group.

2.6. Determination of VEGF Expression and Localization in Palatal Wounds of WT and Hmgb1$^{+/-}$ Mice

The expression of VEGF within the area of granulation tissue was used as read-out for neovascular processes [30]. Real time PCR analysis demonstrated the ratio of VEGF normalized to the glyceraldehyde-3-phosphate dehydrogenase (GAPDH) transcript content in palatal wound site of *Hmgb1*$^{+/-}$ and WT mice (Figure 6A). At three days post-surgery, VEGF mRNA in WT mice wound site (1.4 ± 0.07) were significantly greater (*p* < 0.001) than *Hmgb1*$^{+/-}$ mice (0.5 ± 0.1). There was no statistically significant difference of VEGF values between three days and seven days post-surgery in WT mice (*p* > 0.05). At seven days post-surgery, the VEGF value was significantly higher (*p* < 0.001) in

WT mice (1.3 ± 0.3) than *Hmgb1*$^{+/-}$ mice (0.37 ± 0.05). Next, we examined by immunohistochemistry the presence of VEGF protein within the wound site at three and seven days after surgery. At three days post-surgery, VEGF was detected in the oral epithelia above the basal layer (Figure 6B) to the spindle cell layers with rising density in the wound site of WT mice. In *Hmgb1*$^{+/-}$ group, on the other hand, specific labeling was faint or negative, indicating that VEGF is poorly expressed in these cells. There was also scattered VEGF in capillary endothelial cells, infiltrating inflammatory cells, and fibroblast-like cells in the connective tissue of all WT mice and *Hmgb1*$^{+/-}$ mice wound sites. At seven days post-surgery, VEGF was detected in the whole layer of oral epithelia; whereas in the *Hmgb1*$^{+/-}$ group, VEGF is absent in these cells. No tissues in samples from WT and *Hmgb1*$^{+/-}$ group were stained with isotype-matched control IgG (Figure S2). Collectively, these results demonstrate that HMGB1 ablation contributes to the delayed angiogenic response in heterozygous mouse model.

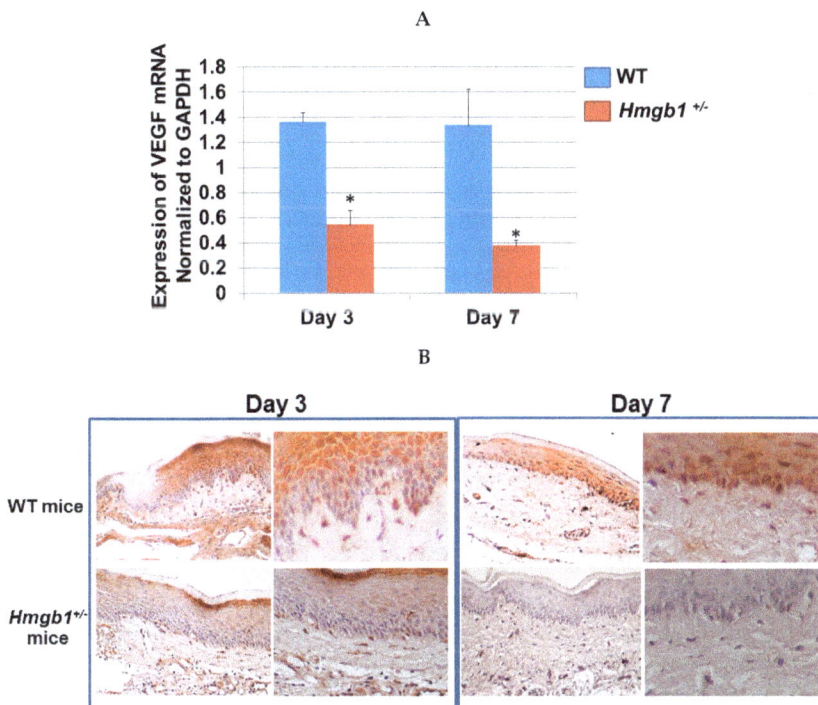

Figure 6. Determination of VEGF mRNA expression and VEGF protein immunolocalization in the palatal wounds of WT and *Hmgb1*$^{+/-}$ mice. (**A**) Expression of VEGF mRNA in palatal wounds on Day 3 and Day 7 post surgery was quantified by real time PCR analysis and reported by normalized to GAPDH. Data are expressed as the mean ± SD of at least three independent determinations. * $p < 0.001$ vs. the WT group; (**B**) Analysis of angiogenic response by immunohistochemistry stained with the anti-VEGF Ab. Original magnifications: left panels 400×, right panels 800×. Experiments were performed in triplicate.

2.7. Efficiency of RAGE Gene Knockdown

The silencing of RAGE gene expression was carried out following the transfection of RAGE-specific siRNA in gingival epithelial cells (Figure 7A). Western blot analysis showed that both 150 and 300 nM of RAGE siRNA significantly decreased the expression of RAGE protein by 50% and 70%, respectively, without affecting house-keeping gene (β-actin) expression or any toxicities after transfection (MTT assay; Figure 7B).

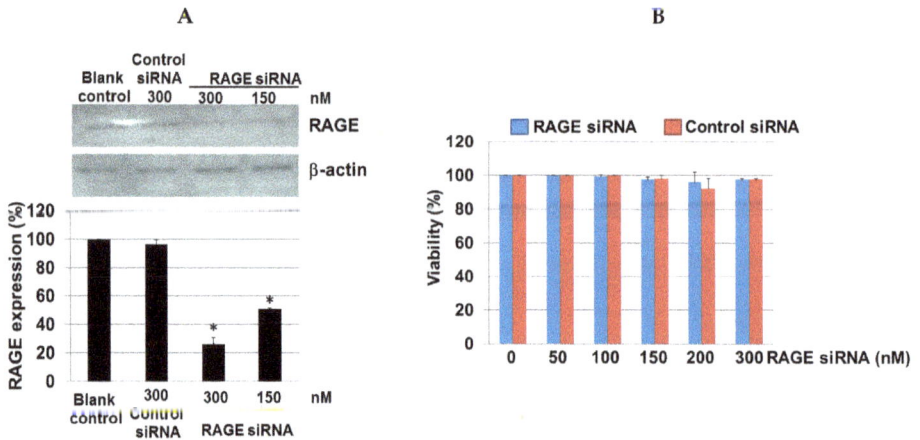

Figure 7. Efficiency of RAGE gene knockdown in gingival epithelial cells. (**A**) A representative Western blot using antibodies against RAGE and β-actin is shown. The results showed decreased expression of RAGE protein in the siRNA group compared to the control siRNA and blank control groups. The intensity of the protein bands in Western blotting was quantified by densitometry and normalized to β-actin. Three independent measurements were performed. Data are expressed as mean ± SD. * $p < 0.001$ vs. the control group; (**B**) Cytotoxicity of the siRNA complexes was assessed by MTT assay. There was no evidence of cell toxicity found in the RAGE siRNA transfected cells. Data are expressed as the mean ± SD. Error bars indicate SD of at least three independent determinations.

2.8. RAGE Silencing Reduces Closure of an In Vitro Scratch Wound and PCNA Expression

A cell monolayer scratch assay is used to evaluate keratinocyte proliferation and migration during re-epithelialization [31]. To assess whether delayed in the wound healing at the wound sites of the $Hmgb1^{+/-}$ mice is accompanied by retarded re-epithelialization mechanism, gingival epithelial cells were subjected to wound-healing (scratch) assay, in which the ability of cells to migrate and cover the cell-free space is monitored. Effects of recombinant HMGB1 (rHMGB1) on the cell migration were studied. Cells were either exposed to rHMGB1 at the concentration of 50 and 100 ng/mL or pretreated with RAGE siRNA. As shown in Figure 8A,B, untreated cells tend to close the wound by about 8% within 48 h. Interestingly, treatment with 100 ng/mL rHMGB1 enhanced cell migration by 32% and 100% at 24 and 48 h, respectively. Previous study reported that HMGB1 binds RAGE and promotes NF-κB activation [32]. In the present study, transfection of gingival epithelial cells with RAGE siRNA suppressed rHMGB1-induced cell migration by ~85% compared with cells treated with 100 ng/mL rHMGB1 for 48 h (Figure 8B). PCNA is considered to be a marker of cell proliferation [33]. In our study, immunohistochemical staining for PCNA in the wound section revealed that PCNA expression was significantly lower in the $Hmgb1^{+/-}$ mice compared with WT mice. We therefore sought to determine whether retarded in the wound healing of $Hmgb1^{+/-}$ mice is related to cell proliferation mechanism. Cells were treated with rHMGB1 at the concentration 100 ng/mL or pretreated with 300 nM RAGE siRNA (Figure 8C). Transcript levels of PCNA were significantly higher in the rHMGB1-treated cell than in the control group ($p < 0.001$). Knockdown of RAGE prevents the effect of HMGB1-induced PCNA expression in gingival epithelial cells by ~70% compared with cells treated with rHMGB1. No effects were observed on control siRNA-treated cells. These results thus suggested that knockdown of RAGE prevents the effect of HMGB1-induced cell migration and cell proliferation in gingival epithelial cells. In addition, HMGB1 promotes molecular signaling, leading to cell migration and proliferation via RAGE.

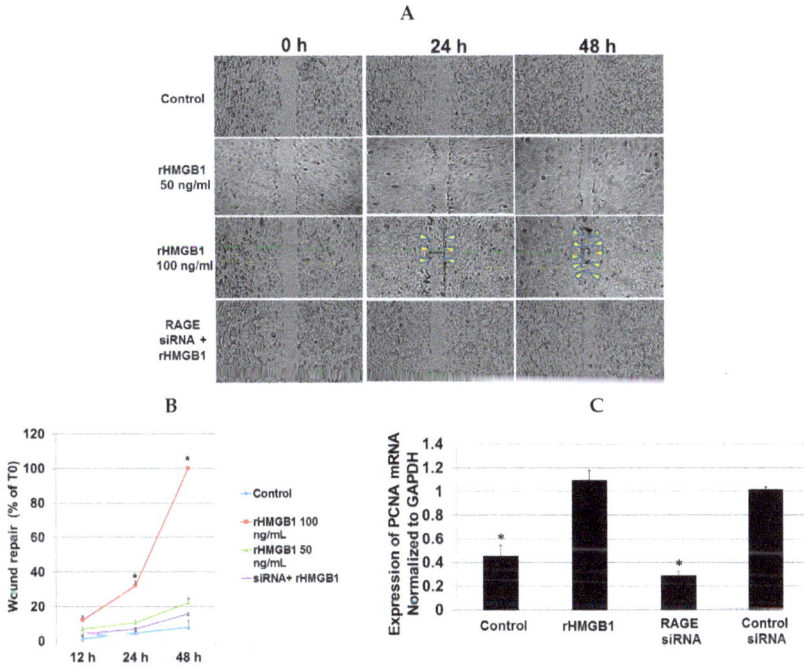

Figure 8. RAGE silencing reduces closure of an in vitro scratch wound, gingival epithelial cell migration and PCNA expression. (**A**) Representative photomicrographs of a cell migration assay. Cells were pretreated with 300 nM siRNA against RAGE and stimulated with rHMGB1 at different concentrations in culture media containing 2% FBS for the cell migration assay with 20× magnification. Treatment with rHMGB1 (100 ng/mL) increases the ability of cells to migrate into the empty area and to repair the wound at all-time points (24 and 48 h) examined. Yellow arrows indicate migration into the cell free-space; (**B**) The ability of cells to migrate and cover the empty space was determined as a wound repair by percent of 0 h (T0). Data are mean ± SD of 3–5 independent experiments. * $p < 0.001$ vs. the control siRNA and blank control groups; (**C**) Expression of PCNA mRNA in the cells after addition of 300 nM RAGE siRNA or control siRNA and stimulation by 100 ng/mL rHMGB1 was quantified by real time PCR analysis and reported by normalized to GAPDH. Data are expressed as the mean ± SD of three independent determinations. * $p < 0.001$ vs. the rHMGB1 group.

3. Discussion

The present study is, as far as we know, the first in which the regulatory role of HMGB1 and its potential involved signal pathways in palatal wound healing has been examined. We show that wound healing is attenuated in *Hmgb1*$^{+/-}$ mice compared with WT mice. In vitro studies demonstrate cell mobility and cell proliferation in gingival epithelial cells after rHMGB1 stimulation. HMGB1 protein is a multifunctional cytokine involved in tissue inflammation and regeneration. Recent studies have documented an important role of HMGB1 in mediating tissue repair [15–17]. Straino et al. has shown that HMGB1 administration promotes wound healing in diabetic mice [26]. Furthermore, HMGB1 had a chemotactic effect on skin fibroblasts and keratinoyctes during wound healing process [18]. Additionally, HMGB1 was able to significantly induce proliferation of human gingival fibroblast and induce cell migration [27]. With these outcomes, it is interesting that HMGB1 may be important for wound healing and promote evidences which support our study.

Although the distribution of HMGB1 in relation to wound healing has been observed in human skin [26] and cholesteatoma [28], studies dealing with palatal wound healing and the target mutation of *Hmgb1* gene in vivo have not been previously performed. In our study, we found that HMGB1 expression in unwounded and wounded site of *Hmgb1*$^{+/-}$ mice is very low compared to WT mice. In addition, a delayed epithelialization contributed to the attenuation of *Hmgb1*$^{+/-}$ mice wound closure and was significantly difference from WT mice at three days post-surgery. These results imply that HMGB1 is responsible for palatal wound repair. HMGB1 is translocalized and released after inflammatory stimuli [29,30] and is mobilized from gingival epithelial cells in response to periodontal inflammation [31]. Our findings demonstrate that at seven days after surgery, tissues from the WT mice presented HMGB1 protein translocation from the nucleus to the cytoplasm of basal epithelial cells and cells of connective tissue, which is in accordance with results obtained in other cell types [29–31].

Palatal wound healing involves a complex, well-orchestrated series of events like hemostasis, inflammatory response and collagen synthesis. A previous study has demonstrated that treatment with 100 ng/mL has a chemotactic effect on fibroblasts and keratinocytes. Following a single dose of 200 and 400 ng HMGB1, wound closure was accelerated in diabetes mice compared to wounds treated either with saline or 800 ng of this protein [26]. In our in vitro study, rHMGB1 at a concentration of 100 ng/mL induces gingival epithelial cells migration and PCNA mRNA overexpression. Accordingly, it is likely that HMGB1 promotes palatal wound healing by accelerating the re-epithelialization and proliferation of gingival epithelial cells, may occur in vivo. Nevertheless, another study showed that treatment with 100 ng/mL of HMGB1 protein impairs fibroblast collagen synthesis in rats undergoing full-thickness incisional wound. Reduction of HMGB1 levels in the wound using ethyl pyruvate leads to significant increases in reparative collagen deposition [32]. The difference in the results of these studies may be associated with the variations in fibroblast subpopulations and the proliferative potential of the cells in the lesion.

VEGF appears to play a crucial role in the proliferative phase of wound healing, promote migration, differentiation and tube formation of endothelial cells which are key elements in early stages of angiogenesis [33]. Our study demonstrated new vessels appear as early as three days after wounding and staining for VEGF was most prominent in keratinocytes at seven days post-surgery, consistent with VEGF expression results of previous reports [34–36]. HMGB1 administration restored the blood flow recovery in the ischemic muscle of diabetic mice is associated with the increased expression of VEGF [37]. In our study, we show that decreased VEGF expression in keratinocytes is associated with impaired wound healing in the *Hmgb1*$^{+/-}$ mice. One explanation for this result is that about three to ten days after the wound occurs, macrophages are abundant in the wound tissue and new blood vessels are formed [38]. Activated macrophages and monocytes secrete HMGB1 [39] and this HMGB1 may in turn induce VEGF expression in keratinocytes. However, wound healing is a complex organized process including a variety of cell types (keratinocytes, endothelial cells, fibroblasts, inflammatory cells and epidermal cells) and angiogenesis inducers other than VEGF. Thus, despite an impairment of wound healing in *Hmgb1*$^{+/-}$ mice in this study, further studies are required to rule out crucial biological processes involved in the early and late phase of wound repair in this model.

A cell monolayer scratch assay is used to assess re-epithelialization [40]. In the present study, we demonstrated a large induction of cell migration in gingival epithelial cells by rHMGB1 during wound scratch assay after 24–48 h. This finding is consistent with previous report that HMGB1 is a potent cell migration and proliferation promoting agent [41]. Nevertheless, HMGB1-induced increase in human gingival and periodontal ligament fibroblast migration after 16 h stimulation in the transwell chamber assay [27]. It is possible that the difference in the speed of cell migration in our experiment compared to the previous study may depend in part on particular cell type and assay protocol.

HMGB1 binding to RAGE triggers transcription factor NF-κB [42]. Knockdown of HMGB1 or RAGE inhibited NF-κB p50 and p65 expression [43]. Our study demonstrated that NF-κB p50-positive epithelial cells of WT mice were significantly higher than in *Hmgb1*$^{+/-}$ mice. This data correlated with previous studies in which reduction in HMGB1-RAGE [43] or Toll-like receptor [44] axis reduced NF-κB activation and could not promote cellular proliferation. To elucidate the molecular mechanisms of HMGB1-induced wound re-epithelialization in vivo, we performed in vitro cell migration assay and PCNA transcription in gingival epithelial cells-treated with rHMGB1 in the presence or absence of RAGE siRNA. The result of our study demonstrates that HMGB1 signaling promotes wound healing of palatal mucosa via RAGE-dependent. It is noteworthy that intraoral wound healing process involves interaction of other signal molecules such as Smad3 [8] and transforming growth factor beta 1 [45]. Hence, the HMGB1 mouse model in our study may partly explain one of the biological factors that regulate palatal tissue repair and support the view that the local inflammatory response can promote wound closure. At the molecular level, these observations may be explained, at least in part, by the capacity of HMGB1 molecules in keratinocytes on palatal wound repair. As such, further analysis of specific effects of this protein may lead to novel therapies for improved wound healing properties.

Attempts have been made to repair and restore destroyed periodontal tissues, including the use of particular bioactive substances or surgical bone grafts [20,34]. Nowadays, conventional therapies are poorly effective on complex chronic ulcers and tissue lesions occurring in specific diseases, like diabetes. Topical application of HMGB1 to the wounds of streptozotocin-induced diabetic models in mice, enhanced arteriole density, granulation tissue formation, and accelerated wound healing in mice skin [26]. HMGB1-induced human gingival fibroblasts [27] and gingival epithelial cells proliferation and migration. These data suggest that HMGB1 may have potential beneficial effects on tissue remodeling and repair. Therefore, there is a great interest for a new generation of topical chronic wound treatments containing low levels of HMGB1 to accelerate intraoral wound healing and reduce wound-related complications.

4. Materials and Methods

4.1. Reagents

VEGF and HMGB1 Ab were obtained from Abcam (Cambridge, MA, USA). NF-κB p50 and PCNA Ab were purchased from Santa Cruz Biotechnology (Santa Cruz, CA, USA). Unless otherwise stated, all other reagents were supplied by Sigma-Aldrich Inc. (St. Louis, MO, USA).

4.2. Mice

All animal experiments were performed according to protocols approved by the institutional guidelines at Kagoshima University Graduate School of Medical and Dental Science. (Ethic approval number: H26/078, Approval date: 24 October 2014) Generation of Hmgb1 heterozygotes ($Hmgb1^{+/-}$) mice on a pure BALB/c (Wild-type, WT) genetic background used in the present study were described before [46]. All mice were maintained and bred under standard pathogen-free conditions. Northern blot analysis of total RNA extracted from $Hmgb1^{+/-}$ mice newborn liver using cDNA-encoding HMGB1 demonstrated approximately one-half the level of HMGB1 mRNA and protein compared with WT mice [46].

4.3. Polymerase Chain Reaction (PCR) Genotyping Assay

For genotyping PCR analysis, 3 mm sections of tail tip were dissolved in 0.1 mL of 50 mM Tris (pH 8.0), 100 mM EDTA, 0.5% SDS, and 0.5 mg/mL proteinase K solution at 55 °C for 2 to 6 h with vigorous shaking. DNA was prepared using DNeasy Blood Tissue Kit (Qiagen, Redwood City, CA, USA) and subjected to reverse transcription (RT)-PCR. We routinely genotyped newborn mice RT-PCR analysis of tail DNA (100 ng) using a commercial RT-PCR kit (Takara Biomedicals, Shiga, Japan), and the reaction was performed following the manufacturer's instructions. The resulting cDNA mixture was amplified with Taq polymerase, and the following specific primers (Sigma, St. Louis, MO, USA) were used: WT allele, 5'-GCAGGCTTCGTTGTTTTCATACAG-3' and 5'-TCAAAGAGTAATACTGCCACCTTC-3', which generate a 495-bp fragment. The mutant *Hmgb1* allele was detected by using primer complementary to the neomycin resistance gene, 5'-TGGTTTGCAGTGTTCTGCCTAGC-3' and 5'-CCCAGTCATAGCCGAATAGCC-3' which generate a

336-bp fragment [11]. Amplification conditions were 1 cycle of 95 °C for 5 min, 35 cycles of 95 °C for 45 s, 60 °C for 30 s, 72 °C for 30 s with primer extension time at 72 °C for 5 min.

4.4. Palatal Wound Healing Model and Histological Analysis

The criteria used for assessing wound healing included; the epithelialization (as assessed by macroscopic evaluation for wound closure) [47], the degree of inflammation in the tissues [22] and collagen formation, which is responsible for tissue repair [48]. The macroscopic finding of continuous wound closure was confirmed by histology (H & E stain) and Mallory Azan. We determined the expression of cell proliferation (PCNA) and angiogenesis (VEGF) in wound tissue of the mice studied by immunohistochemistry. Eight- to fifteen-week-old mice were used in this study (5–7 mice per group). All animals were anaesthetized with isoflurane. A full-thickness incision wound, 1.0-mm width and 2.0-mm length, was made in the mucoperiosteum of the hard palate under sterile condition based on previous study with a slight modification [49]. After the incision was made, animals were sacrificed on Days 0, 3 and 7. No medication was used throughout the experiment. At the time points indicated, the wound kinetics were determined by using image-processing software (ImageJ, US National Institutes of Health, Bethesda, MD, USA) to measure the wound area; wound area was demonstrated as a percentage of the initial wound area. The wound closure was considered complete when the entire surface area was covered with tissue. The measurements were performed by three different examiners and blinded to each other. Pearson's correlation coefficient was used to represent inter-examiner reliability.

To assess the wound healing process by microscopic examination, maxillary tissues after wounding were harvested. Tissues were fixed in 4% freshly made paraformaldehyde in 100 mM sodium phosphate buffer (pH 7.0) overnight at 4 °C. The samples were then decalcified with 19% EDTA in 100 mM PBS for 3 weeks, dehydrated, and embedded in paraffin. Serial sections, 4 μm thick, were cut in the frontal plane through the midpoints of the bilateral second molars. Slides were stained with H&E or Mallory Azan, and evaluated for the histological changes under light microscope (Jenamed II, Carl Zeiss, Gottingen, Germany).

4.5. Immunohistochemistry

Paraffin-embedded sections were deparaffinized in xylene and rehydrated through a series of decreasing concentrations of ethanol. Staining was carried out using indirect immunoperoxidase diaminobenzidine (DAB). Endogenous peroxidase was blocked by 0.3% H_2O_2 for 5 min. Sections were incubated for 1 h at room temperature with polyclonal anti-PCNA (1:200), NF-κB p50 (1:500), HMGB1 (1:200) and VEGF Ab (1:200) in Ab diluent with background reducing components (DakoCytomation, Carpinteria, CA, USA). As a negative control, the IgG isotype control was employed at the same time and concentration as the test antibodies. After rinsing with PBS, sections were finally developed with a DAKO LSAB+ System, horseradish peroxidase (HRP) (DakoCytomation; KO679), and immunostaining was visualized with substrate solution (DAB). Counterstaining was performed with Mayer's hematoxylin. Immunostaining of PCNA and NF-κB p50 were localized in the nucleus of epithelial cells. VEGF staining was noticed in the cytoplasm. The number of PCNA and NF-κB p50 immunopositive cells from six slides in experimental and control samples were selected randomly and evaluated by two blinded observers and scored.

4.6. Quantitative Real-Time PCR Analysis of VEGF Expression in the Wound Tissues

Total RNA was extracted from palatal wound tissues of WT mice or $Hmgb1^{+/-}$ mice using TRIzol Reagent (Invitrogen, Carlsbad, CA, USA). Total RNA samples (2 μg) were reverse transcribed using a First Strand complementary DNA synthesis kit for RT-PCR (Roche, Indianapolis, IN, USA). cDNA was augmented by real-time RT-PCR (C_t value 20–30 s cycles) using a 7300 Real-Time PCR System (Applied Biosystems, Foster City, CA, USA). The primers for gene amplification were as follows: VEGF sense, 5'-AACGATGAAGCCCTGGAGTG-3'; VEGF antisense, 5'-GACAAACAAATGCTTTCTCCG-3' (accession number MN_00441242); GAPDH sense, 5'-TGTGTCCGTCGTGGATCTGA-3'; GAPDH antisense, 5'-CCTGCTTCACCACCTTCTTGAT-3' (accession number NM_008084.3). The PCR conditions were 95 °C for 2 min followed by 40 cycles of 95 °C for 15 s, annealing at 58 °C for 30 s, and extension at 72 °C for 15 s. All reactions were run in triplicate. VEGF expression was defined on the basis of the threshold cycle (C_t value) and normalized to the GAPDH expression.

4.7. Primary Cell Cultures

Palatal gingival explants were prepared for the cell culture according to our previous study [50] with some modification. Briefly, the gingival tissues were surgically removed from the animal, placed in tissue culture plates and soaked in Dulbecco's-modified Eagle Medium (DMEM; Sigma-Aldrich, St. Louis, MO, USA) containing 10% fetal bovine serum (FBS). After 2 weeks, gingival epithelial cells were harvested from the culture medium and further incubated in Keratinocyte-Serum Free Medium (Life Technologies, Rockville, MD, USA) supplemented with epidermal growth factor (5 ng/mL) and bovine pituitary extract (30–50 µg/mL). The cells of passages 4–6 were used for experiments. All cells were cultured in serum-free Opti-MEM-I medium (Gibco, Grand Island, NY, USA) for at least 15 h before treatment to eliminate the possible side effect of growth factors.

4.8. Silencing of RAGE Gene Expression and Western Blot Detection for RAGE Protein Expression

RNA silencing was performed with siRNA targeting RAGE mRNA or control siRNA (Santa Cruz, catalog: sc-36375) prepared according to the method described in our previous study [29]. Cells were transfected with siRNA duplexes suspended in lipofectamine reagent (Life Technologies) following the manufacturer's protocol. Briefly, cells were cultured in 6-well plates until 60% confluence. Cells were washed with serum-free Opti-MEM I Reduced Serum Medium (Gibco, Grand Island, NY, USA). RAGE siRNA and negative control siRNA were gently premixed with lipofectamine reagent in Opti-MEM medium for 20 min at RT. The siRNA (final concentration 150–300 nM)/lipofectamine reagent complex was overlaid onto the washed cells for an additional 24 h at 37 °C in 5% CO_2. The efficacy of gene silencing was evaluated using Western blot analysis. Protein concentrations were determined by Bradford protein assay using bovine serum albumin as standard (Bio-Rad, Hercules, CA, USA). Samples were mixed with $2\times$ electrophoresis sample buffer solution with bromophenol blue (Santa Cruz Biotechnology) before being subjected to 12% SDS-polyacrylamide gel electrophoresis (PAGE) and transferred onto nitrocellulose membranes (Schleicher & Schuell, Dassel, Germany). Samples containing 10 or 15 µg of total protein were used. To prevent nonspecific binding, the membrane was blocked with a solution containing 5% (w/v) nonfat dry milk with 1% (v/v) Tween 20 in PBS for 1 h at RT. Rabbit anti-RAGE primary Ab was incubated for overnight at 4 °C. Then the membranes were washed and incubated with HRP-conjugated anti-rabbit polyclonal IgG (MP Biomedicals Inc., Solon, OH, USA) at RT for 1 h. Labeled bands were visualized using an enhanced chemiluminescence system (GE Healthcare Bio-Science, Pittsburgh, PA, USA) and exposed to high-performance chemiluminescence film (GE Healthcare). The intensity of the protein bands in Western blotting was quantified using National Institutes of Health Image 1.63 software.

4.9. Cell Viability Test

Cell viability was monitored after incubation with siRNA for 48 h by MTT (3-[4,5]-2,5-diphenyltetrazolium bromide) assay. Briefly, cells were harvested and incubated with MTT (0.5 mg/mL; final concentration) for 3 h. Formazan crystal was solubilized by adding dimethyl sulfoxide for 16 h. Dehydrogenase activity was expressed as absorbance at 570 and 630 nm.

4.10. In Vitro Scratch Assay

Cells were scratched with a sterile P200 pipette tip according to a method previously described [51]. After repeated washes to remove the resulting debris, cells were either exposed to rHMGB1 at the concentration of 50 and 100 ng/mL for 0, 12, 24, and 48 h or pretreated with RAGE siRNA. Control siRNA were used as a negative control. The wound closure in scratch assay was monitored by phase microscopy capturing images of the same field with a 20× objective at different times, as specified. Migration of cells into the cell-free space was determined by the digital image processing software "Image J" developed by NIH. In some experiments, the cells were harvested for RNA extraction.

4.11. Quantitative Real-Time PCR Analysis of PCNA Expression in the Gingival Epithelial Cells

After the induction of the cell damage for 48 h with or without 300 nM of RAGE siRNA, cells were collected and lysed with TRIzol Reagent according to the manufacturer's instruction and performed with nearly the same protocol with VEGF mRNA expression analysis. The selected specific primer was purchased from Biomol International (Plymouth Meeting, PA, USA). The PCR conditions were 95 °C for 30 s, followed by 1 min annealing at 58 °C and then followed by 1 min extension at 72 °C for a total of 45 cycles.

4.12. Statistical Analysis

Experimental values are given as mean ± S.D. Statistical significances between different groups were assessed by one-way analysis of variance (ANOVA) test or Student's paired t-test. The correlations among three examiners and the wound area measurement were calculated using Pearson's correlation coefficient. All calculations were performed employing Sigma Stat for Windows, version 3.5 (Systat Software, Inc., Chicago, IL, USA). p values < 0.05 were considered statistically significant.

Supplementary Materials: Supplementary materials can be found at www.mdpi.com/1422-0067/17/11/1961/s1.

Acknowledgments: The authors thank Dr. Marco E. Bianchi (San Raffaele Institute, Italy) for providing the heterozygous HMGB1 mice. This work was supported by the Faculty of Dentistry Grant (2013), Mahidol University.

Author Contributions: Salunya Tancharoen and Satoshi Gando conceived and designed the experiments; Salunya Tancharoen, Shrestha Binita, Tomoka Nagasato, Kiyoshi Kikuchi and Pornpen Dararat performed the experiments; Yuko Nawa and Mika Yamamoto analyzed the data; Somphong Narkpinit and Ikuro Maruyama contributed reagents, materials and analysis tools; and Salunya Tancharoen wrote the paper.

Conflicts of Interest: The authors declare no conflict of interest.

References

1. Guo, S.; di Pietro, L.A. Factors Affecting Wound Healing. *J. Dent. Res.* **2010**, *89*, 219–229. [CrossRef] [PubMed]
2. Leoni, G.; Neumann, P.A.; Sumagin, R.; Denning, T.L.; Nusrat, A. Wound repair: Role of immune-epithelial interactions. *Mucosal Immunol.* **2015**, *8*, 959–968. [CrossRef] [PubMed]
3. Martin, P. Wound healing—Aiming for perfect skin regeneration. *Science* **1997**, *276*, 75–81. [CrossRef] [PubMed]
4. Werner, S.; Grose, R. Regulation of wound healing by growth factors and cytokines. *Physiol. Rev.* **2003**, *83*, 835–870. [PubMed]
5. Mori, R.; Ohshima, T.; Ishida, Y.; Mukaida, N. Accelerated wound healing in tumor necrosis factor receptor p55-deficient mice with reduced leukocyte infiltration. *FASEB. J.* **2002**, *16*, 963–974. [CrossRef] [PubMed]
6. Ishida, Y.; Kondo, T.; Takayasu, T.; Iwakura, Y.; Mukaida, N. The essential involvement of cross-talk between IFN-gamma and TGF-β in the skin wound-healing process. *J. Immunol.* **2004**, *172*, 1848–1855. [CrossRef] [PubMed]
7. Gallucci, R.M.; Simeonova, P.P.; Matheson, J.M. Impaired cutaneous wound healing in interleukin-6-deficient and immunosuppressed mice. *FASEB. J.* **2000**, *14*, 2525–2531. [CrossRef] [PubMed]
8. Jinno, K.; Takahashi, T.; Tsuchida, K.; Tanaka, E.; Moriyama, K. Acceleration of palatal wound healing in Smad3-deficient mice. *J. Dent. Res.* **2009**, *88*, 757–761. [CrossRef] [PubMed]
9. Eming, S.A.; Werner, S.; Bugnon, P. Accelerated wound closure in mice deficient for interleukin-10. *Am. J. Pathol.* **2007**, *170*, 188–202. [CrossRef] [PubMed]

10. Bianchi, M.E. HMG proteins: Dynamic players in gene regulation and differentiation. *Curr. Opin. Genet. Dev.* **2005**, *15*, 496–506. [CrossRef] [PubMed]

11. Oyama, Y.; Hashiguchi, T.; Taniguchi, N. High-mobility group box-1 protein promotes granulomatous nephritis in adenine-induced nephropathy. *Lab. Investig.* **2010**, *90*, 853–866. [CrossRef] [PubMed]

12. Taniguchi, N.; Kawahara, K.; Yone, K. High mobility group box chromosomal protein 1 plays a role in the pathogenesis of rheumatoid arthritis as a novel cytokine. *Arthritis Rheum.* **2003**, *48*, 971–981. [CrossRef] [PubMed]

13. Kikuchi, K.; Uchikado, H.; Miura, N. HMGB1 as a therapeutic target in spinal cord injury: A hypothesis for novel therapy development. *Exp. Ther. Med.* **2011**, *2*, 767–770. [PubMed]

14. Morimoto, Y.; Kawahara, K.I.; Tancharoen, S. Tumor necrosis factor-α stimulates gingival epithelial cells to release high mobility-group box 1. *J. Periodontal Res.* **2008**, *43*, 76–83. [CrossRef] [PubMed]

15. Huttunen, H.J.; Kuja-Panula, J.; Rauvala, H. Receptor for advanced glycation end products (RAGE) signaling induces CREB-dependent chromogranin expression during neuronal differentiation. *J. Biol. Chem.* **2002**, *277*, 38635–38646. [CrossRef] [PubMed]

16. Palumbo, R.; Sampaolesi, M.; de Marchis, F. Extracellular HMGB1, a signal of tissue damage, induces mesoangioblast migration and proliferation. *J. Cell Biol.* **2004**, *164*, 441–449. [CrossRef] [PubMed]

17. Limana, F.; Germani, A.; Zacheo, A. Exogenous high-mobility group box 1 protein induces myocardial regeneration after infarction via enhanced cardiac C-kit⁺ cell proliferation and differentiation. *Circ. Res.* **2005**, *97*, 73–83. [CrossRef] [PubMed]

18. Ranzato, E.; Patrone, M.; Pedrazzi, M.; Burlando, B. HMGb1 promotes scratch wound closure of HaCaT keratinocytes via ERK1/2 activation. *Mol. Cell. Biochem.* **2009**, *332*, 199–205. [CrossRef] [PubMed]

19. Ito, Y.; Bhawal, U.K.; Sasahira, T. Involvement of HMGB1 and RAGE in IL-1β-induced gingival inflammation. *Arch. Oral Biol.* **2012**, *57*, 73–80. [CrossRef] [PubMed]

20. Schmidt, A.M.; Hofmann, M.; Taguchi, A.; Du, Y.S.; Stern, D.M. RAGE: A Multiligand Receptor Contributing to the Cellular Response in Diabetic Vasculopathy and Inflammation. *Semin. Thromb. Hemost.* **2000**, *26*, 485–494. [CrossRef] [PubMed]

21. Schmidt, S.D. Receptor for age (RAGE) is a gene within the major histocompatibility class III region: implications for host response mechanisms in homeostasis and chronic disease. *Front. Biosci.* **2001**, *6*, D1151–D1160. [CrossRef] [PubMed]

22. Hori, B.J.; Slattery, T.; Cao, R.; Zhang, J.; Chen, J.X.; Nagashima, M.; Lundh, E.R.; Vijay, S.; Nitecki, D. The receptor for advanced glycation end products (RAGE) is a cellular binding site for amphoterin. Mediation of neurite outgrowth and co-expression of rage and amphoterin in the developing nervous system. *J. Biol. Chem.* **1995**, *270*, 25752–25761. [CrossRef] [PubMed]

23. Hofmann, S.D.; Fu, C. RAGE mediates a novel proinflammatory axis: A central cell surface receptor for S100/calgranulin polypeptides. *Cell* **1999**, *97*, 889–901. [CrossRef]

24. Bierhaus, H.P.; Morcos, M.; Wendt, T.; Chavakis, T.; Arnold, B.; Stern, D.M.; Nawroth, P.P. Understanding RAGE, the receptor for advanced glycation end products. *J. Mol. Med. (Berl.)* **2005**, *83*, 876–886. [CrossRef] [PubMed]

25. Ranzato, P.M.; Pedrazzi, M.; Burlando, B. HMGB1 promotes wound healing of 3T3 mouse fibroblasts via RAGE-dependent ERK1/2 activation. *Cell. Biochem. Biophys.* **2010**, *57*, 9–17. [CrossRef] [PubMed]

26. Straino, S.; di, C.; Mangoni, A. High-mobility group box 1 protein in human and murine skin: Involvement in wound healing. *J. Investig. Dermatol.* **2008**, *128*, 1545–1553. [CrossRef] [PubMed]

27. Chitanuwat, A.; Laosrisin, N.; Dhanesuan, N. Role of HMGB1 in proliferation and migration of human gingival and periodontal ligament fibroblasts. *J. Oral Sci.* **2013**, *55*, 45–50. [CrossRef] [PubMed]

28. Szczepanski, M.J.; Luczak, M.; Olszewska, E. Molecular signaling of the HMGB1/RAGE axis contributes to cholesteatoma pathogenesis. *J. Mol. Med.* **2015**, *93*, 305–314. [CrossRef] [PubMed]

29. Tancharoen, S.; Tengrungsun, T.; Suddhasthira, T.; Kikuchi, K.; Vechvongvan, N.; Tokuda, M.; Maruyama, I. Overexpression of Receptor for Advanced Glycation End Products and High-Mobility Group Box 1 in Human Dental Pulp Inflammation. *Mediat. Inflamm.* **2014**. [CrossRef] [PubMed]

30. Chaichalotornkul, S.; Nararatwanchai, T.; Narkpinit, S. Secondhand smoke exposure-induced nucleocytoplasmic shuttling of HMGB1 in a rat premature skin aging model. *Biochem. Biophys. Res. Commun.* **2015**, *456*, 92–97. [CrossRef] [PubMed]

31. Ebe, N.; Iwasaki, K.; Iseki, S.; Okuhara, S.; Podyma-Inoue, K.A.; Terasawa, K.; Watanabe, A.; Akizuki, T.; Watanabe, H.; Yanagishita, M.; et al. Pocket epithelium in the pathological setting for HMGB1 release. *J. Dent. Res.* **2011**, *90*, 235–240. [CrossRef] [PubMed]

32. Zhang, Q.; O'Hearn, S.; Kavalukas, S.L.; Barbul, A. Role of high mobility group box 1 (HMGB1) in wound healing. *J. Surg. Res.* **2012**, *176*, 343–347. [CrossRef] [PubMed]

33. Bates, D.O.; Jones, R.O. The role of vascular endothelial growth factor in wound healing. *Int. J. Lower Extremity Wounds* **2003**, *2*, 107–120. [CrossRef] [PubMed]

34. Bueno, F.G.; Moreira, E.A.; Morais, G.R.; Pacheco, I.A.; Baesso, M.L.; Leite-Mello, E.V.; Mello, J.C. Enhanced cutaneous wound healing in vivo by standardized crude extract of *poincianella pluviosa*. *PLoS ONE* **2016**, *11*, e0149223. [CrossRef] [PubMed]

35. Jacobi, J.; Tam, B.Y.; Sundram, U.; von Degenfeld, G.; Blau, H.M.; Kuo, C.J.; Cooke, J.P. Discordant effects of a soluble VEGF receptor on wound healing and angiogenesis. *Gene Ther.* **2004**, *11*, 302–409. [CrossRef] [PubMed]

36. Jain, V.; Triveni, M.G.; Kumar, A.B.; Mehta, D.S. Role of platelet-rich-fibrin in enhancing palatal wound healing after free graft. *Contemp. Clin. Dent.* **2012**, *3*, S240–S243. [PubMed]

37. Biscetti, F.; Straface, G.; de, C.R. High-Mobility Group Box-1 protein promotes angiogenesis after peripheral ischemia in diabetic mice through a VEGF-dependent mechanism. *Diabetes* **2010**, *59*, 1496–1505. [CrossRef] [PubMed]

38. Matthias, S. Cancer as an overhealing wound: An old hypothesis revisited. *Nat. Rev. Mol. Cell Biol.* **2008**, *9*, 628–638.

39. Tang, D.; Shi, Y.; Kang, R. Hydrogen peroxide stimulates macrophages and monocytes to actively release HMGB1. *J. Leukoc. Biol.* **2007**, *81*, 741–747. [CrossRef] [PubMed]

40. Mohammedsaeed, W.; Cruickshank, S.; McBain, A.J.; O'Neill, C.A. Lactobacillus rhamnosus GG Lysate Increases Re-Epithelialization of Keratinocyte Scratch Assays by Promoting Migration. *Sci. Rep.* **2015**, *5*, 16147. [CrossRef] [PubMed]

41. Rossini, A.; Zacheo, A.; Mocini, D. HMGB1-stimulated human primary cardiac fibroblasts exert a paracrine action on human and murine cardiac stem cells. *J. Mol. Cell. Cardiol.* **2008**, *44*, 683–693. [CrossRef] [PubMed]

42. Luan, Z.G.; Zhang, H.; Yang, P.T. HMGB1 activates nuclear factor-kappaB signaling by RAGE and increases the production of TNF-α in human umbilical vein endothelial cells. *Immunobiology* **2010**, *215*, 956–962. [CrossRef] [PubMed]

43. Chen, R.; Yi, P.P.; Zhou, R.R. The role of HMGB1-RAGE axis in migration and invasion of hepatocellular carcinoma cell lines. *Mol. Cell. Biochem.* **2014**, *390*, 271–280. [CrossRef] [PubMed]

44. Ruochan, C.; Qiuhong, Z.; Rui, K.; Xue-Gong, F.; Daolin, T. Emerging roles for HMGB1 protein in immunity, inflammation, and cancer. *Mol. Med.* **2013**, *19*, 357–366.

45. Diabetic, R.; Dorria, M.; Elbardisey, H.M.; Eltokhy, M.; Doaa, T. Effect of transforming growth factor β 1 on wound healing in induced diabetic rats. *Int. J. Health Sci.* **2013**, *7*, 160–172.

46. Calogero, S.; Grassi, F.; Aguzzi, A.; Voigtländer, T.; Ferrier, P.; Ferrari, S.; Bianchi, M.E. The lack of chromosomal protein HMG1 does not disrupt cell growth but causes lethal hypoglycaemia in newborn mice. *Nat. Genet.* **1999**, *22*, 276–280. [PubMed]

47. Silva, C.O.; Ribeiro Edel, P.; Sallum, A.W.; Tatakis, D.N. Free gingival grafts: Graft shrinkage and donor-site healing in smokers and non-smokers. *J. Periodontol.* **2010**, *81*, 692–701. [CrossRef] [PubMed]

48. Boucek, R.J. Factors affecting wound healing. *Otolaryngol. Clin. N. Am.* **1984**, *17*, 243–264.

49. Jettanacheawchankit, S.; Sasithanasate, S.; Sangvanich, P.; Banlunara, W.; Thunyakitpisal, P. Acemannan stimulates gingival fibroblast proliferation; expressions of keratinocyte growth factor-1, vascular endothelial growth factor, and type I collagen; and wound healing. *J. Pharm. Sci.* **2009**, *109*, 525–531. [CrossRef]

50. Tancharoen, S.; Matsuyama, T.; Kawahara, K. Cleavage of host cytokeratin-6 by lysine-specific gingipain induces gingival inflammation in periodontitis patients. *PLoS ONE* **2015**, *10*, e0117775. [CrossRef] [PubMed]
51. Merlo, S.; Canonico, P.L.; Sortino, M.A. Differential involvement of estrogen receptorα and estrogen receptor β in the healing promoting effect of estrogen in human keratinocytes. *J. Endocrinol.* **2009**, *200*, 189–197. [CrossRef] [PubMed]

© 2016 by the authors. Licensee MDPI, Basel, Switzerland. This article is an open access article distributed under the terms and conditions of the Creative Commons Attribution (CC BY) license (http://creativecommons.org/licenses/by/4.0/).

International Journal of
Molecular Sciences

MDPI

Article

Broad-Spectrum Inhibition of the CC-Chemokine Class Improves Wound Healing and Wound Angiogenesis

Anisyah Ridiandries [1,2], Christina Bursill [1,2,*,†] and Joanne Tan [1,2,*,†]

[1]   Heart Research Institute, 7 Eliza Street, Newtown, Sydney 2042, NSW, Australia;
     anisyah.ridiandries@hri.org.au
[2]   Sydney Medical School, University of Sydney, Camperdown, Sydney 2050, NSW, Australia
*   Correspondence: christina.bursill@hri.org.au (C.B.); joanne.tan@hri.org.au (J.T.);
     Tel.: +61-2-8208-8905 (C.B.); +61-2-8208-8900 (J.T.); Fax: +61-2-9565-5584 (C.B. & J.T.)
†   These authors contributed equally to this work.

Academic Editors: Allison Cowin and Terrence Piva
Received: 17 August 2016; Accepted: 10 January 2017; Published: 13 January 2017

Abstract: Angiogenesis is involved in the inflammation and proliferation stages of wound healing, to bring inflammatory cells to the wound and provide a microvascular network to maintain new tissue formation. An excess of inflammation, however, leads to prolonged wound healing and scar formation, often resulting in unfavourable outcomes such as amputation. CC-chemokines play key roles in the promotion of inflammation and inflammatory-driven angiogenesis. Therefore, inhibition of the CC-chemokine class may improve wound healing. We aimed to determine if the broad-spectrum CC-chemokine inhibitor "35K" could accelerate wound healing in vivo in mice. In a murine wound healing model, 35K protein or phosphate buffered saline (PBS, control) were added topically daily to wounds. Cohorts of mice were assessed in the early stages (four days post-wounding) and in the later stages of wound repair (10 and 21 days post-wounding). Topical application of the 35K protein inhibited CC-chemokine expression (CCL5, CCL2) in wounds and caused enhanced blood flow recovery and wound closure in early-mid stage wounds. In addition, 35K promoted neovascularisation in the early stages of wound repair. Furthermore, 35K treated wounds had significantly lower expression of the p65 subunit of NF-κB, a key inflammatory transcription factor, and augmented wound expression of the pro-angiogenic and pro-repair cytokine TGF-β. These findings show that broad-spectrum CC-chemokine inhibition may be beneficial for the promotion of wound healing.

Keywords: chemokine; angiogenesis; wound; healing; inflammation

1. Introduction

Wound healing is a complex multistep process comprised of three overlapping but distinct phases: inflammation, proliferation and remodelling. It begins with the inflammation phase, where neutrophils and macrophages are recruited to the wound site to remove cell debris and phagocytose infectious pathogens. This is followed by the proliferation stage, where re-epithelialization and the formation of granulation tissue begin to close the wound. Angiogenesis, the formation of new blood vessels, is important during both the inflammation and proliferation phases, forming angiogenic capillary sprouts that allow for recruitment of inflammatory cells to the wound for debris removal, invasion of the fibrin/fibronectin-rich wound clot, and reorganization of a microvascular network to maintain the new granulation tissue being formed [1]. Finally, during the remodelling phase, fibroblasts reorganize the collagen matrix, forming a closed wound [2]. All three phases must occur in the proper sequence and time frame in order for a wound to heal successfully. Excessive macrophage accumulation during the inflammation and proliferation stages may prolong the inflammatory response leading to delayed wound healing and scar formation at the injury site. Macrophage phenotype has also been reported to change throughout the wound healing process, alternating between inflammatory M1 and repair M2 phenotypes [3]. The regulation of these phenotypes has been shown to affect wound functionality and the extent of scar tissue [4]. Wounds that exhibit impaired healing frequently enter a state of pathological inflammation, with most chronic wounds developing into ulcers. The incidence of chronic wounds is associated with ischemia, diabetes mellitus, venous stasis disease or pressure [5]. Chronic, non-healing wounds greatly impact a person's quality of life and, when left untreated, sometimes lead to amputation, resulting in an enormous health care burden [6,7]. While current therapies provide some relief, there is a need for the continued development of novel therapies that address the debilitating effects of impaired wound healing.

CC-chemokines are small inflammatory cytokines that have been shown to play a key role in the induction of inflammation and inflammation-mediated angiogenesis. In the wound healing process, macrophage infiltration is highly regulated by CC-chemokine gradients released by hyper-proliferating keratinocytes, fibroblasts and other macrophages [8,9]. In human wounds, a host of CC-chemokines including CCL1, CCL2, CCL3, CCL4, CCL5 and CCL7 are expressed during the first week after injury and high levels of CCL2 have been found in human burns [10,11]. Additionally, studies in excisional wounds have found localized expression of CCL2 and CCL3 in the epidermis, while CCL3 is expressed in follicular epithelium and sebaceous glands [12]. CC-chemokines regulate key angiogenic processes such as the recruitment of inflammatory cells to the wound, to provide support for proliferating and migrating cells [13], and the formation of granulation tissue [9]. Furthermore, incubation with the recombinant CC-chemokines CCL2 and CCL5 stimulate angiogenic functions such as endothelial cell tubulogenesis and migration [14–18]. Therefore, manipulation of the CC-chemokine family may modulate key wound healing processes to confer benefits. To date, studies have inhibited a single CC-chemokine and shown modest or no effect on wound healing. CCL2 knockout mice have delayed re-epithelialization and reduced angiogenesis in the early stages of wound repair, whilst the CCL3 knockout mice have normal wound healing [19]. Similarly, CCR1 knockout models have no alteration in wound healing [20]. Inhibition of a single CC-chemokine, however, may not be as effective in modulating wound healing due to redundancies in CC-chemokine signalling [21,22]. A broad-spectrum CC-chemokine inhibition approach may therefore have increased efficacy.

The broad-spectrum CC-chemokine inhibitor "35K" is a 35 kDa soluble protein produced by the Vaccinia virus that uniquely inhibits only the CC-chemokine class [23]. It recognizes and binds to common structural features shared by most CC-chemokines, preventing binding to their cognate receptors [24]. Broad-spectrum CC-chemokine inhibition using 35K inhibits a host of inflammatory diseases including atherosclerosis, acute peritonitis, hepatitis and liver fibrosis [25–28]. Given the inhibitory effects of 35K on inflammatory-mediated diseases and the role of CC-chemokines in angiogenesis, we sought to elucidate the effect of broad-spectrum CC-chemokine inhibition on wound healing. We found that topical application of 35K enhanced wound closure, blood flow perfusion and neovascularisation at the early-mid stages of wound healing. Augmentation of wound angiogenesis by 35K was mediated via an increase in the pro-angiogenic and pro-repair cytokine TGF-β. In addition, 35K treatment inhibited wound levels of inflammatory transcription factor NF-κB and reduced collagen deposition at the late stages of wound healing in the 35K-treated wounds, suggesting a reduction in scar formation. Taken together, these findings show that broad-spectrum inhibition of CC-chemokines may be beneficial for the promotion of wound healing by suppressing inflammation and promoting wound closure and neovascularisation and reducing scar formation.

2. Results

2.1. Topical 35K Increases Wound Blood Perfusion and the Rate of Wound Closure in Early-Mid Stage Wound Repair

Topical treatment of wounds with 35K protein caused significantly faster wound closure throughout the early to mid-stages of wound healing at Day 7 (42%, $p < 0.05$) and Day 8 (33%, $p < 0.05$) compared to PBS treated wounds (Figure 1a). Using Laser Doppler imaging, it was revealed that 35K protein augmented wound blood perfusion (Figure 1b). The Laser Doppler Index (LDI), used as a marker of wound angiogenesis and determined as the ratio of 35K:PBS wounds, was significantly elevated in the 35K treated wounds in the early stages following wounding (Day 3, 4), compared to PBS control wounds (values above 100% indicate increased wound blood perfusion with 35K treatment relative to PBS control). In the later stages of the wound healing process, at seven days post-wounding, blood perfusion started to decline in wounds treated with 35K, compared to the PBS treated control wounds. After Day 7, blood perfusion fluctuated along the baseline and very little blood perfusion was detected. There were no significant increases or decreases to the end point.

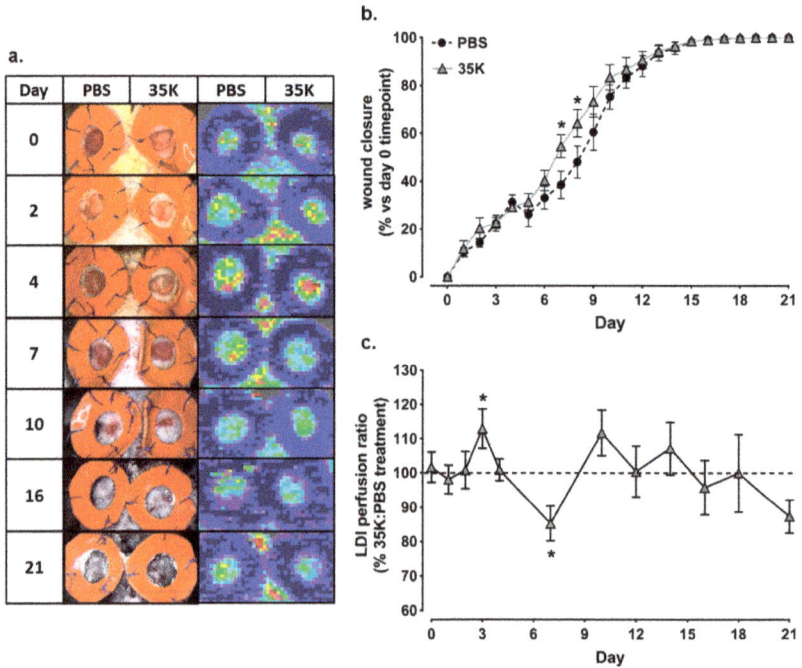

Figure 1. Topical 35K increases the rate of wound closure and wound blood perfusion in early-mid stage wound repair. Two full thickness wounds were created on C57Bl/6J mice ($n = 7$–12). Mice received daily topical application of 35K protein (200 nM) or PBS (vehicle). (**a**) representative images of the wounds and wound blood perfusion using Laser Doppler imaging (high (red) to low (blue) blood flow); (**b**) wound area was calculated from the average of three daily diameter measurements along the x-, y- and z-axes. Wound closure is expressed as a percentage of initial wound area at Day 0. Black circles are PBS treated wounds; grey triangles are 35K treated wounds; and (**c**) the 35K:PBS wound blood flow perfusion ratio was determined using Laser Doppler imaging (LDI). Data is represented as mean ± SEM. Points above the dotted line represent an improvement with 35K treatment. Statistical analysis was performed by an unpaired two-tailed t-test. * $p < 0.05$ compared to PBS treated wounds.

2.2. Inhibition of CC-Chemokines by 35K Increases Neovessel Formation in Early Stage Wounds but Decreases Neovessels in Late Stage Wounds

To investigate wound angiogenesis, neovessels and arterioles were assessed. In the early stages of wound healing (Day 4), there was an increase in the presence of wound neovessels in the 35K treated wounds as determined by CD31+ staining (182%, $p < 0.05$, Figure 2a). However, at Day 10 post-wounding, there were significantly fewer wound neovessels (39% decrease, $p < 0.05$) following 35K treatment. At Day 21, there were no differences in neovessels between 35K and PBS treated wounds. A similar biphasic pattern was also seen with the arterioles (α-actin+ staining), although, at Day 4, the trend for an increase in arterioles did not reach significance, but there was a significant 48% decrease in arterioles ($p < 0.05$) in the 35K treated wounds at Day 10 post-wounding (Figure 2b). No CD31+ and α-actin+ staining is detected in IgG controls (Figure S1).

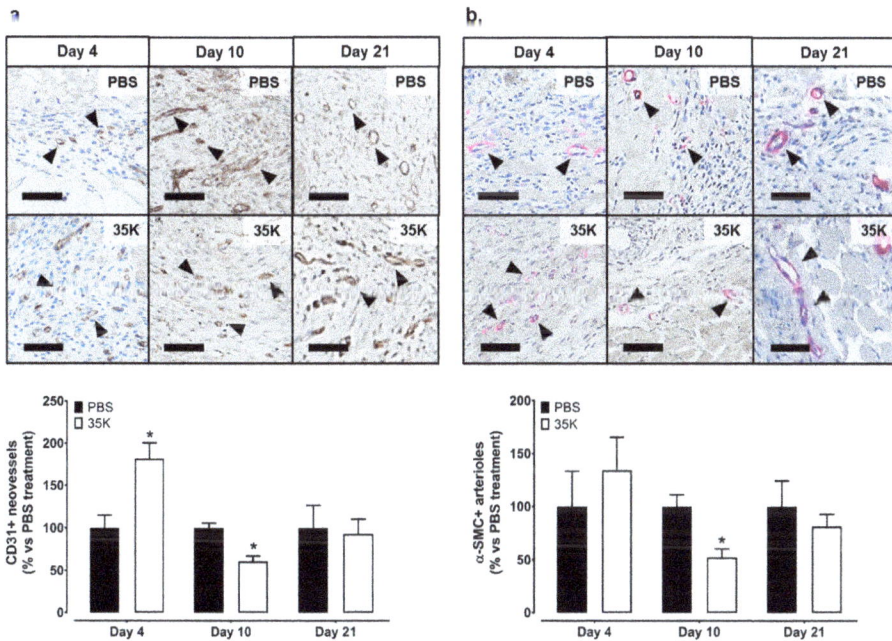

Figure 2. Inhibition of CC-chemokines by 35K increases neovessel formation in the early stages of wound repair but decreases neovessels in later stages. Immunocytochemistry was used to detect the presence of neovessels and arterioles. Photomicrographs represent wounds stained for (**a**) CD31+ neovessels (stained brown, noted by black arrows) and (**b**) α-SMC+ arterioles (stained pink, noted by black arrows). Three tiled images were taken per wound ($n = 7$–12/treatment). Scale bars represent 50 µm. Data is represented as mean $\pm$ SEM. Statistical analysis was performed by an unpaired two-tailed t-test. * $p < 0.05$ compared to PBS treated wounds.

2.3. Inhibition of CC-Chemokines by 35K Modulates Pro-Angiogenic Markers TGF-β and Vascular Endothelial Growth Factor (VEGF) in Early and Late Stage Wounds

The expression of key angiogenic markers TGF-β and vascular endothelial growth factor (VEGF), known to be involved in wound healing, were measured next. Interestingly, TGF-β expression was significantly higher in 35K-treated wounds in the early stages of wound repair (Day 4 post-wounding, 88%, $p < 0.05$, Figure 3a). However, in the later stages post-wounding (Day 10), the expression of TGF-β had normalised back to control levels. There were no differences in VEGF protein levels at Day 4 post-wounding between treatment groups. In the later stages, wounds from both treatment groups had significantly lower levels of VEGF protein, compared to wounds at Day 10 ($p < 0.01$). The protein levels of fibroblast growth factor-2 (FGF-2), were also measured, but no differences were detected (Figure S2a). Neither were there changes in wound protein levels of HIF-1α, a regulator of VEGF (Figure S2b).

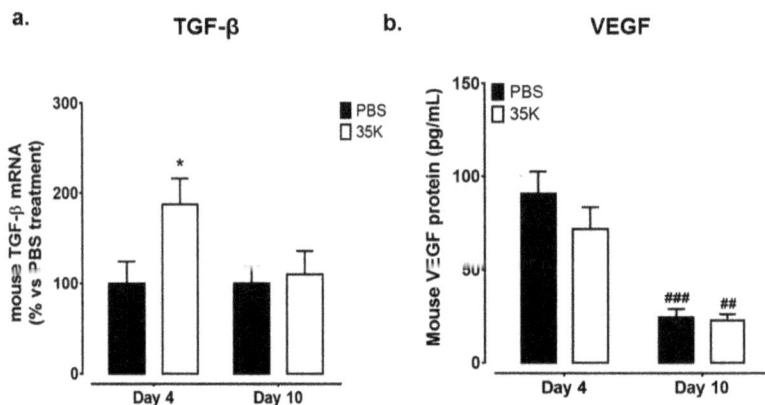

Figure 3. Modulation of pro-angiogenic markers following inhibition of CC-chemokines by 35K in early and late stage wounds. RNA and protein were isolated from PBS and 35K treated wounds at both the early (Day 4) and late (Day 10) time points ($n = 12$/time point); (**a**) real-time PCR was used to measure mRNA levels of TGF-β; (**b**) vascular endothelial growth factor (VEGF) protein levels were detected by ELISA. Data is represented as mean ± SEM. Statistical analysis was performed by an unpaired two-tailed t-test. ## $p < 0.01$, ### $p < 0.001$ compared to respective treatment group wounds at Day 4. * $p < 0.05$ compared to PBS treated wounds.

2.4. Inhibition of CC-Chemokines by 35K Reduces Inflammation but Has No Effect on Wound Macrophage Content

Broad-spectrum CC-chemokine inhibition suppressed the mRNA levels of p65, the active subunit of NF-κB, with 34% and 62% decreases seen in Day 4 and 10 in 35K treated wounds respectively, compared to PBS treated wounds (all $p < 0.05$, Figure 4a). A similar trend was observed for mRNA levels of the macrophage marker CD68, which were also lower in the 35K treated wounds but did not reach significance (Day 4: 39% ns; Day 10: 30% ns, Figure 4b). Despite this, histological analysis of macrophages in wound sections did not find a difference in CD68+ cells in 35K and PBS treated wounds at Day 4, 10, and 21 post-wounding (Figure 4c). Analysis of wound macrophage phenotype revealed that there were no significant changes in the mRNA levels of M2 macrophage marker CD206 in the 35K treated wounds (Figure S3a). Interestingly, M1 macrophage markers CD80 and CD86 were unchanged at Day 4 but elevated in Day 10 35K treated wounds (278% and 342% respectively, $p < 0.01$, Figure S3b,c). No CD68+ staining is detected in IgG control (Figure S1).

2.5. Inhibition of CC-Chemokines by 35K Decreases Collagen Formation in Late Stage Wounds

Milligan's trichrome staining was used to detect the collagen content in the wounds as a marker of tissue remodelling. While no differences in collagen deposition were observed at Day 4 and 21 between 35K and PBS wounds, a 25% decrease ($p < 0.05$) in collagen content was seen in the 35K treated wounds at Day 10 (Figure 5a). The picrosirius red stain was imaged under polarized light to differentiate between type III and type I collagen. There was 10–100 times more type I collagen to type III collagen (across Day 4–21), but there were no significant differences between 35K and PBS treated wounds at each time point (Figure 5b).

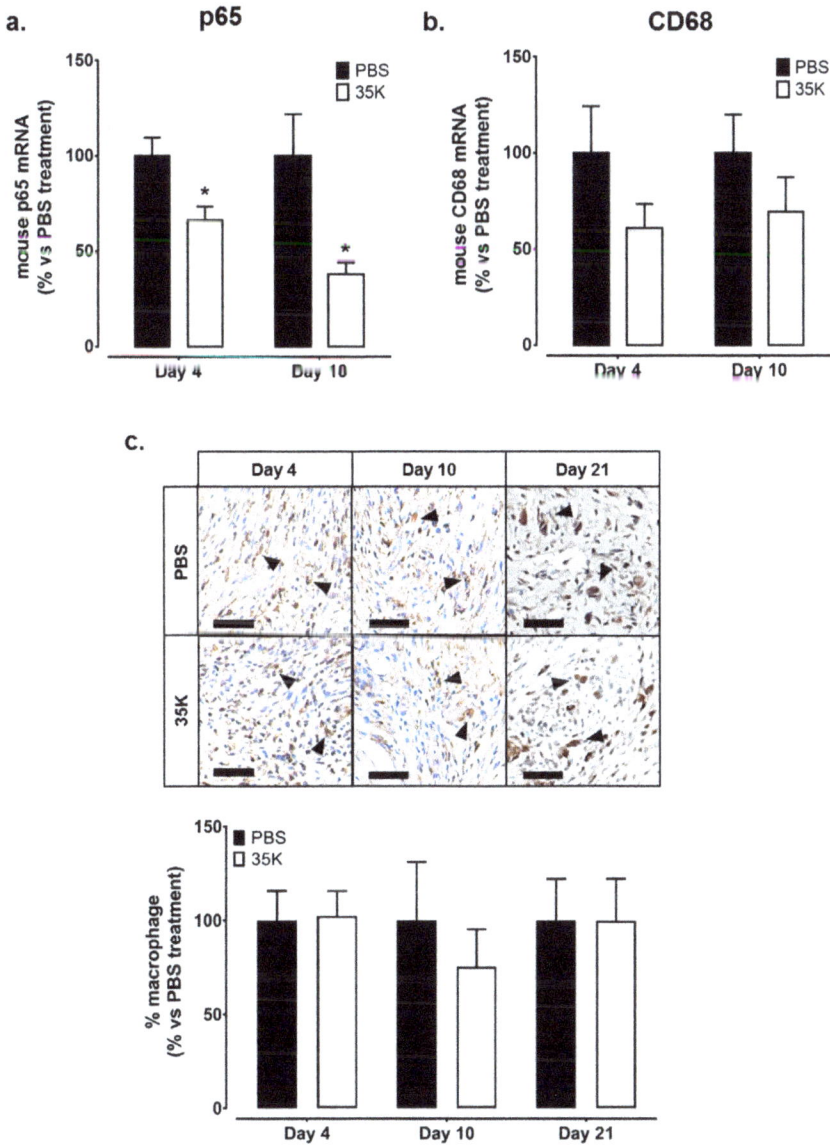

Figure 4. Inhibition of CC-chemokines by 35K reduces inflammation but has no effect on wound macrophage content. Total RNA was isolated from PBS and 35K treated wounds at both the early (Day 4) and late (Day 10) time points ($n = 12$/time point). Real-time PCR was used to measure mRNA levels of (**a**) p65, the active subunit of the key inflammatory transcription factor NF-κB and (**b**) macrophage marker CD68; (**c**) immuno-histochemistry was used to detect the presence of wound macrophages. Photomicrographs represent wound sections stained for CD68+ macrophages (stained brown, noted by black arrows). Three tiled images were taken per wound ($n = 7$–12/treatment). Scale bars represent 50 μm. Data is represented as mean ± SEM. Statistical analysis was performed by an unpaired two-tailed *t*-test. * $p < 0.05$ compared to PBS treated wounds.

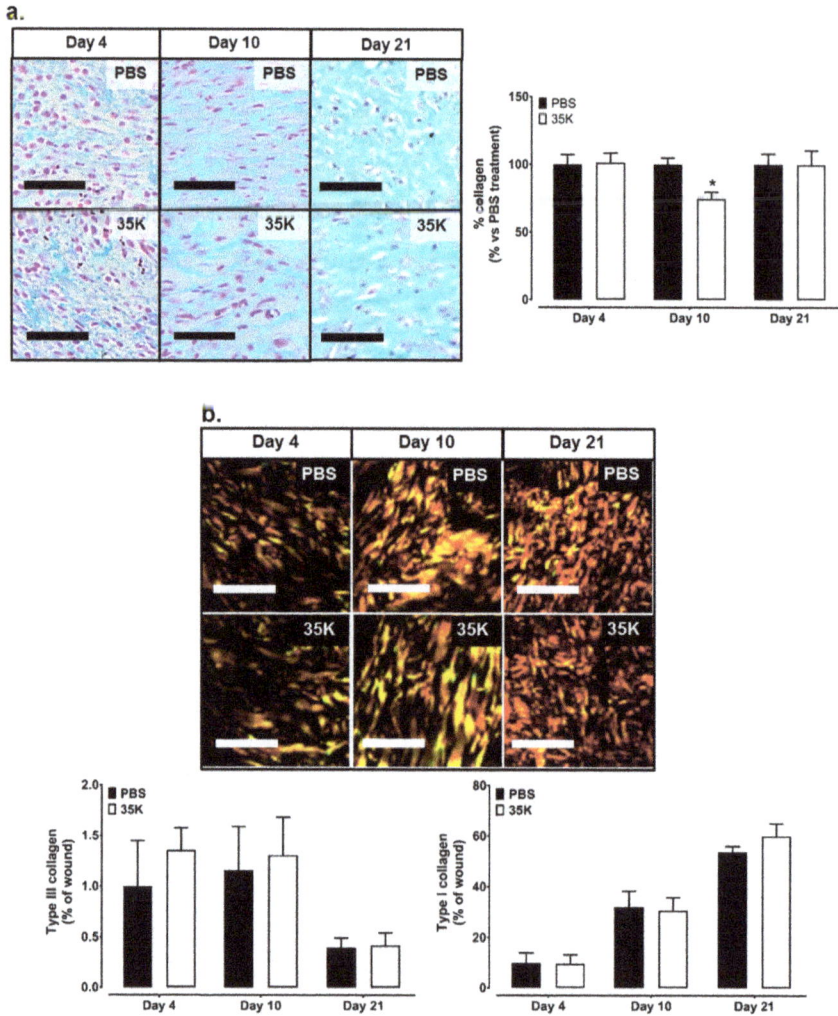

Figure 5. Inhibition of CC-chemokines by 35K decreases collagen formation in Day 10 wounds. Masson's trichrome staining was used to detect the collagen content in the wounds. (**a**) green staining in wound sections is collagen, purple staining are nuclei; (**b**) to differentiate collagen type, wound sections were stained red with picrosirius and imaged under polarized light. Green represents Type III collagen, yellow to red represents Type I collagen. Three tiled images were taken per wound (*n* = 12 or 7/treatment). Scale bars represent 50 μm. Data is represented as mean ± SEM. Statistical analysis was performed by an unpaired two-tailed *t*-test. * *p* < 0.05 compared to PBS treated wounds.

2.6. Topical 35K Suppresses CC-Chemokine Expression in the Wounds

The effect of 35K on wound CC-chemokine protein and gene expression was measured. There was a significant decrease in CCL5 protein levels in 35K treated wounds at both the Day 4 (36%, *p* < 0.05) and Day 10 (66%, *p* < 0.05) time points (Figure 6a). There were also similar decreases in CCL5 mRNA levels in wounds treated with 35K (Day 4: 32%, ns and Day 10: 62%, *p* < 0.05, Figure 6b). CCL2 protein levels were lower in wounds at the Day 4 time point (41%, *p* < 0.05, Figure 6c), but the trend for a decline at the Day 10 time point did not reach significance following 35K treatment. This was consistent for wound CCL2 mRNA levels in which 35K treatment decreased CCL2 wound mRNA at the Day 4 time point (36%, $p < 0.05$, Figure 6d), with no significant decrease in Day 10 wounds.

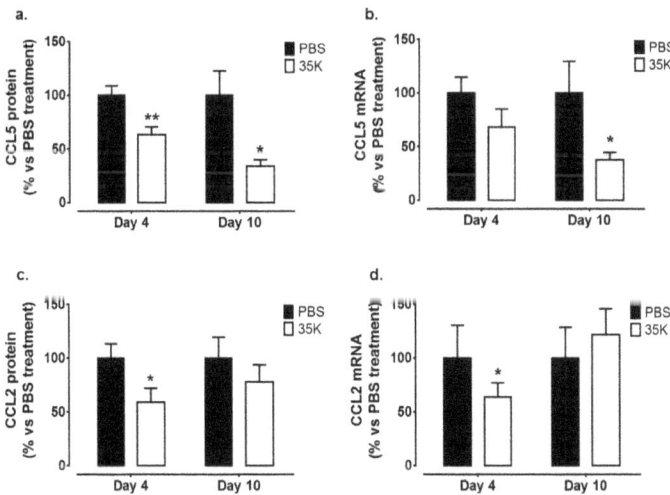

Figure 6. Topical 35K suppresses CC-chemokines in wound tissues. Protein and RNA were isolated from PBS and 35K treated wounds at both the early (Day 4) and late (Day 10) time points ($n = 12$/time point). ELISAs and real-time PCR was used to measure protein and mRNA levels of (**a**,**b**) CCL5 and (**c**,**d**) CCL2. Data is represented as mean ± SEM. Statistical analysis was performed by an unpaired two-tailed t-test. * $p < 0.05$, ** $p < 0.01$ compared to PBS treated wounds.

3. Discussion

The stages of the wound healing process encompass the most complex biological processes that occur in human life. An imbalance can greatly affect the outcome of wound repair, with impaired wound healing often leading to severe unfavourable outcomes such as amputation. This highlights the need to develop novel agents that promote wound recovery. CC-chemokines have been shown to play a key role in the promotion of inflammation and inflammatory-induced angiogenesis, two important processes involved in wound healing. To date, only single CC-chemokine intervention studies have been done, with minimal to no benefit, suggesting that a broad-spectrum inhibition approach may have improved efficacy. In this study, we report that topical application of the broad-spectrum CC-chemokine inhibitor "35K" decreased wound CCL5 and CCL2 protein levels, leading to enhanced wound healing and wound neovascularisation during the early stages of wound healing. Histological analysis of wounds showed that 35K treatment significantly increased neovessels in the early stages of wound healing, when angiogenesis is most important. Mechanistically, we found that the classical VEGF-mediated angiogenic pathway was not responsible for the augmentation of angiogenesis by 35K, but, instead, 35K elevated the pro-angiogenic and pro-repair cytokine TGF-β. Furthermore, 35K treated wounds had significantly lower mRNA levels of p65, the active subunit of the inflammatory transcription factor NF-κB. In addition, 35K treated wounds at Day 10 had lower collagen content, suggesting a reduction in scar formation. Taken together, these findings suggest that broad-spectrum inhibition of CC-chemokines may be an alternate therapeutic approach to improve wound healing by simultaneously suppressing inflammation while augmenting neovascularisation at the critical early stages of wound repair.

Macroscopic wound measurements revealed that wound closure was faster in 35K treated wounds compared to PBS treated control wounds. This was supported by the increased blood flow perfusion observed during the early stages post-wounding in 35K treated wounds. At this stage (~Day 3), perfusion is critical to support the wound healing process to supply the wound with nutrients and growth factors, and accelerate debris removal [2,29]. Increased blood flow perfusion also allows for increased oxygenation to help maintain the newly forming wound tissue [30]. This early boost in perfusion may support the increased closure seen at Day 7 and 8. By Day 7, wound blood perfusion then started to decline in the 35K treated wounds. At this stage, remodelling is commencing, and, whilst neovessels are still present, the impetus for new wound neovessels has started to decline, supporting the decrease in wound blood perfusion at this stage. After Day 7 post-wounding, there is very little detectable blood perfusion using the Laser Doppler for both treatments. Consistent with our blood perfusion measurements, the current study also found that topical application of 35K increased the presence of wound neovessels at Day 4, and, at Day 10, neovessels and arterioles were reduced in the 35K treated wounds. At this late stage, the wound has nearly healed, re-epithelialization is complete, inflammation has ceased, a scar has formed and collagen remodelling from type III to type I collagen begins [31]. Importantly, as the wound healing process is near-completion, angiogenesis may decline to allow for continued remodelling [9,32]. The lower number of neovessels and arterioles seen at Day 10 suggest that these wounds have healed quicker and have now entered the remodelling phase. However, at Day 21, the wounds are further in the remodelling phase, bringing stability to neovessel content. The reduction in collagen content of the 35K treated wounds at Day 10 may further lead to reduced scar formation, as scars are mainly comprised of collagen [33]. At Day 21, the collagen content returns to a stable level, when compared to PBS wounds. This suggests that whilst the wounds may have less scar the stability of the skin is not compromised. Interestingly, at Day 4 there is 10 times more type I collagen than type III, and, when the wound is completely healed and undergoing remodelling, there is an increase of 50 times in type I collagen in wounds. These findings are consistent with previous studies where reduced fibrosis was detected in the kidneys and lungs of mice following inhibition of the chemokine receptors CCR1 or CCR5 [34–36].

Angiogenic mediators such as VEGF and FGF-2 are released by various cell types including endothelial cells, fibroblasts, keratinocytes and macrophages. VEGF is primarily produced by endothelial cells, whilst FGF-2 is produced by fibroblasts. Interestingly, 35K had no effect on these key angiogenic mediators involved in wound healing. Furthermore, we found no change in wound HIF-1α expression, which regulates VEGF and is triggered in response to tissue ischemia. This suggests that there was sufficient blood perfusion and oxygenation to the wounds. VEGF is involved in angiogenesis and synthesis of collagen in the wound [8,37]. It is produced by wound fibroblasts, keratinocytes and macrophages [8,32,38]. VEGF levels remained unchanged in 35K treated wounds, suggesting that VEGF is not involved in the augmentation of wound healing and wound angiogenesis by 35K. An alternative suggestion is that VEGF expression increased very early post-wounding with 35K, which was missed at the Day 4 time point. Studies have shown increases in VEGF just 24 h post-wounding [39], which then return to baseline once angiogenesis is underway. There was, however, an increase in wound neovessels with 35K treatment at the Day 4 time point, despite no changes in VEGF or HIF-1α. Interestingly, both VEGF and HIF-1α are induced by NF-κB in inflammatory pathological angiogenesis. This study found that 35K significantly inhibited p65 (active NF-κB subunit) mRNA levels. This reduction in NF-κB may be part of the reason for the lack of change in VEGF and HIF-1α and suggests that 35K was preventing pathological excessive wound angiogenesis. Furthermore, the reduction in wound CC-chemokine levels by 35K would also contribute to the suppression of pathological angiogenesis that can compromise wound repair if it persists in the later stages of healing. The increase in angiogenesis seen during early wound healing, where it is critical, is likely to be the result of increased TGF-β expression in the Day 4 35K treated wounds. TGF-β increases key angiogenic functions including endothelial cell proliferation and migration [40,41] TGF-β also has a number of roles in the wound repair process including fibroblast proliferation, collagen formation and remodelling of the extracellular matrix [42] and the initiation in the formation of granulation tissue [43]. TGF-β stimulates contraction of fibroblasts [44] and promotes the migration of keratinocytes in the wound [45] which promotes wound closure. Taken together, our results demonstrate that 35K does not augment wound angiogenesis via the classical HIF-1α/VEGF angiogenic pathway but rather through the induction of TGF-β, and may in fact suppress excessive pathological angiogenesis in the mid-late stages of wound repair via inhibition of NF-κB and CC-chemokine levels.

Despite the reduction in NF-κB by 35K, there were no changes in the number of wound macrophages as may have been expected. Macrophages can be roughly divided into pro- and anti-inflammatory phenotypes (M1 and M2) and both are involved in different stages of wound healing [46]. Our results indicate an increase in the M2 macrophage phenotype during the early wound healing stage (TGF-β, CD206), which is known to be associated with improved wound repair, sufficient formation of granulation tissue and new vascular networks [46]. Interestingly, there was a surprising increase in M1 macrophage markers at Day 10 (CD80, CD86), which is inconsistent with the reduction in NF-κB at this time point. However, previous studies have shown that the presence of M1 macrophages at this late-stage does not affect the wound healing process [4].

There was a decrease in CCL5 and CCL2 protein in the tissues of wounds treated with the CC-chemokine inhibitor 35K. This would typically be associated with a decrease in macrophages; however, no significant reductions in macrophages were observed. A previous study reported that despite dominant expression of CCL2 in dermal mouse wounds, a relatively low CCL2 level correlated with optimal monocyte accumulation [12], indicating that a reduction, but not complete ablation of CCL2, may not affect macrophage recruitment in the early stages of wound healing. However, at the later stage (i.e., Day 10), the wound is preparing to enter the remodelling phase in which macrophages undergo apoptosis and chemokines and angiogenesis are subsequently reduced. Additionally, chemokines from other classes and other cytokines that are not affected by 35K (for example CX_3CL1, IL-1, IFN-γ) are released from the wound site and may have assisted in maintaining macrophage infiltration [47], even with lower CCL2 and CCL5 levels. This may explain the consistent macrophage levels seen in our study despite the reduction in CC-chemokine levels.

4. Materials and Methods

4.1. Generation and Isolation of 35K Protein

A recombinant adenovirus overexpressing 35K (Ad35K) was generated as described previously [48]. To isolate 35K protein, 20 large flasks of Ad293 cells (293AD cell line, AD-100, Cell Biolabs Inc., San Diego, CA USA) were infected with 5×10^{11} Ad35K virus particles. After 24–48 h, cells were microscopically examined for evidence of complete cytopathic effect (CPE). 35K protein was isolated from Ad35K viral media using anti-HA tagged agarose-conjugated beads (A2095, Sigma-Aldrich, St. Louis, MO, USA). The media was run through a column packed with anti-HA agarose beads to bind 35K protein, which was then eluted from the column with 3 M sodium thiocyanate into 1 M Trizma Base (pH 8). Isolated 35K protein was dialyzed into sterile PBS and then filtered through 0.45 μm low protein binding syringe filter (Pall Corporation, Port Washington, NY, USA). The concentration of 35K protein was determined by measuring absorbance at 280 nm and aliquoted to 200 nM before freezing at −80 °C.

4.2. Murine Wound Healing Model

All experimental procedures and protocols were conducted with approval from the Sydney Local Health District Animal Welfare Committee (#2013/027A), and conformed to the Guide for the Care and Use of Laboratory Animals (United States National Institute of Health, Bethesda, MD, USA). To explore the effect of broad-spectrum CC-chemokine inhibition on wound healing, a murine model that closely mimics the human wound healing process was used [49]. Briefly, C57Bl6/J wildtype mice were anesthetized by inhalation of methoxyflurane. Then, a 6 mm biopsy punch was used to outline two circular full-thickness excisions, 5 mm in diameter, that include the panniculus carnosus created on the dorsum, one on each side of the midline of the mouse. A 0.5 mm thick silicone splint (Life Technologies, Carlsbad, CA, USA) was then placed around the wound and secured with interrupted sutures. For each mouse, one wound received purified 35K protein (200 nM in 50 µL PBS) and the other endotoxin-free PBS (50 µL, vehicle control), topically applied daily beginning on the day of wounding (Day 0). A transparent occlusive dressing (Opsite™ Flexifix™, Smith & Nephew, London, UK) was then applied. Mice were given carprofen (5 mg/kg) daily 3 days post-surgery to alleviate any pain. Digital images were taken and micro-callipers were used to measure wound area daily along the x-, y- and z-axes. Blood perfusion in wound areas was determined using Laser Doppler Perfusion Imaging (moorLDI2-IR, Moor Instruments, Devon, UK). For this study, two cohorts were taken at 4 (n = 12), 10 (n = 12), or 21 (n = 7) days after wounding. Both PBS and 35K treated wounds were collected for histological, protein and gene analysis.

4.3. Immunohistochemistry

Wound tissues were fixed in 4% (v/v) paraformaldehyde overnight, and then embedded in paraffin. Furthermore, 5 µm wound sections were taken from the midpoint. Wounds were assessed for neovessels by detection with rabbit polyclonal anti-CD31 antibody (1:100, ab28364, Abcam, Cambridge, UK), arterioles by detection with mouse monoclonal smooth muscle α-actin antibody (1:100, A5691, Sigma-Aldrich) and for macrophages by detection with mouse monoclonal anti-CD68 antibody (1:100, ab31630, Abcam) staining. Secondary antibodies were pre-diluted horseradish peroxidase (HRP) secondary antibody α-rabbit (K4011, Dako) or pre-diluted HRP secondary antibody α-mouse (K4007, Dako, Glostrup, Denmark) for CD31 and CD68, respectively. The staining was visualized using 3,3′-Diaminobenzidine (DAB) (Dako) for CD31 and CD68 or Vector Red alkaline phosphatase substrate (SK-5100, Vector Laboratories, Burlingame, CA, USA) for smooth muscle α-actin. Appropriate IgG controls were used for all antibodies. Collagen was measured following Milligan's trichrome staining. Picrosirius red staining, imaged under polarized light, was used to differentiate type I and type III collagen. For all histological quantification, three sections were imaged at 10× magnification, the images were tiled and "stitched" to obtain the entire wound cross section for each treatment group per mouse (n = 7–12) and assessed using Image-Pro® Premier 9.0 software (Media Cybernetics, Rockville, MD, USA).

4.4. Protein Expression

Wound tissues were homogenized in lysis buffer (80 mM Tris HCl, 10 mM NaCl, 50 mM NaF, 5 mM $Na_4P_2O_7$, 15 mM Triton-X 100). Wound tissue lysates (50 µg) were subjected to Western immunoblotting and probed for HIF-1α (1:1000, NB100-105, Novus Biologicals, Littleton, CO, USA) α Tubulin (1:5000, AB40742, Abcam) was used to confirm even protein loading. Commercially available ELISA kits (Quantikine, RnD Systems, Minneapolis, MN, USA) were used to determine protein levels of VEGF, CCL2 (MCP-1) and CCL5 (RANTES) in 100 µg of wound tissue lysates, while FGF-2 protein levels were measured in 50 µg of wound tissue lysates.

4.5. Gene Expression

Total RNA was isolated from wound samples using TRI reagent (Sigma-Aldrich). Furthermore, 300 ng total RNA was reverse transcribed using the iScript cDNA synthesis kit (Bio-Rad, Hercules, CA, USA) before amplification using the iQ SYBR Supermix (Bio-Rad) in a Bio-Rad Cfx384 thermocycler. The following mouse primers were used to probe for HIF-1α (F 5′-TCCCTTGCTCTTTGTGGTTGGGT-3′, R 5′-AACGTAAGCGCTGACCCAGG-3′), VEGF (F 5′-GAGTACCCCGACGAGATAGAGT-3′, R 5′-GGT GAGGTTTGATCCGCATGA-3′), p65 (F 5′-AGTATCCATAGCTTCCAGAACC-3′, R 5′-ACTGCATTCAA GTCATAGTCC-3′), CD68 (F 5′-GGGGCTCTTGGGAACTACAC-3′, R 5′-GTACCGTCACAACCTC CCTG-3′), TGF-β (F 5′-GGATACCAACTATTGCTTCAGCTCC-3′, R 5′-AGGCTCCAAATATAGGGG CAGGGTC-3′), CD206 (F 5′-CAGGTGTGGGCTCAGGTAGT-3′, R 5′-TGTGGTGAGCTGAAAGGT GA-3′), CD80 (F 5′-ACCCCCAACATAACTGAGTCT-3′, R 5′-TTCCAACCAAGAGAAGCGAGG-3′), CD86 (F 5′-CAGCTCACTCAGGCTTATGTTT-3′, R 5′-TGTTTCCGTGGAGACGCAAG-3′), CCL2 (F 5′-GCTGGAGCATCCACGTGTT-3′, R 5′-ATCTTGCTGGTGAATGAGTAGCA-3′), CCL5 (F 5′-GCA

AGTGCTCCAATCTTGCA-3′, R 5′-CTTCTCTGGGTTGGCACACA-3′) and 36B4 (F 5′-CAACGCAGCA
TTTATAACCC-3′, R 5′-CCCATTGATGATGGAGTGTGG-3′). Relative changes in gene expression were
normalized using the $\Delta\Delta C_t$ method to 36B4 as the housekeeping gene.

4.6. Statistics

All results are expressed as mean ± SEM. All data were compared using an unpaired two-tailed
t-test. Significance was set at a value of $p < 0.05$.

5. Conclusions

In conclusion, our findings show that broad-spectrum CC-chemokine inhibition via topical
application of 35K enhanced wound closure through the early promotion of neovascularisation.
Furthermore, inhibiting the CC-chemokine class resulted in a reduced inflammatory state as indicated
by reduced NF-κB. Mechanistically, 35K augmented wound angiogenesis via the induction of the
pro-angiogenic cytokine TGF-β, rather than the HIF-1α/VEGF angiogenic pathway. Taken together,
our findings (summarized in Figure 7) demonstrate that broad-spectrum CC-chemokine inhibition
may improve wound healing by suppressing the inflammatory state of the wound while augmenting
wound angiogenesis and suppressing scar formation.

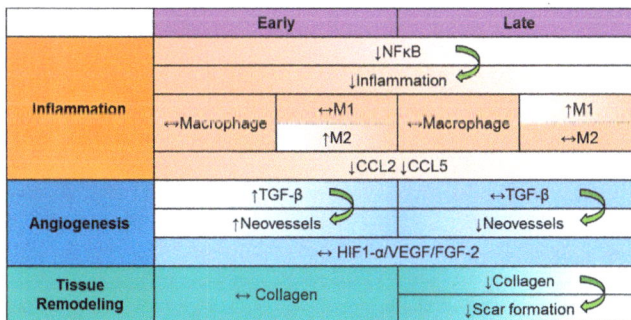

	Early		Late	
Inflammation		↓NFκB		
		↓Inflammation		
	↔Macrophage	↔M1	↔Macrophage	↑M1
		↑M2		↔M2
		↓CCL2 ↓CCL5		
Angiogenesis		↑TGF-β	↔TGF-β	
		↑Neovessels	↓Neovessels	
		↔ HIF1-α/VEGF/FGF-2		
Tissue Remodeling	↔ Collagen		↓Collagen	
			↓Scar formation	

Figure 7. Summary of proposed 35K action in wound healing. Broad-spectrum CC-chemokine
inhibition by 35K has a significant impact on key mechanisms involved in wound healing. In addition,
35K enhances wound closure through the early promotion of neovascularisation via the induction of
TGF-β, rather than the HIF-1α/VEGF angiogenic pathway. 35K treatment reduces inflammation via
inhibition of NF-κB, which may help to stave off pathological angiogenesis in the later stages of wound
repair, along with the reductions in CCL5 and CCL2 but with no change in macrophages. Finally, 35K
suppresses collagen deposition at the late stage of wound healing, but not early, indicative of reduced
scar formation.

Supplementary Materials: Supplementary materials can be found at www.mdpi.com/1422-0067/18/1/155/s1.

Acknowledgments: This work was supported by the Heart Research Institute (PhD scholarship to
Anisyah Ridiandries) and the Heart Foundation of Australia (Career Development Fellowship #CR07S3331 to
Christina A. Bursill). The authors acknowledge the facilities as well as the scientific and technical assistance of
the Australian Microscopy & Microanalysis Research Facility (ammrf.org.au) node at the University of Sydney:
Sydney Microscopy & Microanalysis.

Author Contributions: Joanne Tan and Christina Bursill developed the study and designed the experiments.
Anisyah Ridiandries and Joanne Tan performed the experiments. Anisyah Ridiandries, Joanne Tan and
Christina Bursill interpreted the data and wrote the manuscript. Anisyah Ridiandries, Joanne Tan and Christina
Bursill reviewed and edited the manuscript.

Conflicts of Interest: The authors declare no conflict of interest.

References

1. Tonnesen, M.G.; Feng, X.; Clark, R.A.F. Angiogenesis in wound healing. *J. Investig. Dermatol. Symp. Proc.* **2000**, *5*, 40–46. [CrossRef] [PubMed]
2. Gurtner, G.C.; Werner, S.; Barrandon, Y.; Longaker, M.T. Wound repair and regeneration. *Nature* **2008**, *453*, 314–321. [CrossRef] [PubMed]
3. Daley, J.M.; Brancato, S.K.; Thomay, A.A.; Reichner, J.S.; Albina, J.E. The phenotype of murine wound macrophages. *J. Leukoc. Biol.* **2010**, *87*, 59–67. [CrossRef] [PubMed]
4. Lucas, T.; Waisman, A.; Ranjan, R.; Roes, J.; Krieg, T.; Müller, W.; Roers, A.; Eming, S.A. Differential roles of macrophages in diverse phases of skin repair. *J. Immunol.* **2010**, *184*, 3964–3977. [CrossRef] [PubMed]
5. Guo, S.; DiPietro, L.A. Factors affecting wound healing. *J. Dent. Res.* **2010**, *89*, 219–229. [CrossRef] [PubMed]
6. Mathieu, D.; Linke, J.-C.; Wattel, F. Non-Healing Wounds. In *Handbook on Hyperbaric Medicine*; Mathieu, D., Ed.; Springer: Dordrecht, The Netherlands, 2006; pp. 401–428.
7. Menke, N.B.; Ward, K.R.; Witten, T.M.; Bonchev, D.G.; Diegelmann, R.F. Impaired wound healing. *Clin. Dermatol.* **2007**, *25*, 19–25. [CrossRef] [PubMed]
8. Eming, S.A.; Krieg, T.; Davidson, J.M. Inflammation in wound repair: Molecular and cellular mechanisms. *J. Investig. Dermatol.* **2007**, *127*, 514–525. [CrossRef] [PubMed]
9. Gillitzer, R.; Goebeler, M. Chemokines in cutaneous wound healing. *J. Leukoc. Biol.* **2001**, *69*, 513–521. [PubMed]
10. Engelhardt, E.; Toksoy, A.; Goebeler, M.; Debus, S.; Brocker, E.B.; Gillitzer, R. Chemokines IL-8, GROα, MCP-1, IP-10, and Mig are sequentially and differentially expressed during phase-specific infiltration of leukocyte subsets in human wound healing. *Am. J. Pathol.* **1998**, *153*, 1849–1860. [CrossRef]
11. Gibran, N.S.; Ferguson, M.; Heimbach, D.M.; Isik, F.F. Monocyte chemoattractant protein-1 mRNA expression in the human burn wound. *J. Surg. Res.* **1997**, *70*, 1–6. [CrossRef] [PubMed]
12. Jackman, S.H.; Yoak, M.B.; Keerthy, S.; Beaver, B.L. Differential expression of chemokines in a mouse model of wound healing. *Ann. Clin. Lab. Sci.* **2000**, *30*, 201–207. [PubMed]
13. Folkman, J.; Shing, Y. Angiogenesis. *J. Biol. Chem.* **1992**, *267*, 10931–10934. [PubMed]
14. Stamatovic, S.M.; Keep, R.F.; Mostarica-Stojkovic, M.; Andjelkovic, A.V. CCL2 regulates angiogenesis via activation of Ets-1 transcription factor. *J. Immunol.* **2006**, *177*, 2651–2661. [CrossRef] [PubMed]
15. Salcedo, R.; Young, H.A.; Ponce, M.L.; Ward, J.M.; Kleinman, H.K.; Murphy, W.J.; Oppenheim, J.J. Eotaxin (CCL11) induces in vivo angiogenic responses by human CCR3+ endothelial cells. *J. Immunol.* **2001**, *166*, 7571–7578. [CrossRef] [PubMed]
16. Strasly, M.; Doronzo, G.; Cappello, P.; Valdembri, D.; Arese, M.; Mitola, S.; Moore, P.; Alessandri, G.; Giovarelli, M.; Bussolino, F. CCL16 activates an angiogenic program in vascular endothelial cells. *Blood* **2004**, *103*, 40–49. [CrossRef] [PubMed]
17. Galvez, B.G.; Genis, L.; Matias-Roman, S.; Oblander, S.A.; Tryggvason, K.; Apte, S.S.; Arroyo, A.G. Membrane type 1-matrix metalloproteinase is regulated by chemokines monocyte-chemoattractant protein-1/CCL2 and interleukin-8/CXCL8 in endothelial cells during angiogenesis. *J. Biol. Chem.* **2005**, *280*, 1292–1298. [CrossRef] [PubMed]
18. Ishida, Y.; Kimura, A.; Kuninaka, Y.; Inui, M.; Matsushima, K.; Mukaida, N.; Kondo, T. Pivotal role of the CCL5/CCR5 interaction for recruitment of endothelial progenitor cells in mouse wound healing. *J. Clin. Investig.* **2012**, *122*, 711–721. [CrossRef] [PubMed]
19. Low, Q.E.H.; Drugea, I.A.; Duffner, L.A.; Quinn, D.G.; Cook, D.N.; Rollins, B.J.; Kovacs, E.J.; DiPietro, L.A. Wound healing in MIP-1α$^{-/-}$ and MCP-1$^{-/-}$ mice. *Am. J. Pathol.* **2001**, *159*, 457–463. [CrossRef]
20. Kaesler, S.; Bugnon, P.; Gao, J.-L.; Murphy, P.M.; Goppelt, A.; Werner, S. The chemokine receptor CCR1 is strongly up-regulated after skin injury but dispensable for wound healing. *Wound Repair Regen.* **2004**, *12*, 193–204. [CrossRef] [PubMed]
21. Mantovani, A. The chemokine system: Redundancy for robust outputs. *Immunol. Today* **1999**, *20*, 254–257. [CrossRef]
22. Devalaraja, M.N.; Richmond, A. Multiple chemotactic factors: Fine control or redundancy? *Trends Pharmacol. Sci.* **1999**, *20*, 151–156. [CrossRef]
23. Zhang, L.; Derider, M.; McCornack, M.A.; Jao, S.C.; Isern, N.; Ness, T.; Moyer, R.; LiWang, P.J. Solution structure of the complex between poxvirus-encoded CC chemokine inhibitor vCCI and human MIP-1B. *Proc. Natl. Acad. Sci. USA* **2006**, *103*, 13985–13990. [CrossRef] [PubMed]

24. Carfí, A.; Smith, C.A.; Smolak, P.J.; McGrew, J.; Wiley, D.C. Structure of a soluble secreted chemokine inhibitor VCCI (P35) from cowpox virus. *Proc. Natl. Acad. Sci. USA* **1999**, *96*, 12379–12383. [CrossRef] [PubMed]
25. Ali, Z.A.; Bursill, C.A.; Hu, Y.; Choudhury, R.P.; Xu, Q.; Greaves, D.R.; Channon, K.M. Gene transfer of a broad-spectrum CC-chemokine inhibitor reduces vein graft atherosclerosis in apolipoprotein E-knockout mice. *Circulation* **2005**, *112*, I-235–I-241. [PubMed]
26. Bursill, C.A.; Cash, J.L.; Channon, K.M.; Greaves, D.R. Membrane-bound CC chemokine inhibitor 35K provides localized inhibition of CC-chemokine activity in vitro and in vivo. *J. Immunol.* **2006**, *177*, 5567–5573. [CrossRef] [PubMed]
27. Bursill, C.A.; Choudhury, R.P.; Ali, Z.; Greaves, D.R.; Channon, K.M. Broad-spectrum CC-chemokine blockade by gene transfer inhibits macrophage recruitment and atherosclerotic plaque formation in apolipoprotein E-knockout mice. *Circulation* **2004**, *110*, 2460–2466. [CrossRef] [PubMed]
28. Bursill, C.A.; McNeill, E.; Wang, L.; Hibbitt, O.C.; Wade-Martins, R.; Paterson, D.J.; Greaves, D.R.; Channon, K.M. Lentiviral gene transfer to reduce atherosclerosis progression by long-term CC-chemokine inhibition. *Gene Ther.* **2009**, *16*, 93–102. [CrossRef] [PubMed]
29. Sunderkotter, C.; Steinbrink, K.; Goebeler, M.; Bhardwaj, R.; Sorg, C. Macrophages and angiogenesis. *J. Leukoc. Biol.* **1994**, *55*, 410–422. [PubMed]
30. Jonsson, K.; Jensen, J.A.; Goodson, W.H.; Scheuenstuhl, H.; West, J.; Hopf, H.W.; Hunt, T.K. Tissue oxygenation, anemia, and perfusion in relation to wound healing in surgical patients. *Ann. Surg.* **1991**, *214*, 605–613. [CrossRef] [PubMed]
31. Lovvorn, H.N.; Cheung, D.T.; Nimni, M.E.; Perelman, N.; Estes, J.M.; Adzick, N.S. Relative distribution and crosslinking of collagen distinguish fetal from adult sheep wound repair. *J. Pediatr. Surg.* **1999**, *34*, 218–223. [CrossRef]
32. Singer, A.J.; Clark, R.A.F. Cutaneous wound healing. *N. Engl. J. Med.* **1999**, *341*, 738–746. [PubMed]
33. Werner, S.; Krieg, T.; Smola, H. Keratinocyte-fibroblast interactions in wound healing. *J. Investig. Dermatol.* **2007**, *127*, 998–1008. [CrossRef] [PubMed]
34. Schuh, J.M.; Blease, K.; Hogaboam, C.M. The role of CC-chemokine receptor 5 (CCR5) and RANTES/CCL5 during chronic fungal asthma in mice. *FASEB J.* **2002**, *16*, 228–230. [PubMed]
35. Anders, H.J.; Vielhauer, V.; Frink, M.; Linde, Y.; Cohen, C.D.; Blattner, S.M.; Kretzler, M.; Strutz, F.; Mack, M.; Grone, H.J.; et al. A chemokine receptor CCR-1 antagonist reduces renal fibrosis after unilateral ureter ligation. *J. Clin. Investig.* **2002**, *109*, 251–259. [CrossRef] [PubMed]
36. Seki, E.; Minicis, S.D.; Gwak, G.-Y.; Kluwe, J.; Inokuchi, S.; Bursill, C.A.; Llovet, J.M.; Brenner, D.A.; Schwabe, R.F. CCR1 and CCR5 promote hepatic firbosis in mice. *J. Clin. Investig.* **2009**, *119*, 1858–1870. [PubMed]
37. Bao, P.; Kodra, A.; Tomic-Canic, M.; Golinko, M.S.; Ehrlich, H.P.; Brem, H. The role of vascular endothelial growth factor in wound healing. *J. Surg. Res.* **2009**, *153*, 347–358. [CrossRef] [PubMed]
38. Martin, P. Wound healing-aiming for perfect skin regeneration. *Science* **1997**, *276*, 75–81. [CrossRef] [PubMed]
39. Bates, D.O.; Jones, R.O. The role of vascular endothelial growth factor in wound healing. *Int. J. Lower Extremity Wounds* **2003**, *2*, 107–120. [CrossRef] [PubMed]
40. Hyman, K.M.; Seghezzi, G.; Pintucci, G.; Stellari, G.; Kim, J.H.; Grossi, E.A.; Galloway, A.C.; Mignatti, P. Transforming growth factor-β1 induces apoptosis in vascular endothelial cells by activation of mitogen-activated protein kinase. *Surgery* **2002**, *132*, 173–179. [CrossRef] [PubMed]
41. Choi, M.E.; Ballermann, B.J. Inhibition of capillary morphogenesis and associated apoptosis by dominant negative mutant transforming growth factor-β receptors. *J. Biol. Chem.* **1995**, *270*, 21144–21150. [CrossRef] [PubMed]
42. Penn, J.W.; Grobbelaar, A.O.; Rolfe, K.J. The role of the TGF-β family in wound healing, burns and scarring: A review. *Int. J. Burns Trauma* **2012**, *2*, 18–28. [PubMed]
43. Behm, B.; Babilas, P.; Landthaler, M.; Schreml, S. Cytokines, chemokines and growth factors in wound healing. *J. Eur. Acad. Dermatol. Venereol.* **2012**, *26*, 812–820. [CrossRef] [PubMed]
44. Desmouliere, A.; Geinoz, A.; Gabbiani, F.; Gabbiani, G. Transforming growth factor-β1 induces α-smooth muscle actin expression in granulation tissue myofibroblasts and in quiescent and growing cultured fibroblasts. *J. Cell Biol.* **1993**, *122*, 103–111. [CrossRef] [PubMed]
45. Gailit, J.; Welch, M.P.; Clark, R.A. TGF-β1 stimulates expression of keratinocyte integrins during re-epithelialization of cutaneous wounds. *J. Investig. Dermatol.* **1994**, *103*, 221–227. [CrossRef] [PubMed]

46. Brancato, S.K.; Albina, J.E. Wound macrophages as key regulators of repair: Origin, phenotype, and function. *Am. J. Pathol.* **2011**, *178*, 19–25. [CrossRef] [PubMed]

47. Gregory, J.L.; Morand, E.F.; McKeown, S.J.; Ralph, J.A.; Hall, P.; Yang, Y.H.; McColl, S.R.; Hickey, M.J. Macrophage migration inhibitory factor induces macrophage recruitment via CC-chemokine ligand 2. *J. Immunol.* **2006**, *177*, 8072–8079. [CrossRef] [PubMed]

48. Bursill, C.A.; Cai, S.; Channon, K.M.; Greaves, D.R. Adenoviral-mediated delivery of a viral chemokine binding protein blocks CC-chemokine activity in vitro and in vivo. *Immunobiology* **2003**, *207*, 187–196. [CrossRef] [PubMed]

49. Dunn, L.; Prosser, H.C.G.; Tan, J.T.M.; Vanags, L.Z.; Ng, M.K.C.; Bursill, C.A. Murine model of wound healing. *J. Vis. Exp.* **2013**. [CrossRef] [PubMed]

© 2017 by the authors. Licensee MDPI, Basel, Switzerland. This article is an open access article distributed under the terms and conditions of the Creative Commons Attribution (CC BY) license (http://creativecommons.org/licenses/by/4.0/).

International Journal of
Molecular Sciences

MDPI

Article

Keratinocyte Growth Factor Combined with a Sodium Hyaluronate Gel Inhibits Postoperative Intra-Abdominal Adhesions

Guangbing Wei [1], Cancan Zhou [2], Guanghui Wang [1], Lin Fan [1], Kang Wang [1] and Xuqi Li [1,*]

[1] Department of General Surgery, the First Affiliated Hospital of Xi'an Jiaotong University, Xi'an 710061, Shaanxi, China; weiguangbing1208@163.com (G.W.); amon@mail.xjtu.edu.cn (G.W.); linnet@mail.xjtu.edu.cn (L.F.); wangkang5754@163.com (K.W.)

[2] Department of Hepatobiliary Surgery, the First Affiliated Hospital of Xi'an Jiaotong University, Xi'an 710061, Shaanxi, China; trytofly@stu.xjtu.edu.cn

* Correspondence: lixuqi@163.com; Tel./Fax: +86-29-8532-3899

Academic Editor: Allison Cowin
Received: 18 July 2016; Accepted: 16 September 2016; Published: 22 September 2016

Abstract: Postoperative intra-abdominal adhesion is a very common complication after abdominal surgery. One clinical problem that remains to be solved is to identify an ideal strategy to prevent abdominal adhesions. Keratinocyte growth factor (KGF) has been proven to improve the proliferation of mesothelial cells, which may enhance fibrinolytic activity to suppress postoperative adhesions. This study investigated whether the combined administration of KGF and a sodium hyaluronate (HA) gel can prevent intra-abdominal adhesions by improving the orderly repair of the peritoneal mesothelial cells. The possible prevention mechanism was also explored. The cecum wall and its opposite parietal peritoneum were abraded after laparotomy to induce intra-abdominal adhesion formation. Animals were randomly allocated to receive topical application of HA, KGF, KGF + HA, or normal saline (Control). On postoperative day 7, the adhesion score was assessed with a visual scoring system. Masson's trichrome staining, picrosirius red staining and hydroxyproline assays were used to assess the magnitude of adhesion and tissue fibrosis. Cytokeratin, a marker of the mesothelial cells, was detected by immunohistochemistry. The levels of tissue plasminogen activator (tPA), interleukin-6 (IL-6), and transforming growth factor β1 (TGF-β1) in the abdominal fluid were determined using enzyme-linked immunosorbent assays (ELISAs). Western blotting was performed to examine the expression of the TGF-β1, fibrinogen and α-smooth muscle actin (α-SMA) proteins in the rat peritoneal adhesion tissue. The combined administration of KGF and HA significantly reduced intra-abdominal adhesion formation and fibrin deposition and improved the orderly repair of the peritoneal mesothelial cells in the rat model. Furthermore, the combined administration of KGF and HA significantly increased the tPA levels but reduced the levels of IL-6, tumor necrosis factor α (TNF-α) and TGF-β1 in the abdominal fluid. The expression levels of TGF-β1, fibrinogen and α-SMA protein and mRNA in the rat peritoneum or adhesion tissues were also down-regulated following the combined administration of KGF and HA. The combined administration of KGF and HA can significantly prevent postoperative intra-abdominal adhesion formation by maintaining the separation of the injured peritoneum and promoting mesothelial cell regeneration. The potential mechanism may be associated with rapid mesothelial cell repair in the injured peritoneum. This study suggests that combined administration of KGF and HA may be a promising pharmacotherapeutic strategy for preventing abdominal adhesions, which is worth further study, and has potential value in clinical applications.

Keywords: peritoneum; mesothelial cells; regeneration; postoperative adhesions; keratinocyte growth factor

---

Int. J. Mol. Sci. **2016**, *17*, 1611

1. Introduction

Postoperative intra-abdominal adhesions are a very common complication of abdominal surgery that may occur in 90% to 95% of patients undergoing abdominal surgery [1–3]. These unavoidable postoperative adhesions can cause a series of clinical problems, such as intestinal obstruction, postoperative abdominal and pelvic pain, female infertility, and difficult access in a subsequent surgery [4,5]. Ten percent of patients with intestinal obstructions caused by intra-abdominal adhesions must undergo adhesiolysis, for which the mortality rate is between 5% and 20%. The recurrent rate is very high after surgical treatment of adhesive intestinal obstructions. Without effective prevention of adhesions, the recurrence rate is 12% within 41 months after surgery; the recurrent risk is still considerable even 20 years after surgery. Studies of postoperative intra-abdominal adhesions have been ongoing since the emergence of surgery [6], but an effective medication that has a reliable effect and fewer side effects has not been identified. Clinically, many medications and approaches have been tested; however, the results have been inconclusive. Thus, a current clinical problem that remains to be solved is to identify an ideal strategy that prevents intra-abdominal adhesions [7].

Intra-abdominal adhesion formation results from fibrin exudation and deposition caused by a series of inflammatory processes in the injured peritoneum; furthermore, a decreased ability to dissolve fibrin at the injured sites can also lead to adhesion formation [8]. Peritoneal injury can induce the production of a fibrinogen-rich serous exudate, which activates the coagulation system and converts fibrinogen into fibrin to form deposits. Fibrin deposits can degrade into fibrin degradation products, and then, the injured peritoneum can be repaired by mesothelial cell regeneration. The degeneration of fibrosis is mainly completed by the peritoneal fibrinolytic system. Tissue plasminogen activator (tPA) and plasminogen activator inhibitor type 1 (PAI-1) of mesothelial cells, the main components of the peritoneal fibrinolytic system, play key roles in adhesion formation and development. Fibrin deposition can persist and form permanent fibrous adhesions when fibrinolytic activity is decreased due to poor mesothelial cell regeneration.

To this end, mesothelial injury is considered to be one of the causes of peritoneal adhesion formation [9]. The integrity of the peritoneal mesothelial cell layer is closely associated with fibrinolytic activity. Moreover, the loss of or poor proliferation and migration of mesothelial cells is another factor that promotes adhesion formation [10]. Peritoneal mesothelial cells not only form the smooth surface of the peritoneum but also exhibit fibrinolytic activity. Mesothelial cells are considered as the primary source of tPA in the serous cavity; their fibrinolytic activity in serous cavities plays a decisive role in local fibrin deposition and removal [10]. Normally, the balance between fibrin formation and dissolution of peritoneal mesothelial cells is maintained. However, when some pathological factors, such as peritoneal inflammation, mechanical damage, tissue ischemia, and foreign body implantation, are present, the mesothelial cells can be injured, reducing their fibrinolytic capacity. An imbalance between fibrin exudation and fibrinolysis can subsequently lead to tissue organization and adhesion formation.

The target of keratinocyte growth factor (KGF) is epithelial cells (including mesothelial cells). KGF can promote epithelial cell proliferation and growth and increase the rate of epithelialization [11,12]. KGF has rarely been reported to prevent intra-abdominal adhesions by promoting mesothelial cell regeneration. KGF has been shown to improve mesothelial cell proliferation, which may enhance fibrinolytic activity to potentially suppress postoperative adhesions [13]. This study investigated whether the combined administration of KGF and sodium hyaluronate (HA) can prevent intra-abdominal adhesions in rats by promoting the orderly repair of the injured peritoneal mesothelial cell layer. The possible prevention mechanisms were also explored.

2. Results

2.1. The Combined Administration of Keratinocyte Growth Factor (KGF) and Sodium Hyaluronate (HA) Significantly Reduced the Abdominal Adhesion Score in the Rat Model Using the Visual Scoring System

Animal death was not observed, and all animals completed the entire experimental protocol. No wound disruption, wound infection or intra-abdominal infections were observed in any of the groups. The magnitude of abdominal adhesions between groups was different (Figure 1). Adhesion formation was rarely observed in the sham laparotomy group, and the surface of the parietal peritoneum was smooth. Sheet-like adhesions were observed in the control group, which were difficult to separate. In the HA group, the jelly-like HA gel on the animal's cecal surface was completely absorbed, and the partially injured visceral surface was repaired. Strip-like adhesions were observed between the parietal peritoneum and cecum wound bed or between the omentums. The magnitude of adhesion information was milder than the control group. The magnitude of adhesion formation in the KGF group was similar to the HA group. In the animals of the KGF and HA groups, there was a low magnitude of adhesion formation, which appeared loose and had a thin thread-like morphology.

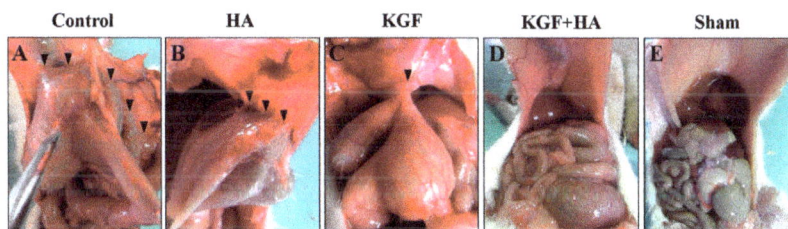

Figure 1. The combined administration of keratinocyte growth factor (KGF) and sodium hyaluronate (HA) prevented postoperative abdominal adhesion formation in rats. Representative images showing the intra-abdominal adhesions (black triangle) or the injured areas on the opposite parietal peritoneum on postoperative day 8 in each group of rats. (**A**) A large area of dense adhesions formed and the adhesions were difficult to separate in the control group; (**B**) Moderate adhesions were less severe in the HA group of rats than in the control group; (**C**) The relatively significant adhesions in the KGF group of rats were similar to those in the HA group in magnitude; (**D**) Very mild adhesions in some rats in the KGF plus HA group; (**E**) No adhesions were observed in the sham group. The surfaces of the parietal peritoneum and cecum were smooth.

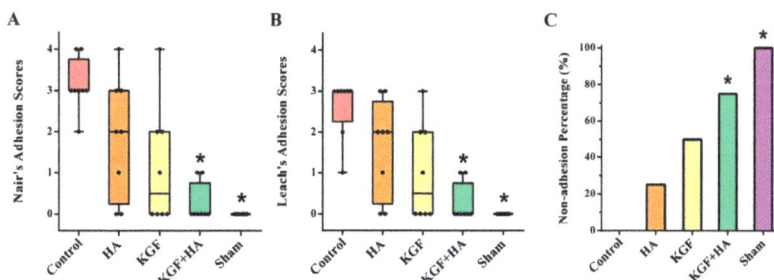

Figure 2. Comparison of intra-abdominal adhesion score in each group (*n* = 8). HA or KGF has a non-obvious trend to reduce adhesions in the rats. The magnitude of adhesion was significantly decreased in the KGF plus HA group. (**A**) Intra-abdominal adhesion score based on Nair's classification (* $p < 0.05$, compared with the control group); (**B**) Intra-abdominal adhesion score based on Leach's classification (* $p < 0.05$ compared with the control group); (**C**) Comparison of the non-adhesion rate (* $p < 0.05$, Fisher's exact test).

The magnitude of intra-abdominal adhesions was scored according to Nair's classification, and there was a significant difference among the five groups of rats ($p < 0.05$) (Figure 2A). Compared to the control group, HA or KGF reduced the postoperative adhesions in the rats respectively. The differences were not significant. In contrast, the magnitude of intra-abdominal adhesion formation was significantly decreased in the KGF plus HA group of rats ($p < 0.05$). In addition, the adhesion score was the lowest in the KGF plus HA group when using the scale described by Leach et al. (Figure 2B). The proportion of rats without abdominal adhesions in all groups was analyzed; we found that the rate of animals without adhesion formation was 75% in the KGF plus HA group and was significantly increased compared to the HA and KGF groups. The above-reported results indicated that the combined administration of KGF and HA can significantly prevent postoperative intra-abdominal adhesion formation in rats.

2.2. The Combined Administration of KGF and HA Decreased Collagen Deposition in the Injured Peritoneum of the Rat Model

Collagen fibers are stained blue by Masson's trichrome staining (Figure 3). The structure of the collagen fibers was intact, and collagen fiber proliferation was not observed in the sham operation group, whereas collagen fibers were stained dark blue in the control group, representing obvious hyperplasia. In the HA and KGF groups, collagen fibers were stained different shades of blue, which was somewhat lighter than the staining in the control group. More importantly, the area size and density of collagen fibers were not only significantly reduced in the KGF plus HA group compared with the control group but also significantly reduced compared with the HA and KGF groups. In addition, we examined the thickness of the abdominal adhesions between the injured rat peritoneum using picrosirius red staining of collagen fibers (Figure 4A). Compared with the HA and KGF groups, the collagen fibers in the rat adhesion tissue appear more loose, and the thickness of the abdominal adhesions was significantly decreased in the KGF plus HA group ($p < 0.05$) (Figure 4B). Furthermore, the combined administration significantly reduced the hydroxyproline content of the adhesion tissue in the animals of the KGF plus HA group ($p < 0.05$) (Figure 4C). Thus, our data suggested that the combined administration of KGF plus HA could decrease collagen deposition during adhesion formation in the rat model.

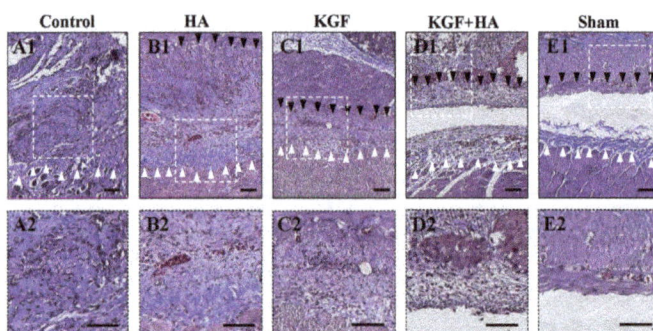

Figure 3. Masson's trichrome-stained images showing the intra-abdominal adhesions (black triangle) or injured areas on the opposite parietal peritoneum (white triangle) in each group of rats. (100× magnification pictures in upper plate, 200× magnification pictures in below plate; the black scale bar represents 100 μm). (**A1,2**) Image showing the compact structure and significant hyperplasia of the collagen fibers (dark blue staining) in the control group; (**B1,2**) Image showing the collagen fibers (dark to light blue staining) in the HA group, in which the staining is somewhat lighter than the control group; (**C1,2**) The abundance of collagen fiber hyperplasia in the KGF group was similar to the HA group; (**D1,2**) The collagen fiber was significantly impoverished in the KGF plus HA group; (**E1,2**) The structure of the collagen fibers was clear and intact with a very loose arrangement in the control group.

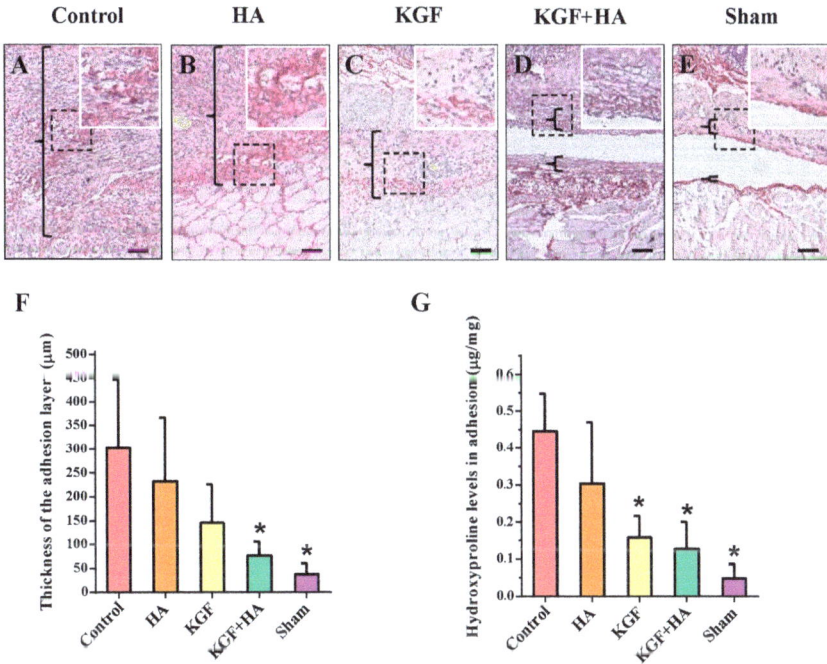

Figure 4. The magnitude of fibrosis in the intra-abdominal adhesions or the injured areas on the opposite parietal peritoneum in each group of rats ($n = 8$). (**A–E**) Representative images of picrosirius red staining ($100\times$; insets, $200\times$) in each group. The adhesive tissues are located between the black brackets. The black scale bar represents 100 μm; (**F**) The thickness of the collagen deposits in the adhesive tissue of each group of rats (* $p < 0.05$ compared with the control group); (**G**) Hydroxyproline content in the adhesion tissues of each group of rats (* $p < 0.05$ compared with the control group).

2.3. The Combined Administration of KGF and HA Can Reduce Inflammatory Infiltration in the Injured Rat Peritoneum

We utilized the hematoxylin and eosin (HE) staining technique to compare the magnitude of inflammatory cell infiltration in the peritoneum (Figure 5A–E). The criteria for the inflammation score were listed in Table S3. In the control group, the accumulation of a number of inflammatory cells and some small abscesses were noted in the peritoneal tissue. The same inflammatory cells were also noted in the HA and KGF groups; however, the magnitude of the accumulation was less severe than that in the control group. Only a few scattered inflammatory cells were noted in the KGF plus HA group, indicating a significant decrease in inflammatory cell accumulation compared to that in the HA or KGF group. The inflammation scores were significantly different among the five groups of rats (Figure 5F), in which the highest score was observed in the control group. The score in the KGF plus HA group was not only significantly reduced compared with the control group ($p < 0.05$) but also significantly reduced compared with the HA and KGF groups ($p < 0.05$). Thus, the results have shown that the combined administration of KGF and HA significantly inhibited the accumulation of inflammatory cells in the peritoneum.

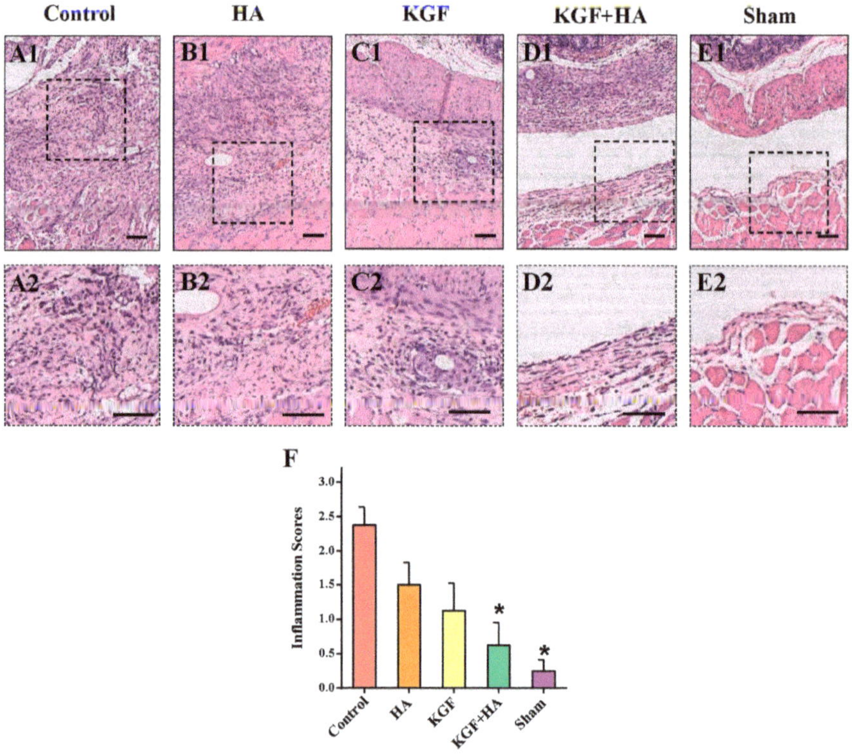

Figure 5. Inflammatory infiltration in the intra-abdominal adhesions or the injured areas on the opposite parietal peritoneum in each group of rats (*n* = 8). (**A1,2–E1,2**) Representative images of HE staining in the intra-abdominal adhesions or the injured areas on the opposite parietal peritoneum in each group of rats (100× magnification pictures in upper plate, 200× magnification pictures in below plate; the black scale bar represents 100 μm); (**F**) Inflammatory infiltration score for the adhesive tissue in each group of rats (* $p < 0.05$ compared with the control group).

2.4. The Combined Administration of KGF and HA Promotes Mesothelial Cell Repair on the Injured Peritoneal Surface

Immunohistochemical staining for cytokeratin (CK), a marker of mesothelial cells, was used to assess the continuity of the mesothelial cell layer of the injured rat peritoneum (Figure 6). The shape of representative mesothelial cells resembles cobblestones, and the cytokeratin staining is positive [14]. Our studies have observed the presence of a continuous and intact cell layer on the surface of the parietal and visceral peritoneum in the sham group. In contrast, mesothelial cells were not present on the surface of the injured parietal and visceral peritoneum in the control group. Similarly, mesothelial cell repair was not evident in the HA group. Mesothelial cell regeneration was noted in the KGF or KGF plus HA group, but, more importantly, a well-repaired mesothelial cell layer was noted in the KGF plus HA group, and its continuity was similar to the sham group. Therefore, the combined administration KGF and HA can significantly contribute to mesothelial cell repair on the surface of the injured peritoneum.

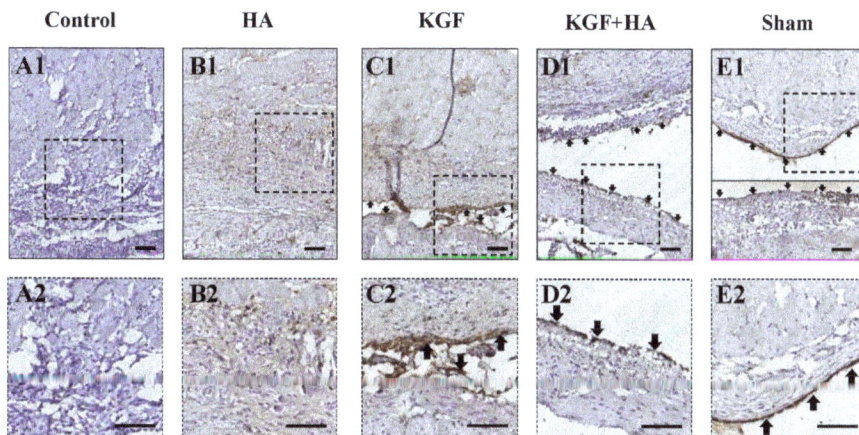

Figure 6. Immunohistochemical staining of cytokeratin in the intra-abdominal adhesions or the injured areas on the opposite parietal peritoneum in each group of rats (100× magnification pictures in upper plate, 200× magnification pictures in below plate; the black scale bar represents 100 μm). (**A1,2**) In the control group, the brown thin continuous line was completely interrupted; the interrupted brown line indicated the injured mesothelial cell layer; (**B1,2, C1,2**) The mesothelial cell layer (brown thin continuous line) was interrupted in the KGF and HA groups; (**D1,2**) In the KGF plus HA group, continuous brown lines were observed, indicating that the peritoneal mesothelial cell layer was completely repaired by regeneration; (**E1,2**) In the sham operation group, the parietal peritoneum and visceral peritoneum of the intestine surface were stained brown. A brown thin continuous line was formed, indicating that the peritoneal mesothelial cell layer was intact. Arrow: peritoneal mesothelial cell.

2.5. The Combined Administration of KGF and HA Inhibited the Severity of the Fibrous Changes in the Injured Peritoneum and/or Adhesion Tissue in the Rat Model

Using immunohistochemical staining for α-smooth muscle actin (α-SMA), an activated fibroblast marker, the degree of fibrous changes in the injured rat peritoneum and/or adhesion tissue was examined (Figure 7A–E). In the sham laparotomy group, the peritoneum was intact, and no positive staining was observed. In the control group, a large amount of fusiform fibroblasts with positive brown staining were observed in the thick adhesive tissue. In the HA and KGF groups, the number of fibroblasts with positive α-SMA expression was slightly less than the number in the control group. However, in the KGF plus HA group, the degree of fibrous changes in the injured peritoneum and/or adhesion tissue was significantly reduced. Immunohistochemical staining for collagen I deposition (Figure 7F–J), a marker of fibrosis, showed that the amount of collagen I deposition on a scale from high to low was the control group, the HA group, the KGF group, the KGF plus HA group and the sham operation group. Thus, the results showed that the combined administration of KGF and HA could effectively inhibit fibrosis of the injured or adhesive peritoneum in a rat model.

2.6. The Combined Administration of KGF and HA Suppressed the Abdominal Fluid Levels of Tissue Plasminogen Activator (tPA) and the Pro-Inflammatory Cytokines Transforming Growth Factor β1 (TGF-β1) and Interleukin 1β (IL-1β) in the Rat Model

On postoperative day 7, the abdominal fluid levels of tPA, TGF-β1 and IL-1β were measured by ELISAs (Figure 8). The results revealed that the tPA levels in the abdominal fluid were significantly reduced in the control group compared with the sham operation group, but the levels of the pro-inflammatory cytokines TGF-β1 and IL-1β were significantly increased compared with the sham operation group, suggesting that surgical injury-induced adhesion formation resulted in decreased local fibrinolytic activity and significant inflammatory responses. Moreover, the KGF plus HA intervention could significantly inhibit the decrease in the tPA levels and the increase in the TGF-β1 and IL-1β levels ($p < 0.05$). This inhibitory effect was not significantly different in the HA group and the KGF group ($p > 0.05$). These results suggested that the combined administration inhibited the down-regulation of fibrinolytic activity and the significant release of inflammatory factors induced by the injured peritoneum during adhesion formation.

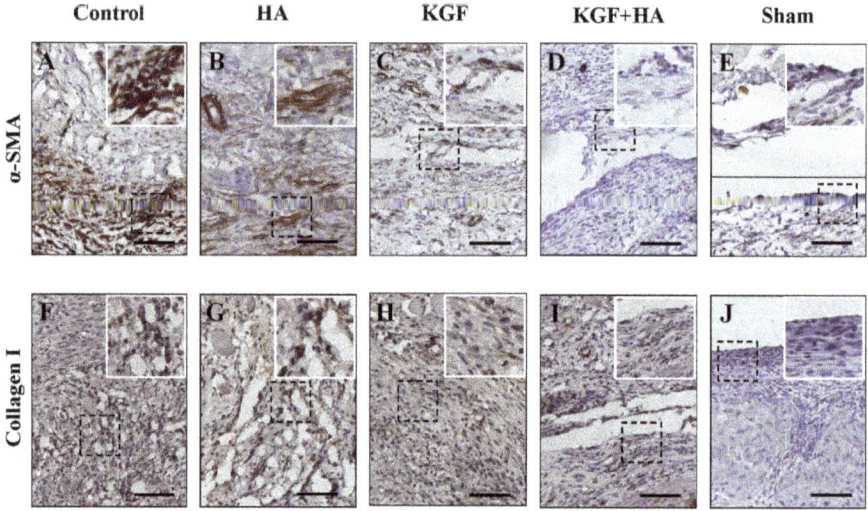

Figure 7. Immunohistochemical staining of α-smooth muscle actin (α-SMA), vimentin and collagen I in the intra-abdominal adhesions or the injured areas on the opposite parietal peritoneum in each group of rats (200×; insets, 400×). The black scale bar represents 100 μm. (**A–E**) α-SMA staining; (**F–J**) Collagen I staining.

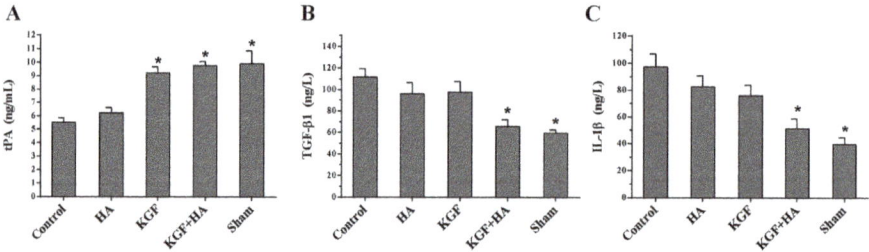

Figure 8. Abdominal fluid levels of tissue plasminogen activator (tPA) (**A**); transforming growth factor β1 (TGF-β1) (**B**); and interleukin 1β (IL-1β) (**C**) in each group ($n = 8$) (* $p < 0.05$ compared with the control group).

2.7. The Combined KGF and HA Treatment Inhibited Src Phosphorylation and Expression of TGF-β1, Fibrinogen, and α-Smooth Muscle Actin (α-SMA) in the Injured Peritoneum and/or Adhesion Tissue in the Rat Model

The Western blot analysis showed that Src phosphorylation was significantly increased on postoperative day 7 in the KGF group and KGF plus HA group, suggesting that KGF can promote Src phosphorylation and activation (Figure 9). In addition, TGF-β1 expression in the abdominal adhesion tissue was significantly increased in the control group compared with the sham operation group. Compared with the control group, the HA and KGF groups did not show apparent TGF-β1 down-regulation. However, in the KGF plus HA group, expression of the TGF-β1 protein in the injured peritoneum and/or adhesion tissue was significantly decreased compared with that in the control, HA, and KGF groups (Figure 9). Moreover, we examined the expression of the fibrinogen and α-SMA proteins. With a similar pattern as the TGF-β1 levels in the groups, the expression levels of fibrinogen and α-SMA protein were reduced by the combined administration of KGF plus HA in the injured peritoneum and/or adhesion tissue (Figure 9).

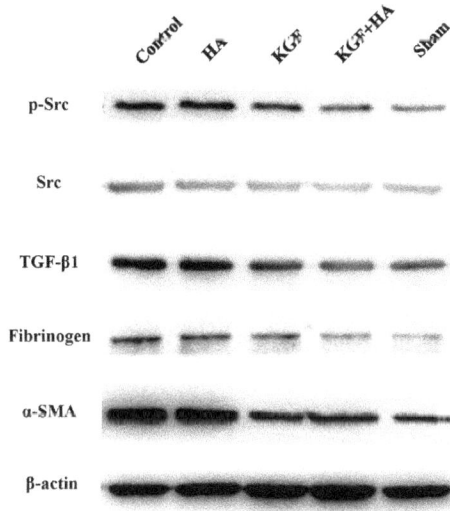

Figure 9. Expression of phosphorylated Src, Src, TGF-βl, fibrinogen and α-SMA protein in the intra-abdominal adhesions or the injured areas on the opposite parietal peritoneum by Western blotting.

The expression levels of TGF-β1, fibrinogen and α-SMA mRNA in the injured peritoneum and/or adhesion tissue on postoperative day 8 were determined using the real-time RT-PCR technique (Figure 10). We found that the expression levels of TGF-β1, fibrinogen and α-SMA mRNA in the injured peritoneum and/or adhesion tissue were the lowest in the KGF plus HA group ($p < 0.05$). The mRNA expression levels were consistent with the protein expression levels shown in the Western blot analysis.

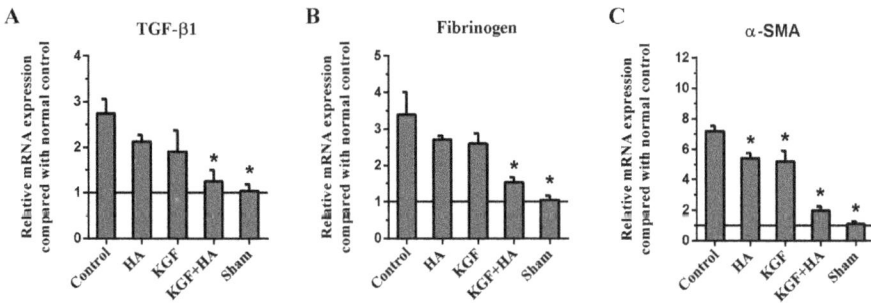

Figure 10. Expression of TGF-βl (**A**); fibrinogen (**B**) and α-SMA (**C**) mRNA in the intra-abdominal adhesions or the injured areas on the opposite parietal peritoneum by real-time RT-PCR ($n = 8$). (* $p < 0.05$ compared with the control group)

3. Discussion

The results of this study have shown that, compared to the administration of either KGF or HA alone, the combined administration of KGF and HA can more effectively prevent intra-abdominal adhesion formation in a rat model. This good preventative effect may be associated with the synergistic effect of the two agents, which can promote the repair of the mesothelial cell layer. Therefore, this study showed that the combined administration of KGF plus HA is a promising initial strategy for the prevention of intra-abdominal adhesions. It is worth continuing the study of its mechanism and translating the results into clinical practice.

At present, the following pharmacological preventive strategies against postoperative intra-abdominal adhesions should be considered [15–17]: (1) minimizing the initial inflammatory reaction and exudate; (2) inhibiting agglutination of the exudate; (3) promoting fibrinolysis; (4) using a physical barrier that can isolate the injured surface coated with fibrin; and (5) inhibiting fibroblast proliferation. Unfortunately, most of the regimens that have been used to prevent adhesion formation have limited effectiveness and considerable side effects [18]. It has been reported that intra-abdominal perfusion of crystalloid solutions and dextran injections are ineffective at preventing adhesions; moreover, the use of anti-inflammatory drugs has increased the risk of gastrointestinal anastomotic leakage. Currently, the most widely accepted method for the prevention of intra-abdominal adhesions is to place a medication barrier or gel between the injured peritoneal surfaces for at least 5 to 7 days. Some animal and clinical trials have shown that HA, chitosan and polyethylene glycol/polylactic acid films can indeed reduce adhesions [1–3], but others reported that the effectiveness of these medications is not significant [19]. The reason that the above-mentioned medications are not ideally effective in adhesion prevention may be because researchers have overlooked the function of mesothelial cells on the surface of the peritoneum.

Mesothelial cells are specialized epithelial cells that line the internal organs and body wall in the peritoneal, pleural, and pericardial cavities [20,21]. The surface of the serous peritoneum is an intact mesothelial cell layer. Intra-abdominal adhesions form between the injured surfaces of the two mesothelial cell layers [22]. Surgical injury may cause local tissue ischemia and hypoxia in the peritoneum [23], which trigger the inflammatory responses involving polymorphonuclear leukocytes, macrophages, fibroblasts and neovascularization and result in fibrin deposition for the initial adhesion formation [24,25]. Thereafter, the activation of the fibrinolytic system can degrade the fibrin deposits in the abdominal cavity. If fibrinolytic activation did not occur, fibroblasts would produce large amounts of collagen to form adhesions in 5 to 6 days after the injury of peritoneal mesothelial cell layer. Then, the mesothelial cells would completely cover the injured surface and adhesion area by proliferating to form fibrous adhesions, leading to permanent adhesion formation [26].

Mature adhesions can be effectively prevented if the intended techniques are used to enhance the regeneration ability of mesothelial cells and to rebuild the intact mesothelial cell layer in early intra-abdominal adhesions [27]. However, strategies to protect mesothelial cell regeneration are rarely reported, even though there have been a number of approaches described for the prevention of intra-abdominal adhesions. Bertram et al. [28] reported that intraperitoneal transplantation of isologous mesothelial cells resulted in a significant reduction of adhesion formation. Guo and colleagues [29] showed that high expression of sphingosine kinase (SPK1) can enhance both the proliferation and migration of mesothelial cells to accentuate the repair of mesothelial cells; SPK1 can also induce the cells to secrete plasminogen activator to enhance the peritoneal fibrinolytic capacity and ultimately prevent adhesion formation. These results suggest that the rapid enhancement of mesothelial cell regeneration is a possible strategy to prevent adhesion formation.

KGF is a single-chain polypeptide that was first isolated from the culture medium of human embryonic lung fibroblasts by Finch and Rubin in 1989 [30]. Its molecular weight is 26 to 28 kDa. The gene sequence analysis indicates that KGF is a member of the fibroblast growth factor (FGF) family and thus is known as FGF-7. KGF targets epithelial cells (including mesothelial cells) and may have potential mitogen activity and promotes proliferation in a variety of epithelial cells [31].

KGF can prevent and repair epithelium injuries of the skin, cornea, bladder, lungs, intestines, and liver through multiple approaches and multiple levels [32]. Lopes et al. [13] employed KGF to prevent swine pericardial adhesions in their study and suggested that the use of growth factors targeting mesothelial cell proliferation or regeneration can reduce the severity of adhesions. However, it has not been reported that KGF can prevent peritoneal adhesions. This study assessed the effect of KGF on adhesion formation in rats. The results have shown that KGF indeed reduced the severity of peritoneal adhesions and that the administration of KGF alone had approximately the same effect as the HA gel.

KGF function depends on the phosphorylation of the tyrosine kinase Src. Src can be phosphorylated by the KGF receptor (KGFR) after KGF binds KGFR [33] and thus exert its biological effects on downstream targets. In our study, we evaluated Src phosphorylation levels after the administration of KGF. We found that the levels of the phosphorylated Src protein were up-regulated in the peritoneal tissue after the administration of KGF to rats. This result suggests that KGF indeed plays a role in the local repair of the injured peritoneum.

KGFR is only distributed in epithelial cells (including mesothelial cells), and thus, KGF has no effect on fibroblasts and endothelial cells [11]. Therefore, the use of KGF can reduce adhesions without stimulating fibroblast proliferation and activation and thus can avoid aggravating adhesion fibrosis. Our experimental results showed that the degrees of collagen deposition and fibrosis in the adhesive tissue were reduced in the rats of the KGF intervention group compared with the control group.

Although KGF reduced the peritoneal adhesions, its effect was similar to the effect of the HA gel in clinical practice. The proportion of rats without adhesion formation did not reach 50% of the total number of rats. This result indicated that a considerable number of rats did not obtain a benefit from the KGF intervention. The possible reason is that the timing and speed of mesothelial cell repair is crucial for peritoneal adhesion prevention. Typically, mesothelial cell repair begins at 1 or 2 days and is completed in approximately 8–10 days after peritoneal injury [34] because peritoneal injury causes a large amount of fibrin deposition in the injured sites during this period. If the regenerated mesothelial cells cover the surface of the fibers and collagen, they can instead promote adhesion maturation and lead to permanent adhesion formation. Therefore, we suspect that the inhibition of fibrin deposition on the injured peritoneal surface during KGF-promoted mesothelial cell repair can greatly increase the effect on adhesion prevention.

HA gels are a commonly used barrier-like material for the clinical prevention of postoperative abdominal adhesions [1]. Hyaluronic acid is an acidic mucopolysaccharide. It is a normal glycosaminoglycan that is distributed in the extracellular matrix of the connective tissue in the human body. At physiological concentrations, hyaluronic acid molecules are entangled and form a disordered network of fibers that comprise its unique three-dimensional structure. It is difficult for some macromolecules, such as fibrinogen, collagen and proteoglycan, to enter this network structure. Hyaluronic acid enters the abdominal cavity in a colloidal state to form a jelly-like substance and covers the surface of the injured peritoneum to avoid the direct and effective contact between the surfaces. HA is a form of the sodium salt of hyaluronic acid, a naturally degraded and absorbable biomedical material [35]. Furthermore, HA can reduce the exudation of inflammatory cells, reduce the deposition of collagen at injured sites, and promote the physiological repair of injured tissue [36]. HA definitely plays a beneficial role in isolating the injured surface. This study showed that the barrier effect of HA also mildly or moderately reduced the magnitude of abdominal adhesions.

In theory, better prevention of adhesion formation can be achieved if a barrier material is used to maintain the separation between the parietal and visceral peritoneum before KGF accentuates the restoration of the integrity of the mesothelial cell layer of the peritoneum. The mechanisms by which KGF and HA prevent adhesions are different. Thus, we speculated that, theoretically, the combined administration of the two agents may be more efficacious in preventing adhesions. Our study also showed that the combined administration of KGF and HA can achieve better results in terms of adhesion severity and reducing the inflammatory cytokine levels compared to each treatment alone. The combined administration of KGF and HA significantly reduced the magnitude of inflammatory infiltration and collagen deposition at the injured peritoneal adhesions in rats. The combined administration can take advantage of the synergistic effects to achieve better results in the prevention of peritoneal adhesions (Figure 11).

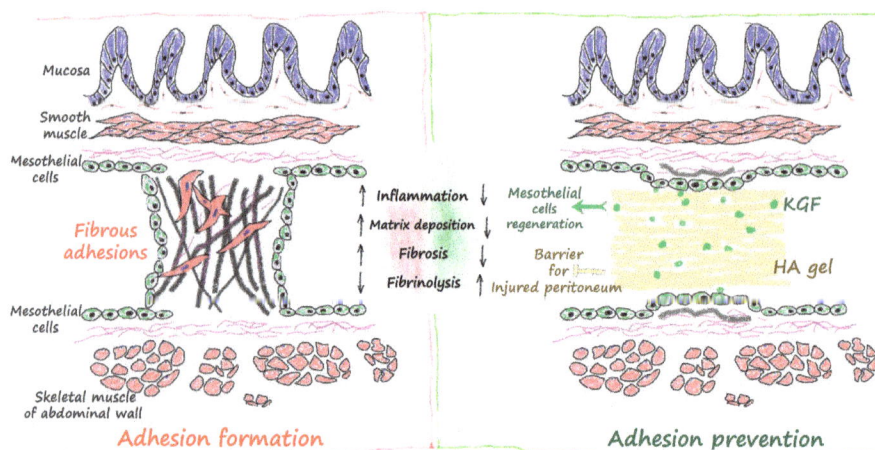

Figure 11. Mechanism by which the combined administration of KGF and HA prevents intra-abdominal adhesions. The HA gel can maintain the separation of the parietal and visceral peritoneum; KGF accentuates the regeneration and repair of the peritoneal mesothelial cell layer and thus the orderly repair of the integrity of the injured peritoneal mesothelial cell layer.

There are some limitations and concerns with this study. KGF should be used with caution to prevent postoperative abdominal adhesions in the setting of abdominal tumor surgery because experimental studies suggested that KGF, a growth factor, may have the potential to promote tumor growth [36] or increase tumor resistance to chemotherapy [37]. Oelmann et al. [38] have proven that under the effects of recombinant KGF (r-KGF), the majority of tumor cell lines did not exhibit meaningful proliferation in vitro among 35 cancer cell lines of epithelial origin and 22 lymphoma and leukemia cell lines. Only five cancer cell lines (two lung cancer, one gastric cancer, one colon cancer and one breast cancer) exhibited statistically significant proliferation in a dose-dependent manner. Although KGF has been used at non-tumor sites to treat disease [39], people are still concerned about whether its use will cause tumor growth. Another issue of this present study regarded the lasting effective period of KGF in the body. KGF has poor stability and a short biological half-life; the biological activity of KGF is highly susceptible to environmental changes. Mesothelial cells require 8 days to repair the injured surface; however, the activity of intraperitoneally administered KGF cannot last for a long time due to the impact of a variety of biological enzymes. Thus, further improvements in the KGF dosage form are needed to maintain stable pharmacological effects on the body.

In summary, the present study has shown that the combined administration of KGF and HA can effectively prevent postoperative intra-abdominal adhesion formation in a rat model by maintaining the separation of the injured peritoneum and promoting mesothelial cell regeneration. The combined administration of KGF and HA has shown considerable advantages from the aspects of the inhibition of collagen deposition, tissue fibrosis, and inflammation. This study suggests that the combined administration of KGF and HA is a promising pharmacotherapeutic strategy in intra-abdominal adhesion prevention and also provides more ideas for the development of novel drugs to prevent adhesion formation.

4. Materials and Methods

4.1. Agents

The r-KGF was purchased from Prospec (Rehovot, Israel; CAT#: cyt-219) in a highly purified state (95%). Stock solutions were made after an initial dilution with sterile phosphate-buffered saline and stored at −20 °C at a 1000 ng/mL concentration. At the time of use, the stock solutions were thawed and diluted in sterile water. The sample for the KGF group consisted of a 15 mL solution containing 25 ng/mL r KGF. The medical HA gel (10 mg/mL) was produced by Hangzhou Singclean Medical Products Co., Ltd., China. KGF at a 1000 ng/mL concentration was thawed and dissolved in the sterile hyaluronate gel. The final sample of KGF + HA gel contains 25 ng/mL r-KGF.

4.2. Surgical Procedures

Male Sprague-Dawley rats weighing 200 to 250 g were purchased from the Experimental Animal Center of Xi'an Jiaotong University. These animals were housed at room temperature (22 ± 2 °C), with free access to water and standard rat chow. All animal experiment protocols were approved by Xi'an Jiaotong University Experimental Animal Ethics Committee (No. 2015-156, 6 March 2015). The animals were anesthetized through the inhalation of methoxyflurane. The abdominal skin was prepared and disinfected with the povidone-iodine prior to the procedure. As previously described in the literature [40], a 2- to 3-cm-long lower abdominal midline incision was used to access the abdominal cavity. The pouch-like cecum was located in the right iliac fossa. The cecum wall and its opposite parietal peritoneum were abraded with sterile gauze until spot bleeding was observed. The abraded area was approximately 2–3 cm^2 and was exposed to air for approximately 5 min. The bowels were arranged to ensure that the abraded cecum wall was opposite the abraded peritoneum. In the group with a sham operation, the abrasion and exposure was not performed. In the Control, HA, KGF or KGF + HA groups, a 1 mL normal saline, HA gel, r-KGF or HA gel containing r-KGF was applied to the abraded peritoneum and its surrounding areas, respectively. Interrupted 3-0 Vicryl sutures were used to close the peritoneum, the abdominal muscles, and the skin in 2 layers.

4.3. Adhesion Grade and Assessment

On day 8 after surgery, all rats were anesthetized, and an inverted "U" shape incision was used to open the abdomen. The magnitude of the intra-abdominal adhesions was assessed according to the adhesion grade criteria reported by Nair et al. [41] (Table S1) or Leach et al. [42] (Table S2). The investigators who assessed the adhesion grade were independent researchers and were blinded to the protocol. The rats were sacrificed after the assessment, and the specimens were collected for the subsequent studies.

4.4. Hematoxylin and Eosin (HE) Staining and Microscopic Histological Grading of Inflammation

The injured peritoneum and adhesion tissues were excised. Specimen fixation and section preparation were carried out. Then, HE staining was performed. The tissues were evaluated under a microscope in regards to the severity of inflammatory cell reaction by using the classification described by Mahdy et al. [43]. The standard was as follows: degree of inflammation (grade 0: absent or normal in number; grade 1: mild increase giant cells, occasional scattered lymphocytes and plasma cells; grade 2: moderate infiltration, giant cells with increased numbers of admixed lymphocytes, plasma cells, eosinophils, neutrophils; grade 3: massive infiltration, many admixed inflammatory cells, microabscesses present).

4.5. Picrosirius Red Staining for Collagen

Picrosirius red staining for collagen was achieved using 0.1% picrosirius red (Direct Red 80; Sigma-Aldrich, St. Louis, MO, USA) and counterstained with Weigert's hematoxylin. The percentage of the positively stained area in eight randomly selected fields was evaluated using ImagePro Plus 5.0 software (Leica Qwin. Plus, Leica Microsystem Imaging Solutions Ltd., Cambridge, UK), and the average of the eight values was taken as the collagen content in the adhesions.

4.6. Immunohistochemistry

Immunohistochemical staining was performed using the SABC kit (Maxim, Fuzhou, China), according to the manufacturer's instructions. The tissue sections were incubated with primary antibodies for cytokeratin AE1/AE3 (1:400 dilution), alpha smooth muscle actin (α-SMA) (1:500 dilution), and collagen I (1:300 dilution) overnight at 4 °C and incubated with the appropriate biotinylated secondary antibody for 30 min at room temperature, followed by a 30 min incubation with streptavidin peroxidase (Dako LSAB + HRP kit). After rinsing, the results were visualized using Diaminobenzidine tetrahydrochloride (DAB), and the slides were counterstained with hematoxylin.

4.7. Western Blot

Total proteins were extracted from the tissues using RIPA lysis buffer as previously described [44]. Cell lysates were resolved on 10% sodium dodecyl sulfate polyacrylamide gels and transferred to PVDF membranes. The membranes were blocked with 5% skim milk and incubated with primary antibodies overnight at 4 °C, which included an anti-phospho Src antibody (ab185617, Abcam, Cambridge, UK, 1:1000 dilution), anti-Src antibody (ab47405, Abcam, 1:600 dilution), anti-TGF-β1 antibody (sc-146, Santa Cruz Biotechnology, Dallas, TX, USA, 1:400 dilution), anti-fibrinogen antibody (sc-18029, Santa Cruz Biotechnology, 1:800 dilution), anti-α-SMA antibody (sc-53015, Santa Cruz Biotechnology, 1:800 dilution), and anti-β-actin antibody (sc-47778, Santa Cruz Biotechnology, 1:1000 dilution). The membranes were washed and incubated with the secondary antibodies for 2 h at room temperature. Protein expression was detected using a chemiluminescence system (Millipore, Billerica, MA, USA) according to the manufacturer's specifications.

4.8. Real-Time RT-PCR

Real-time RT-PCR was performed to determine the messenger RNA (mRNA) levels of TGF-β1, fibrinogen, α-SMA and GAPDH. Total RNA was extracted using TRIzol reagent (Invitrogen, Carlsbad, CA, USA), and reverse transcription was performed using a PrimeScript RT reagent Kit (TaKaRa, Dalian, China). The real-time experiments were conducted on an iQ5 Multicolor Real-Time PCR Detection System (Bio-Rad, Hercules, CA, USA) using a SYBR Green Real-time PCR Master Mix (TaKaRa). The PCR reactions consisted of 5 s at 94 °C followed by 40 cycles at 94 °C for 30 s, 60 °C for 30 s, and 72 °C for 30 s. The PCR primer sequences are listed in Table S4. The comparative C(T) method was used to quantitate the expression of each target gene using GAPDH as the normalization control [45].

4.9. ELISA Quantification of Abdominal Fluid Levels of IL-6, Tumor Necrosis Factor α (TNF-α), TGF-β1

Abdominal fluid samples were collected from the abdomens of the animals and centrifuged at 3000 rpm for 30 min. The supernatant was stored at −20 °C. The concentrations of IL-6, tumor necrosis factor α (TNF-α), and TGF-β1 were measured using ELISA kits (R & D Systems, Minneapolis, MN, USA) according to the manufacturer's instructions.

4.10. Hydroxyproline Determination

Hydroxyproline content was determined using a Hydroxyproline Assay Kit (Sigma-Aldrich) according to the manufacturer's instructions. The tissue hydroxyproline levels, which were used as an indicator of the adhesion severity, are presented as micrograms of hydroxyproline per gram of protein.

4.11. Statistical Analyses

Categorical variables are listed as medians (minimum to maximum), and continuous variables are listed as averages and standard deviations. The categorical variables were evaluated by the Kruskal-Wallis H variance analysis for independent samples and the post hoc Mann-Whitney U test for multiple matches. Continuous variables were analyzed using one-way analysis of variance followed by the least significant different (LSD) test. The SPSS 15.0 software package (SPSS, Chicago, IL, USA) was used for data analysis. $p < 0.05$ was considered significant.

Supplementary Materials: Supplementary materials can be found at www.mdpi.com/1422-0067/17/10/1611/s1. References [41,42] are cited in the supplementary materials.

Acknowledgments: This study was supported by the National Natural Science Foundation of China (No. 81572734) and the Fundamental Research Funds for the Central Universities in Xi'an Jiaotong University (No. 2013jdhz33).

Author Contributions: Xuqi Li conceived and designed the experiments; Guangbing Wei, Cancan Zhou, and Guanghui Wang performed the experiments; Xuqi Li, Kang Wang and Lin Fan contributed to the interpretation of data; Xuqi Li and Guangbing Wei wrote the manuscript.

Conflicts of Interest: The authors declare no conflict of interest.

References

1. Hu, J.; Fan, D.; Lin, X.; Wu, X.; He, X.; He, X.; Wu, X.; Lan, P. Safety and efficacy of sodium hyaluronate gel and chitosan in preventing postoperative peristomal adhesions after defunctioning enterostomy: A prospective randomized controlled trials. *Medicine* **2015**, *94*, e2354. [CrossRef] [PubMed]

2. Yang, B.; Gong, C.; Zhao, X.; Zhou, S.; Li, Z.; Qi, X.; Zhong, Q.; Luo, F.; Qian, Z. Preventing postoperative abdominal adhesions in a rat model with PEG-PCL-PEG hydrogel. *Int. J. Nanomed.* **2012**, *7*, 547–557.

3. Yuan, F.; Lin, L.X.; Zhang, H.H.; Huang, D.; Sun, Y.L. Effect of carbodiimide-derivatized hyaluronic acid gelatin on preventing postsurgical intra-abdominal adhesion formation and promoting healing in a rat model. *J. Biomed. Mater. Res. A* **2016**, *104*, 1175–1181. [CrossRef] [PubMed]

4. Ten Broek, R.P.; Issa, Y.; van Santbrink, E.J.; Bouvy, N.D.; Kruitwagen, R.F.; Jeekel, J.; Bakkum, E.A.; Rovers, M.M.; van Goor, H. Burden of adhesions in abdominal and pelvic surgery: Systematic review and met-analysis. *BMJ* **2013**, *347*, f5588. [CrossRef] [PubMed]

5. Miller, G.; Boman, J.; Shrier, I.; Gordon, P.H. Natural history of patients with adhesive small bowel obstruction. *Br. J. Surg.* **2000**, *87*, 1240–1247. [CrossRef] [PubMed]

6. Alpay, Z.; Saed, G.M.; Diamond, M.P. Postoperative adhesions: From formation to prevention. *Semin. Reprod. Med.* **2008**, *26*, 313–321. [CrossRef] [PubMed]

7. Becker, J.M.; Stucchi, A.F. Intra-abdominal adhesion prevention: Are we getting any closer? *Ann. Surg.* **2004**, *240*, 202–204. [CrossRef] [PubMed]

8. Beyene, R.T.; Kavalukas, S.L.; Barbul, A. Intra-abdominal adhesions: Anatomy, physiology, pathophysiology, and treatment. *Curr. Probl. Surg.* **2015**, *52*, 271–319. [CrossRef] [PubMed]

9. Liu, H.J.; Wu, C.T.; Duan, H.F.; Wu, B.; Lu, Z.Z.; Wang, L. Adenoviral-mediated gene expression of hepatocyte growth factor prevents postoperative peritoneal adhesion in a rat model. *Surgery* **2006**, *140*, 441–447. [CrossRef] [PubMed]

10. Uguralp, S.; Akin, M.; Karabulut, A.B.; Harma, B.; Kiziltay, A.; Kiran, T.R.; Hasirci, N. Reduction of peritoneal adhesions by sustained and local administration of epidermal growth factor. *Pediatr. Surg. Int.* **2008**, *24*, 191–197. [CrossRef] [PubMed]

11. Yen, T.T.; Thao, D.T.; Thuoc, T.L. An overview on keratinocyte growth factor: From the molecular properties to clinical applications. *Protein Pept. Lett.* **2014**, *21*, 306–317. [CrossRef] [PubMed]

12. Wang, X.; Yu, M.; Zhu, W.; Bao, T.; Zhu, L.; Zhao, W.; Zhao, F.; Wang, H. Adenovirus-mediated expression of keratinocyte growth factor promotes secondary flap necrotic wound healing in an extended animal model. *Aesthet. Plast. Surg.* **2013**, *37*, 1023–1033. [CrossRef] [PubMed]

13. Lopes, J.B.; Dallan, L.A.; Campana-filho, S.P.; Lisboa, L.A.; Gutierrez, P.S.; Moreira, L.F.; Oliveira, S.A.; Stolf, N.A. Keratinocyte growth factor: A new mesothelial targeted therapy to reduce postoperative pericardial adhesions. *Eur. J. Cardiothorac. Surg.* **2009**, *35*, 313–318. [CrossRef] [PubMed]

14. Shen, J.; Xu, Z.W. Combined application of acellular bovine pericardium and hyaluronic acid in prevention of postoperative pericardial adhesion. *Artif. Organs* **2014**, *38*, 224–230. [CrossRef] [PubMed]

15. Ward, B.C.; Panitch, A. Abdominal adhesions: Current and novel therapies. *J. Surg. Res.* **2011**, *165*, 91–111. [CrossRef] [PubMed]

16. Wei, G.; Chen, X.; Wang, G.; Jia, P.; Xu, Q.; Ping, G.; Wang, K.; Li, X. Inhibition of cyclooxygenase-2 prevents intra-abdominal adhesions by decreasing activity of peritoneal fibroblasts. *Drug Des. Dev. Ther.* **2015**, *9*, 3083–3098.

17. Wei, G.; Chen, X.; Wang, G.; Fan, L.; Wang, K.; Li, X. Effect of resveratrol on the prevention of intra-abdominal adhesion formation in a rat model. *Cell. Physiol. Biochem.* **2016**, *39*, 33–46. [CrossRef] [PubMed]

18. Robb, W.B.; Mariette, C. Strategies in the prevention of the formation of postoperative adhesions in digestive surgery: A systematic review of the literature. *Dis. Colon Rectum* **2014**, *57*, 1228–1240. [CrossRef] [PubMed]

19. Fayez, J.A.; Schneider, P.J. Prevention of pelvic adhesion formation by different modalities of treatment. *Am. J. Obstet. Gynecol.* **1987**, *157*, 1184–1188. [CrossRef]

20. Yung, S.; Chan, T.M. Mesothelial cells. *Perit. Dial. Int.* **2007**, *27*, S110–S115. [PubMed]

21. Mutsaers, S.E. Mesothelial cells. Their structure, function and role in serosal repair. *Respirology* **2002**, *7*, 171–191. [CrossRef] [PubMed]

22. Haney, A.F.; Doty, E. The formation of coalescing peritoneal adhesions requires injury to both contacting peritoneal surfaces. *Fertil. Steril.* **1994**, *61*, 767–775. [CrossRef]

23. Fletcher, N.M.; Awonuga, A.O.; Abusamaan, M.S.; Saed, M.G.; Diamond, M.P.; Saed, G.M. Adhesion phenotype manifests an altered metabolic profile favoring glycolysis. *Fertil. Steril.* **2016**, *105*, 1628–1637. [CrossRef] [PubMed]

24. Saed, G.M.; Fletcher, N.M.; Diamond, M.P. The creation of a model for ex vivo development of postoperative adhesions. *Reprod. Sci.* **2016**, *23*, 610–612. [CrossRef] [PubMed]

25. Saed, G.M.; Kruger, M.; Diamond, M.P. Expression of transforming growth factor-β and extracellular matrix by human peritoneal mesothelial cells and by fibroblasts from normal peritoneum and adhesions: Effect of Tisseel. *Wound Repair Regen.* **2004**, *12*, 557–564. [CrossRef] [PubMed]

26. Chegini, N. Peritoneal molecular environment, adhesion formation and clinical implication. *Front. Biosci.* **2002**, *7*, e91–e115. [CrossRef] [PubMed]

27. Kawanishi, K.; Nitta, K. Cell sheet-based tissue engineering for mesothelial cell injury. *Contrib. Nephrol.* **2015**, *185*, 66–75. [PubMed]

28. Bertram, P.; Tietze, L.; Hoopmann, M.; Treutner, K.H.; Mittermayer, C.; Schumpelick, V. Intraperitoneal transplantation of isologous mesothelial cells for prevention of adhesions. *Eur. J. Surg.* **1999**, *165*, 705–709. [CrossRef] [PubMed]

29. Guo, Q.; Li, Q.F.; Liu, H.J.; Li, R.; Wu, C.T.; Wang, L.S. Sphingosine kinase 1 gene transfer reduces postoperative peritoneal adhesion in an experimental model. *Br. J. Surg.* **2008**, *95*, 252–258. [CrossRef] [PubMed]

30. Finch, P.W.; Rubin, J.S. Keratinocyte growth factor expression and activity in cancer: Implications for use in patients with solid tumors. *J. Natl. Cancer Inst.* **2006**, *98*, 812–824. [CrossRef] [PubMed]

31. Kovacs, D.; Raffa, S.; Flori, E.; Aspite, N.; Briganti, S.; Cardinali, G.; Torrisi, M.R.; Picardo, M. Keratinocyte growth factor down-regulates intracellular ROS production induced by UVB. *J. Dermatol. Sci.* **2009**, *54*, 106–113. [CrossRef] [PubMed]

32. Abo, T.; Nagayasu, T.; Hishikawa, Y.; Tagawa, T.; Nanashima, A.; Yamayoshi, T.; Matsumoto, K.; An, S.; Koji, T. Expression of keratinocyte growth factor and its receptor in rat tracheal cartilage: Possible involvement in wound healing of the damaged cartilage. *Acta Histochem. Cytochem.* **2010**, *43*, 89–98. [CrossRef] [PubMed]

33. Belleudi, F.; Scrofani, C.; Torrisi, M.R.; Mancini, P. Polarized endocytosis of the keratinocyte growth factor receptor in migrating cells: Role of SRC-signaling and cortactin. *PLoS ONE* **2011**, *6*, e29159. [CrossRef] [PubMed]

34. Cheong, Y.C.; Laird, S.M.; Li, T.C.; Shelton, J.B.; Ledger, W.L.; Cooke, I.D. Peritoneal healing and adhesion formation/reformation. *Hum. Reprod. Update* **2001**, *7*, 556–566. [CrossRef] [PubMed]

35. Krüger-Szabó, A.; Aigner, Z.; Balogh, E.; Sebe, I.; Zelkó, R.; Antal, I. Microstructural analysis of the fast gelling freeze-dried sodium hyaluronate. *J. Pharm. Biomed. Anal.* **2015**, *104*, 12–16. [CrossRef] [PubMed]

36. Yates, A.C.; Stewart, A.A.; Byron, C.R.; Pondenis, H.C.; Kaufmann, K.M.; Constable, P.D. Effects of sodium hyaluronate and methylprednisolone acetate on proteoglycan metabolism in equine articular chondrocytes treated with interleukin-1. *Am. J. Vet. Res.* **2006**, *67*, 1980–1986. [CrossRef] [PubMed]

37. Rotolo, S.; Ceccarelli, S.; Romano, F.; Frati, L.; Marchese, C.; Angeloni, A. Silencing of keratinocyte growth factor receptor restores 5-fluorouracil and tamoxifen efficacy on responsive cancer cells. *PLoS ONE* **2008**, *3*, e2528. [CrossRef] [PubMed]

38. Oelmann, E.; Haghgu, S.; Kulimova, E.; Mesters, R.M.; Kienast, J.; Herbst, H.; Schmitmann, C.; Kolkmeyer, A.; Serve, H.; Berdel, W.E. Influence of keratinocyte growth factor on clonal growth of epithelial tumor cells, lymphoma and leukemia cells and on sensitivity of tumor cells towards 5-fluorouracil in vitro. *Int. J. Oncol.* **2004**, *25*, 1001–1012. [PubMed]

39. Kanuga, S. Cryotherapy and keratinocyte growth factor may be beneficial in preventing oral mucositis in patients with cancer, and sucralfate is effective in reducing its severity. *J. Am. Dent. Assoc.* **2013**, *144*, 928–929. [CrossRef] [PubMed]

40. Peyton, C.C.; Keys, T.; Tomblyn, S.; Burmeister, D.; Beumer, J.H.; Holleran, J.L.; Sirintrapun, J.; Washburn, S.; Hodges, S.J. Halofuginone infused keratin hydrogel attenuates adhesions in a rodent cecal abrasion model. *J. Surg. Res.* **2012**, *178*, 545–552. [CrossRef] [PubMed]

41. Nair, S.K.; Bhat, I.K.; Aurora, A.L. Role of proteolytic enzyme in the prevention of postoperative intraperitoneal adhesions. *Arch. Surg.* **1974**, *108*, 849–853. [CrossRef] [PubMed]

42. Leach, R.E.; Burns, J.W.; Dawe, E.J.; SmithBarbour, M.D.; Diamond, M.P. Reduction of postsurgical adhesion formation in the rabbit uterine horn model with use of hyaluronate/carboxymethylcellulose gel. *Fertil. Steril.* **1998**, *69*, 415–418. [CrossRef]

43. Mahdy, T.; Mohamed, G.; Elhawary, A. Effect of methylene blue on intra-abdominal adhesion formation in rats. *Int. J. Surg.* **2008**, *6*, 452–455. [CrossRef] [PubMed]

44. Lei, J.; Ma, J.; Ma, Q.; Li, X.; Liu, H.; Xu, Q.; Duan, W.; Sun, Q.; Xu, J.; Wu, Z.; et al. Hedgehog signaling regulates hypoxia induced epithelial to mesenchymal transition and invasion in pancreatic cancer cells via a ligand-independent manner. *Mol. Cancer* **2013**, *12*, 66. [CrossRef] [PubMed]

45. Schmittgen, T.D.; Livak, K.J. Analyzing real-time PCR data by the comparative C(T) method. *Nat. Protoc.* **2008**, *3*, 1101–1108. [CrossRef] [PubMed]

© 2016 by the authors. Licensee MDPI, Basel, Switzerland. This article is an open access article distributed under the terms and conditions of the Creative Commons Attribution (CC BY) license (http://creativecommons.org/licenses/by/4.0/).

International Journal of
Molecular Sciences

MDPI

Article

Early Healing Events after Periodontal Surgery: Observations on Soft Tissue Healing, Microcirculation, and Wound Fluid Cytokine Levels

Doğan Kaner [1,*], Mouaz Soudan [1], Han Zhao [2], Georg Gaßmann [3], Anna Schönhauser [1] and Anton Friedmann [1]

[1]   Department of Periodontology, Witten/Herdecke University, 58455 Witten, Germany;
       mouaz.soudan@uni-wh.de (M.S.); anna-frederike.schoenhauser@uni-wh.de (A.S.);
       anton.friedmann@uni-wh.de (A.F.)
[2]   Multi-Disciplinary Treatment Center, Beijing Stomatological Hospital, Capital Medical University,
       Beijing 100050, China; 775301704@163.com
[3]   praxisHochschule, University of applied sciences, 50670 Cologne, Germany;
       g.gassmann@praxishochschule.de
*   Correspondence: dogan.kaner@uni-wh.de; Tel.: +49-2302-926-656

Academic Editor: Allison Cowin
Received: 10 October 2016; Accepted: 19 January 2017; Published: 27 January 2017

Abstract: Early wound healing after periodontal surgery with or without enamel matrix derivative/biphasic calcium phosphate (EMD/BCP) was characterized in terms of soft tissue closure, changes of microcirculation, and expression of pro- and anti-inflammatory cytokines in gingival crevicular fluid/wound fluid (GCF/WF). Periodontal surgery was carried out in 30 patients (18 patients: application of EMD/BCP for regeneration of bony defects; 12 patients: surgical crown lengthening (SCL)). Healthy sites were observed as untreated controls. GCF/WF samples were collected during two post-surgical weeks. Flap microcirculation was measured using laser Doppler flowmetry (LDF). Soft tissue healing was evaluated after two weeks. GCF/WF levels of interleukin 1β (IL-1β), tumour necrosis factor (TNF-α), IL-6, and IL-10 were determined using a multiplex immunoassay. Surgery caused similar reductions of flap microcirculation followed by recovery within two weeks in both EMD/BCP and SCL groups. GCF/WF and pro-inflammatory cytokine levels were immediately increased after surgery, and returned only partially to baseline levels within the two-week observation period. Levels of IL-10 were temporarily reduced in all surgical sites. Flap dehiscence caused prolonged elevated levels of GCF/WF, IL-1β, and TNF-α. These findings show that periodontal surgery triggers an immediate inflammatory reaction corresponding to the early inflammatory phase of wound healing, and these inflammation measures are temporary in case of maintained closure of the flap. However, flap dehiscence causes prolonged inflammatory exudation from the periodontal wound. If the biological pre-conditions for periodontal wound healing are considered important for the clinical outcome, care should be taken to maintain primary closure of the flap.

Keywords: periodontal surgery; periodontal regeneration; surgical crown lengthening; wound healing; cytokines

1. Introduction

Periodontal surgery is routinely performed in order to remove microbial deposits from root surfaces and for corrective measures such as pocket elimination, removal of roots, or contouring of bony defects. Further, periodontal surgery allows placing of biomaterials or grafts for regeneration of lost periodontal structures [1]. Surgical lengthening of the clinical crown in order to increase retention of prospective restorations or to avoid subgingival restoration margins is another common periodontal surgical procedure [2].

Irrespective of the intention and the modalities of surgery, periodontal wound healing always begins with a blood clot in the space maintained by the closed flap after suturing [3]. In the early inflammation phase of wound healing, inflammatory cells are attracted by platelet and complement derived mediators and aggregate around the blood clot. While polymorphnuclear neutrophil granulocytes (PMN) dominate initially, monocytes and macrophages emerge within the first days [4]. The blood clot also provides a provisional matrix for cells originating from the surrounding tissues (i.e., gingiva, periodontal ligament (PDL), cementum, and alveolar bone) [5]. Thus, gingival fibroblasts, endothelial cells, osteoblasts, and special fibroblast populations originating from the PDL proliferate into the wound area. Wound healing progresses consequently through several phases from inflammation to cell proliferation and matrix formation and repair; then, these stages are followed by remodelling and maturation [4]. Reparative healing with formation of a long junctional epithelium and only little reorganization of connective tissue attachment is the expected outcome in periodontal wound healing without regenerative measures [6,7].

Clinical and human histological data show the efficacy of enamel matrix derivative (EMD) for inducement of periodontal regeneration [8–10]. These effects are mainly attributed to the proven stimulatory effects of EMD on PDL fibroblasts, cementoblasts, and osteoblasts [11–13], as well as to inhibitory effects on the competing epithelial cells [14]. However, other mechanisms may also contribute to the beneficial effects of EMD. Since a high microbial load negatively affects "gain" of clinical attachment after periodontal surgery [15], antibacterial effects of the EMD preparation may add to the stimulation of regeneration [16,17]. As wound healing is regulated by endogenous substances such as pro- and anti-inflammatory cytokines, the resolution of inflammation appears important for the outcome of wound healing [18]. For example, tissue destruction due to high levels of pro-inflammatory mediators like interleukin 1β (IL-1β), IL-6, and tumour necrosis factor (TNF-α) are reversed by the anti-inflammatory cytokine IL-10 [19]. Suppression of various pro-inflammatory cytokines and anti-inflammatory modulation of macrophages have been confirmed for EMD in vitro [20,21]. Direct influences have been shown on T helper lymphocyte migration, CD25 activation, and apoptosis in a three-dimensional collagen matrix migration model [22]. Correspondingly, these immunomodulatory effects of EMD may promote favourable conditions for periodontal wound healing.

Regardless of with or without application of bioactive substances, the course of early healing after periodontal surgery is scarcely investigated in humans, although the first post-surgical weeks are considered important [23]. The aim of this descriptive study was to characterize early wound healing after periodontal surgery with or without application of EMD combined with granular biphasic calcium phosphate (BCP) in terms of soft tissue closure, changes of microcirculation, and expression of pro- and anti-inflammatory cytokines.

2. Results

2.1. Patients

Thirty patients were recruited, treated, and analysed (EMD/BCP group: 18 patients; SCL group: 12 patients). Demographic and clinical characteristics are shown in Table 1.

Table 1. Demographic and clinical patient characteristics at baseline. EMD/BCP: enamel matrix derivative/biphasic calcium phosphate; SCL: surgical crown lengthening. IQ: inter-quartiles. n.d.: Not Detected.

Patient Data	EMD/BCP 18 Patients	SCL 12 Patients	p
Age (years, median, IQ)	58 (47, 71)	51 (48, 58)	>0.05
Gender (male/female)	9/9	7/5	>0.05
Cigarette smokers (yes/no)	3/15	1/11	>0.05
Probing depth (mm, median, IQ) Min/max	7 (6.3, 9) 6, 11	n.d.	-
median tooth mobility (0–III) Min/max	0 (0, 1) 0, 2	0	>0.05
Radiographic defect depth (mm, median, IQ) Min/max	6 (5, 7) 3, 9	0	<0.001

2.2. Microcirculation

In each group, untreated control sites did not show any significant changes of perfusion throughout all 14 days of observation. Directly after surgery (D0b), both EMD/BCP and SCL groups showed significantly reduced microcirculation values at the papillary base compared to baseline at D0a (Figure 1A). Although perfusion increased significantly from termination of surgery (D0b) to D1 in both surgery groups, microcirculation was significantly lower than at untreated control sites not only after surgery (D0b), but also at D1. The ongoing significant recovery until 14 days after surgery (D14) was similar in both surgery groups.

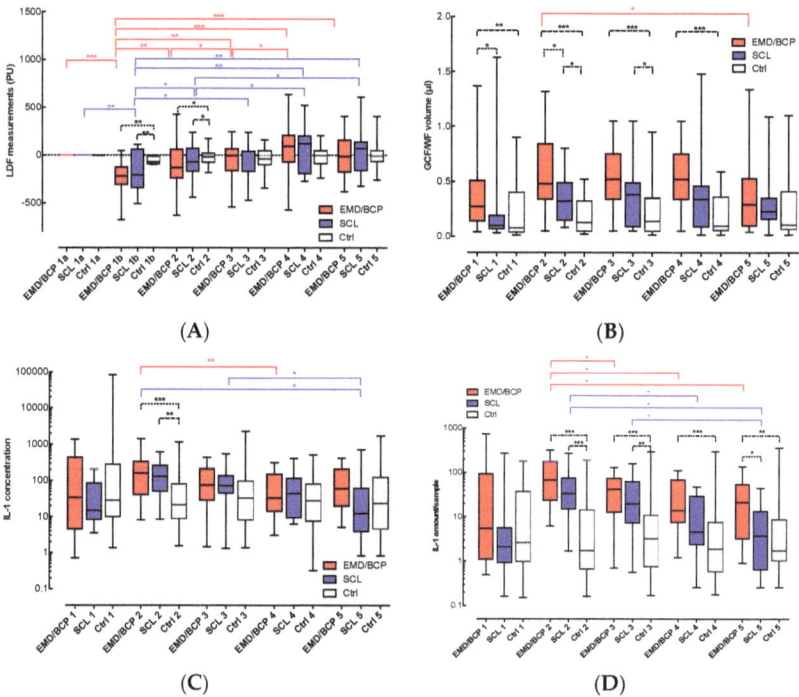

Figure 1. (A) LDF measurements of microcirculation; (B) Measurements of GCF/WF volume; (C) Concentration of IL-1β in GCF/WF samples; and (D) Total amount of IL-1β in GCF/WF samples. EMD/BCP (enamel matrix derivative/biphasic calcium phosphate): red, SCL (surgical crown lengthening): blue, Ctrl (control): white; LDF: laser Doppler flowmetry; GCF/WF: gingival crevicular fluid/wound fluid; IL-1: Interleukin 1β; *: $p < 0.05$; **: $p < 0.01$; ***: $p < 0.001$.

Int. J. Mol. Sci. **2017**, 18, 283

2.3. GCF/WF

The results for analyses of GCF/WF samples and their changes are shown in Figure 1B. At baseline (D0a), GCF sample volume was significantly greater in EMD/BCP sites, when compared to SCL and control sites. One and three days after surgery (D1, D3), both surgical groups showed significantly higher sample volumes than found in control sites. In addition, WF sample volume at D1 was significantly greater in EMD/BCP sites than in SCL sites. Further, EMD/BCP sites showed higher volumes of WF at days three and seven (D3, D7), when compared to control sites.

2.4. Cytokine Levels at EMD/BCP, SCL, and Control Sites

All three groups showed similar concentrations of IL-1β at baseline (Figure 1C). One day after surgery (D1), IL-1β concentrations were similarly elevated over healthy control sites in both surgical groups (EMD/BCP: 7.5-fold, SCL: 6-fold; $p < 0.05$). In both EMD/BCP and SCL sites, the concentrations of IL-1β decreased significantly again until the end of the observation period (from D3 to D7).

At baseline (D0a), all three groups showed similar amounts of IL-1β/sample (Figure 1D). At day one (D1) and at day three (D3), the amounts of IL-1β were significantly enhanced in EMD/BCP and SCL samples, when compared to control samples (39-fold and 19-fold, respectively, $p < 0.001$). The amount of IL-1β/sample decreased significantly for both surgical groups from D1 to D14. However, IL-1β levels detected in EMD/BCP sites after 14 days (D14) were still significantly elevated, when compared to both other groups (6-fold and 12-fold, respectively, $p < 0.05$).

At baseline (D0a), control sites showed a significantly higher concentration of TNF-α than EMD/BCP sites. No further differences or significant changes were found (Figure 2A).

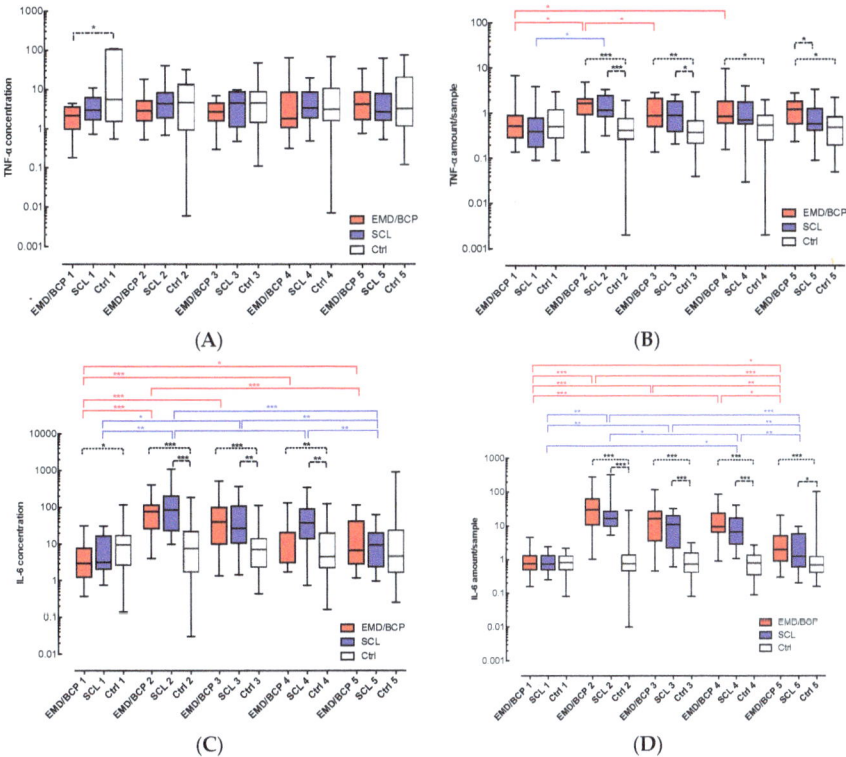

Figure 2. *Cont.*

87

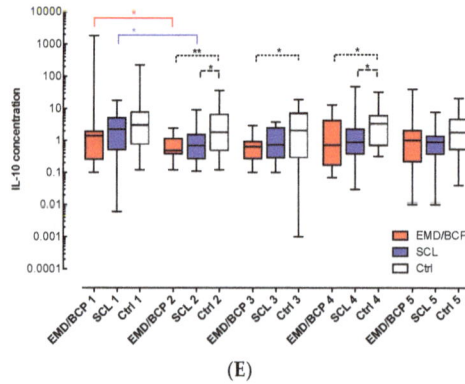

(E)

Figure 2. (**A**) Concentration of TNF-α in GCF/WF samples; (**B**) Total amount of TNF-α in GCF/WF samples; (**C**) Concentration of IL-6 in GCF/WF samples; (**D**) Total amount of IL-6 in GCF/WF samples; (**E**) Concentration of IL-10 in GCF/WF samples. EMD/BCP (enamel matrix derivative/biphasic calcium phosphate): red, SCL (surgical crown lengthening): blue, Ctrl (control): white; GCF/WF: gingival crevicular fluid/wound fluid; TNF-α: Tumour necrosis factor α; IL-6: Interleukin 6; IL-10: Interleukin 10; *: $p < 0.05$; **: $p < 0.01$, ***: $p < 0.001$.

All three groups showed similar amounts of TNF-α/sample prior to surgery at D0a (Figure 2B). After one and three days (D1, D3), the amount of TNF-α had significantly increased in both surgical groups and exceeded the levels found in control sites significantly. In the EMD/BCP group, the amount of TNF-α/sample remained high, and a significantly greater amount of TNF-α was found in EMD/BCP sites after 14 days (D14) in comparison to both SCL and healthy sites (2- and 2.5-fold, respectively, $p < 0.05$).

Prior to surgery (D0a), the concentration of IL-6 in EMD/BCP sites was significantly lower than in control sites (Figure 2C). In both surgical groups, the concentration of IL-6 increased significantly after surgery, and decreased again from D1 to D7. Accordingly, the EMD/BCP and SCL groups showed significantly higher IL-6 concentrations than controls at D1, D3, and D7.

The amounts of IL-6/sample were similar in all groups at baseline (D0a) (Figure 2D). At D1, the amounts of IL-6 were significantly increased by a factor of 39 for EMD/BCP and 22 for SCL, respectively ($p < 0.001$), and decreased significantly again over the course of 14 days (D3 to D14, $p < 0.05$).

However, when compared to controls, both surgical groups showed significantly higher amounts of IL-6 continuously from D1 to D14 ($p < 0.05$).

The concentration of IL-10 decreased similarly and significantly between D0a and D1 in both EMD/BCP and SCL groups; accordingly, significantly reduced concentrations of IL-10 were found in both surgical groups at D1, D3, and at D7, when compared to healthy sites (0.25- to 0.4-fold in both surgery groups, $p < 0.05$; Figure 2E).

The amount of IL-10/sample did not change significantly over time within all groups, and significant differences among the groups were not detected (data not shown).

2.5. Soft Tissue Healing

Two weeks after surgery (D14), the SCL group presented significantly better EHI values compared to the EMD/BCP group ($p = 0.032$, Table 2, Figure 3). While 13 EMD/BCP sites presented inter-proximal dehiscence of the flap, only three sites with dehiscence were observed in SCL patients ($p = 0.024$, Table 3, Figure 3).

Table 2. Contingency table for wound healing in both groups, evaluated using the Early Healing Index (EHI). Significant difference favouring the SCL group (p = 0.032, Pearson's Chi^2 test). EMD/BCP: enamel matrix derivative/biphasic calcium phosphate; SCL: surgical crown lengthening.

EHI	1	2	3	4	5
EMD/BCP	2 (11%)	2 (11%)	1 (6%)	4 (22%)	9 (50%)
SCL	0 (0%)	5 (42%)	4 (33%)	1 (8%)	2 (17%)

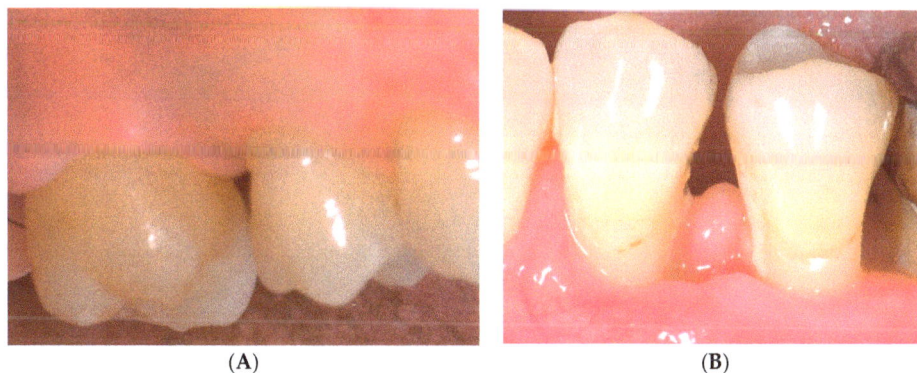

(A)   (B)

Figure 3. Soft tissue healing two weeks after surgery (EMD/BCP group): (**A**) EHI 1 and closed flap; (**B**) EHI 5 and flap dehiscence. EMD/BCP: enamel matrix derivative/biphasic calcium phosphate; EHI: early healing index.

Table 3. Contingency table for flap dehiscences (yes/no) noted two weeks after surgery. Significant difference favouring the SCL group (p = 0.024, Pearson's Chi^2 test). EMD/BCP: enamel matrix derivative/biphasic calcium phosphate; SCL: surgical crown lengthening.

	EMD/BCP	SCL
flap dehiscence	13 (72%)	3 (25%)
no dehiscence	5 (28%)	9 (75%)

2.6. GCF/WF/Cytokine Levels at Sites with or without Flap Dehiscence

Sixteen of thirty surgical sites displayed dehiscent inter-proximal flaps after two weeks, when the results for all sites of both groups were pooled. In these sites, the amount of GCF/WF was 1.5-fold elevated (p = 0.049, Figure 4A). In GCF/WF of sites with dehiscence, the total amounts/sample of IL-1β and of TNF-α were 4-fold and 2.5-fold increased, when compared to closed sites (p = 0.033 and p = 0.021, Figure 4B). Other cytokine parameters were unaffected by the course of soft tissue healing (p > 0.05, data not shown).

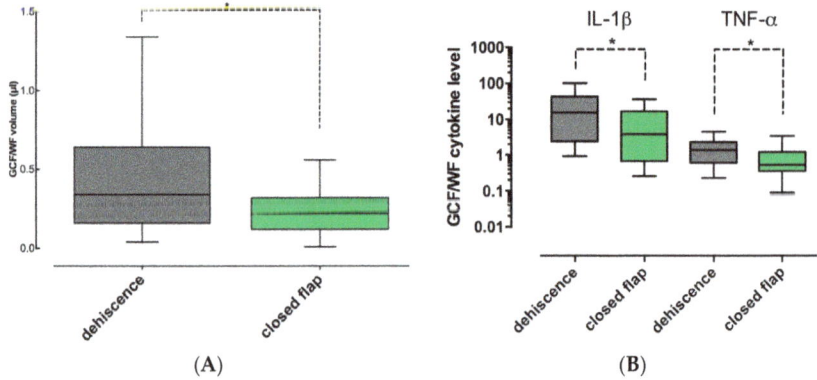

Figure 4. GCF/WF and cytokine levels in sites with/without dehiscence (pooled results for all sites/both groups): (**A**) Significantly higher GCF/WF volume were found in sites with flap dehiscence, when compared to closed flaps (*: $p < 0.05$); (**B**) Significantly higher total amounts/sample of IL-1β and TNF-α were found in sites with flap dehiscence, when compared to closed flaps (*: $p < 0.05$). Dehiscence: grey, closed flap: green.

3. Discussion

The main events of periodontal wound healing are completed within two to three weeks of wound closure, followed by tissue maturation and remodelling [4]. The aim of this descriptive study was to characterize this early phase of wound healing in terms of soft tissue closure, changes of microcirculation, and expression of pro- and anti-inflammatory cytokines.

Maintenance of inter-proximal flap closure was assessed using both the Early Healing Index (EHI) and a dichotomous classification (dehiscence yes/no). EHI values ranging from 2–3 (complete flap closure with presence of a fibrin line or fibrin clot) were found in 75% of patients treated with SCL, while 25% of SCL sites showed EHI values of 4–5 (incomplete flap closure with partial or complete inter-proximal necrosis), or dehiscence of the flap. Suchlike inter-proximal flap dehiscences despite initial primary closure occur frequently after resective periodontal surgeries and heal by second intention [24]. Healing by second intention causes formation of gingival clefts and craters, but these soft tissue deformities may even out over time and are of limited clinical relevance after resective surgery [25].

Significantly greater proportions of dehiscent flaps were found in EMD/BCP patients (72%), which is an unexpected finding. The frequency of incomplete flap closure ranged from 0%–10% in similar studies combining minimally invasive techniques with EMD [26–28], and therefore, the high proportion of inter-proximal dehiscence is surprising. Generally, local anaesthesia and flap elevation disturb microcirculation and induce ischemia [29], while maintenance of blood flow is essential for survival of the operated tissue [30]. In our study, the effects of surgery on flap perfusion were evaluated using Laser Doppler flowmetry (LDF). LDF measurements of microcirculation detect disturbances in blood flow caused by surgical trauma and are able to discriminate surgical techniques according to different extents of tissue traumatisation [31,32]. Further, LDF has been shown to predict flap dehiscence after surgery [32]. Both groups showed significant reductions of microcirculation directly after surgery and after one day, when compared to untreated control sites. However, microcirculation returned quickly to unimpaired levels, and no significant difference with regard to reduction of blood flow was found between EMD/BCP and SCL sites. Since suchlike alterations of microcirculation are normal responses to surgery, the poor outcome of soft tissue healing in the EMD/BCP group should not be attributed to insufficient blood flow caused by allegedly excessive tissue traumatisation during treatment. Instead, it should be scrutinised that the surgeon attempted to re-establish the normal convex inter-proximal bony contour by adding grafting material supra-crestally beyond the remaining margins of the bone defect. The ideal amount of granular materials for periodontal defects is not known. Enhancement of flap support at non-containing defects is an important rationale for implantation of defect fillers; however, the placement of bone substitutes and the selection of graft type is considered subordinate to primary closure and wound stability [33]. In our study, overfilling of the bony defects may have promoted flap dehiscence in EMD sites, especially since coronal advancement of the flaps had not been carried out. Consequently, clinicians applying bone substitutes during regenerative periodontal surgery should always consider resultant difficulties in maintaining primary closure of the flap

Irrespective of the treatment applied, surgery resulted in elevated volumes of GCF/WF samples and the levels of IL-1β, TNF-α, and IL-6 were markedly increased as early as after one day. The early peak of these pro-inflammatory cytokines relates to the physiologic reaction to tissue injury, the early inflammatory phase of wound healing [34]; IL-1β and TNF-α are necessary for inducement of expression of adhesion molecules and chemokines, secretion of other inflammatory mediators, and of matrix metalloproteinases [35]. Lack of these pro-inflammatory cytokines in the early phase, however, delays or even impairs wound healing [36–38]. After the initial peak, an incremental reduction of the amount of GCF/WF and the levels of IL-1β and TNF-α was found at most sites during 14 days, which may correspond to the transition between different phases of wound healing [34]. In contrast to IL-1β and TNF-α, the elevated levels of IL-6 were maintained over the entire observation period of 14 days. Levels of IL-6 are generally elevated during the early phase of cutaneous wound healing, since IL-6 has been shown to be involved in regulation of leukocyte infiltration and angiogenesis [37,39]. Unlike IL-1β, TNF-α, and IL-6, IL-10 is an immunosuppressive cytokine and regulates innate and adaptive immune responses [34]. Both surgical groups showed significant reductions of IL-10 concentrations in GCF/WF one day after surgery and the reduced levels were maintained for one week. Considering the contrasting levels of IL-1β and TNF-α, the decreased levels of Il-10 may truly reflect the first inflammatory phase of wound healing [34].

Given the well-known anti-inflammatory effects of EMD found in vitro, reduced levels of pro-inflammatory cytokines or enhanced levels of IL-10 were to be expected in GCF/WF samples of EMD/BCP sites. Interestingly, we failed to reproduce the effects of EMD found in vitro in our clinical study. This is in line with a recent report by Villa et al., who found a similar discrepancy between cytokine levels in clinical samples of wound fluid after periodontal surgery and in vitro measurements [40]. However, in contrast to "clean" in vitro assays focusing on one particular cell type such as PDL fibroblasts, osteoblasts, or keratinocytes, a wide variety of different cell types and exudates will always contribute to GCF/WF samples, which may obscure the detectable effects of biological mediators on a given cell population within the periodontal wound. Further, in the study by Villa et al. [40], conventional macro-surgical flap designs and suture materials were used instead of micro-surgical or minimally invasive techniques and materials, which may have negatively affected the parameters assessed for characterization of early wound healing. Similarly, the poor outcome of soft tissue healing in EMD/BCP sites in our study may have contributed to the lack of detectable effects of EMD on cytokine levels. Inflammation, as characterized by high levels of IL-1β, reduces the beneficial effects of EMD on wound healing: wound fill rate, cell proliferation and adhesion, synthesis of growth factors and collagen as well as mineralization were negatively affected under concomitant challenge with IL-1β in vitro [41]. In our study, a suchlike interrelation between inflammation and scarce effect of EMD may be clinically reflected by the fact that significantly higher levels of GCF/WF, IL-1β, and TNF-α were found in sites with inter-proximal flap dehiscence, despite application of EMD. Flap dehiscence always leads to bacterial colonization of the surgical site and especially of implanted materials [42,43]. Subsequently, the early physiologic inflammatory response to wounding cannot be resolved and becomes chronic due to the bacterial infection, with detrimental effects on the attempted regeneration [44]. For example, prolonged high levels of IL-1β and TNF-α impair wound healing by inhibition of collagen synthesis [45,46]. In contrast, inflammation at non-infected sites with maintained flap closure can be reduced promptly by pro-resolving mediators, and normal healing can occur [44].

In conclusion, periodontal surgery immediately increased the amount of GCF/WF and the levels of pro-inflammatory cytokines, which was reflected by contrasting effects on the anti-inflammatory cytokine IL-10. Increased measures of inflammation were temporary in case of maintained closure of the flap. However, flap dehiscence caused prolonged inflammatory exudation from the periodontal wound. If the biological pre-conditions for periodontal wound healing are considered important for the clinical outcome, care should be taken to maintain primary closure of the flap.

4. Materials and Methods

4.1. Treatment Groups

The study protocol was approved by the institutional Ethics Committee of the Witten/Herdecke University (no. 39/2011; 24 April 2011). The study was conducted in accordance with the guidelines of Good Clinical Practice (GCP-ICH) and the principles of the Helsinki Declaration of 1975, as revised in 2008. Written informed consent was obtained from all patients. Two groups of patients were treated with periodontal surgery and monitored during the first two weeks of healing:

(1) regenerative periodontal surgery of intrabony defects, using enamel matrix derivative and a granular bone substitute (EMD/BCP group).
(2) surgical crown lengthening prior to prosthetic treatment (SCL group).

4.2. Study Patients

Participants were recruited from the patient pool of the Witten/Herdecke University's Dental School (first patient in: 17 May 2011; last patient out: 27 September 2012). Patients presenting a periodontal site in need of regenerative therapy (probing depth (PD) $\geq$ 6mm, and radiographic evidence of a vertical bone defect of at least 4 mm, assessed at a supportive periodontal therapy visit three months after completion of non-surgical therapy for moderate to advanced generalized chronic periodontitis) were assigned to the EMD/BCP group. A minimal width of 2 mm of keratinized gingiva at the prospective surgical site and an inter-proximal control site considered "clinically healthy" (PD $\leq$ 3 mm, negative for bleeding on probing and suppuration) were also required.

Periodontally healthy patients (no site with PD > 3 mm, full-mouth bleeding on probing score <25%, mean full-mouth plaque score <25%) were recruited to the SCL group in case of presence of a site in need of surgical crown lengthening for prosthetic reasons (distance of a planned restoration margin to bone crest $\leq$2 mm). A minimal width of 2 mm of keratinized gingiva at the prospective surgical site and presence of a control site considered "clinically healthy" (PD < 3 mm, negative for bleeding on probing and suppuration) were also required.

Exclusion criteria included pregnancy, lactation period, use of antibiotics or anti-inflammatory drugs in the previous six months, systemic diseases and medications affecting periodontal inflammation or surgical procedures, and any condition requiring premedication before dental treatment.

4.3. Surgery

All surgeries were carried out by the same experienced surgeon (GG). Patients were advised to abstain from brushing the surgical area, and to use a 0.2% chlorhexidine mouth rinse twice daily for 1 min until removal of sutures.

4.3.1. Access Flap with EMD and Granular Bone Substitute (EMD/BCP)

The simplified papilla preservation flap technique (SPPF) [47] was used in case of an interdental width of $\leq$2 mm, as measured at the papilla base. A modified papilla preservation flap (MPPF) [48] was performed at sites presenting with an interdental width of >2 mm. Vertical releasing incisions were avoided. After flap elevation, thorough debridement of the root surface, and degranulation, the defect was rinsed with sterile saline. A 24% ethylenediamine tetraacetic acid gel (PrefGel, Institut Straumann AG, Basel, Switzerland) was applied to the exposed root surface and removed after 2 min using saline. Then, EMD (Emdogain, Institut Straumann AG, Basel, Switzerland) was administered to the root surface. The bone defect was filled with a granular bone substitute (biphasic calcium phosphate (BCP); Bone Ceramic, Institut Straumann AG, Basel, Switzerland) mixed with EMD, using a sterile amalgam gun. BCP was not only applied into the bone defect for flap support, but also used for vertical augmentation beyond the remaining defect margins in order to compensate for both vertical and horizontal bone loss and to re-establish the physiologic convex inter-proximal bone contour (Figure 5A,B). Then, the flaps were closed with modified vertical mattress sutures, using fine monofilament suture material (6.0). Sutures were removed after two weeks.

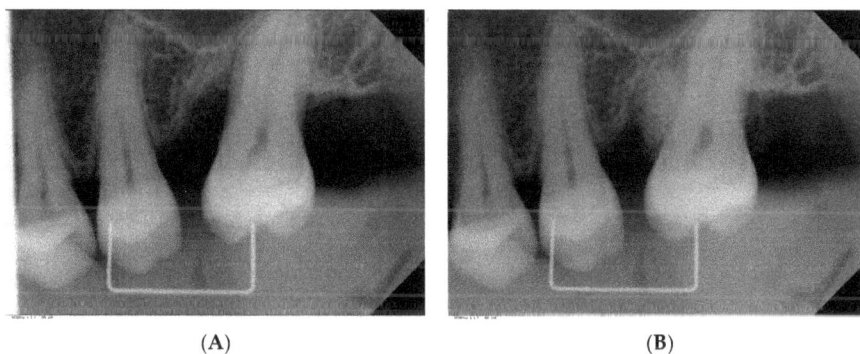

(A)   (B)

Figure 5. Pre- and post-surgical radiographs (EMD/BCP group). (**A**) Deep vertical infra-bony defect is visible on the mesial aspect of tooth 27; (**B**) The bone defect is filled with radio-opaque bone substitute (BCP) beyond the residual contour of the alveolar crest.

4.3.2. Surgical Crown Lengthening (SCL)

After placement of bevelled internal and sulcular incisions, a mucoperiosteal flap was reflected and the bone was exposed. Vertical releasing incisions were avoided. Ostectomy and osteoplasty were carried out with burs and hand instruments in order to establish a distance of up to 3 mm from the projected restoration margin to the new bony crest, as described previously [2,49]. Then, the flaps were closed with vertical mattress sutures with periosteal anchorage for apical repositioning, using fine monofilament suture material (6.0). Sutures were removed after two weeks.

4.4. Assessment of Microcirculation

A laser Doppler flowmeter (Periflux 5010, Perimed AB, Jarfalla, Sweden) equipped with a PF 416 probe (outside diameter 1.0 mm, fibre separation 0.25 mm; wavelength 780 nm) was used for assessing changes of microcirculation at surgical and control sites before surgery (D0a), directly after completion of surgery (D0b), and one (D1), three (D3), seven (D7), and 14 days (D14) after surgery. The LDF probe was aligned at a custom-made acrylic stent and the measurements were carried out always at the same position (base of the buccal papilla) for 1 min, perpendicular to the tissue and at a distance of 0.5 mm to the flap. The flowmeter recordings were monitored using the Perisoft software (Perisoft 2.10, Perimed AB, Jarfalla, Sweden), measured in perfusion units (PU), and analysed as changes relative to the baseline value defined as zero.

4.5. Sampling of Gingival Crevicular Fluid/Wound Fluid (GCF/WF)

Samples of gingival crevicular fluid or wound fluid (GCF/WF) were harvested at one surgical and one healthy control site/patient before surgery (D0a), and after one (D1), three (D3), seven (D7), and 14 days (D14). For sampling, sites were isolated with cotton rolls and gently air-dried. A paper strip (Perio paper, Oraflow, Amityville, NY, USA) was inserted into the site until mild resistance was felt and left in place for 30 s. The GCF/WF sample volume was immediately determined with a micro-moisture meter (Periotron 8000, Oraflow, Amityville, NY, USA) and calculated in microliters from a standard curve. Samples were stored at −80 °C in dry vials until further processing.

4.6. Determination of GCF/WF Cytokine Levels

The GCF/WF levels of cytokines (IL-1β, TNF-α, IL-6, IL-10) were determined with a multiplex immunoassay (Magpix, Luminex, Austin, TX, USA) following the manufacturer's instructions. The samples were thawed in assay buffer, shaken for 1 min with a vortex mixer, and centrifuged for 5 min at $3000 \times g$ for recovery of GCF. Data are presented in pg/mL for concentration and in pg/sample for total amount of cytokine per sample.

4.7. Assessment of Clinical Soft Tissue Healing

Pictures were taken from all surgical sites two weeks after surgery (D14). Using the photographs, wound closure was independently evaluated by two calibrated examiners blinded to the group allocation, using the Early Healing Index (EHI; 1: complete flap closure—no fibrin line in the inter-proximal area; 2: complete flap closure—fine fibrin line in the inter-proximal area; 3: complete flap closure—fibrin clot in the inter-proximal area; 4: incomplete flap closure—partial necrosis of the inter-proximal tissue; 5: incomplete flap closure—complete necrosis of the inter-proximal tissue) [28]; in addition, wound closure was assessed dichotomously (soft tissue dehiscence yes/no).

4.8. Statistical Analysis

Statistics were calculated with the patient as the unit of analysis (split-mouth design, using one surgical site and one control site/patient). Non-parametric statistical tests were used according to the non-normal data distribution (confirmed with Kolmogorov-Smirnov test, data not shown). Medians and inter-quartile ranges (IQ) were calculated for metric parameters (LDF, GCF/WF volume, cytokine concentration, and total amount/sample). Wilcoxon's signed-rank test was used for longitudinal comparisons of repeated measurements within the groups. Cross-sectional analyses were carried out with the Mann–Whitney U test. Categorical data were analysed using Pearson's Chi^2 test. A statistical software program was used for all calculations (SPSS 22 for OSX, SPSS Inc., Chicago, IL, USA). Statistical significance was defined as $p < 0.05$.

Acknowledgments: The study was supported by Institut Straumann AG, Basel, Switzerland. The Open Access publication fee was covered by the Witten/Herdecke University's Open Access Publication Fund.

Author Contributions: Anton Friedmann and Georg Gaßmann conceived and designed the study protocol; Georg Gaßmann performed all surgeries; Mouaz Soudan performed all sampling and clinical measurement procedures; Han Zhao and Anna Schönhauser conducted the multiplex assays; Doğan Kaner and Anton Friedmann analysed the data; Doğan Kaner wrote the paper.

Conflicts of Interest: The authors declare no conflict of interest. The funding sponsors had no role in the design of the study; in the collection, analyses, or interpretation of data; in the writing of the manuscript, and in the decision to publish the results.

References

1. Heitz-Mayfield, L.J.; Lang, N.P. Surgical and nonsurgical periodontal therapy. Learned and unlearned concepts. *Periodontol. 2000* **2013**, *62*, 218–231. [CrossRef] [PubMed]
2. Bragger, U.; Lauchenauer, D.; Lang, N.P. Surgical lengthening of the clinical crown. *J. Clin. Periodontol.* **1992**, *19*, 58–63. [CrossRef] [PubMed]

3. Wikesjo, U.M.; Crigger, M.; Nilveus, R.; Selvig, K.A. Early healing events at the dentin-connective tissue interface. Light and transmission electron microscopy observations. *J. Periodontol.* **1991**, *62*, 5–14. [CrossRef] [PubMed]

4. Susin, C.; Fiorini, T.; Lee, J.; de Stefano, J.A.; Dickinson, D.P.; Wikesjo, U.M. Wound healing following surgical and regenerative periodontal therapy. *Periodontol. 2000* **2015**, *68*, 83–98. [CrossRef] [PubMed]

5. Melcher, A.H. On the repair potential of periodontal tissues. *J. Periodontol.* **1976**, *47*, 256–260. [CrossRef] [PubMed]

6. Yukna, R.A. A clinical and histologic study of healing following the excisional new attachment procedure in rhesus monkeys. *J. Periodontol.* **1976**, *47*, 701–709. [CrossRef] [PubMed]

7. Listgarten, M.A.; Rosenberg, M.M. Histological study of repair following new attachment procedures in human periodontal lesions. *J. Periodontol.* **1979**, *50*, 333–344. [CrossRef] [PubMed]

8. Sculean, A.; Donos, N.; Windisch, P.; Brecx, M.; Gera, I.; Reich, E.; Karring, T. Healing of human intrabony defects following treatment with enamel matrix proteins or guided tissue regeneration. *J. Periodontal. Res.* **1999**, *34*, 310–322. [CrossRef] [PubMed]

9. Heijl, L.; Heden, G.; Svardstrom, G.; Ostgren, A. Enamel matrix derivative (emdogain) in the treatment of intrabony periodontal defects. *J. Clin. Periodontol.* **1997**, *24*, 705–714. [CrossRef] [PubMed]

10. Meyle, J.; Hoffmann, T.; Topoll, H.; Heinz, B.; Al-Machot, E.; Jervoe-Storm, P.M.; Meiss, C.; Eickholz, P.; Jepsen, S. A multi-centre randomized controlled clinical trial on the treatment of intra-bony defects with enamel matrix derivatives/synthetic bone graft or enamel matrix derivatives alone: Results after 12 months. *J. Clin. Periodontol.* **2011**, *38*, 652–660. [CrossRef] [PubMed]

11. Ashkenazi, M.; Shaked, I. In vitro clonogenic capacity of periodontal ligament fibroblasts cultured with emdogain. *Dent. Traumatol.* **2006**, *22*, 25–29. [CrossRef] [PubMed]

12. Cattaneo, V.; Rota, C.; Silvestri, M.; Piacentini, C.; Forlino, A.; Gallanti, A.; Rasperini, G.; Cetta, G. Effect of enamel matrix derivative on human periodontal fibroblasts: Proliferation, morphology and root surface colonization. An in vitro study. *J. Periodontal. Res.* **2003**, *38*, 568–574. [CrossRef] [PubMed]

13. Berry, J.E.; Zhao, M.; Jin, Q.; Foster, B.L.; Viswanathan, H.; Somerman, M.J. Exploring the origins of cementoblasts and their trigger factors. *Connect. Tissue Res.* **2003**, *44*, 97–102. [CrossRef] [PubMed]

14. Kawase, T.; Okuda, K.; Yoshie, H.; Burns, D.M. Anti-TGF-β antibody blocks enamel matrix derivative-induced upregulation of p21WAF1/CIP1 and prevents its inhibition of human oral epithelial cell proliferation. *J. Periodontal. Res.* **2002**, *37*, 255–262. [CrossRef] [PubMed]

15. Heitz-Mayfield, L.; Tonetti, M.S.; Cortellini, P.; Lang, N.P. Microbial colonization patterns predict the outcomes of surgical treatment of intrabony defects. *J. Clin. Periodontol.* **2006**, *33*, 62–68. [CrossRef] [PubMed]

16. Walter, C.; Jawor, P.; Bernimoulin, J.P.; Hagewald, S. Moderate effect of enamel matrix derivative (emdogain gel) on porphyromonas gingivalis growth in vitro. *Arch. Oral. Biol.* **2006**, *51*, 171–176. [CrossRef] [PubMed]

17. Spahr, A.; Lyngstadaas, S.P.; Boeckh, C.; Andersson, C.; Podbielski, A.; Haller, B. Effect of the enamel matrix derivative emdogain on the growth of periodontal pathogens in vitro. *J. Clin. Periodontol.* **2002**, *29*, 62–72. [CrossRef] [PubMed]

18. Kantarci, A.; Hasturk, H.; van Dyke, T.E. Host-mediated resolution of inflammation in periodontal diseases. *Periodontol. 2000* **2006**, *40*, 144–163. [CrossRef] [PubMed]

19. De Waal Malefyt, R.; Abrams, J.; Bennett, B.; Figdor, C.G.; de Vries, J.E. Interleukin 10(IL-10) inhibits cytokine synthesis by human monocytes: An autoregulatory role of IL-10 produced by monocytes. *J. Exp. Med.* **1991**, *174*, 1209–1220. [CrossRef] [PubMed]

20. Myhre, A.E.; Lyngstadaas, S.P.; Dahle, M.K.; Stuestol, J.F.; Foster, S.J.; Thiemermann, C.; Lilleaasen, P.; Wang, J.E.; Aasen, A.O. Anti-inflammatory properties of enamel matrix derivative in human blood. *J. Periodontal. Res.* **2006**, *41*, 208–213. [CrossRef] [PubMed]

21. Fujishiro, N.; Anan, H.; Hamachi, T.; Maeda, K. The role of macrophages in the periodontal regeneration using emdogain gel. *J. Periodontal. Res.* **2008**, *43*, 143–155. [CrossRef] [PubMed]

22. Gassmann, G.; Schwenk, B.; Entschladen, F.; Grimm, W.D. Influence of enamel matrix derivative on primary CD4+ t-helper lymphocyte migration, cd25 activation, and apoptosis. *J. Periodontol.* **2009**, *80*, 1524–1533. [CrossRef] [PubMed]

23. Dickinson, D.P.; Coleman, B.G.; Batrice, N.; Lee, J.; Koli, K.; Pennington, C.; Susin, C.; Wikesjo, U.M. Events of wound healing/regeneration in the canine supraalveolar periodontal defect model. *J. Clin. Periodontol.* **2013**, *40*, 527–541. [CrossRef] [PubMed]

24. Takei, H.H.; Han, T.J.; Carranza, F.A., Jr.; Kenney, E.B.; Lekovic, V. Flap technique for periodontal bone implants. Papilla preservation technique. *J. Periodontol.* **1985**, *56*, 204–210. [CrossRef] [PubMed]
25. Jenkins, W.M.; Wragg, P.F.; Gilmour, W.H. Formation of interdental soft tissue defects after surgical treatment of periodontitis. *J. Periodontol.* **1990**, *61*, 564–570. [CrossRef] [PubMed]
26. Al Machot, E.; Hoffmann, T.; Lorenz, K.; Khalili, I.; Noack, B. Clinical outcomes after treatment of periodontal intrabony defects with nanocrystalline hydroxyapatite (ostim) or enamel matrix derivatives (emdogain): A randomized controlled clinical trial. *BioMed Res. Int.* **2014**, *2014*, 786353. [CrossRef] [PubMed]
27. Jepsen, S.; Topoll, H.; Rengers, H.; Heinz, B.; Teich, M.; Hoffmann, T.; Al-Machot, E.; Meyle, J.; Jervoe-Storm, P.M. Clinical outcomes after treatment of intra-bony defects with an EMD/synthetic bone graft or EMD alone: A multicentre randomized-controlled clinical trial. *J. Clin. Periodontol.* **2008**, *35*, 420–428. [CrossRef] [PubMed]
28. Wachtel, H.; Schenk, G.; Bohm, S.; Weng, D.; Zuhr, O.; Hurzeler, M.B. Microsurgical access flap and enamel matrix derivative for the treatment of periodontal intrabony defects: A controlled clinical study. *J. Clin. Periodontol.* **2003**, *30*, 496–504. [CrossRef] [PubMed]
29. McLean, T.N.; Smith, B.A.; Morrison, E.C.; Nasjleti, C.E.; Caffesse, R.G. Vascular changes following mucoperiosteal flap surgery: A fluorescein angiography study in dogs. *J. Periodontol.* **1995**, *66*, 205–210. [CrossRef] [PubMed]
30. Nakayama, Y.; Soeda, S.; Kasai, Y. The importance of arterial inflow in the distal side of a flap: An experimental investigation. *Plast. Reconstr. Surg.* **1982**, *69*, 61–67. [CrossRef] [PubMed]
31. Retzepi, M.; Tonetti, M.; Donos, N. Comparison of gingival blood flow during healing of simplified papilla preservation and modified widman flap surgery: A clinical trial using laser doppler flowmetry. *J. Clin. Periodontol.* **2007**, *34*, 903–911. [CrossRef] [PubMed]
32. Kaner, D.; Zhao, H.; Terheyden, H.; Friedmann, A. Improvement of microcirculation and wound healing in vertical ridge augmentation after pre-treatment with self-inflating soft tissue expanders—A randomized study in dogs. *Clin. Oral Implants Res.* **2015**, *26*, 720–724. [CrossRef] [PubMed]
33. Cortellini, P.; Tonetti, M.S. Clinical concepts for regenerative therapy in intrabony defects. *Periodontol. 2000* **2015**, *68*, 282–307. [CrossRef] [PubMed]
34. Eming, S.A.; Krieg, T.; Davidson, J.M. Inflammation in wound repair: Molecular and cellular mechanisms. *J. Investig. Dermatol.* **2007**, *127*, 514–525. [CrossRef] [PubMed]
35. Morand, D.N.; Davideau, J.L.; Clauss, F.; Jessel, N.; Tenenbaum, H.; Huck, O. Cytokines during periodontal wound healing: Potential application for new therapeutic approach. *Oral Dis.* **2016**, *12469*. [CrossRef] [PubMed]
36. Graves, D.T.; Nooh, N.; Gillen, T.; Davey, M.; Patel, S.; Cottrell, D.; Amar, S. IL-1 plays a critical role in oral, but not dermal, wound healing. *J. Immun.* **2001**, *167*, 5316–5320. [CrossRef] [PubMed]
37. Lin, Z.Q. Essential involvement of IL-6 in the skin wound-healing process as evidenced by delayed wound healing in IL-6-deficient mice. *J. Leukoc. Biol.* **2003**, *73*, 713–721. [CrossRef] [PubMed]
38. Heo, S.C.; Jeon, E.S.; Lee, I.H.; Kim, H.S.; Kim, M.B.; Kim, J.H. Tumor necrosis factor-α-activated human adipose tissue-derived mesenchymal stem cells accelerate cutaneous wound healing through paracrine mechanisms. *J. Investig. Dermatol.* **2011**, *131*, 1559–1567. [CrossRef] [PubMed]
39. Gallucci, R.M.; Simeonova, P.P.; Matheson, J.M.; Kommineni, C.; Guriel, J.L.; Sugawara, T.; Luster, M.I. Impaired cutaneous wound healing in interleukin-6-deficient and immunosuppressed mice. *FASEB J.* **2000**, *14*, 2525–2531. [CrossRef] [PubMed]
40. Villa, O.; Wohlfahrt, J.C.; Koldsland, O.C.; Brookes, S.J.; Lyngstadaas, S.P.; Aass, A.M.; Reseland, J.E. EMD in periodontal regenerative surgery modulates cytokine profiles: A randomised controlled clinical trial. *Sci. Rep.* **2016**, *6*, 23060. [CrossRef] [PubMed]
41. Nokhbehsaim, M.; Winter, J.; Rath, B.; Jager, A.; Jepsen, S.; Deschner, J. Effects of enamel matrix derivative on periodontal wound healing in an inflammatory environment in vitro. *J. Clin. Periodontol.* **2011**, *38*, 479–490. [CrossRef] [PubMed]
42. Costerton, J.W.; Montanaro, L.; Arciola, C.R. Biofilm in implant infections: Its production and regulation. *Int. J. Artif. Organs* **2005**, *28*, 1062–1068. [PubMed]
43. Sela, M.N.; Steinberg, D.; Klinger, A.; Krausz, A.A.; Kohavi, D. Adherence of periodontopathic bacteria to bioabsorbable and non-absorbable barrier membranes in vitro. *Clin. Oral Implants Res.* **1999**, *10*, 445–452. [CrossRef] [PubMed]

44. Thomas, M.V.; Puleo, D.A. Infection, inflammation, and bone regeneration: A paradoxical relationship. *J. Dent. Res.* **2011**, *90*, 1052–1061. [CrossRef] [PubMed]
45. Harrison, J.R.; Kleinert, L.M.; Kelly, P.L.; Krebsbach, P.H.; Woody, C.; Clark, S.; Rowe, D.W.; Lichtler, A.C.; Kream, B.E. Interleukin-1 represses colia1 promoter activity in calvarial bones of transgenic colcat mice in vitro and in vivo. *J. Bone Miner. Res.* **1998**, *13*, 1076–1083. [CrossRef] [PubMed]
46. Rapala, K. The effect of tumor necrosis factor-α on wound healing. An experimental study. *Ann. Chir. Gynaecol. Suppl.* **1996**, *211*, 1–53. [PubMed]
47. Cortellini, P.; Pini-Prato, G.; Tonetti, M.S. The simplified papilla preservation flap. A novel surgical approach for the management of soft tissues in regenerative procedures. *Int. J. Periodontics Restor. Dent.* **1999**, *19*, 589–599.
48. Cortellini, P.; Pini-Prato, G.; Tonetti, M.S. The modified papilla preservation technique. A new surgical approach for interproximal regenerative procedures. *J. Periodontol.* **1995**, *66*, 261–266. [CrossRef] [PubMed]
49. Ingber, J.S.; Rose, L.F.; Coslet, J.G. The "biologic width"—A concept in periodontics and restorative dentistry. *Alpha Omegan* **1977**, *70*, 62–65. [PubMed]

© 2017 by the authors. Licensee MDPI, Basel, Switzerland. This article is an open access article distributed under the terms and conditions of the Creative Commons Attribution (CC BY) license (http://creativecommons.org/licenses/by/4.0/).

International Journal of
Molecular Sciences

MDPI

Article

Effects of Remote Ischemic Preconditioning on Heme Oxygenase-1 Expression and Cutaneous Wound Repair

Niels A. J. Cremers [1,2,4], Kimberley E. Wever [3,5,†], Ronald J. Wong [6,†], René E. M. van Rheden [1,4],
Eline A. Vermeij [2,4], Gooitzen M. van Dam [7], Carine E. Carels [1,4,8,9], Ditte M. S. Lundvig [1,4]
and Frank A. D. T. G. Wagener [1,4,*]

[1]   Department of Orthodontics and Craniofacial Biology, Radboud University Medical Center,
      Nijmegen 6500HB, The Netherlands; niels.cremers@radboudumc.nl (N.A.J.C.);
      rene.vanrheden@radboudumc.nl (R.E.M.v.R.); carine.carels@radboudumc.nl (C.E.C.);
      dittelundvig@hotmail.com (D.M.S.L.)
[2]   Department of Rheumatology, Radboud University Medical Center, Nijmegen 6500HB, The Netherlands;
      eline.vermeij@radboudumc.nl
[3]   Central Animal Laboratory, Radboud University Medical Center, Nijmegen 6500HB, The Netherlands;
      kim.wever@radboudumc.nl
[4]   Radboud Institute for Molecular Life Sciences, Nijmegen 6500HB, The Netherlands
[5]   Radboud Institute for Health Sciences, Nijmegen 6500HB, The Netherlands
[6]   Department of Pediatrics, Stanford University School of Medicine, Stanford, CA 94305, USA;
      rjwong@stanford.edu
[7]   Department of Surgery, University Medical Center Groningen, Groningen 9700RB, The Netherlands;
      g.m.van.dam@umcg.nl
[8]   Department of Human Genetics, Radboud University Medical Center, Nijmegen 6500HB, The Netherlands
[9]   Department of Oral Health Sciences, Faculty of Medicine, KU Leuven, 3000 Leuven, Belgium
*   Correspondence: frank.wagener@radboudumc.nl; Tel.: +31-24-3614082
†   These authors contributed equally to this work.

Academic Editor: Allison Cowin
Received: 20 December 2016; Accepted: 13 February 2017; Published: 17 February 2017

Abstract: Skin wounds may lead to scar formation and impaired functionality. Remote ischemic preconditioning (RIPC) can induce the anti-inflammatory enzyme heme oxygenase-1 (HO-1) and protect against tissue injury. We aim to improve cutaneous wound repair by RIPC treatment via induction of HO-1. RIPC was applied to HO-1-*luc* transgenic mice and HO-1 promoter activity and mRNA expression in skin and several other organs were determined in real-time. In parallel, RIPC was applied directly or 24h prior to excisional wounding in mice to investigate the early and late protective effects of RIPC on cutaneous wound repair, respectively. HO-1 promoter activity was significantly induced on the dorsal side and locally in the kidneys following RIPC treatment. Next, we investigated the origin of this RIPC-induced HO-1 promoter activity and demonstrated increased mRNA in the ligated muscle, heart and kidneys, but not in the skin. RIPC did not change HO-1 mRNA and protein levels in the wound 7 days after cutaneous injury. Both early and late RIPC did not accelerate wound closure nor affect collagen deposition. RIPC induces HO-1 expression in several organs, but not the skin, and did not improve excisional wound repair, suggesting that the skin is insensitive to RIPC-mediated protection.

Keywords: remote ischemic preconditioning; heme oxygenase-1; tissue injury; wound repair

1. Introduction

Severe skin wounds following burns, trauma, or surgery often lead to scar formation and impaired functionality [1]. Cutaneous wound repair is a dynamic and highly regulated process, involving several overlapping phases: inflammation, proliferation, and remodelling [2]. Aberrant wound repair and scarring occurs following prolongation of the inflammatory phase that together with oxidative stress fuels (myo)fibroblast proliferation and interferes with myofibroblast apoptosis [2]. This leads to excessive deposition of extracellular matrix proteins, subsequently promoting excessive scar formation [3,4]. Unfortunately, conventional therapies to accelerate wound repair and to prevent scarring are insufficient [5–7]. Therefore, adjuvant therapies aimed at resolving inflammation are warranted. Pharmacological preconditioning has been shown to improve wound repair, as exemplified by heme and curcumin that also induce the cytoprotective protein heme oxygenase-1 (HO-1) [8–14]. HO-1 is one of the most important enzymes protecting against oxidative and inflammatory insults [15]. HO catabolizes heme to biliverdin, free iron, and carbon monoxide (CO). Biliverdin is then rapidly converted to the antioxidant bilirubin by biliverdin reductase [16,17]. The iron scavenger ferritin is co-induced by HO-1 and renders iron inactive [18]. Recent studies have shown that induction of HO-1 expression attenuates the inflammatory response and accelerates wound healing in HO-1-deficient mice; whereas, decreased HO activity in mice results in slower cutaneous wound closure [9,19]. In addition, intraperitoneal administration of the HO-effector molecule bilirubin accelerates wound repair [20]. Since increased HO-1 expression improves wound repair, its induction may be a good candidate for preventing aberrant cutaneous wound repair.

A promising novel preconditioning strategy is ischemic preconditioning (IPC), hereby, short cycles of ischemia/reperfusion to an organ protects against subsequent more harmful insults to the same organ. In remote ischemic preconditioning (RIPC), the target organ is not subjected to the initial stress, but a remote organ, e.g., the hind limb, is exposed. [21,22]. Interestingly, RIPC protects against injury in the liver [23–26], lung [27], intestines [28], heart [29,30], and kidneys [31,32] often via the induction of HO-1, since inhibition of HO-activity abrogates the protective effects of RIPC [23,26,33]. Following RIPC, there exists both a rapid phase of protection initiated within 1h after preconditioning, and a later phase after one to several days [21,34]. In addition, different modes of action have been reported between single and repeated RIPC procedures, as demonstrated by differential expression of genes involved in autophagy, endoplasmic reticulum stress, mitochondrial oxidative metabolism, and cell survival [35,36].

Successful translation towards its clinical use was recently established by inducing temporary occlusion and restoration of blood flow in arm or thigh of patients [29,30,37]. Patient outcome after myocardial surgery was significantly improved when RIPC was applied before surgery [27]. However, recently conflicting results have been reported showing that RIPC does not always mediate protection [38–41]. Data from animal and human studies demonstrated the need for careful interpretation because of translational differences [38–43]. RIPC improves microcirculation by an increase in tissue oxygenation and capillary blood flow in the skin [44] and skin flaps [45], and forms a novel target for skin flap transplantation [46] and the healing of diabetic foot ulcers [47,48]. Although RIPC has been shown to protect in several models of tissue injury, its role in cutaneous excisional wound healing is still unclear. We postulated that RIPC induces HO-1 expression and improves skin repair following excisional skin injury.

2. Results

2.1. Effects of RIPC on HO-1 Promoter Activity and HO-1 mRNA Expression in Mice

RIPC can induce HO-1 expression in different organs. In order to evaluate if RIPC can induce HO-1 expression, we used a combination of HO-1 promoter activity and HO-1 mRNA analyses in HO-1-*luc* transgenic (Tg) mice. We previously demonstrated that treatment with cadmium chloride (CdCl$_2$) potently induced HO-1 promoter activity in the liver and kidney using the HO-1-*luc* Tg model [49]. To validate the RIPC model, we corroborated that blood flow was indeed hampered after applying elastic rings (Figure A1). After RIPC treatment, HO-1 promoter activity was measured using the In Vivo Imaging System at 1, 6, and 24 h. Measurements of HO-1 promoter activity were acquired at the dorsal aspect of each mouse (Figure 1a). Because of variations in HO-1 promoter activity, each mouse served as its own control. Relative HO-1 promoter activity of the dorsal side of the mice after RIPC treatment is shown over time (Figure 1b). We found a significant increase in HO-1 promoter activity after 6 and 24 h of RIPC treatment compared to 1 h after RIPC. HO-1 promoter activity was strongly observed in the renal area, suggesting RIPC induced HO-1 promoter activity in an organ-specific manner 6 and 24 h after RIPC treatment when compared to 1 h after RIPC treatment (Figure 1c).

To discriminate whether the skin or underlying organs were responsible for the increase in HO-1 promoter activity, and to test whether RIPC induced HO-1 expression in a tissue-specific manner, we measured HO-1 mRNA expression levels in the skin and several organs 1, 6 and 24 h after RIPC using RT-PCR and then compared these results to untreated controls. First, HO-1 mRNA expression in the hind limb muscles that had been exposed to repeated ischemia/ reperfusion cycles were measured and significantly increased 6 h after RIPC treatment when compared to untreated controls and other time points after RIPC (Figure 1d). We then analyzed the effects of RIPC on HO-1 mRNA levels in remote organs like kidney, heart, and skin. We also observed a significant increase in HO-1 mRNA expression in the kidney 6h after RIPC treatment when compared to untreated controls and other time points after RIPC (Figure 1e). Similarly, in the heart, HO-1 mRNA levels significantly increased 6h after RIPC treatment when compared to untreated controls (Figure 1f). No significant induction was found at other time points compared to untreated controls. After 24 h, HO-1 mRNA levels returned to control levels. However, no significant induction in HO-1 mRNA expression was detected in the skin at any time point (Figure 1g). In conclusion, we demonstrated a tissue- and time-specific induction of HO-1 promoter activity and HO-1 mRNA expression after RIPC treatment.

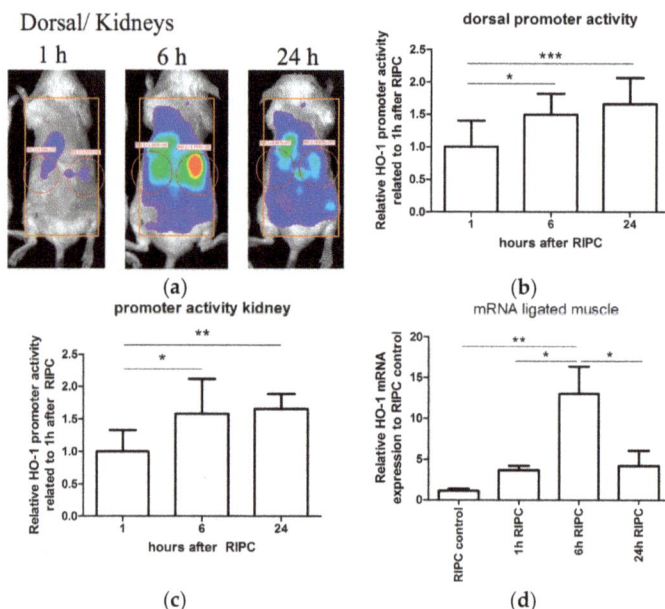

Figure 1. *Cont.*

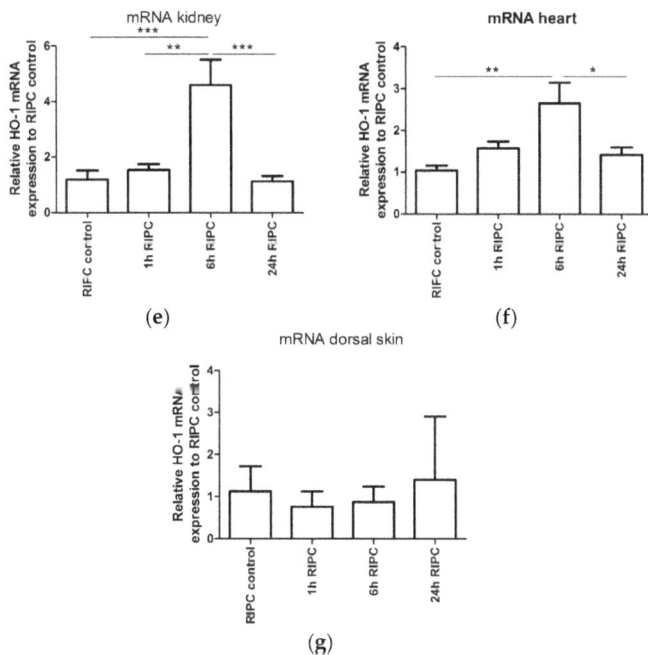

Figure 1. Heme oxygenase-1 (HO-1) promoter activity (**a–b**) and mRNA expression (**b–g**) after remote ischemic preconditioning (RIPC) treatment. (**a**) Representative dorsal images of HO-1 promoter activity after RIPC over time. Both the overall dorsal side (inserted orange rectangles) and the specific regions of the kidneys (inserted red circles) were analyzed and the total flux of emitted photons per second was quantified. Quantification of HO-1 promoter activity in the overall dorsal area after RIPC treatment (**b**); and locally in the regions of the kidneys (**c**), and HO-1 mRNA expression in muscle at the place of ligation (**d**); kidney (**e**); heart (**f**); and dorsal skin (**g**), 1, 6 and 24 h after RIPC treatment compared to untreated controls ($n = 6$ animals per group). Data are expressed as mean $\pm$ SD of six individual mice. * $p < 0.05$, ** $p < 0.01$, *** $p < 0.001$.

2.2. Effects of Early or Late RIPC on Excisional Cutaneous Wound Closure in Mice

Since RIPC was observed to improve the cutaneous microcirculation [44], we next investigated whether RIPC could also modulate cutaneous wound repair. RIPC induced HO-1 expression in several organs, but not in the skin. Interestingly, HO-1 can also promote regeneration in a paracrine fashion via its versatile effector molecules biliverdin/bilirubin, CO, and ferritin [20,50–54], which are increased following (R)IPC [55–58]. Moreover, RIPC-mediated protection can act via various alternative signaling pathways, including humoral, neuronal, and systemic mechanisms [22,59]. Since there are early and late protective effects of RIPC, we evaluated whether early (5 min) and/or late (24 h) RIPC treatment before wounding improved full-thickness excisional wound closure. Examples of untreated excisional wounds are shown in Figure 2a. Wound sizes were normalized to the wound size at day 0 (Figure 2b). As expected, after quantification of the wound surface area, we found a reduction in the wound size over time. However, no significant differences were observed in wound closures between early or late RIPC treatment mice and controls.

(a)

(b)

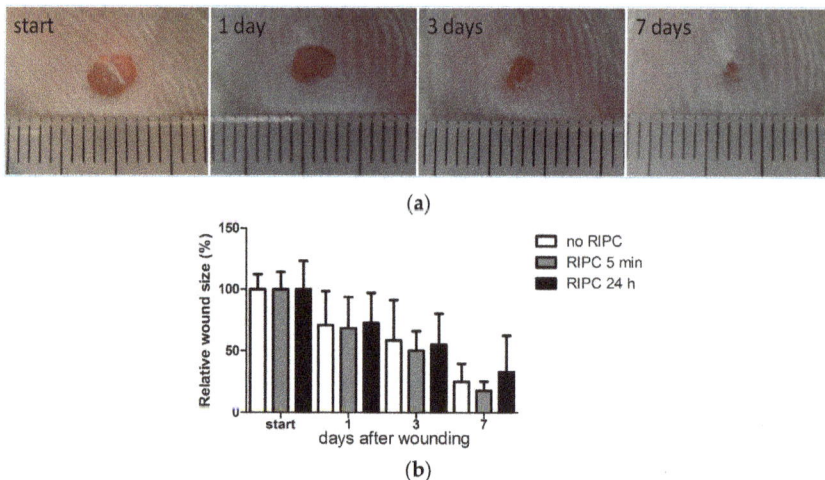

Figure 2. Excisional wound closure in time after RIPC treatment 24 h and directly after wounding compared to the control group. Representative images of the wounds of a single mouse receiving no RIPC treatment at day 0, and at 1, 3, and 7 days after wounding (ruler is incorporated in the pictures and each bar represent 1 mm) (**a**) and their relative wound sizes after different treatments in time, compared to control group at day 0 (**b**). Data are expressed as mean $\pm$ SD. No significant differences were observed between the different groups: no RIPC ($n = 8$), RIPC 5 min ($n = 6$) and RIPC 24 h ($n = 6$).

2.3. HO-1 mRNA and Protein Expression in Wounds After RIPC

To further elucidate the role of RIPC-induced HO-1 expression during wound repair, we investigated whether RIPC modulates HO-1 mRNA and protein expression in day-7 wounds. Using RT-PCR, HO-1 mRNA expression was assessed in the wounds and compared to non-wounded day-0 skin (Figure 3a). Here, we found no significant differences between the wounds and their corresponding control skins for both RIPC-treated and control groups as well as between the different treatment groups.

HO-1 protein was found in the epithelial cells of the epidermis and in recruited leukocytes in the dermis (Figure 3b). In the epidermis, HO-1-positive cells were clustered in the re-epithelialized tissue underneath the wound crust and were likely newly-formed keratinocytes [60,61]. In the dermis, HO-1-positive cells in inflamed tissues were individually spread, and based on their location and morphology, they appear to be macrophages [60–63]. Moreover, in unpublished data from a previous experiment on excisional wound healing in C57Bl/6 mice at day-2 post-wounding, fluorescent staining for HO-1 (red) and F4/80 (green) clearly showed co-localization (orange) of HO-1 and macrophages in a majority of cells (Figure A2).

The wounds were scored for the levels of HO-1 expression in the epidermal and dermal regions, and compared between the different treatment groups (Figure 3c). RIPC treatment did not modulate HO-1 protein expression in either region of day-7 wounds when compared to controls. Variations in HO-1 expression was found between animals, but was independent of RIPC treatment. In summary, RIPC treatment does not appear to alter HO-1 mRNA and protein expression in day-7 wounds.

(a)

(b)

(c)

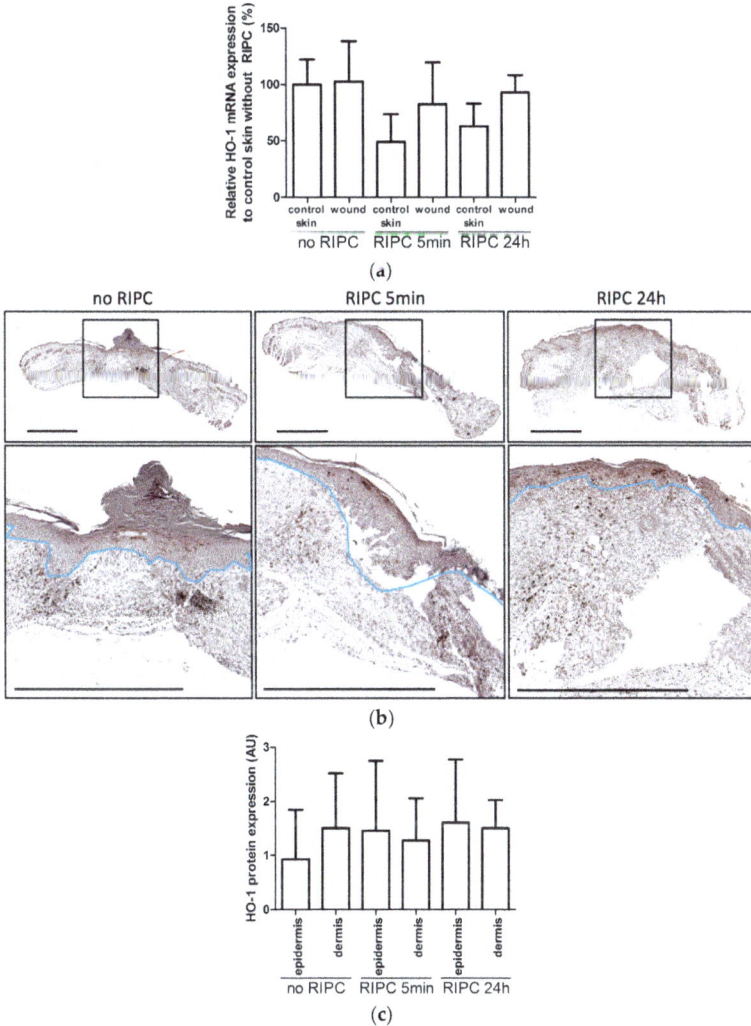

Figure 3. HO-1 expression in wounds. (**a**). HO-1 mRNA expression in unwounded (control) skin at day 0 and wounds after 7 days for the different treatments compared to control skin; (**b**) HO-1 protein expression in control, and early and late RIPC-treated wounds after 7 days of healing. Region above the marked blue line is the epidermis and underneath the blue line is the dermal layer (bars represent 1 mm); (**c**) Scored HO-1 protein staining in epidermis and dermis of the wounds after 7 days in arbitrary units (AU). Data are expressed as mean ± SD. No significant differences were observed between the different groups: no RIPC ($n = 8$), RIPC 5 min ($n = 6$) and RIPC 24 h ($n = 6$).

2.4. Effects of RIPC on Wound Morphology and Collagen Deposition

To determine if RIPC modulates other processes during wound repair, we performed H&E staining to examine wound morphology, and AZAN staining to investigate the effects on collagen deposition.

H&E staining revealed that the wound area could be easily distinguished from non-injured skin by a disruption of the epidermis, subcutaneous fat, and muscle layers. Figure 4a shows H&E staining of day-7 wounds from mice treated with early and late RIPC, and control mice (left). At the surface of the wound, the re-epithelialized tissue had marked epithelial hyperplasia under the crust of the wound. More distally, the highly cellularized granulation tissue was less organized, and consisted of inflammatory cells, such as macrophages, granulocytes, and (myo)fibroblasts. Variations in the thickness and size of the wounds were observed between the tissue sections of the animals. When comparing the different treatment groups, no differences were found in morphology and in the presence of different cell types in day-7 wounds.

Figure 4a (right) shows collagen deposition by AZAN staining. The wound regions were marked after which the level of collagen deposition in the wounds were measured and corrected for the total wound area (Figure 4b). No significant differences were observed between the different groups. Summarizing, RIPC did not affect wound morphology and collagen deposition of day-7 wounds.

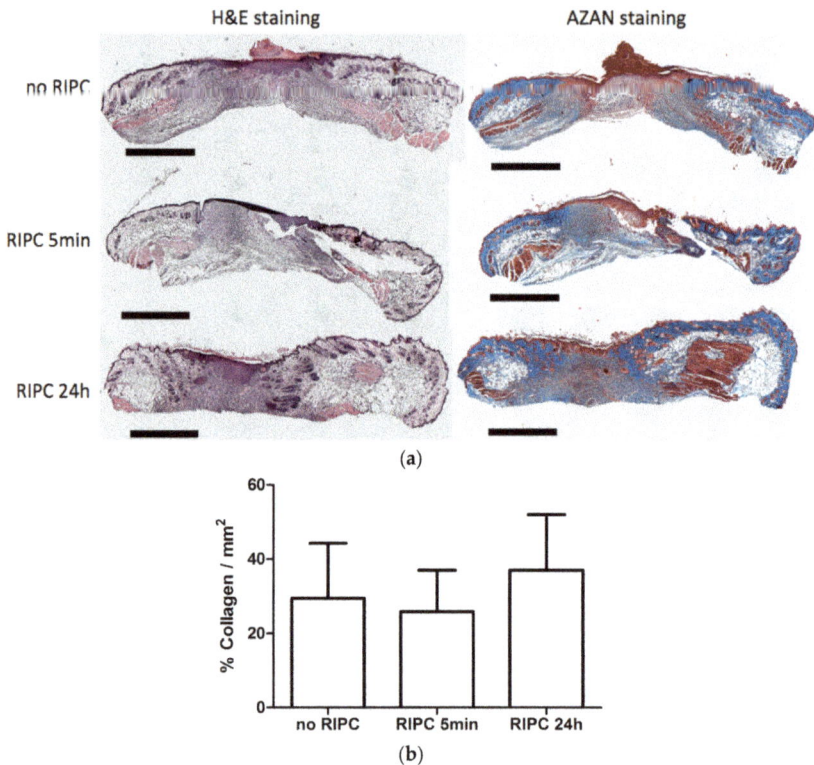

Figure 4. Effects of early and late RIPC on morphology of 7-day-old excisional wounds. (**a**) H&E and AZAN staining were performed to evaluate wound morphology (bars represent 1 mm); (**b**) Quantification of collagen deposition to assess the level of wound remodeling using AZAN staining. Data are expressed as mean ± SD. No significant differences were observed between the different groups: no RIPC ($n = 8$), RIPC 5 min ($n = 6$) and RIPC 24 h ($n = 6$).

3. Discussion

We postulated that RIPC increases HO-1 induction and improves cutaneous wound repair. RIPC induced HO-1 in kidney, heart, and skeletal muscles, but not in the skin. Although RIPC had previously been shown to target the skin [44–46,48], both early and late RIPC did not affect cutaneous wound closure. In addition, skin morphology and collagen deposition at day-7 wounds did not change after early or late RIPC.

RIPC-mediated protection to organs seems therefore tissue-specific and/or dependent on the insult. Organ- and time-specific protective effects of RIPC have also previously been demonstrated. For example, RIPC does not improve wound healing in small bowel anastomoses [64,65]. Although late RIPC (24 h) attenuates ischemia/reperfusion injury (IRI) in muscle flaps, it is ineffective in adipocutaneous flaps [66]. In contrast, early RIPC (30 min) enhances adipocutaneous flap survival [67]. IPC improves the survival of myocutaneous and skin flaps subjected to secondary ischemia of 1h in rats [68,69]. Since these RIPC protocols vary from ours, and IPC and RIPC are different procedures, these studies cannot be directly extrapolated to our study. The protective actions of RIPC are thus dependent on the targeted organ and the type of RIPC treatment [70]. Remote preconditioning by trauma (RPCT) by abdominal incision, has previously been reported to improve cardiac outcome following induced heart infarcts by coronary artery occlusion in murine and canine models [71–73]. Similarly, the inflicted injuries in our study could have led to the induction of overlapping cytoprotective pathways. When these RPCT-induced protective pathways were stronger than the effects by RIPC or used similar pathways, this could explain in part the observed lack of protective effects by RIPC on cutaneous wound repair. Similar protective pathways of RIPC and RPCT include the activation of protein kinase C, mitogen-activated protein kinases and mitochondrial potassium ATP channels, bradykinin and adenosine [71,73,74].

Previously, the stress enzyme HO-1 was found to be important in wound repair and is generally expressed at wound sites [60,75,76]. HO-1 and HO-2 knockout mice showed a delayed wound repair; whereas, induction of HO-1 or administration of its effector molecule bilirubin accelerated wound repair [19,20,77]. Since some of the protective effects of RIPC were shown to be dependent on HO-1 expression in IRI of diverse organs, like the liver [23,24,26], lungs [33], and intestines [28], we further evaluated the role of RIPC on HO-1 in excisional wound healing. HO-1-*luc* Tg mice allowed monitoring the effects of RIPC on HO-1 promoter activity levels in different organs in real-time. 6 and 24 h after RIPC treatment, HO-1 promoter activity was significantly induced compared to 1h after RIPC. This correlates well with our RT-PCR data showing significantly increased endogenous HO-1 mRNA levels in muscle, heart, and kidney 6h after RIPC treatment. However, despite RIPC improving cutaneous microcirculation [44], no HO-1 mRNA induction was found in the skin, and may thus not affect the skin directly. HO-1 induction in the skin is possible using pharmacological preconditioning since we previously observed that i.p. administration of the HO-1 inducer cobalt protoporphyrin at a concentration of 25 mg/kg body weight in HO-1 *luc* Tg mice induced HO-1 mRNA specifically in the skin after 24 h (Figure A3). This underscores the tissue- and time-dependent effects of RIPC, which is probably due to the structural and physiological differences between different organs. Since organs that have induce HO-1 expression upon RIPC treatment correlate with the organs that are protected by RIPC, it is tempting to speculate that HO-1 facilitates these protective effects.

Also the long-term protective effects of RIPC via activation of HO-1 were absent in the skin. Both HO-1 mRNA and protein expression levels were observed to be independent of RIPC treatment in the epidermal and dermal regions of day-7 wounds. Like others, we found HO-1-positive keratinocytes in the hyperproliferative epithelia of the wound margins covering the wound [60,61,78]. In the dermis, we also observed HO-1-positive infiltrating leukocytes that are likely macrophages [60,62,79]. HO-1-positive macrophages are thought to protect the wound environment against oxidative stress [60,80]. The pro-inflammatory HO substrate heme is abundantly released at the edges of the wound site and stimulates recruitment of leukocytes [60,63,78]. HO-1 is thought to attenuate these inflammatory and oxidative triggers at the wound site.

The effect of both early and late RIPC on wound closure was monitored regularly. However, the (immuno)histochemical and PCR analysis was only performed on day 7 wounds, which limits our insight in wound repair processes, like inflammatory signaling, during the first days. Using the HO-1-*luc* Tg mice we previously found that HO-1 promoter activity was indeed significantly induced on day 3 post-wounding, however the level of HO-1 promoter activity did not decline significantly at day 7 compared to day 3 [79]. Although no effects in wound closure, collagen deposition, or HO-1 expression in the skin were observed, we cannot exclude that paracrine effects of HO-1 or other protective signaling pathways may have been triggered by RIPC. HO-effector molecules biliverdin, bilirubin, CO, and ferritin, have all shown to improve wound repair [20,51–54], suggesting that RIPC-induced HO-1 induction in various organs stimulate wound repair in a paracrine fashion. Interestingly, increased systemic levels of bilirubin augment vascular function [81,82] and may contribute to the reported RIPC induced improvement of cutaneous microcirculation. In a previous study, no adverse effects of HO inhibition following RIPC were observed in a kidney injury model, suggesting that other mediators may have been protective [31]. Alternative protective pathways triggered by RIPC include humoral, neural, or systemic anti-inflammatory, anti-apoptotic responses [22,59]. In addition, RIPC may have more effects in more stringent wound repair models where there is a shortage of cytoprotective molecules, such as in diabetic wound repair models and pressure ulcers [47]. Although our method has shown to be effective in diverse animal models [32], other RIPC regimens may enhance these protective effects such as combinations with remote ischemic postconditioning [35,36,47]. Recently, it was found that the sex of the animal may play a role in the efficacy of RIPC treatment, and was observed to be lower in experimental groups of mixed sexes, which we also used in our wound repair study [83].

4. Materials and Methods

4.1. Animals

The Committee for Animal Experiments of the Radboud University Nijmegen approved all procedures involving animals (RU-DEC 2010-248) on 1 February 2011. Fifty mice (strain: HO-1-*luc* FVB/N-Tg background; see Table 1) of 4–5 months in age, and weighing 21–35 g were provided with food and water ad libitum and maintained on a 12 h light/dark cycle and specific pathogen-free housing conditions at the Central Animal Facility Nijmegen. The transgene consists of the full-length mouse HO-1 promoter fused to the reporter gene luciferase (*luc*). More details on the housing conditions are previously described [84]. Mice were originally derived from Stanford University (Stanford, CA, USA) as previously described [85]. An overview of the animals used for the different experiments can be found in Table 1. No animals died during the experiments and no animals were excluded during the experiments or data analyses. All mice were randomly divided over the experiments, and split evenly over their sex and age. All outcomes were measured by an observer who was blinded for the allocation of the animals to the experimental groups, when possible.

Table 1. Overview animal experiments.

Aim Experiment	Read Out	Animals (*n*: ♂/♀)
Investigate the effects of RIPC on HO-1 promoter activity	HO-1 promoter activity at 1, 6 and 24 h after RIPC treatment	6: 0/6
Investigate the effects of RIPC on HO-1 gene expression in different organs during time	HO-1 mRNA levels at 0, 1, 6, and 24 h after RIPC treatment	24: 0/24 (6 per time point)
Investigate the effects of RIPC on dermal wound healing	Early (5 min before wounding) and late (24 h before wounding) effects of RIPC on wound healing compared to controls without receiving RIPC treatment (endpoint: day 7)	6: 4/2 (early RIPC) 6: 4/2 (late RIPC) 8: 4/4 (controls)

4.2. RIPC Treatment

RIPC by brief hind limb ischemia was induced by applying elastic latex-free O-rings (Miltex Integra: 28–155) using a hemorrhoidal ligator (Miltex McGivney: 26–154B) bilaterally around the most upper position of the proximal thigh (Figure A1). Reperfusion was accomplished by cutting the elastic rings with scissors, confirmed by the disappearance of blue color to the limbs (Figure A1) as described previously [86–88]. The mice were anesthetized with isoflurane in O_2/N_2O (5% isoflurane for induction and 2%–3% to maintain anesthesia) during RIPC treatment and treatment consisted of three cycles of 4-min ischemia interspersed with 4-min reperfusion. This RIPC regime is based on a previous study in which we found that bilateral repetitive (3 times 4 min) ischemia/reperfusion gave the most potent protection in a kidney injury model [32].

4.3. Measuring of HO-1 Promoter Activity

In order to monitor HO-1 promoter activity after RIPC treatment in time, HO-1-*luc* Tg mice underwent RIPC treatment as described above. HO-1-*luc* expression was measured in vivo and the mice were sacrificed at 24 h. In vivo bioluminescence imaging was performed as described before on the IVIS Lumina System (Caliper Life Sciences, Hopkinton, MA, USA) [89]. Images taken were quantified using Living Image 3.0 software (Caliper Life Sciences) by selecting the regions of interest (ROI). ROIs included both the dorsal images encompassing the back region below the head and above the tail to cover the area where the wounds would be made and the renal area. Emitted photons per second (or total flux) per region of interest (ROI) was measured, and then calculated as fold change from baseline levels and related to 1 h after RIPC.

4.4. Excisional Wound Model

Wounds were made 24 h or 5 min after RIPC. Control mice did not receive RIPC, but underwent the same anesthetic procedure 1h before wounding. Two full-thickness excisional wounds of 4 mm in diameter were made on the shaved dorsal side of anesthetized mice using a sterile disposable biopsy punch (Kai Medical, Seki City, Japan), as previously described by our group [77]. The wounds were created on the dorsum to either side of the midline, with approximately 1cm between the wounds, and just below the shoulders and pelvis. Skin biopsies taken to create the 4-mm wounds served as control skin. Wounds were photographically documented immediately, and 1, 3 and 7 days after wounding with a ruler placed perpendicular to the wounds for wound size normalization. The area of the wounds was blindly measured twice using ImageJ v1.44p software (http://imagej.nih.gov/ij; NIH, Bethesda, MD, USA).

4.5. Sample Collection

At day 7, the mice were anesthetized with 5% isoflurane in O_2/N_2O and sacrificed by exsanguination, followed by cervical dislocation. Kidney and muscle (*m. quadriceps femoris*) were dissected, and wound tissue was collected using a 4-mm biopsy punch. Half of the tissue was fixated with 4% paraformaldehyde and processed for paraffin embedding and (immuno)histochemistry, and the other half was snap frozen in liquid nitrogen and stored at −80 °C until use for RT-PCR.

4.6. (Immuno-)histochemical Staining and Analyses

Standard H&E, Weigert-AZAN staining (azo carmine and aniline blue), and immunohistochemical HO-1 staining were performed on paraffin sections of the wounds as previously described [77]. Stained sections were analyzed and photographed using the Zeiss Imager Z1 microscope (Zeiss, Sliedrecht, The Netherlands) and Axiovision software version 4.8 (Zeiss).

Analysis of collagen deposition in AZAN stained wound sections was performed by image analysis using a macro built in Image J [90]. The wound area was manually defined using the edges of the *panniculus carnosus* and epithelium as boundaries before running the macro. Measurements were averaged per mouse and mean intensity/mm^2 was used for further analysis.

HO-1 immunoreactivity was evaluated by blindly scoring the epidermal zone and the dermal region of the wounds. A single section per wound of each animal was semi-quantitatively scored as previously described using the following scale: 0 (minimal), 1 (mild), 2 (moderate), and 3 (marked).

4.7. RNA Isolation and Quantitative-RT-PCR

Tissue was pulverized in TRIzol (Invitrogen, Carlsbad, CA, USA) using a micro-dismembrator (Sartorius BBI Systems GmbH, Melsungen, Germany) and RNA was further extracted as previously described [13]. All values were normalized to the household gene *gapdh*, which is often used in RIPC experiments [91,92] according to the comparative method ($2^{-\Delta\Delta Ct}$). *Gapdh* mRNA expression levels were stable and were not affected by RIPC treatment. The sequences of the mouse-specific primers for *gapdh* are forward 5′GGCAAATTCAACGGCACA3′, and reverse 5′GTTAGTGGGGTCTCGC TCCTG3′, and for *Hmox1* (*HO-1*) forward 5′CAACATTGAGCTGTTTGAGG3′, and reverse 5′TGGTCTTTGTGTTCCTCTGTC3′.

4.8. Statistics

Data were analyzed using GraphPad Prism 5.01 software (San Diego, CA, USA). Outliers were tested using the Grubbs' test, but no outliers were found (except for the data in Figure A3 where one outlier was found in the skin and one in the kidney group). Data were analyzed using two-sided *t*-tests to compare two variables or a one-way analysis of variance when comparing multiple variables. Bonferroni's multiple comparison *post hoc* test was applied as correction for multiple comparisons when investigating multiple dependent research questions. Results were considered significantly different at $p < 0.05$ (* $p < 0.05$, ** $p < 0.01$, and *** $p < 0.001$).

5. Conclusions

RIPC treatment induced HO-1 mRNA expression in kidney, heart, and ligated muscle, and may therefore directly contribute to enhanced protection to injurious stressors and/or microcirculation in these tissues. However, RIPC did not alter HO-1 in the skin and was not modulated in day-7 skin wounds, demonstrating organ- and time-specific effects. Both early and late RIPC treatment did not affect dermal wound closure time, collagen deposition, or wound morphology. A better understanding of the mechanistic insight by which RIPC mediates organ protection is needed.

Acknowledgments: This study was supported by grants from Radboudumc and the Dutch Burns Foundation (#09.110). The Committee for Animal Experiments of the Radboud University Nijmegen approved all procedures involving animals (RU-DEC 2010-248). No financial disclosures were reported by the authors of this paper.

Author Contributions: Niels A. J. Cremers, Ronald J. Wong, Ditte M. S. Lundvig, and Frank A. D. T. G. Wagener conceived and designed the experiments; Niels A. J. Cremers, René E. M. van Rheden, and Kimberley E. Wever performed the experiments; Niels A. J. Cremers, Ronald J. Wong, Ditte M. S. Lundvig, and Frank A. D. T. G. Wagener analyzed the data; Eline A. Vermeij, Gooitzen M. van Dam, and Kimberley E. Wever contributed reagents/materials/analysis tools; Niels A. J. Cremers, Ronald J. Wong, Carine E. Carels, Ditte M. S. Lundvig, and Frank A. D. T. G. Wagener wrote the paper.

Conflicts of Interest: The authors declare no conflict of interest.

Abbreviations

H&E	hematoxylin and eosin
HO-1	heme oxygenase-1
HO-1-*luc* Tg	HO-1-*luciferase* transgenic
I/R	Ischemia/Reperfusion
IPC	Ischemic preconditioning
IRI	Ischemia/Reperfusion injury
RIPC	Remote ischemic preconditioning
RPCT	Remote preconditioning by trauma
ROS	Reactive oxygen species

Appendix A

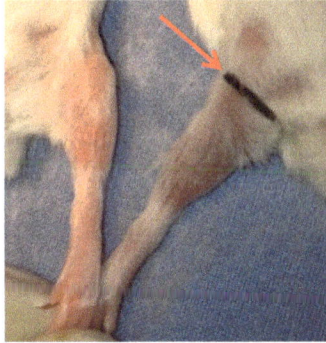

Figure A1. Hind limb ischemia by ligation using elastic ring (**red arrow**). Note the difference in the color of the legs after obstruction of the blood flow, confirming RIPC treatment was successful.

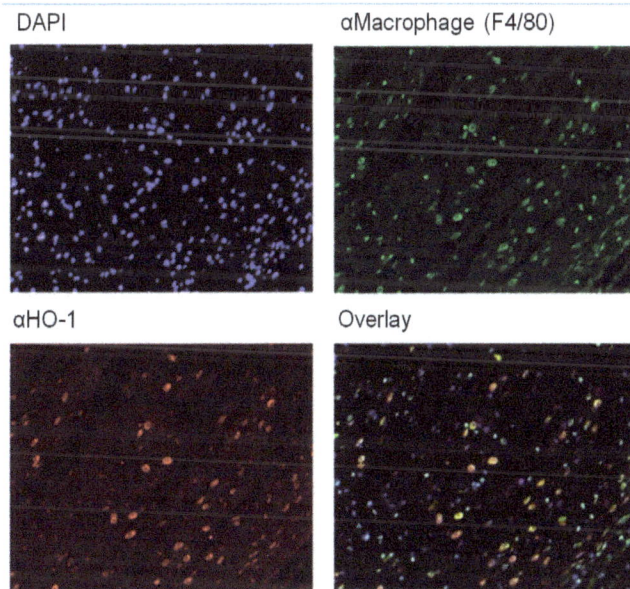

Figure A2. Co-localization of HO-1 and macrophages during excisional wound healing. Fluorescent staining from a previous experiment on excisional wound healing in C57Bl/6 mice at day 2 post wounding, showing cell nuclei stained with DAPI (**blue**), HO-1 (**red**), macrophages stained for F4/80 (**green**), and an overlay. The overlay picture clearly shows co-localization (**orange**) of HO-1 and macrophages.

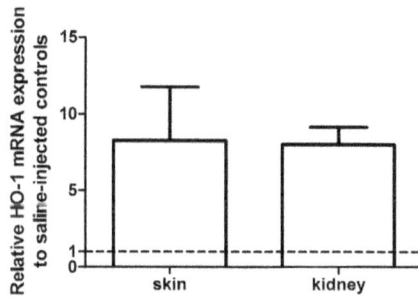

Figure A3. Pharmacological preconditioning with cobalt protoporphyrin (25 mg/kg body weight) in HO-1 *luc* Tg mice (*n* = 6) induced HO-1 mRNA expression in both the skin and kidney 24 h after treatment when compared to saline-injected control mice (*n* = 6). Dashed line represents the mRNA expression levels in the corresponding control organs after saline injection, which are set at 1. Data represents the relative mean ± SD. One significant outlier was found in both skin and kidney group and therefore excluded in the graph.

References

1.   Rabello, F.B.; Souza, C.D.; Farina Junior, J.A. Update on hypertrophic scar treatment. *Clinics* **2014**, *69*, 565–573. [CrossRef]
2.   Eming, S.A.; Martin, P.; Tomic-Canic, M. Wound repair and regeneration: Mechanisms, signaling, and translation. *Sci. Transl. Med.* **2014**, *6*, 265sr266. [CrossRef] [PubMed]
3.   Sidgwick, G.P.; Bayat, A. Extracellular matrix molecules implicated in hypertrophic and keloid scarring. *J. Eur. Acad. Dermatol. Venereol.* **2012**, *26*, 141–152. [CrossRef] [PubMed]
4.   Grice, E.A.; Segre, J.A. Interaction of the microbiome with the innate immune response in chronic wounds. *Adv. Exp. Med. Biol.* **2012**, *946*, 55–68. [PubMed]
5.   Pereira, R.F.; Bartolo, P.J. Traditional therapies for skin wound healing. *Adv. Wound Care* **2016**, *5*, 208–229. [CrossRef] [PubMed]
6.   Aarabi, S.; Longaker, M.T.; Gurtner, G.C. Hypertrophic scar formation following burns and trauma: New approaches to treatment. *PLoS Med.* **2007**, *4*, e234. [CrossRef] [PubMed]
7.   Tziotzios, C.; Profyris, C.; Sterling, J. Cutaneous scarring: Pathophysiology, molecular mechanisms, and scar reduction therapeutics part II. Strategies to reduce scar formation after dermatologic procedures. *J. Am. Acad. Dermatol.* **2012**, *66*, 13–24. [CrossRef] [PubMed]
8.   Chen, Q.Y.; Wang, G.G.; Li, W.; Jiang, Y.X.; Lu, X.H.; Zhou, P.P. Heme oxygenase-1 promotes delayed wound healing in diabetic rats. *J. Diabetes Res.* **2016**, *2016*, 9726503. [CrossRef] [PubMed]
9.   Ahanger, A.A.; Prawez, S.; Leo, M.D.; Kathirvel, K.; Kumar, D.; Tandan, S.K.; Malik, J.K. Pro-healing potential of hemin: An inducer of heme oxygenase-1. *Eur. J. Pharmacol.* **2010**, *645*, 165–170. [CrossRef] [PubMed]
10. Panchatcharam, M.; Miriyala, S.; Gayathri, V.S.; Suguna, L. Curcumin improves wound healing by modulating collagen and decreasing reactive oxygen species. *Mol. Cell. Biochem.* **2006**, *290*, 87–96. [CrossRef] [PubMed]
11. Tejada, S.; Manayi, A.; Daglia, M.; Nabavi, S.F.; Sureda, A.; Hajheydari, Z.; Gortzi, O.; Pazoki-Toroudi, H.; Nabavi, S.M. Wound healing effects of curcumin: A short review. *Curr. Pharm. Biotechnol.* **2016**, *17*, 1002–1007. [CrossRef] [PubMed]
12. Kant, V.; Gopal, A.; Kumar, D.; Pathak, N.N.; Ram, M.; Jangir, B.L.; Tandan, S.K.; Kumar, D. Curcumin-induced angiogenesis hastens wound healing in diabetic rats. *J. Surg. Res.* **2015**, *193*, 978–988. [CrossRef] [PubMed]
13. Cremers, N.A.; Lundvig, D.M.; van Dalen, S.C.; Schelbergen, R.F.; van Lent, P.L.; Szarek, W.A.; Regan, R.F.; Carels, C.E.; Wagener, F.A. Curcumin-induced heme oxygenase-1 expression prevents H_2O_2-induced cell death in wild type and heme oxygenase-2 knockout adipose-derived mesenchymal stem cells. *Int. J. Mol. Sci.* **2014**, *15*, 17974–17999. [CrossRef] [PubMed]

14. Akbik, D.; Ghadiri, M.; Chrzanowski, W.; Rohanizadeh, R. Curcumin as a wound healing agent. *Life Sci.* **2014**, *116*, 1–7. [CrossRef] [PubMed]

15. Wagener, F.A.; Volk, H.D.; Willis, D.; Abraham, N.G.; Soares, M.P.; Adema, G.J.; Figdor, C.G. Different faces of the heme-heme oxygenase system in inflammation. *Pharmacol. Rev.* **2003**, *55*, 551–571. [CrossRef] [PubMed]

16. Gozzelino, R.; Jeney, V.; Soares, M.P. Mechanisms of cell protection by heme oxygenase-1. *Annu. Rev. Pharmacol. Toxicol.* **2010**, *50*, 323–354. [CrossRef] [PubMed]

17. Morse, D.; Choi, A.M. Heme oxygenase-1: From bench to bedside. *Am. J. Respir. Crit. Care Med.* **2005**, *172*, 660–670. [CrossRef] [PubMed]

18. Gozzelino, R.; Soares, M.P. Coupling heme and iron metabolism via ferritin h chain. *Antioxid. Redox. Signal.* **2014**, *20*, 1754–1769. [CrossRef] [PubMed]

19. Grochot-Przeczek, A.; Lach, R.; Mis, J.; Skrzypek, K.; Gozdecka, M.; Sroczynska, P.; Dubiel, M.; Rutkowski, A.; Kozakowska, M.; Zagorska, A.; et al. Heme oxygenase-1 accelerates cutaneous wound healing in mice. *PLoS ONE* **2009**, *4*, e5803. [CrossRef] [PubMed]

20. Ahanger, A.A.; Leo, M.D.; Gopal, A.; Kant, V.; Tandan, S.K.; Kumar, D. Pro-healing effects of bilirubin in open excision wound model in rats. *Int. Wound J.* **2014**. [CrossRef] [PubMed]

21. Souza Filho, M.V.; Loiola, R.T.; Rocha, E.L.; Simao, A.F.; Gomes, A.S.; Souza, M.H.; Ribeiro, R.A. Hind limb ischemic preconditioning induces an anti-inflammatory response by remote organs in rats. *Braz. J. Med. Biol. Res.* **2009**, *42*, 921–929. [CrossRef] [PubMed]

22. Hausenloy, D.J.; Yellon, D.M. Remote ischaemic preconditioning: Underlying mechanisms and clinical application. *Cardiovasc. Res.* **2008**, *79*, 377–386. [CrossRef] [PubMed]

23. Lai, I.R.; Chang, K.J.; Chen, C.F.; Tsai, H.W. Transient limb ischemia induces remote preconditioning in liver among rats: The protective role of heme oxygenase-1. *Transplantation* **2006**, *81*, 1311–1317. [CrossRef] [PubMed]

24. Tapuria, N.; Junnarkar, S.P.; Dutt, N.; Abu-Amara, M.; Fuller, B.; Seifalian, A.M.; Davidson, B.R. Effect of remote ischemic preconditioning on hepatic microcirculation and function in a rat model of hepatic ischemia reperfusion injury. *HPB (Oxford)* **2009**, *11*, 108–117. [CrossRef] [PubMed]

25. Kageyama, S.; Hata, K.; Tanaka, H.; Hirao, H.; Kubota, T.; Okamura, Y.; Iwaisako, K.; Takada, Y.; Uemoto, S. Intestinal ischemic preconditioning ameliorates hepatic ischemia reperfusion injury in rats: Role of heme oxygenase-1 in the second-window of protection. *Liver Transpl.* **2014**. [CrossRef] [PubMed]

26. Wang, Y.; Shen, J.; Xiong, X.; Xu, Y.; Zhang, H.; Huang, C.; Tian, Y.; Jiao, C.; Wang, X.; Li, X. Remote ischemic preconditioning protects against liver ischemia-reperfusion injury via heme oxygenase-1-induced autophagy. *PLoS ONE* **2014**, *9*, e98834. [CrossRef] [PubMed]

27. Thielmann, M.; Kottenberg, E.; Kleinbongard, P.; Wendt, D.; Gedik, N.; Pasa, S.; Price, V.; Tsagakis, K.; Neuhauser, M.; Peters, J.; et al. Cardioprotective and prognostic effects of remote ischaemic preconditioning in patients undergoing coronary artery bypass surgery: A single-centre randomised, double-blind, controlled trial. *Lancet* **2013**, *382*, 597–604. [CrossRef]

28. Saeki, I.; Matsuura, T.; Hayashida, M.; Taguchi, T. Ischemic preconditioning and remote ischemic preconditioning have protective effect against cold ischemia-reperfusion injury of rat small intestine. *Pediatr. Surg. Int.* **2011**, *27*, 857–862. [CrossRef] [PubMed]

29. Kharbanda, R.K.; Mortensen, U.M.; White, P.A.; Kristiansen, S.B.; Schmidt, M.R.; Hoschtitzky, J.A.; Vogel, M.; Sorensen, K.; Redington, A.N.; MacAllister, R. Transient limb ischemia induces remote ischemic preconditioning in vivo. *Circulation* **2002**, *106*, 2881–2883. [CrossRef] [PubMed]

30. Lim, S.Y.; Hausenloy, D.J. Remote ischemic conditioning: From bench to bedside. *Front. Physiol.* **2012**, *3*, 27. [CrossRef] [PubMed]

31. Wever, K.E.; Masereeuw, R.; Wagener, F.A.; Verweij, V.G.; Peters, J.G.; Pertijs, J.C.; van der Vliet, J.A.; Warle, M.C.; Rongen, G.A. Humoral signalling compounds in remote ischaemic preconditioning of the kidney, a role for the opioid receptor. *Nephrol. Dial. Transplant.* **2013**, *28*, 1721–1732. [CrossRef] [PubMed]

32. Wever, K.E.; Warle, M.C.; Wagener, F.A.; van der Hoorn, J.W.; Masereeuw, R.; van der Vliet, J.A.; Rongen, G.A. Remote ischaemic preconditioning by brief hind limb ischaemia protects against renal ischaemia-reperfusion injury: The role of adenosine. *Nephrol. Dial. Transplant.* **2011**, *26*, 3108–3117. [CrossRef] [PubMed]

33. Jan, W.C.; Chen, C.H.; Tsai, P.S.; Huang, C.J. Limb ischemic preconditioning mitigates lung injury induced by haemorrhagic shock/resuscitation in rats. *Resuscitation* **2011**, *82*, 760–766. [CrossRef] [PubMed]

34. Narayanan, S.V.; Dave, K.R.; Perez-Pinzon, M.A. Ischemic preconditioning and clinical scenarios. *Curr. Opin. Neurol.* **2013**, *26*, 1–7. [CrossRef] [PubMed]

35. Depre, C.; Park, J.Y.; Shen, Y.T.; Zhao, X.; Qiu, H.; Yan, L.; Tian, B.; Vatner, S.F.; Vatner, D.E. Molecular mechanisms mediating preconditioning following chronic ischemia differ from those in classical second window. *Am. J. Physiol. Heart. Circ. Physiol.* **2010**, *299*, H752–H762. [CrossRef] [PubMed]

36. Shen, Y.T.; Depre, C.; Yan, L.; Park, J.Y.; Tian, B.; Jain, K.; Chen, L.; Zhang, Y.; Kudej, R.K.; Zhao, X.; et al. Repetitive ischemia by coronary stenosis induces a novel window of ischemic preconditioning. *Circulation* **2008**, *118*, 1961–1969. [CrossRef] [PubMed]

37. Hausenloy, D.J.; Boston-Griffiths, E.; Yellon, D.M. Cardioprotection during cardiac surgery. *Cardiovasc. Res.* **2012**, *94*, 253–265. [CrossRef]

38. Hausenloy, D.J.; Candilio, L.; Evans, R.; Ariti, C.; Jenkins, D.P.; Kolvekar, S.; Knight, R.; Kunst, G.; Laing, C.; Nicholas, J.; et al. Remote ischemic preconditioning and outcomes of cardiac surgery. *N. Engl. J. Med.* **2015**, *373*, 1408–1417. [CrossRef] [PubMed]

39. Sukkar, L.; Hong, D.; Wong, M.G.; Badve, S.V.; Rogers, K.; Perkovic, V.; Walsh, M.; Yu, X.; Hillis, G.S.; Gallagher, M.; et al. Effects of ischaemic conditioning on major clinical outcomes in people undergoing invasive procedures: Systematic review and meta-analysis. *BMJ* **2016**, *355*, i5599. [CrossRef] [PubMed]

40. Garratt, K.N.; Whittaker, P.; Przyklenk, K. Remote ischemic conditioning and the long road to clinical translation: Lessons learned from ericca and ripheart. *Circ. Res.* **2016**, *118*, 1052–1054. [CrossRef] [PubMed]

41. King, N.D.G.; Smart, N.A. Remote ischaemic pre-conditioning does not affect clinical outcomes following coronary artery bypass grafting. A systematic review and meta-analysis. *Clin. Trials Regul. Sci. Cardiol.* **2016**, *17*, 1–8. [CrossRef]

42. Przyklenk, K. Ischaemic conditioning: Pitfalls on the path to clinical translation. *Br. J. Pharmacol.* **2015**, *172*, 1961–1973. [CrossRef] [PubMed]

43. Dorresteijn, M.J.; Paine, A.; Zilian, E.; Fenten, M.G.; Frenzel, E.; Janciauskiene, S.; Figueiredo, C.; Eiz-Vesper, B.; Blasczyk, R.; Dekker, D.; et al. Cell-type-specific downregulation of heme oxygenase-1 by lipopolysaccharide via bach1 in primary human mononuclear cells. *Free Radic. Biol. Med.* **2015**, *78*, 224–232. [CrossRef] [PubMed]

44. Kraemer, R.; Lorenzen, J.; Kabbani, M.; Herold, C.; Busche, M.; Vogt, P.M.; Knobloch, K. Acute effects of remote ischemic preconditioning on cutaneous microcirculation–a controlled prospective cohort study. *BMC Surg.* **2011**, *11*, 32. [CrossRef] [PubMed]

45. Kolbenschlag, J.; Sogorski, A.; Kapalschinski, N.; Harati, K.; Lehnhardt, M.; Daigeler, A.; Hirsch, T.; Goertz, O. Remote ischemic conditioning improves blood flow and oxygen saturation in pedicled and free surgical flaps. *Plast. Reconstr. Surg.* **2016**, *138*, 1089–1097. [CrossRef] [PubMed]

46. Masaoka, K.; Asato, H.; Umekawa, K.; Imanishi, M.; Suzuki, A. Value of remote ischaemic preconditioning in rat dorsal skin flaps and clamping time. *J. Plast. Surg. Hand. Surg.* **2015**, *50*, 107–110. [CrossRef] [PubMed]

47. Shaked, G.; Czeiger, D.; Abu Arar, A.; Katz, T.; Harman-Boehm, I.; Sebbag, G. Intermittent cycles of remote ischemic preconditioning augment diabetic foot ulcer healing. *Wound Repair Regen.* **2015**, *23*, 191–196. [CrossRef]

48. Epps, J.A.; Smart, N.A. Remote ischaemic conditioning in the context of type 2 diabetes and neuropathy: The case for repeat application as a novel therapy for lower extremity ulceration. *Cardiovasc. Diabetol.* **2016**, *15*, 130. [CrossRef] [PubMed]

49. Zhang, W.; Contag, P.R.; Hardy, J.; Zhao, H.; Vreman, H.J.; Hajdena-Dawson, M.; Wong, R.J.; Stevenson, D.K.; Contag, C.H. Selection of potential therapeutics based on in vivo spatiotemporal transcription patterns of heme oxygenase-1. *J. Molecular Med.* **2002**, *80*, 655–664. [CrossRef] [PubMed]

50. Zarjou, A.; Kim, J.; Traylor, A.M.; Sanders, P.W.; Balla, J.; Agarwal, A.; Curtis, L.M. Paracrine effects of mesenchymal stem cells in cisplatin-induced renal injury require heme oxygenase-1. *Am. J. Physiol. Ren. Physiol.* **2011**, *300*, F254–262. [CrossRef] [PubMed]

51. Halilovic, A.; Patil, K.A.; Bellner, L.; Marrazzo, G.; Castellano, K.; Cullaro, G.; Dunn, M.W.; Schwartzman, M.L. Knockdown of heme oxygenase-2 impairs corneal epithelial cell wound healing. *J. Cell. Physiol.* **2011**, *226*, 1732–1740. [CrossRef] [PubMed]

52. Dulak, J.; Deshane, J.; Jozkowicz, A.; Agarwal, A. Heme oxygenase-1 and carbon monoxide in vascular pathobiology: Focus on angiogenesis. *Circulation* **2008**, *117*, 231–241. [CrossRef]

53. Ahanger, A.A.; Prawez, S.; Kumar, D.; Prasad, R.; Amarpal; Tandan, S.K.; Kumar, D. Wound healing activity of carbon monoxide liberated from co-releasing molecule (co-rm). *Naunyn Schmiedebergs Arch. Pharmacol.* **2011**, *384*, 93–102. [CrossRef] [PubMed]

54. Coffman, L.G.; Parsonage, D.; D'Agostino, R., Jr.; Torti, F.M.; Torti, S.V. Regulatory effects of ferritin on angiogenesis. *Proc. Natl. Acad. Sci. USA* **2009**, *106*, 570–575. [CrossRef] [PubMed]

55. Penna, C.; Granata, R.; Tocchetti, C.G.; Gallo, M.P.; Alloatti, G.; Pagliaro, P. Endogenous cardioprotective agents: Role in pre and postconditioning. *Curr. Drug Target.* **2015**, *16*, 843–867. [CrossRef]

56. Guimaraes Filho, M.A.; Cortez, E.; Garcia-Souza, E.P.; Soares Vde, M.; Moura, A.S.; Carvalho, L.; Maya, M.C.; Pitombo, M.B. Effect of remote ischemic preconditioning in the expression of IL-6 and IL-10 in a rat model of liver ischemia-reperfusion injury. *Acta Cir. Bras.* **2015**, *30*, 152–160. [CrossRef]

57. Chevion, M.; Leibowitz, S.; Aye, N.N.; Novogrodsky, O.; Singer, A.; Avizemer, O.; Bulvik, B.; Konijn, A.M.; Berenshtein, E. Heart protection by ischemic preconditioning: A novel pathway initiated by iron and mediated by ferritin. *J. Mol. Cell. Cardiol.* **2008**, *45*, 839–845. [CrossRef] [PubMed]

58. Andreadou, I.; Iliodromitis, E.K.; Rassaf, T.; Schulz, R.; Papapetropoulos, A.; Ferdinandy, P. The role of gasotransmitters NO, H2S and CO in myocardial ischaemia/reperfusion injury and cardioprotection by preconditioning, postconditioning and remote conditioning. *Br. J. Pharmacol.* **2015**, *172*, 1587–1606. [CrossRef] [PubMed]

59. Tapuria, N.; Kumar, Y.; Habib, M.M.; Abu Amara, M.; Seifalian, A.M.; Davidson, B.R. Remote ischemic preconditioning: A novel protective method from ischemia reperfusion injury—A review. *J. Surg. Res.* **2008**, *150*, 304–330. [CrossRef]

60. Hanselmann, C.; Mauch, C.; Werner, S. Haem oxygenase-1: A novel player in cutaneous wound repair and psoriasis? *Biochem. J.* **2001**, *353*, 459–466. [CrossRef] [PubMed]

61. Kampfer, H.; Kolb, N.; Manderscheid, M.; Wetzler, C.; Pfeilschifter, J.; Frank, S. Macrophage-derived heme-oxygenase-1: Expression, regulation, and possible functions in skin repair. *Mol. Med.* **2001**, *7*, 488–498. [PubMed]

62. Schurmann, C.; Seitz, O.; Klein, C.; Sader, R.; Pfeilschifter, J.; Muhl, H.; Goren, I.; Frank, S. Tight spatial and temporal control in dynamic basal to distal migration of epithelial inflammatory responses and infiltration of cytoprotective macrophages determine healing skin flap transplants in mice. *Ann. Surg.* **2009**, *249*, 519–534. [CrossRef] [PubMed]

63. Wagener, F.A.; van Beurden, H.E.; von den Hoff, J.W.; Adema, G.J.; Figdor, C.G. The heme-heme oxygenase system: A molecular switch in wound healing. *Blood* **2003**, *102*, 521–528. [CrossRef] [PubMed]

64. Holzner, P.A.; Kulemann, B.; Kuesters, S.; Timme, S.; Hoeppner, J.; Hopt, U.T.; Marjanovic, G. Impact of remote ischemic preconditioning on wound healing in small bowel anastomoses. *World J. Gastroenterol.* **2011**, *17*, 1308–1316. [CrossRef] [PubMed]

65. Colak, T.; Turkmenoglu, O.; Dag, A.; Polat, A.; Comelekoglu, U.; Bagdatoglu, O.; Polat, G.; Kanik, A.; Akca, T.; Aydin, S. The effect of remote ischemic preconditioning on healing of colonic anastomoses. *J. Surg. Res.* **2007**, *143*, 200–205. [CrossRef]

66. Kuntscher, M.V.; Kastell, T.; Engel, H.; Gebhard, M.M.; Heitmann, C.; Germann, G. Late remote ischemic preconditioning in rat muscle and adipocutaneous flap models. *Ann. Plastic Surg.* **2003**, *51*, 84–90. [CrossRef] [PubMed]

67. Kuntscher, M.V.; Schirmbeck, E.U.; Menke, H.; Klar, E.; Gebhard, M.M.; Germann, G. Ischemic preconditioning by brief extremity ischemia before flap ischemia in a rat model. *Plast. Reconstr. Surg.* **2002**, *109*, 2398–2404. [CrossRef]

68. Shah, A.A.; Arias, J.E.; Thomson, J.G. The effect of ischemic preconditioning on secondary ischemia in myocutaneous flaps. *J. Reconstr. Microsurg.* **2009**, *25*, 527–531. [CrossRef] [PubMed]

69. Zahir, K.S.; Syed, S.A.; Zink, J.R.; Restifo, R.J.; Thomson, J.G. Ischemic preconditioning improves the survival of skin and myocutaneous flaps in a rat model. *Plast. Reconstr. Surg.* **1998**, *102*, 140–152. [CrossRef] [PubMed]

70. Kolh, P. Remote ischaemic pre-conditioning in cardiac surgery: Benefit or not? *Eur. Heart J.* **2014**, *35*, 141–143. [PubMed]

71. Gross, G.J.; Baker, J.E.; Moore, J.; Falck, J.R.; Nithipatikom, K. Abdominal surgical incision induces remote preconditioning of trauma (RPCT) via activation of bradykinin receptors (BK2R) and the cytochrome p450 epoxygenase pathway in canine hearts. *Cardiovasc. Drugs Ther.* **2011**, *25*, 517–522. [CrossRef] [PubMed]

72. Gross, G.J.; Hsu, A.; Gross, E.R.; Falck, J.R.; Nithipatikom, K. Factors mediating remote preconditioning of trauma in the rat heart: Central role of the cytochrome p450 epoxygenase pathway in mediating infarct size reduction. *J. Cardiovasc. Pharmacol. Ther.* **2013**, *18*, 38–45. [CrossRef] [PubMed]

73. Jones, W.K.; Fan, G.C.; Liao, S.; Zhang, J.M.; Wang, Y.; Weintraub, N.L.; Kranias, E.G.; Schultz, J.E.; Lorenz, J.; Ren, X. Peripheral nociception associated with surgical incision elicits remote nonischemic cardioprotection via neurogenic activation of protein kinase c signaling. *Circulation* **2009**, *120*, S1–S9. [CrossRef] [PubMed]

74. Chai, Q.; Liu, J.; Hu, Y. Cardioprotective effect of remote preconditioning of trauma and remote ischemia preconditioning in a rat model of myocardial ischemia/reperfusion injury. *Exp. Ther. Med.* **2015**, *9*, 1745–1750. [CrossRef] [PubMed]

75. Wagener, F.A.; Scharstuhl, A.; Tyrrell, R.M.; Von den Hoff, J.W.; Jozkowicz, A.; Dulak, J.; Russel, F.G.; Kuijpers-Jagtman, A.M. The heme-heme oxygenase system in wound healing; implications for scar formation. *Curr. Drug Target* **2010**, *11*, 1571–1585. [CrossRef]

76. Lundvig, D.M.; Immenschuh, S.; Wagener, F.A. Heme oxygenase, inflammation, and fibrosis: The good, the bad, and the ugly? *Front. Pharmacol.* **2012**, *3*, 81. [CrossRef] [PubMed]

77. Lundvig, D.M.; Scharstuhl, A.; Cremers, N.A.; Pennings, S.W.; te Paske, J.; van Rheden, R.; van Run-van Breda, C.; Regan, R.F.; Russel, F.G.; Carels, C.E.; et al. Delayed cutaneous wound closure in ho-2 deficient mice despite normal ho-1 expression. *J. Cell. Mol. Med.* **2014**, *18*, 2488–2498. [CrossRef] [PubMed]

78. Auf dem Keller, U.; Kumin, A.; Braun, S.; Werner, S. Reactive oxygen species and their detoxification in healing skin wounds. *J. Investig. Dermatol. Symp. Proc.* **2006**, *11*, 106–111. [CrossRef] [PubMed]

79. Cremers, N.A.; Suttorp, M.; Gerritsen, M.M.; Wong, R.J.; van Run-van Breda, C.; van Dam, G.M.; Brouwer, K.M.; Kuijpers-Jagtman, A.M.; Carels, C.E.; Lundvig, D.M.; et al. Mechanical stress changes the complex interplay between HO-1, inflammation and fibrosis, during excisional wound repair. *Front. Med.* **2015**, *2*, 86. [CrossRef] [PubMed]

80. Ishii, T.; Itoh, K.; Sato, H.; Bannai, S. Oxidative stress-inducible proteins in macrophages. *Free Radical Res.* **1999**, *31*, 351–355. [CrossRef]

81. Dekker, D.; Dorresteijn, M.J.; Pijnenburg, M.; Heemskerk, S.; Rasing-Hoogveld, A.; Burger, D.M.; Wagener, F.A.; Smits, P. The bilirubin-increasing drug atazanavir improves endothelial function in patients with type 2 diabetes mellitus. *Arterioscler. Thromb. Vasc. Biol.* **2011**, *31*, 458–463. [CrossRef] [PubMed]

82. Maruhashi, T.; Soga, J.; Fujimura, N.; Idei, N.; Mikami, S.; Iwamoto, Y.; Kajikawa, M.; Matsumoto, T.; Kihara, Y.; Chayama, K.; et al. Hyperbilirubinemia, augmentation of endothelial function, and decrease in oxidative stress in gilbert syndrome. *Circulation* **2012**, *126*, 598–603. [CrossRef] [PubMed]

83. Wever, K.E.; Hooijmans, C.R.; Riksen, N.P.; Sterenborg, T.B.; Sena, E.S.; Ritskes-Hoitinga, M.; Warle, M.C. Determinants of the efficacy of cardiac ischemic preconditioning: A systematic review and meta-analysis of animal studies. *PLoS ONE* **2015**, *10*, e0142021. [CrossRef] [PubMed]

84. Wever, K.E.; Wagener, F.A.; Frielink, C.; Boerman, O.C.; Scheffer, G.J.; Allison, A.; Masereeuw, R.; Rongen, G.A. Diannexin protects against renal ischemia reperfusion injury and targets phosphatidylserines in ischemic tissue. *PLoS ONE* **2011**, *6*, e24276. [CrossRef] [PubMed]

85. Su, H.; van Dam, G.M.; Buis, C.I.; Visser, D.S.; Hesselink, J.W.; Schuurs, T.A.; Leuvenink, H.G.; Contag, C.H.; Porte, R.J. Spatiotemporal expression of heme oxygenase-1 detected by in vivo bioluminescence after hepatic ischemia in HO-1/*luc* mice. *Liver Transplant.* **2006**, *12*, 1634–1639. [CrossRef] [PubMed]

86. Shin, H.J.; Won, N.H.; Lee, H.W. Remote ischemic preconditioning prevents lipopolysaccharide-induced liver injury through inhibition of NF-κB activation in mice. *J. Anesth.* **2014**, *28*, 898–905. [CrossRef] [PubMed]

87. Abu-Amara, M.; Yang, S.Y.; Quaglia, A.; Rowley, P.; Tapuria, N.; Seifalian, A.M.; Fuller, B.J.; Davidson, B.R. Effect of remote ischemic preconditioning on liver ischemia/reperfusion injury using a new mouse model. *Liver Transplant.* **2011**, *17*, 70–82. [CrossRef] [PubMed]

88. Cai, Z.P.; Parajuli, N.; Zheng, X.; Becker, L. Remote ischemic preconditioning confers late protection against myocardial ischemia-reperfusion injury in mice by upregulating interleukin-10. *Basic Res. Cardiol.* **2012**, *107*, 277. [CrossRef] [PubMed]

89. van den Brand, B.T.; Vermeij, E.A.; Waterborg, C.E.; Arntz, O.J.; Kracht, M.; Bennink, M.B.; van den Berg, W.B.; van de Loo, F.A. Intravenous delivery of HIV-based lentiviral vectors preferentially transduces F4/80+ and ly-6c+ cells in spleen, important target cells in autoimmune arthritis. *PLoS ONE* **2013**, *8*, e55356. [CrossRef] [PubMed]

90. Hadi, A.M.; Mouchaers, K.T.; Schalij, I.; Grunberg, K.; Meijer, G.A.; Vonk-Noordegraaf, A.; van der Laarse, W.J.; Belien, J.A. Rapid quantification of myocardial fibrosis: A new macro-based automated analysis. *Cell. Oncol.* **2011**, *34*, 343–354. [CrossRef] [PubMed]

91. Bjornsson, B.; Winbladh, A.; Bojmar, L.; Sundqvist, T.; Gullstrand, P.; Sandstrom, P. Conventional, but not remote ischemic preconditioning, reduces iNOS transcription in liver ischemia/reperfusion. *World J. Gastroenterol.* **2014**, *20*, 9506–9512. [PubMed]

92. He, X.; Zhao, M.; Bi, X.Y.; Yu, X.J.; Zang, W.J. Delayed preconditioning prevents ischemia/reperfusion-induced endothelial injury in rats: Role of ROS and eNOS. *Lab. Investig.* **2013**, *93*, 168–180. [CrossRef] [PubMed]

© 2017 by the authors. Licensee MDPI, Basel, Switzerland. This article is an open access article distributed under the terms and conditions of the Creative Commons Attribution (CC BY) license (http://creativecommons.org/licenses/by/4.0/).

International Journal of

Molecular Sciences

MDPI

Article

The Acute Inflammatory Response to Absorbed Collagen Sponge Is Not Enhanced by BMP-2

Hairong Huang [1], Daniel Wismeijer [1], Ernst B. Hunziker [2] and Gang Wu [1,*]

[1] Department of Oral Implantology and Prosthetic Dentistry, Academic Centre for Dentistry Amsterdam (ACTA), University of Amsterdam and Vrije Universiteit Amsterdam, Amsterdam Movement Sciences, Gustav Mahlerlaan 3004, 1081LA Amsterdam, The Netherlands; hhrstudy@126.com (H.H.); d.wismeijer@acta.nl (D.W.)

[2] Departments of Osteoporosis and Orthopaedic Surgery, Inselspital (DKF), University of Bern, Murtenstrasse 35, 3008 Bern, Switzerland; ernst.hunziker@dkf.unibe.ch

* Correspondence: g.wu@acta.nl; Tel.: +31-20-598-0866; Fax: +31-20-598-0333

Academic Editor: Allison Cowin
Received: 1 January 2017; Accepted: 16 February 2017; Published: 25 February 2017

Abstract: Absorbed collagen sponge (ACS)/bone morphogenetic protein-2 (BMP-2) are widely used in clinical practise for bone regeneration. However, the application of this product was found to be associated with a significant pro-inflammatory response, particularly in the early phase after implantation. This study aimed to clarify if the pro-inflammatory activities, associated with BMP-2 added to ACS, were related to the physical state of the carrier itself, i.e., a wet or a highly dehydrated state of the ACS, to the local degree of vascularisation and/or to local biomechanical factors. ACS (0.8 cm diameter)/BMP-2 were implanted subcutaneously in the back of 12 eight-week-old Sprague Dawley rats. Two days after surgery, the implanted materials were retrieved and analysed histologically and histomorphometrically. The acute inflammatory response following implantation of ACS was dependent of neither the presence or absence of BMP-2 nor the degree of vascularization in the surrounding tissue nor the hydration state (wet versus dry) of the ACS material at the time of implantation. Differential micro biomechanical factors operating at the implantation site appeared to have an influence on the thickness of inflammation. We conclude that the degree of the early inflammatory response of the ACS/BMP-2 may be associated with the physical and chemical properties of the carrier material itself.

Keywords: bone morphogenetic protein-2 (BMP-2); absorbed collagen sponge (ACS); inflammation; vascularization; biomechanical

1. Introduction

Recombinant human bone morphogenetic protein-2 (BMP-2), a member of the transforming growth factor beta (TGF-β) superfamily, is in clinical use since more than a decade [1,2]. It is used in clinical practice for spinal fusion [3] and for treatment of non-unions to enhance bone formation processes and to accelerate the bony healing response; in dental practice it is used for oral and maxillofacial reconstruction [4,5].

Even though the clinical use of BMP-2 is very successful, its clinical application is associated with some serious unwanted effects such as heterotopic bone formation [6], bone resorption (by osteoclast activation) and formation of cyst-like bone voids [7], as well as postoperative inflammatory swelling [8,9] and neurological symptoms, etc.

BMP-2 is clinically applied as a free factor (Infuse® (USA), Inductos® (Europe)) together with an ACS as a carrier. BMP-2 of this product is used in very high dosage, and it is believed that it is this high dosage level of BMP-2 that leads to extensive inflammatory responses. This use-associated inflammation is one of the main reasons why several of the above-described unwanted effects do occur. It is also believed by many authors that BMP-2 itself contributes significantly to the enhancing of the inflammatory response during and after the implantation of the construct in this kind of a tissue engineering approach. Indeed, several publications report that BMP-2 itself enhances the swelling and the inflammatory response in the surrounding soft tissues in conjunction with the carrier material (ACS) [10].

Seroma formation is, for example, a frequently observed side effect of BMP-2-use, encountered most commonly in the first week postoperatively, as described in several studies [8,11]. Rihnet et al. [12] found that lumbar seromas occurred in 1.2% of rhBMP-2 treated patients compared to 0% in the control patient population. Robin et al. [8] described postoperative seroma formations associated with BMP-2 use in the surrounding soft tissues in the cervical region that led to bilateral paresthesia of the upper extremities. In clinical cases with BMP-2-induced seromas, elevated serum levels of inflammatory cytokines were found, such as those of IL-6, IL-8, and TNF-α [13], as well as those of IL-10 [10].

Indeed in the publication of Lee et al. [10], a dose-dependency of the inflammatory response to high dosage levels of BMP-2 was found. However, in a report of Wu et al. [14] it was described that BMP-2, in particular when delivered in a slow release system, is able to attenuate inflammatory responses. In other in vivo animal experiments [15], microcomputed tomography and histological analyses confirmed that PCL/PLGA/collagen/rhBMP-2 scaffolds (long-term delivery mode) showed the best bone healing quality at both weeks 4 and 8 after implantation without inflammatory response. Thus, conflicting data are encountered in the scientific literature respecting the role of BMP-2 and it use-associated inflammation.

The purpose of this study was to investigate if the use of BMP-2, when applied at high concentrations as a free factor together with a carrier material (ACS), is indeed associated with a pro-inflammatory response in the acute phase of the body response, i.e., in the initial two days after implantation of this growth factor with the carrier material. It is, indeed, conceivable that it is not the BMP-2 itself that triggers the intensive inflammatory response, but that the inflammation may be elicited by a number of other factors operating in close topographical vicinity to the deposited collagen carrier. Such candidate factors may be the degree of tissue vascularity, or the local micromechanical conditions of different physiological stress fields, i.e., depend on differences in the local biological environment (differential niche biology). Another role may be played by the physical state in which the collagen carrier itself is deposited, i.e., inserted in a dry state or in a wet state into the living tissue spaces. Since burst release of BMP-2 (in surgical practise poured onto the ACS sponge) does readily occur, among others due to mechanical manipulation of the construct itself during surgical implantation [16], we set up in our experiments a specific control group in which the collagen carrier was kept in a dry state to assess the possible role of such mechanical stress-modulated release profiles of BMP-2 in the inflammatory response.

In order to clarify the possible role of these various candidate factors, the Sprague Dawley (SD) rat was used as the animal model. ACS carrier material was implanted in the subcutaneous space in the back area (lumbar level). By this set up the deposited collagen carrier patch is exposed on one side towards the skin, where the skin muscles of the rat generate a continuous instability situation, i.e., a high biomechanical instability [17]. On the opposite side of the collagen patch, facing the large underlying lumber muscle package, a relatively stable micromechanical environment is present. In addition, the two different biomechanical niches around these implants are also characterized by specific differential densities of blood vessels. The differential blood vessel densities at these two opposite locations (skin side versus lumbar body side) were quantified in this study in order to elucidate their possible proinflammatory contribution.

2. Results

Figure 1A–D illustrated that already on the 2nd day after implantation, all collagen implants were surrounded by a capsule of inflamed tissue (delineated by a red line), and was highly vascularized. The inflammatory response involved large numbers of macrophages around each of the implanted collagen sponges (Figure 1E). The outer border of the inflammation border of the collagen implant was delineated by a red line and the inner border of the inflammatory zone by a yellow line (Figure 1A–D).

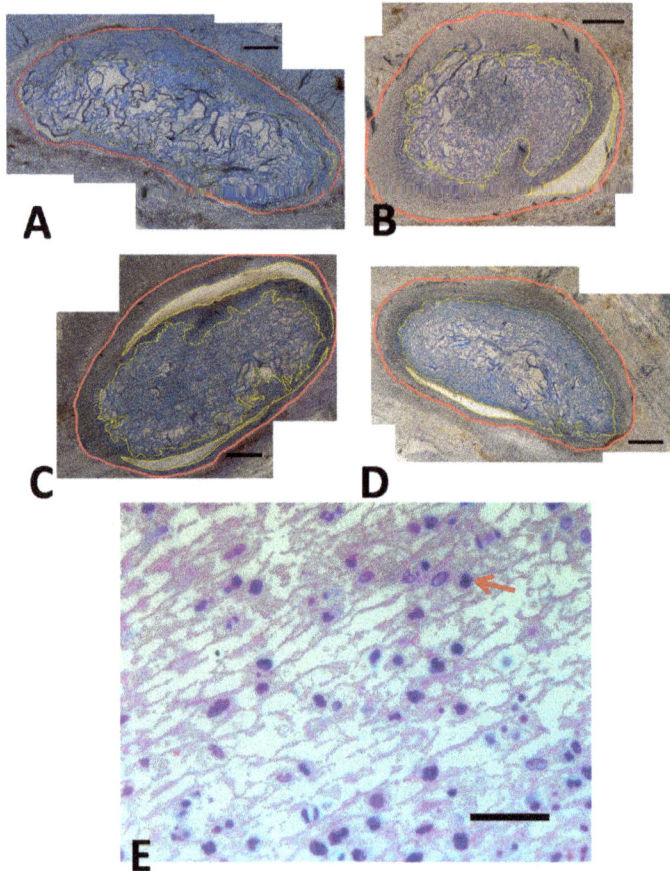

Figure 1. Microscopic findings following subcutaneous implantation of: (**A**) dry Absorbed Collagen Sponge (ACS); (**B**) dry bone morphogenetic protein-2 (BMP-2)/ACS; (**C**) wet ACS; (**D**) wet ACS/BMP-2; (**E**) high magnification of inflamed zone. Red arrow: macrophage. The inflammatory zone was delineated by two different lines: the outer border in red, the inner border in yellow. Bar = 500 μm (in **A**–**D**). The upper side is skin side and the lower side is lumber body side. Numerous macrophages were identified in the highly vascularized inflamed zone (cf. 1E, bar = 20 μm).

The degree of inflammation activity was gauged by estimation of the volume of the implanted sample and the volume of the inflamed tissue. As Figures 2 and 3 showed, the volumes of the implanted collagen sample and the inflammation area were increased when the carrier (ACS) was loaded with BMP-2. However, there were no significant differences observed between the collagen sponge volumes in the presence or absence of BMP-2, nor if implanted in a wet or a dry (dehydrated) state.

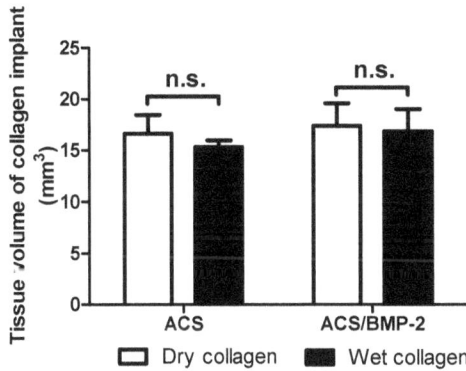

Figure 2. Mean volumes of collagen implants. No significant differences were found between dry ACS and dry ACS/BMP-2 nor between wet ACS and wet ACS/BMP-2. Data were present as Means ± SEM. n.s.: no significant difference.

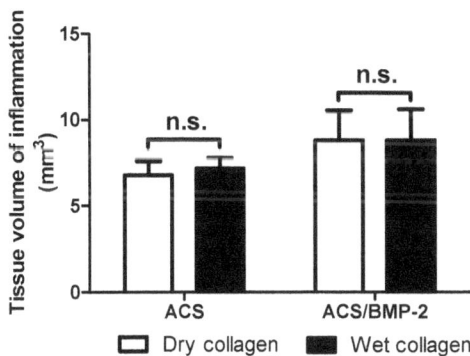

Figure 3. Periimplant inflammation volume. No significant differences were found between dry ACS and dry ACS/BMP-2 nor between wet ACS and wet ACS/BMP-2. Data were present as Means ± SEM. n.s.: no significant difference.

As Figure 4 illustrates, the mean thickness of the inflamed tissue at the skin side and the lumbar body side is different, and significant differences were indeed found around the dry ACS implants in the absence of BMP-2 ($p = 0.001$), and in the wet ACS groups in the presence ($p = 0.0009$) or absence ($p = 0.009$) of BMP-2.

The differential blood vessel densities at these two opposite locations (skin side versus lumbar body side) were quantified in this study in order to elucidate their possible role to contribute to the proinflammatory response. As Figure 5 shows, the area density of blood vessels on both sides were different, the area density of blood vessels in the dry group without BMP-2 on the lumbar body side was significantly higher than that on the skin side ($p = 0.014$) but no significance was found in the wet group. In the group with BMP-2, the area density of blood vessels in the dry group was found to be higher on the lumbar body side than on the skin side, but was not significantly different (due to a high degree of variation; cf. SEM-error bar in Figure 5), in the wet group, the area density of blood vessels on the lumbar body side was significantly higher than that on the skin side ($p = 0.032$). Figure 6 illustrates typical areas and blood vessel densities as encountered on the skin side (Figure 6A,C) and the lumbar body side (Figure 6B,D).

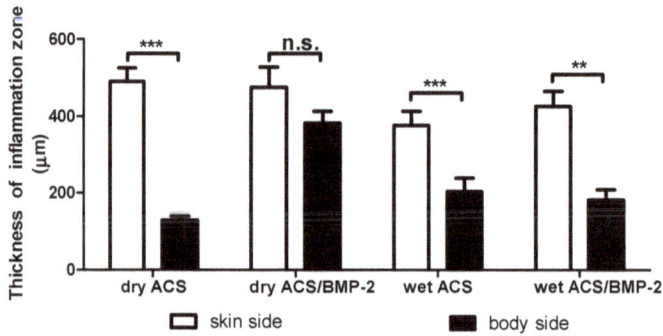

Figure 4. Comparison of the mean thickness of the inflammation zone on the skin side and the body side. There are significant differences in the thickness of the inflammatory zones between the skin side and the lumbar body side in the dry ACS implant group without BMP-2, and in both the wet ACS groups with or without BMP-2. **: $p < 0.01$, ***: $p < 0.001$, n.s.: no significant difference.

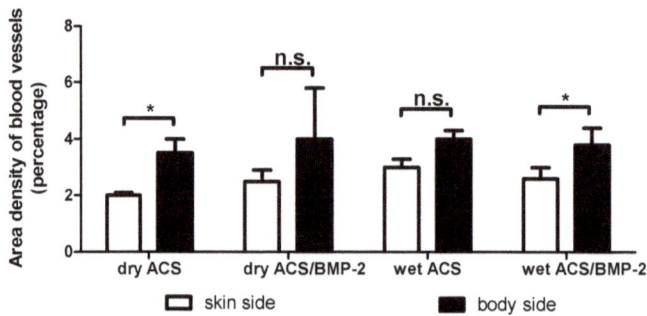

Figure 5. Area density of blood vessels in the dry and wet ACS implant groups, comparing the skin side blood vessel density with the lumbar body side blood vessel density. The data reveal that the density is significantly different physiologically. *: $p < 0.05$, n.s.: no significant difference.

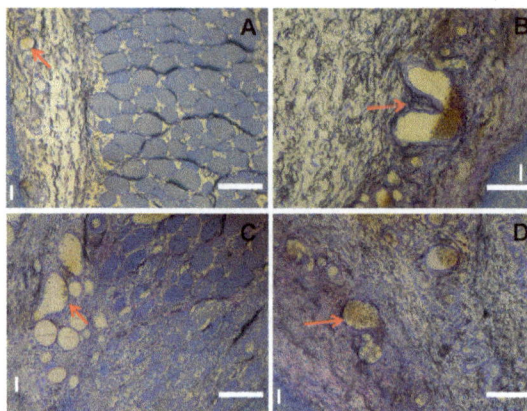

Figure 6. Illustration of blood vessel density in the inflamed area at the skin side (**A,C**) and the lumbar body side (**B,D**) from the dry (**A,B**) and wet (**C,D**) ACS. Arrows point to selected blood vessels. I: inflammation area. Bar = 100 μm. Red arrows: blood vessels.

3. Discussion

This study is focusing on the initial response of the tissue to the implantation of a sterile scaffold i.e., collagen matrix scaffold, available commercially for use in human patients.

The acute phase of inflammation within the two initial days after implantation is a sterile type of inflammation in the absence of an infection. It is a non-specific tissue response to the foreign body material implanted (carriers, biomaterials) [18]. Moreover, it is associated with tissue swelling, formation of edema as well as the influx of a cell population of the acute inflammatory response type, represented mainly by macrophages, and later on by foreign body giant cells [19]. This inflammatory response is not to be confused with infection, which is caused by foreign agents such as bacteria, viruses, etc. In this study, no infection was observed, and the inflammatory responses were all sterile in nature.

The comparison between wet ACS and dry ACS implanted in the subcutaneously space of rats revealed no difference in extent of inflammation in the acute phase (Figure 3). In addition, the sample size of the ACS, implanted the same way in all experimental groups, exhibited no differences occurring during these two early postimplantation days, i.e., no differences in early degradation activities (Figure 2); also the degree of inflammation, quantified by the inflammation volume around the implanted materials (Figure 3) during this acute inflammation phase did not reveal any significant differences between the control group and ACS/BMP-2 groups. These findings indicate that the acute inflammatory response in such cases is most likely based on the non-specific tissue reactions to foreign materials placed into the body, and it is not dependent on other factors in its extent.

In particular, the comparison between the extent of inflammation in topographically different areas such as the skin area compared to the lumbar body area, which are subjected to different biomechanical stress fields [17], and also to different degrees of vascularity (Figures 5 and 6), that both physiologically do occur at these sites, revealed no differences in the extent of the inflammatory response (Figure 3). This basically implies that the degree of vascularity is irrelevant respecting the extent of the acute inflammation response that can be expected following implantation of foreign materials into the body. The same applies to the state of the hydration of the implant material which is similarly irrelevant to the acute inflammation response with these materials i.e., implanted in a wet hydrated state or implanted into the body in a dry state. Due to the absence of the difference in the inflammatory response in 2 days it is probably implied that the dry material implanted get hydrated very rapidly inside the body so that no difference in inflammatory response can be monitored. However, the thickness of the local inflammation appeared quite irregular in the groups carrying BMP-2, represented by larger coefficient of variation (Figure 5) (dry ACS/BMP-2 group: coefficient of variation (CV) = 100%, coefficient of error (CE) = 45%) The thickness of the local inflammatory response was thus the only factor identified to show any differences between the two chosen topographical locations (skin versus lumbar body), and was thus associated with an asymmetrical response and a high degree of variation (dry ACS/BMP-2 group: CV = 100%, CE = 45%). This finding maybe a consequence of the angiogenetic activity of BMP-2 that has been proved previously by various authors [20–22], and may be related to a more rapid formation of blood vessels during the inflammation response when BMP-2 is present, and thus lead to the observed high irregularity of the extent of the inflammatory response. However, as a whole, the total inflammatory response remains the same in all experimental groups (Figure 3).

In the literature, it is described that in the subcutaneous tissue of rats, close to the skin, this area is biomechanically very instable, due to continuous skin muscle activities which are associated with irregular mechanical forces to occur, whereas in deeper areas near the lumbar spine muscles, less biomechanical instability is present in the associated tissues [17]; thus, the implanted materials are physiologically exposed at the skin side and at the lumbar body side to differential mechanical force fields with differential instability conditions. However, no major difference were observed respecting the extent of inflammation around the implanted materials at the different site, minor differences respecting thickness of local inflammation and its variance was found to be different. The most surprising finding in this study is the fact that the presence or absence of BMP-2 has no effect on the extent of the initial acute inflammatory response.

From studies in various animal models, BMPs are known to have species-specific osteoinductive dose requirements [23]. For example, in 2002, ACS/rhBMP-2 was FDA-approved as an autograft replacement for interbody spinal fusion procedures in human patients (at a concentration of 1.5 mg/cc) [24]. The BMP-2 concentration necessary for inducing consistent bone formation is substantially higher in nonhuman primates (0.75–2.0 mg/mL) than in rodents (0.02–0.4 mg/mL) [23]. In a recent publication, Luginbuehl et al. [25] found that 25 µg/mL in rodents, 50 µg/mL in dogs, 100 µg/mL in non-human primates and 800 µg/mL in humans, are quite different optimal osteoinductive BMP-2 concentrations, compared to the presently use clinical setting (0.75 and 1.5 mg/mL BMP-2) [26].

In a study of Lee et al. [10], the total amounts of BMP-2 used were 10 and 20 µg, and were diluted to 1 and 2 mg/mL, for addition to the ACS carrier, and resulted in a final BMP-2 /ACS carrier concentration of 3.3 and 6.67 mg/g for use. These authors found the inflammatory response to this construct not only to be dependent on the presence of BMP-2, but also proportionally related to its concentration. In our study, we used a total BMP-2 amount of 20 µg, dissolved and diluted to 1 mg/mL, and resulting in an ACS/BMP-2 carrier concentration of 10 mg/g, i.e., used BMP-2 in the same order of magnitude. However, we were unable to observe any additional pro-inflammatory response by the presence of BMP-2, as described by other authors [10,27]. Thus we conclude that the primary factors leading to the inflammatory response in the body are actually associated with the carrier itself and its chemical properties, but not to the presence of BMP-2. The materials used and the experimental conditions chosen in our study were the same (BMP-2, collagen) or quite similar (experimental conditions) to these previous studies [10].

It was interesting to find that in the different local areas (skin vs. lumbar body site), the thickness of the inflammatory response was indeed significantly different (Figure 4) and/or of high variability (see discussion above). We hypothesized that at sites of higher blood vessel densities on body side, we would expect more inflammation to occur, since inflammatory responses are dependent on the presence of an extensive blood vasculature, and would expect less inflammation at sites where the blood vessel density is lower. Since this was not the case in our study (see Figure 5), and this factor obviously overpowered by another biological influence, we attribute this finding to a higher biomechanical stability condition on the site with thicker inflammatory response, i.e., on the skin side. As Figure 5 illustrates in the group with dehydrated collagen sponges without BMP-2 and wet collagen sponges with BMP-2, the blood vessel density at the body side is significantly higher than that of the skin side. In the group with a dehydrated collagen sponge with BMP-2 and wet collagen sponges without BMP-2, the thickness of the inflammation zone between these two topographical sites did not show a significant difference, which would not be expected if the suggested hypothesis would be operative. The difference in inflammation thickness may thus be related to other factors, such as discussed above and in a recent review article of James et al. [5], in which the authors describe that specific anatomic locations can be associated with distinctive adverse events to implanted materials.

We thus conclude that according to our experimental findings the use of BMP-2 is not associated with the enhancement of pro-inflammatory effects in the initial phase of scaffold material implantation. The acute inflammatory response appears to be triggered predominately by the carrier material itself, its chemistry and physical properties, irrespective of the presence of BMP-2 or its absence. The aforementioned unwanted side effects, such as postoperative inflammatory swelling [8,9], possibly attribute to the carrier material itself, not by the BMP-2. As for a surgeon, should be very careful to select an optimal carrier for BMP-2. Given the fact that BMP-2 has been described by several authors to have an attenuating effect on inflammatory responses in the later phases of the implantations [14], it is actually not surprising that we are unable to confirm that BMP-2 would have a pro-inflammatory function.

4. Materials and Methods

4.1. Animal Preparation

Twelve eight-week-old male SD rats (mean weight 230 g, range from 190–250 g) were used in this study and divided into 4 experimental groups (*n* = 6 samples per group). ACSs (Medtronic Sofamor Danek, Memphis, TN, USA) were cut into identically sized circular samples (8 mm diameter). The experimental groups were defined as follows: Group1: ACS + 20 µL sterile water, group 2: ACS + 20 µL BMP-2 solution containing 20 µg BMP-2 (The dosage of BMP-2 was determined as previously described [10]); the samples of these two groups were stored under aseptic conditions overnight. Group 3: ACS + 20 µL sterile water and group 4: ACS + 20 µL BMP-2 were prepared freshly before surgery.

For induction of a general anesthesia 3% pentobarbital were intraperitoneally injected. Aseptic techniques were used during the surgical procedures. The iliac crest was used as the landmark for determining the location of the skin incision, a 25 mm posterior longitudinal incision was made bilaterally, 5–10 mm laterally from the midline. ACSs were implanted with or without BMP-2 into the subcutaneous space of the lumbar back. Right after implantation, the soft tissues were repositioned and the wound was closed using standard non-resorbable suture materials. The wound was then disinfected with 10% povidone-iodine. Animals were kept at 23 °C ambient temperature conditions until awakening.

4.2. Animal Husbandry

The SD rats were kept in animal experiment center (Zhejiang Chinese Medical University Laboratory Animal Research Center, Hangzhou, China). Temperature for keeping the SD rats was 18–23 centigrade, day/night light cycle time were 14/10 (h/h), humidity 60%–80%, sterile complete feed(Anlimo, Nanjing, China) and filtered water were freely available.

4.3. Tissue Processing

The rats were sacrificed on postoperative day 2, at which point the collagen samples were retrieved with the adhering/surrounding tissues and chemically fixed in buffered 10% formaldehyde solution [17] for 1 day at ambient temperature, they were rinsed in tap water, dehydrated in ethanol and embedded in methylmethacrylate [14]. Using a Leica diamond saw (Leco VC-50, St. Joseph, MO, USA), the tissue blocks were cut into 5–7 slices, 600-µm-thick and 1mm apart, according to a systematic random sampling protocol [28]. All slices were then glued to plastic specimen holders and ground down to a final thickness of 80–100 µm. They were then surface-polished and surface-stained with McNeal's Tetrachrome, basic Fuchsine and Toluidine blue, according to the publication of Schenk et al. [29].

4.4. Histomorphometry

4.4.1. Sample Volume and Volume of Inflammation

The sections were photographed at a final magnification of ×40 in a Nikon light microscope (Eclipse 50i Microscope, Tokyo, Japan), and photographic subsampling performed according to a systematic random-sampling protocol [28]. Using the photographic prints, the volume of the implants and the inflammation areas (associated with each sample) were determined by point counting [30], respecting stereological principles. The final volumes were estimated using Cavalieri's principle [28].

4.4.2. Thickness of Inflammation Volume

It was visually observed that the inflammation thickness of the periimplant inflammation zone was different when comparing the skin side and lumber body side areas. It therefore was decided to measure the thickness of the skin side and the opposite location at the body side by drawing parallel

Int. J. Mol. Sci. **2017**, *18*, 498

lines across the sample and vertically to its surface; thickness measurements were performed along these lines between the implant surface boundary and the end of the inflammation zone.

4.4.3. Blood Vessel Density

In dry ACS and ACS/BMP-2 group, using the photographic prints (magnification ×40), areas for high magnification imaging (×200) were chosen according to a systematic random protocol to be photographed and for morphometrical determination of the area density of blood vessels, again both on the skin side and on the opposite body side [28].

4.4.4. Statistical Analysis

Independent *t*-tests were applied to the data to obtain specific comparisons between experimental and control groups of the histomorphometrical data. All statistical analyses were performed with SPSS® 21.0 software (SPSS, Chicago, IL, USA), and statistical significance was defined as $p < 0.05$.

5. Conclusions

It is the collagen carrier itself that is the determining factor in eliciting and regulating the degree of the inflammatory response in the acute phase after implantation of an ACS/BMP-2 carrier construct in the bodily environment. This finding suggests that further development and optimization of the carrier material may be a promising way to reduce in the future the incidence and extent of the early inflammatory response as an unwanted side-effect in the soft tissue reactions around this type of implants.

Acknowledgments: This study was supported by the funds of China Scholarship Council, Natural Science Foundation of China (Grant No. 81400475 and No. 81470724) and Zhejiang Provincial Natural Science Foundation of China (No. LY14H140002, No. LY14H140006 and No. Y17H140023).

Author Contributions: Hairong Huang, Daniel Wismeijer, Ernst B. Hunziker and Gang Wu did experimental design; Hairong Huang and Gang Wu performed the experiments and collected the data; Hairong Huang, Daniel Wismeijer, Ernst B. Hunziker and Gang Wu did the data interpretation; Hairong Huang, Ernst B. Hunziker and Gang Wu wrote and revise the manuscript. Hairong Huang, Daniel Wismeijer, Ernst B. Hunziker and Gang Wu gave the final approval.

Conflicts of Interest: The authors declare no conflict of interest.

References

1. Wozney, J.M.; Rosen, V.; Celeste, A.J.; Mitsock, L.M.; Whitters, M.J.; Kriz, R.W.; Hewick, R.M.; Wang, E.A. Novel regulators of bone formation: Molecular clones and activities. *Science* **1988**, *242*, 1528–1534. [CrossRef] [PubMed]
2. Bessa, P.C.; Casal, M.; Reis, R.L. Bone morphogenetic proteins in tissue engineering: The road from the laboratory to the clinic, part I (basic concepts). *J. Tissue Eng. Regen. Med.* **2008**, *2*, 1–13. [CrossRef] [PubMed]
3. Cahill, K.S.; Chi, J.H.; Day, A.; Claus, E.B. Prevalence, complications, and hospital charges associated with use of bone-morphogenetic proteins in spinal fusion procedures. *JAMA* **2009**, *302*, 58–66. [CrossRef] [PubMed]
4. Benglis, D.; Wang, M.Y.; Levi, A.D. A comprehensive review of the safety profile of bone morphogenetic protein in spine surgery. *Neurosurgery* **2008**. [CrossRef] [PubMed]
5. James, A.W.; la Chaud, G.; Shen, J.; Asatrian, G.; Nguyen, V.; Zhang, X.; Ting, K.; Soo, C. A review of the clinical side effects of bone morphogenetic protein-2. *Tissue Eng. Part B Rev.* **2016**, *22*, 284–297. [CrossRef] [PubMed]
6. Shah, R.K.; Moncayo, V.M.; Smitson, R.D.; Pierre-Jerome, C.; Terk, M.R. Recombinant human bone morphogenetic protein 2-induced heterotopic ossification of the retroperitoneum, psoas muscle, pelvis and abdominal wall following lumbar spinal fusion. *Skelet. Radiol.* **2010**, *39*, 501–504. [CrossRef] [PubMed]
7. Balseiro, S.; Nottmeier, E.W. Vertebral osteolysis originating from subchondral cyst end plate defects in transforaminal lumbar interbody fusion using rhBMP-2. Report of two cases. *Spine J.* **2010**, *10*, e6–e10. [CrossRef] [PubMed]

8.   Robin, B.N.; Chaput, C.D.; Zeitouni, S.; Rahm, M.D.; Zerris, V.A.; Sampson, H.W. Cytokine-mediated inflammatory reaction following posterior cervical decompression and fusion associated with recombinant human bone morphogenetic protein-2: A case study. *Spine* **2010**, *35*, E1350–E1354. [CrossRef] [PubMed]

9.   Garrett, M.P.; Kakarla, U.K.; Porter, R.W.; Sonntag, V.K. Formation of painful seroma and edema after the use of recombinant human bone morphogenetic protein-2 in posterolateral lumbar spine fusions. *Neurosurgery* **2010**, *66*, 1044–1049. [CrossRef] [PubMed]

10. Lee, K.B.; Taghavi, C.E.; Song, K.J.; Sintuu, C.; Yoo, J.H.; Keorochana, G.; Tzeng, S.T.; Fei, Z.; Liao, J.C.; Wang, J.C. Inflammatory characteristics of rhBMP-2 in vitro and in an in vivo rodent model. *Spine* **2011**, *36*, E149–E154. [CrossRef] [PubMed]

11. Shahlaie, K.; Kim, K.D. Occipitocervical fusion using recombinant human bone morphogenetic protein-2: Adverse effects due to tissue swelling and seroma. *Spine* **2008**, *33*, 2361–2366. [CrossRef] [PubMed]

12. Rihn, J.A.; Patel, R.; Makda, J.; Hong, J.; Anderson, D.G.; Vaccaro, A.R.; Hilibrand, A.S.; Albert, T.J. Complications associated with single-level transforaminal lumbar interbody fusion. *Spine J.* **2009**, *9*, 623–629. [CrossRef] [PubMed]

13. Shen, J.; James, A.W.; Zara, J.N.; Asatrian, G.; Khadarian, K.; Zhang, J.B.; Ho, S.; Kim, H.J.; Ting, K.; Soo, C. BMP2-induced inflammation can be suppressed by the osteoinductive growth factor NELL-1. *Tissue Eng. Part A* **2013**, *19*, 2390–2401. [CrossRef] [PubMed]

14. Wu, G.; Liu, Y.; Iizuka, T.; Hunziker, E.B. The effect of a slow mode of BMP-2 delivery on the inflammatory response provoked by bone-defect-filling polymeric scaffolds. *Biomaterials* **2010**, *31*, 7485–7493. [CrossRef] [PubMed]

15. Shim, J.H.; Kim, S.E.; Park, J.Y.; Kundu, J.; Kim, S.W.; Kang, S.S.; Cho, D.W. Three-dimensional printing of rhbmp-2-loaded scaffolds with long-term delivery for enhanced bone regeneration in a rabbit diaphyseal defect. *Tissue Eng. Part A* **2014**, *20*, 1980–1992. [CrossRef] [PubMed]

16. Geiger, M.; Li, R.H.; Friess, W. Collagen sponges for bone regeneration with rhBMP-2. *Adv. Drug Deliv. Rev.* **2003**, *55*, 1613–1629. [CrossRef] [PubMed]

17. Hagi, T.T.; Wu, G.; Liu, Y.; Hunziker, E.B. Cell-mediated BMP-2 liberation promotes bone formation in a mechanically unstable implant environment. *Bone* **2010**, *46*, 1322–1327. [CrossRef] [PubMed]

18. Ratner, B.D.; Bryant, S.J. Biomaterials: Where we have been and where we are going. *Annu. Rev. Biomed. Eng.* **2004**, *6*, 41–75. [CrossRef] [PubMed]

19. Rodriguez, A.; Meyerson, H.; Anderson, J.M. Quantitative in vivo cytokine analysis at synthetic biomaterial implant sites. *J. Biomed. Mater. Res. A* **2009**, *89*, 152–159. [PubMed]

20. Deckers, M.M.; van Bezooijen, R.L.; van der Horst, G.; Hoogendam, J.; van Der Bent, C.; Papapoulos, S.E.; Lowik, C.W. Bone morphogenetic proteins stimulate angiogenesis through osteoblast-derived vascular endothelial growth factor A. *Endocrinology* **2002**, *143*, 1545–1553. [CrossRef] [PubMed]

21. De Jesus Perez, V.A.; Alastalo, T.P.; Wu, J.C.; Axelrod, J.D.; Cooke, J.P.; Amieva, M.; Rabinovitch, M. Bone morphogenetic protein 2 induces pulmonary angiogenesis via Wnt-β-catenin and Wnt-RhoA-Rac1 pathways. *J. Cell Biol.* **2009**, *184*, 83–99. [CrossRef] [PubMed]

22. Raida, M.; Clement, J.H.; Leek, R.D.; Ameri, K.; Bicknell, R.; Niederwieser, D.; Harris, A.L. Bone morphogenetic protein 2 (BMP-2) and induction of tumor angiogenesis. *J. Cancer Res. Clin. Oncol.* **2005**, *131*, 741–750. [CrossRef] [PubMed]

23. Bagaria, V. Bone morphogenic protein: Current state of field and the road ahead. *J. Orthop.* **2005**, *2*, e3.

24. McKay, W.F.; Peckham, S.M.; Badura, J.M. A comprehensive clinical review of recombinant human bone morphogenetic protein-2 (infuse bone graft). *Int. Orthop.* **2007**, *31*, 729–734. [CrossRef] [PubMed]

25. Luginbuehl, V.; Meinel, L.; Merkle, H.P.; Gander, B. Localized delivery of growth factors for bone repair. *Eur. J. Pharm. Biopharm.* **2004**, *58*, 197–208. [CrossRef] [PubMed]

26. Govender, S.; Csimma, C.; Genant, H.K.; Valentin-Opran, A.; Amit, Y.; Arbel, R.; Aro, H.; Atar, D.; Bishay, M.; Borner, M.G.; et al. Recombinant human bone morphogenetic protein-2 for treatment of open tibial fractures: A prospective, controlled, randomized study of four hundred and fifty patients. *J. Bone Joint Surg. Am.* **2002**, *84*, 2123–2134. [CrossRef] [PubMed]

27. Zara, J.N.; Siu, R.K.; Zhang, X.; Shen, J.; Ngo, R.; Lee, M.; Li, W.; Chiang, M.; Chung, J.; Kwak, J.; et al. High doses of bone morphogenetic protein 2 induce structurally abnormal bone and inflammation in vivo. *Tissue Eng. Part A* **2011**, *17*, 1389–1399. [CrossRef] [PubMed]

28. Gundersen, H.J.; Bendtsen, T.F.; Korbo, L.; Marcussen, N.; Moller, A.; Nielsen, K.; Nyengaard, J.R.; Pakkenberg, B.; Sorensen, F.B.; Vesterby, A.; et al. Some new, simple and efficient stereological methods and their use in pathological research and diagnosis. *APMIS* **1988**, *96*, 379–394. [CrossRef] [PubMed]
29. Schenk, R.K.; Olah, A.J.; Herrmann, W. Preparation of Calcified Tissues for Light Microscopy. In *Methods of Calcified Tissue Preparation*; Dickson, G.R., Ed.; Elsevier Science Publishers B.V.: Amsterdam, The Netherlands, 1984; pp. 1–56.
30. Cruz-Orive, L.M.; Weibel, E.R. Recent stereological methods for cell biology: A brief survey. *Am. J. Physiol.* **1990**, *258*, L148 L156. [PubMed]

© 2017 by the authors. Licensee MDPI, Basel, Switzerland. This article is an open access article distributed under the terms and conditions of the Creative Commons Attribution (CC BY) license (http://creativecommons.org/licenses/by/4.0/).

International Journal of
Molecular Sciences

MDPI

Article

Inhibition of NLRP3 Inflammasome Pathway by Butyrate Improves Corneal Wound Healing in Corneal Alkali Burn

Fang Bian [1], Yangyan Xiao [1,2], Mahira Zaheer [1], Eugene A. Volpe [1], Stephen C. Pflugfelder [1], De-Quan Li [1] and Cintia S. de Paiva [1,*]

[1] Ocular Surface Center, Department of Ophthalmology, Baylor College of Medicine, Houston, TX 77030, USA; bftongji@hotmail.com (F.B.); yangyanx@bcm.edu (Y.X.); mzaheer@bcm.edu (M.Z.); eugenevolpe@yahoo.com (E.A.V.); stevenp@bcm.edu (S.C.P.); dequanl@bcm.edu (D.-Q.L.)

[2] Department of Ophthalmology, the Second Xiangya Hospital, Central South University, Changsha 410011, Hunan Province, China

* Correspondence: cintiadp@bcm.edu; Tel.: +1-713-798-2124

Academic Editor: Allison Cowin
Received: 26 January 2017; Accepted: 28 February 2017; Published: 5 March 2017

Abstract: Epithelial cells are involved in the regulation of innate and adaptive immunity in response to different stresses. The purpose of this study was to investigate if alkali-injured corneal epithelia activate innate immunity through the nucleotide-binding oligomerization domain-containing protein (NOD)-like receptor family pyrin domain containing 3 (NLRP3) inflammasome pathway. A unilateral alkali burn (AB) was created in the central cornea of C57BL/6 mice. Mice received either no topical treatment or topical treatment with sodium butyrate (NaB), β-hydroxybutyric acid (HBA), dexamethasone (Dex), or vehicle (balanced salt solution, BSS) quater in die (QID) for two or five days (d). We evaluated the expression of inflammasome components including NLRP3, apoptosis-associated speck-like protein (ASC), and caspase-1, as well as the downstream cytokine interleukin (IL)-1β. We found elevation of NLRP3 and IL-1β messenger RNA (mRNA) transcripts, as well as levels of inflammasome component proteins in the alkali-injured corneas compared to naïve corneas. Treatment with NLRP3 inhibitors using NaB and HBA preserved corneal clarity and decreased NLRP3, caspase-1, and IL-1β mRNA transcripts, as well as NLRP3 protein expression on post-injury compared to BSS-treated corneas. These findings identified a novel innate immune signaling pathway activated by AB. Blocking the NLRP3 pathway in AB mouse model decreases inflammation, resulting in greater corneal clarity. These results provide a mechanistic basis for optimizing therapeutic intervention in alkali injured eyes.

Keywords: alkali injury; NLRP3 inflammasome; sodium butyrate; β-hydroxybutyric acid

1. Introduction

Chemical injuries to the eye represent between 11.5% and 22.1% of ocular injuries [1], often resulting in an aggressive inflammatory response that can impair corneal re-epithelization, promote keratolysis, and lead to globe perforation at the acute stage. Despite medical and surgical management, most ocular chemical injuries still result in loss of vision or the eye. It is possible that the poor therapeutic response stems largely from an incomplete understanding of the host inflammatory response.

Although the mechanisms initiated in wound healing have not been fully elucidated, there is increasingly evidence that epithelial cells serve not only as an effective barrier against most microorganisms, but also are central participants in the bridging of innate and adaptive immunity [2,3].

The innate immune response relies on evolutionarily ancient germline-encoded receptors, the pattern-recognition receptors (PRRs). Pattern-recognition receptors are expressed by a variety of cells to identify microbial pathogens known as pathogen-associated molecular patterns (PAMPs) and to recognize cell components that are released during cell damage or death known as damage-associated molecular patterns (DAMPs). The main families of PPRs, Toll like receptors (TLRs) and nucleotide-binding oligomerization domain-containing protein (NOD)-like receptors (NLRs), have been shown to be expressed on certain epithelial cells [4]. Airway epithelial cells express a variety TLRs that help them to sense bacterial, fungal, or viral exposure and respond accordingly by secreting a large array of molecules that initiate adaptive immune response [5]. Human corneal epithelial cells also express a variety of TLRs (TLR1, TLR2, TLR3, TLR4, TLR5, TLR6, and TLR9) that may help them mount an adequate response to microbial exposure and other environment stimuli such as oxidative and desiccating stress [6–8].

NOD-like receptor family pyrin domain containing 3 (NLRP3), a major inflammatory innate defense mechanism, participates in inflammasome formation through the recruitment of the adapter apoptosis-associated speck-like protein (ASC) with subsequent activation of caspase-1, which leads to secretion of interleukin (IL)-1β or IL-18. NOD-like receptor family pyrin domain containing 3 inflammasome (NLRP3) has been described as an innate sensor of host-derived DAMPs that are released following tissue injury or cell death to activate pro-inflammatory pathways [9,10]. The role of the innate immune response of corneal epithelial cells to sterile wounds is still unknown.

Sodium butyrate (NaB) is a fatty acid that is produced by fermentation of dietary fiber by anaerobic bacteria in the colon [11]. Sodium butyrate has been shown to block inflammasome-mediated inflammatory disease in experimental models of obesity-induced inflammation [12]. Ketone body β-hydroxybutyric acid (HBA) has a similar chemical structure as NaB and was also reported to inhibit NLRP3 activation [12], and has been shown to decrease apoptosis of rat corneal epithelia in dry eye conditions [13]. Therefore, we hypothesize that alkali burns may activate the inflammasome and that inhibition of this pathway with NaB and HBA would improve the fate of injured corneas by suppressing inflammation.

Here, we show that a novel innate immune signaling pathway (NLRP3–ASC–caspase-1–IL-1β) is activated in corneal epithelial cells by alkali burn. Blocking the NLRP3 pathway reduced inflammation, leading to improved wound healing and corneal clarity.

2. Results

2.1. Corneal Alkali Burn Upregulates NLRP3 and Adaptor Protein ASC in Corneal Epithelium

To determine the NLRP3 inflammasome expression in the alkali burned cornea, we first assessed the expression of NLRP3 mRNA transcript by real-time PCR using whole corneas harvested from 2 or 5 days post-alkali burns and naïve corneas as a comparator. As shown in Figure 1A, there was a remarkable increase in NLRP3 transcripts at two days (up to a 30-fold) and five days (up to a 10-fold) in wounded corneas compared to normal corneas.

Next, immunoreactivity of corneas to NLRP3 was evaluated by immunostaining (Figure 1B). Minimal levels of NLRP3 were present in the control corneas, while increased immunoreactivity against NLRP3 was observed in the corneal epithelium of wounded corneas at both 2 and 5 days post-alkali burns. Based on the immunostaining results which showed NLRP3 expression exclusively in the corneal epithelial layer, we repeated the experiment, but only collected corneal epithelium for western blot. Consistent with PCR and immunostaining results, the protein production levels of NLRP3 were increased at 2 days post-injury and slightly decreased at 5 days, but remained elevated compared to the naïve corneas (Figure 1C).

Upon activation, NLRP3 associates with the adaptor protein ASC to recruit pro-caspase-1. In order to investigate whether the adapter protein ASC increased in the wounded corneas, we evaluated

ASC expression by immunostaining in frozen sections. In contrast to normal control, ASC expression increased both at 2 and 5 days post-injury (Figure 2).

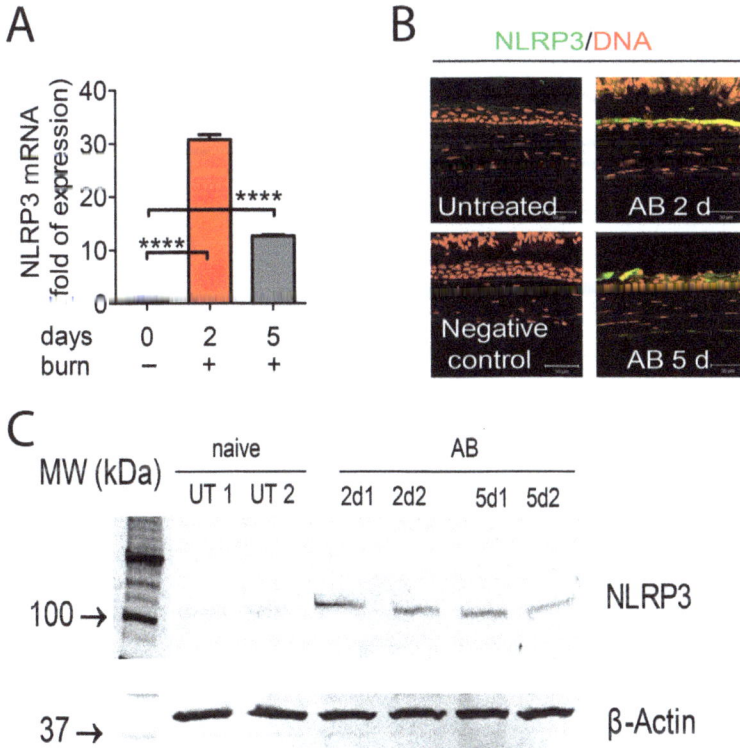

Figure 1. Upregulation of nucleotide-binding oligomerization domain-containing protein (NOD)-like receptor family pyrin domain containing 3 (NLRP3) expression by corneal epithelium in animals subjected to alkali burn (AB) at 2 days (2 d) and 5 days (5 d) post-injury. (**A**) Gene expression analysis of NLRP3 messenger mRNA (mRNA) transcript in whole cornea of animals subjected to alkali burn (n = 8 animals/group). Graphs show means ± standard error of the mean (SEM). **** $p < 0.0001$; (**B**) Representative merged digital images of laser scanning confocal microscopy of corneas cryosections immunostained for NLRP3 (green) with propidium iodide nuclei counterstaining (DNA in red) in corneas subjected to alkali burn (images are representative of n = 6 animals/group). Scale bar: 50 μm; (**C**) Representative digital images of western blot of NLRP3 and β-actin in cornea epithelium of animals subjected to alkali burn for 2 and 5 days. Each lane is a different sample (n = 12 animals/3 samples/group). MW: Molecular weight; UT: untreated control; 2 d and 5 d refer to mice subjected to alkali burn and euthanized 2 or 5 days post-injury, respectively (1 and 2 indicate different samples).

These data demonstrate that both NLRP3 and its adaptor protein ASC in cornea epithelial cells are upregulated following alkali injury, and prompted us to investigate the role of NLRP3 inflammasome in corneal alkali injuries.

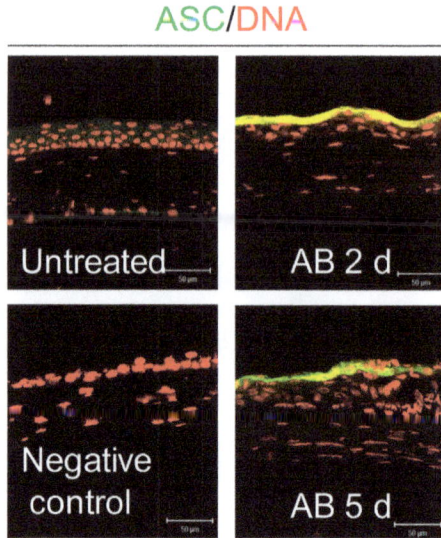

Figure 2. Upregulation of apoptosis-associated speck-like protein (ASC) in corneal epithelium of animals subjected to alkali burns (AB) at 2 days (2 d) and 5 days (5 d) post-injury. Representative merged digital images of laser scanning confocal microscopy in corneal cryosections immunostained for ASC (green) with propidium iodide nuclei counterstaining (DNA in red) in corneas subjected to alkali burn (*n* = 6 animals/group). Scale bar: 50 μm.

2.2. Caspase-1 Activation in Alkali-Injured Corneas

Upon interaction of NLRP3 with ASC, the ASC recruits pro-caspase at the caspase activation and recruitment domain (CARD), thus forming active inflammasome complexes in response to a variety of exogenous and endogenous stressors [14,15]. We evaluated the expression of *caspase-1* gene in injured corneas and compared findings with normal corneas. As shown in Figure 3A, alkali-injured corneas had higher expression of caspase-1 transcripts at both 2 and 5 days compared with normal controls.

Immunoreactivity of corneas to caspase-1 was evaluated by immunostaining (Figure 3B). Although barely detected in naïve corneas, increased reactivity against caspase-1 was noted in cornea epithelia of wounded corneas at 2 and 5 days. Caspase-1 activity in the corneal epithelium demonstrated significantly higher levels of activated caspase-1 at both 2 and 5 days compared to untreated corneas (Figure 3C).

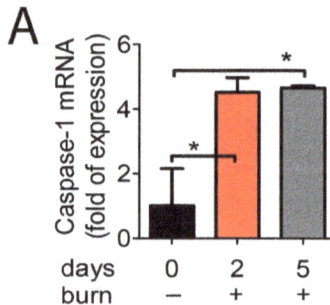

Figure 3. *Cont.*

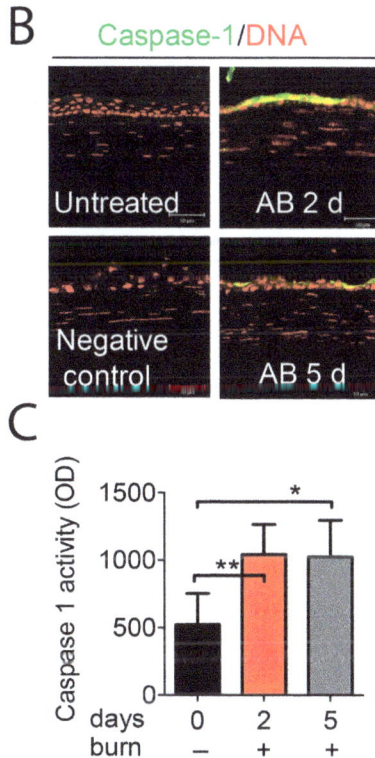

Figure 3. Caspase-1 activation in corneal epithelium in animals subjected to alkali burn (AB) at 2 days (2 d) and 5 days (5 d) post-injury. (**A**) Gene expression analysis of caspase-1 mRNA transcript in whole cornea of animals subjected to alkali burn (n = 8 animals/group). Graphs show mean ± standard error of the mean (SEM); * $p < 0.05$; (**B**) Representative merged digital images of laser scanning confocal microscopy in corneal cryosections immunostained for caspase-1 (green) with propidium iodide nuclei counterstaining (DNA in red) in animals subjected to alkali burn (n = 6 animals/group). Scale bar: 50 μm; (**C**) Caspase-1 activity in whole cornea lysates from eyes subjected to alkali burn (n = 12 animals/3 samples/group). Graphs show mean ± SEM; * $p < 0.05$, ** $p < 0.01$.

These results show that caspase-1 is increased and activated in corneal alkali-burned corneas. Taken together, these findings show that the NLRP3–ASC–caspase-1 inflammasome is assembled in alkali-injured corneas.

2.3. The NLRP3–ASC–Caspase-1 Inflammasome Directs IL-1β Secretion in Alkali-Injured Corneas

The NLRP3 inflammasome is critical for IL-1β production [16,17]. To investigate whether NLRP3 inflammasome formation can lead to IL-1β production in the alkali corneal burn model, we evaluated the expression of IL-1β in wounded corneas by PCR and immunostaining. As shown in Figure 4A, wounded corneas had higher expression of IL-1β mRNA at 2 and 5 days compared to naïve corneas. Consistent with PCR results, increased reactivity to IL-1β in the corneal epithelium was seen in wounded corneas at 2 and 5 days.

Taken together, our results demonstrated that NLRP3 inflammasome is involved in corneal alkali injury, leading to IL-1β production, which may be critical in inducing an inflammatory response.

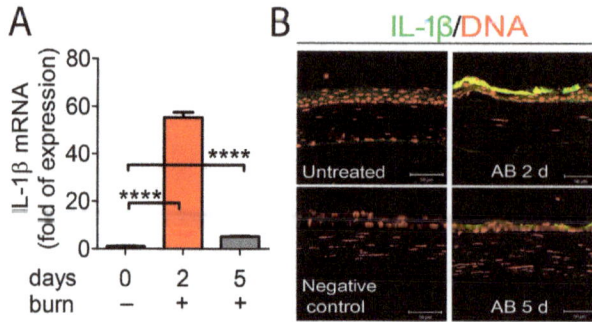

Figure 4. Upregulation of interleukin (IL)-1β in corneal epithelium in animals subjected to alkali burns (AB) at 2 days (2 d) and 5 days (5 d) post-injury. (**A**) Gene expression analysis of IL-1β mRNA transcripts in whole corneas of eyes subjected to alkali burn (n = 8 animals/group). **** p < 0.0001) (**B**) Representative merged digital images of laser scanning confocal microscopy in corneal cryosections immunostained for IL-1β (green) with propidium iodide nuclei counterstaining (DNA in red) in corneas subjected to alkali burn (n = 6 animals/group). Scale bar: 50 μm.

2.4. Sodium Butyrate and β-Hydroxybutyric Acid Topical Treatment Inhibits Activation of NLRP3 Inflammasome in Alkali-Injured Corneas

Recent studies described NaB and HBA as inhibitors of NLRP3 inflammasome that can inhibit IL-1β release and reduce inflammation [12,18,19]. To test the hypothesis that blocking NLRP3 would improve clinical parameters and reduce inflammation, alkali-burned corneas were treated with either NaB, HBA, or vehicle for 2 or 5 days post-injury. To compare the relative potency of these molecules, a group of alkali-injured corneas treated with dexamethasone (Dex) was also included, as our previous study showed that Dex is very efficacious in preserving corneal clarity and suppressing inflammation in the most severe model of alkali burn and dry eye [20].

Since there is no evidence of the safety of NaB and HBA application on the ocular surface, NaB and HBA toxicity to the corneal epithelium was first investigated by topically applying NaB or HBA eyedrops quater in die (QID) in naïve mice. Corneal epithelial integrity was assessed using corneal fluorescein staining after 5 days. As shown in Figure 5A, naïve corneas that received NaB or HBA for 5 days had no corneal epithelial defect, indicating that both HBA and NaB were safe to use in vivo.

Next, mice were subjected to alkali burn (AB) and topically treated with NaB, HBA, or BSS (QID), and clinical parameters of cornea opacity and wound healing were evaluated on a daily basis. Representative color digital images used to score corneal opacity (top row) and fluorescein stained corneas used to measure wound closure rate (bottom row) are shown in Figure 5B. At 5 days post-injury, HBA treated corneas had significantly higher wound closure rate (Figure 5C) and both NaB and HBA treated corneas had lower corneal opacity scores (Figure 5D) compared to vehicle controls, respectively. β-hydroxybutyric acid-treated corneas healed significantly faster than vehicle treated corneas, with 100% wound closure at 4 days post-injury.

To determine if blocking NLRP3 inflammasome can suppress inflammation in alkali-injured corneas, the expression of NLRP3, caspase-1, and IL-1β mRNA transcripts in alkali burned corneas treated with NaB or HBA were evaluated using real-time PCR and compared with vehicle-treated corneas. Polymerase chain reaction results showed that both NaB treatment and HBA treatment after alkali burn significantly decreased NLRP3, caspase-1, and IL-1β mRNA transcripts at both 2 and 5 days compared to BSS-treated corneas (Figure 5E). Western blot confirmed the real-time PCR results with decreased protein expression of NLRP3 in corneal epithelia in AB + HBA and AB + NaB groups compared to AB + vehicle group at 2 days (Figure 5F).

Alkali-burned corneas treated with Dex showed significant improvement in corneal opacity (Figure 5A–C), and impressive decreases in inflammatory cytokines (Figure 5E). Although the anti-inflammatory effect of Dex was greater than NaB and HBA, NaB and HBA treatment can decrease corneal opacity scores to the level of Dex treated groups, indicating that NaB and HBA are very efficacious in preserving corneal clarity. These data indicated that blocking the NLRP3 inflammasome pathway using NaB and HBA drops improves clinical parameters and decreases inflammation in animals subjected to alkali burns at 2 and 5 days post-injury.

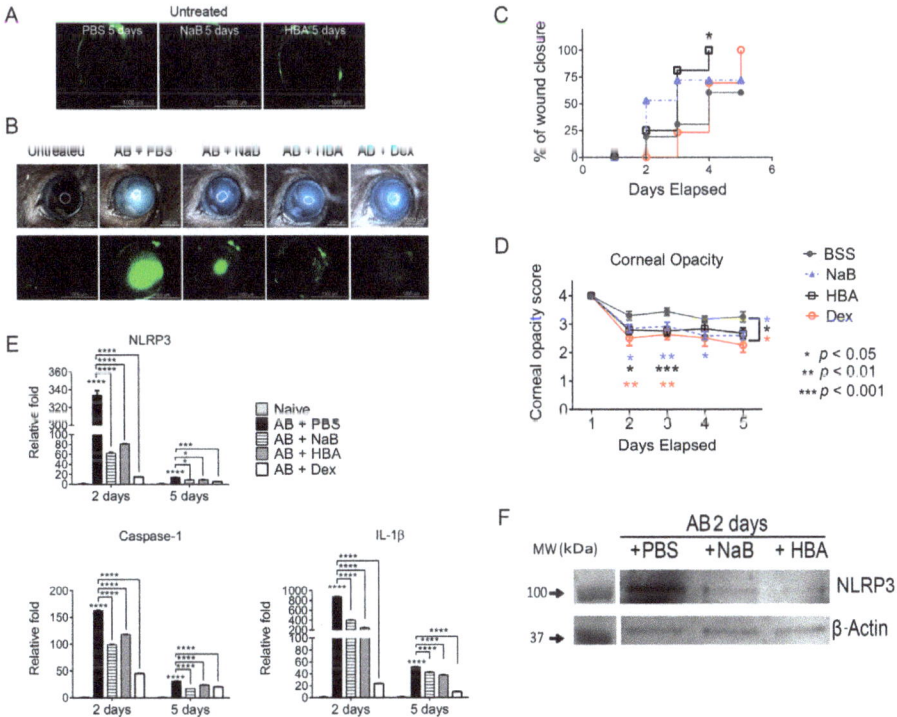

Figure 5. Sodium butyrate (NaB) and hydroxybutyric acid (HBA) eyedrops improve clinical parameters and decrease inflammation in animals subjected to alkali burns (AB). (**A**) Digital images of corneas stained with 0.1% sodium fluorescein demonstrating intact corneal epithelium after topical NaB, HBA or balanced salt solution (BSS) treatment for 5 days quater in die (QID) in untreated mice (*n* = 5 animals/group). Scale bar: 1000 μm; (**B**) Representative color digital images used to score corneal opacity (**top row**) and representative fluorescein stained corneas used to create wound closure rate (**bottom row**) at 5 days (*n* = 15 animals/group) after alkali burn. Scale bar: 1000 μm; (**C**) Corneal opacity in corneas subjected to alkali burn and topically treated with either NaB, HBA, or BSS and compared with dexamethasone (Dex) (*n* = 15 animals/group); (**D**) Wound closure rate in corneas subjected to alkali burn and topically treated with either NaB, HBA, or BSS and compared with Dex (*n* = 15 animals/group); (**E**) Mean ± SEM of results of gene expression analysis of *NLRP3*, *Caspase-1*, and *IL-1β* in whole cornea of animals subjected to alkali burn for 2 or 5 days and topically treated with either NaB, HBA, or BSS and compared with Dex (*n* = 5 animals/group). * $p < 0.05$, *** $p < 0.001$, **** $p < 0.0001$. (**F**) Representative digital images of western blot of NLRP3 and β-actin in cornea epithelium of animals subjected to alkali burn for 2 days and topically treated with either NaB, HBA, or BSS (*n* = 12 animals/3 samples/group).

3. Discussion

In the present study, we demonstrated that corneal epithelial cells participated in innate immunity through the NLRP3–ASC–caspase-1–IL-1β pathway in response to sterile corneal injuries. Blocking the NLRP3 pathway with NaB and HBA in the injured corneas can suppress inflammation and preserve cornea clarity post injury.

Damage to the cornea from an ocular chemical burn is severe, leading to overwhelming sterile inflammatory responses, and corneal epithelial defects or sterile corneal ulceration at the acute stage. The role of corneal epithelium in this inflammatory cascade has not been established. Studies have shown that epithelial cells play a vital role in innate immunity by expressing the main families of PPRs, including TLRs and NLRs, in response to various ocular surface stimuli [6–8]. In the context of infection, *Pseudomonas aeruginosa* flagellin elicits inflammatory response of corneal epithelium through the TLR5–nuclear factor κ-light-chain-enhancer of activated B cells (NF-κB) signaling pathway [7]. Human corneal epithelial cells exposed to *Aspergillus fumigatus* can trigger innate immune response via activating NOD 1 receptors on human corneal epithelial cells, leading to the secretion of IL-6, IL-8, and tumor necrosis factor-α (TNF-α). NOD1 knockdown attenuated *Aspergillus fumigatus* triggered expression of *IL-6*, *IL-8*, and *TNF-α* [4].

It is now evident that PPRs can recognize endogenous molecules that are released during cellular injury termed DAMPs. Damage-associated molecular patterns released by dying cells can stimulate severe inflammation [21]. How corneal epithelial cells response to DAMPs and how this process affects the immune network has not been fully elucidated. The NLRP3 inflammasome has been described as an innate sensor of host-derived DAMPs that are released following tissue injury or cell death to activate pro-inflammatory pathways [9]. NOD-like receptor family pyrin domain containing 3 is expressed by hematopoietic and non-hematopoietic cells, such as keratinocyte and osteoblasts. In this study, we showed that the naïve corneal epithelial cells express low levels of NLRP3, while wounded corneal epithelium upregulated NLRP3 expression, indicating that NLRP3 participates in the wound healing process of a sterile corneal injury in response to DAMPs.

NOD-like receptor family pyrin domain containing 3 contains domains that can interact with adaptor protein ASC which has a CARD to recruit pro-caspase-1, thus forming the inflammasome [22–24]. Once activated, the NLRP3 inflammasome causes the activation of caspase-1 by cleaving pro-IL-1β into biologically active IL-1β. The adaptor protein ASC and pro-caspase-1 are also crucial to induce caspase-1 activation. Shrikant et al. [25] showed that calcium oxalate crystals induce renal inflammation by NLRP3 mediated IL-1β secretion. Interleukin-1β secretion was found to be reduced in NLRP3$^{-/-}$, ASC$^{-/-}$, and Caspase-1$^{-/-}$ mice. In our study, low levels of ASC and caspase-1 were found in naïve mouse corneas, while there was remarkably increased expression of ASC and caspase-1 in the wounded corneal epithelium, suggesting the activation of NLRP3 inflammasome in the corneal epithelia in response to corneal alkali injuries.

NOD-like receptor family pyrin domain containing 3 inflammasome has been implicated in the pathogenesis of many eye diseases. Zheng and colleagues [26] showed that reactive oxygen species induced NLRP3 expression by the corneal epithelium in a dry eye mouse model. We have also reported that NLRP3 and NLRC4 inflammasomes were activated in cultured human corneal epithelial in response to hyperosmolar stress [27]. In the context of infection, *Staphylococcus aureus* activates the NLRP3 inflammasome in human and rat conjunctival goblet cells [28]. In patients with age-related macular degeneration (AMD), increased mRNA levels of NLRP3, IL-1β, and pro-IL-18 in lesions of the retinal pigment epithelium and photoreceptors were detected [29]. Activation of NLRP3 inflammasome also contributed to retinal ganglion cell death following partial optic nerve crush injury [24] in a mouse model, while NLRP3 deficient mice have significantly reduced neuroinflammation and delayed retinal ganglion cell loss. Therefore, inhibiting NLRP3 assembly would be a novel therapeutic approach to limit many diseases, including eye diseases.

Recently, studies of targeted anti-NLRP3 therapy have opened a new chapter in the treatment of inflammatory disease. The ketone body HBA is produced by hepatocytes and astrocytes as an alternative energy source during fasting or exercise. However, in vivo and in vitro studies have demonstrated that HBA is more than just a metabolite. It has important cellular signaling roles as well. Youm et al. [19] reported HBA specifically inhibits NLRP3 inflammasome activation and decreases the production of active IL-1β and IL-18. Sodium butyrate has a similar chemical structure to HBA and also has been reported to inhibit the NLRP3 pathway [12]. To investigate if blocking NLRP3 pathway activation in injured corneas can improve clinical parameters and suppress inflammation, we used NaB and HBA topically in the alkali-burned eyes and evaluated corneal wound healing and corneal clarity, as well as the expression of NLRP3, caspase-1, and IL-1β in the burned corneas, and compared results with vehicle and Dex treated corneas. Dexamethasone therapy has been proven to be very efficacious in preserving corneal clarity and suppressing inflammation [20]. In this study, we observed that HBA and NaB treatment can preserve cornea clarity at a similar level to Dex treated corneas, indicating that HBA and NaB are efficacious in preserving corneal clarity. In agreement with a study by Nakamura et al. [13] who showed that HBA has a protective effect on corneal epithelia in dry eye conditions, our results showed HBA can promote wound healing at 5 days post-injury. In addition to the improvement of clinical parameters, the expression of the key component of NLRP3 pathway also decreased in HBA and NaB treated corneas. Thus, our results suggest that blocking the NLRP3 pathway leads to improved clinical parameters and suppresses inflammation.

In conclusion, this study reveals that the NLRP3 expressed by corneal epithelial cells may respond to DAMPs released after sterile corneal injury, leading to the production of IL-1β via the innate immune signaling pathway (NLRP3–ASC–caspase-1–IL-1β). We have shown for the first time that NLRP3 inflammasome is involved in the sterile corneal injury and provided mechanistic and therapeutic considerations. Inhibiting NLRP3 assembly would be a novel therapeutic approach to limit cornea damage after alkali burns.

4. Materials and Methods

4.1. Animals

All animals were treated in accordance with the Association of Research in Vision and Ophthalmology (ARVO) Statement for the Use of Animals in Ophthalmic and Vision Research, and the protocols were approved by the Baylor College of Medicine Institutional Animal Care and Use Committee (IACUC protocol number AN-5076, first approved on 24 October 2011). Female C57BL/6J mice (6–8 weeks old) were purchased from the Jackson Laboratory (Bar Harbor, ME, USA).

4.2. Unilateral Alkali Burn

After systemic anesthesia with isoflurane using a vaporizer (SomnoSuite; Kent Scientific, Torrington, CT, USA), a unilateral alkali burn was created on the right eye of 6- to 8-week-old C57BL/6 mice. This was achieved by placing one 2-mm diameter filter paper disc that had been presoaked with 1 N NaOH on the central cornea for 10 s, followed by extensive rinsing with balanced salt solution (Alcon, Fort Worth, TX, USA), as previously described [20]. Precautions were taken to avoid damage to the peripheral cornea, conjunctiva, and lids. Alkali burn was created at day 0 and animals were euthanized 2 or 5 days post-injury.

4.3. Histology and Immunostaining

For immunohistochemistry, eyes and adnexae from each group/time point (*n* = 6/group) were excised, embedded in optimal cutting temperature compound (VWR, Suwanee, GA, USA), and flash frozen in liquid nitrogen. Sagittal 8 μm tissue sections were cut with a cryostat (HM 500; Micron, Waldorf, Germany) and placed on glass slides that were stored at −80 °C.

Immunofluorescent staining was performed in frozen tissue sections with rat monoclonal antibody anti-NLRP3 (MAB7578, 10 μg/mL, R&D Systems, Minneapolis, MN, USA), anti-ASC (SC-22514-R, 1 μg/mL, Santa Cruz Biotechnology, Dallas, TX, USA), anti-caspase-1 (SC-56036, 1 μg/mL, Santa Cruz

Biotechnology, Dallas, TX, USA), and goat anti-IL-1β (1:50 dilution, #12426, Cell Signaling Technology, Beverly, MA, USA). Secondary goat anti-rabbit or donkey anti-goat Alexa Fluor 488-conjugated antibodies were used, as previously described [30]. The images were captured and photographed by a laser scanning confocal microscope (LSM 510, with Kr–Ar and He-Ne laser; Carl Zeiss Meditec, Inc., Thornwood, NY, USA).

4.4. RNA Isolation and Quantitative PCR

Whole corneas or corneal epithelium (*n* = 4/group per experiment; total of eight corneas/group) were collected and minced, and total RNA was extracted using a Qiagen MicroPlus RNeasy isolation Kit (Qiagen, Valencia, CA, USA) according to the manufacturer's instructions, quantified by a NanoDrop ND-2000 Spectrophotometer (Thermo Fisher Scientific, Wilmington, DE, USA), and stored at −80 °C. First-strand complement DNA was synthesized with random hexamers by M-MuLV reverse transcription (Ready-To-Go You-Prime First-Strand Beads; GE Healthcare, Inc., Arlington Heights, NJ, USA), as previously described [20].

Real-time PCR was performed with specific Taqman MGB probes (Applied Biosystems, Inc., Foster City, CA, USA) and PCR master mix (Taqman Gene Expression Master Mix), in a commercial thermocycling system (StepOnePlus Real-Time PCR System, Applied Biosystems, Inc.), according to the manufacturer's recommendations. Quantitative real time PCR was performed using gene expression assay primers and probes specific for murine targets, as described in Table 1. The β-2-microglobulin (B2M) gene was used as an endogenous reference for each reaction to correct for differences in the amount of total RNA added. The results of quantitative PCR were analyzed by the comparative cycle threshold (CT) method where the target change equals $2^{-\Delta\Delta CT}$. The results were normalized by the CT value of B2M and the relative mRNA level in the untreated group was used as the calibrator.

Table 1. Oligonucleotide primers used for real-time PCR.

Gene Name	Symbol	Assay ID *
β-2-microglobulin	B2M	Mm00437762
Caspase-1	Caspase-1	Mm00438023
Interleukin-1β	IL-1β	Mm00434228
NLR family, pyrin domain containing 3	NLRP3	Mm00840904

* Identification number from Thermo Fisher Scientific [31].

4.5. Caspase-1 Activation Fluorometric Assays

The activation of caspase-1 was measured in corneal epithelial lysates according to the manufacturer's protocol (K110-200, BioVision, Inc., Mountain View, CA, USA). One sample was pooled from four right corneas and there were a total of three samples per time point. Total protein concentration was measured by the bicinchoninic acid (BCA) protein assay (Thermo Fisher Scientific, Waltham, MA, USA), as previous described [32]. Three samples per group were used. Caspase-1 activities were measured (50 μg/sample) by following the cleavage of the fluorescent substrate analogs in a fluorescent plate reader (Tecan Infinite M200, Magellan V6.55 software, Tecan, Männedorf, Switzerland) with 400 nm excitation filter and 505 nm emission filter. The results were exported and averaged.

4.6. Western Blot

Corneal epithelium was scraped with a scalpel and placed in 100 μL RIPA buffer (R0278; Sigma–Aldrich, St. Louis, MO, USA) with protease inhibitor cocktail (11836170001, Roche, Basel, Switzerland). One sample was pooled from four right corneas and there were a total of three samples per time point. A BCA assay was performed to measure total protein concentration of each sample. Samples (25 μL, equal to a protein concentration 50 μg) were diluted with one part 2X sample buffer

(2X Laemmli Sample Buffer, 161-0737; Bio-Rad Laboratories, Inc., Hercules, CA, USA), boiled for 5 min, and loaded on to the polyacrylamide gel. The gel (mini-PROTEAN TGX stain free Precast Gel 7.5%, 456-8024; Bio-Rad laboratories, Inc., Hercules, CA, USA) was run at constant current at 100V for approximately 90 min at room temperature before the proteins were transferred onto a polyvinylidene difluoride (PVDF) membrane (Immobilion Transfer membranes, IVPH07850; Millipore, Billerica, MA, USA) at 20 V over night at 4 °C. Next, the membranes were incubated in 100 mM Tris-HCl, 0.9% NaCl, 0.1% Tween 20 (TTBS) with 5% fat-free milk for 60 min, followed by incubation in primary antibody NLRP3 (MAB7578, 2 µg/mL, R&D Systems, Minneapolis, MN, USA) or β-actin (#4970, 1 µg/mL, Cell Signaling technology, Beverly, MA, USA) overnight at 4 °C. Subsequently, the membranes were washed in TTBS and incubated in secondary horseradish peroxidase goat anti-rat (629520; Thermo Fisher Scientific), then washed again with TTBS, and finally, developed and photographed.

4.7. Treatment Regimen

Mice subjected to corneal alkali burn were topically treated either with 2 µL of sodium butyrate (0.5 Mm, Sigma–Aldrich), 2 µL of hydroxybutyric acid (80 mM, Sigma–Aldrich), sodium phosphate dexamethasone (0.1%, Spectrum Laboratory, Gardena, CA, USA), or vehicle (balanced salt solution, Alcon, Fort Worth, TX, USA) QID for 2 or 5 days. These doses were chosen based on published manuscripts [12,13,19,33] and also based on our pilot studies.

4.8. Clinical Findings: Opacity Score

Biomicroscopic examination was used to grade corneal edema and opacity by two masked observers in images taken by a color digital camera DS-Fi1 (Melville, NY, USA) by the way described by Yoeruek (2008) [34]. Corneal opacity was scored using a scale of 0–4 where grade 0 represented completely clear conditions; grade 1 slightly hazy, iris, and pupils easily visible; grade 2 slightly opaque, iris, and pupils still detectable; grade 3 opaque, pupils hardly detectable, and grade 4 completely opaque with no view of the pupils.

4.9. Measurement of Corneal Epithelial Defect

Corneal epithelial healing was assessed daily in the experimental groups (four mice per group per experiment; three sets of experiments). Briefly, 1 mL of 0.1% liquid sodium fluorescein was instilled onto the ocular surface. Corneas were rinsed with phosphate-buffered saline and photographed with a stereoscopic zoom microscope (SMZ 1500; Nikon, Melville, NY, USA) under fluorescence excitation at 470 nm (digital camera DS-Qi1Mc, Nikon). Corneal epithelial defect area was graded in digital images by two masked observers in a categorical manner (present/absent) to generate a survival curve [35]. Biological replicate scores were transferred to an Excel database (Microsoft, Redmond, WA, USA) and results analyzed.

4.10. Number of Animals and Statistical Analysis

One hundred and thirty-three C57BL/6J mice were used in this study. Fifty-eight animals were used per time point (2 and 5 days): 6 for histology, 16 for real-time PCR, 12 for caspase-1 activity assay, and 24 for western blot. Contralateral eyes in the alkali burn group were used as untreated controls. Fifteen naïve mice were used to evaluate drug toxicity.

Results are presented as the mean ± SEM. One-way analysis of variance (ANOVA) with Bonferroni post hoc testing was used for statistical comparisons of gene expression. $p \leq 0.05$ was considered statistical significant. These tests were performed using GraphPad Prism 6.0 software (GraphPad Incorporation, San Diego, CA, USA).

Acknowledgments: Supported by W81XWH-12-1-0616 (CSDP), National Institutes of Health (NIH) Training Grant T32-AI053831 (FB), NIH Core Grants (EY002520, EY020799, and CA125123), Research to Prevent Blindness, the Oshman Foundation, William Stamps Farish Fund, and the Hamill Foundation. We would like to thank Kevin Christopher Tesareski for technical assistance. Presented in part as abstract at the annual meeting of the Association for Research in Vision and Ophthalmology, Denver, 2014.

Author Contributions: Cintia S. de Paiva designed the study; Fang Bian, Eugene A. Volpe, Yangyan Xiao and Mahira Zaheer performed the experiments; Fang Bian, Eugene A. Volpe, Stephen C. Pflugfelder, De-Quan Li and Cintia S. de Paiva contributed to manuscript preparation.

Conflicts of Interest: All authors state that they have no conflicts of interest.

Abbreviations

AB	alkali burn
d	days
UT	untreated
IL	interleukin
NaB	Sodium butyrate
HBA	β-hydroxybutyric acid
BSS	balanced salt solution
ASC	apoptosis-associated speck-like protein
B2M	beta-2-microglobulin
BCA	bicinchoninic acid
Dex	dexamethasone
NOD	nucleotide-binding oligomerization domain-containing protein
NLR	NOD-like receptor
PAMPs	pathogen-associated molecular patterns
TLR	Toll like receptor
DAMPs	damage-associated molecular patterns
ANOVA	one-way analysis of variance
CARD	caspase activation and recruitment domain
QID	quater in die (in Latin, four times a day)

References

1. Clare, G.; Suleman, H.; Bunce, C.; Dua, H. Amniotic membrane transplantation for acute ocular burns. *Cochrane Database Syst. Rev.* **2012**, *9*, CD009379.
2. Shao, L.; Kamalu, O.; Mayer, L. Non-classical MHC class i molecules on intestinal epithelial cells: Mediators of mucosal crosstalk. *Immunol. Rev.* **2005**, *206*, 160–176. [CrossRef] [PubMed]
3. Yoshikai, Y. The interaction of intestinal epithelial cells and intraepithelial lymphocytes in host defense. *Immunol. Res.* **1999**, *20*, 219–235. [CrossRef] [PubMed]
4. Zhang, Y.; Wu, J.; Xin, Z.; Wu, X. *Aspergillus fumigatus* triggers innate immune response via NOD1 signaling in human corneal epithelial cells. *Exp. Eye Res.* **2014**, *127*, 170–178. [CrossRef] [PubMed]
5. Bals, R.; Hiemstra, P.S. Innate immunity in the lung: How epithelial cells fight against respiratory pathogens. *Eur. Respir. J.* **2004**, *23*, 327–333. [CrossRef] [PubMed]
6. Rajalakshmy, A.R.; Malathi, J.; Madhavan, H.N. HCV core and NS3 proteins mediate toll like receptor induced innate immune response in corneal epithelium. *Exp. Eye Res.* **2014**, *128*, 117–128. [CrossRef] [PubMed]
7. Zhang, J.; Xu, K.; Ambati, B.; Yu, F.S. Toll-like receptor 5-mediated corneal epithelial inflammatory responses to *Pseudomonas aeruginosa* flagellin. *Investig. Ophthalmol. Vis. Sci.* **2003**, *44*, 4247–4254. [CrossRef]
8. Eslani, M.; Movahedan, A.; Afsharkhamseh, N.; Sroussi, H.; Djalilian, A.R. The role of toll-like receptor 4 in corneal epithelial wound healing. *Investig. Ophthalmol. Vis. Sci.* **2014**, *55*, 6108–6115. [CrossRef] [PubMed]
9. Edye, M.E.; Lopez-Castejon, G.; Allan, S.M.; Brough, D. Acidosis drives damage-associated molecular pattern (damp)-induced interleukin-1 secretion via a caspase-1-independent pathway. *J. Biol. Chem.* **2013**, *288*, 30485–30494. [CrossRef] [PubMed]

10. Iyer, S.S.; Pulskens, W.P.; Sadler, J.J.; Butter, L.M.; Teske, G.J.; Ulland, T.K.; Eisenbarth, S.C.; Florquin, S.; Flavell, R.A.; Leemans, J.C.; et al. Necrotic cells trigger a sterile inflammatory response through the NLRP3 inflammasome. *Proc. Natl. Acad. Sci. USA* 2009, *106*, 20388–20393. [CrossRef] [PubMed]

11. Pajak, B.; Orzechowski, A.; Gajkowska, B. Molecular basis of sodium butyrate-dependent proapoptotic activity in cancer cells. *Adv. Med. Sci.* 2007, *52*, 83–88. [PubMed]

12. Wang, X.; He, G.; Peng, Y.; Zhong, W.; Wang, Y.; Zhang, B. Sodium butyrate alleviates adipocyte inflammation by inhibiting NRLP3 pathway. *Sci. Rep.* 2015, *5*, 12676. [CrossRef] [PubMed]

13. Nakamura, S.; Shibuya, M.; Saito, Y.; Nakashima, H.; Saito, F.; Higuchi, A.; Tsubota, K. Protective effect of D-β-hydroxybutyrate on corneal epithelia in dry eye conditions through suppression of apoptosis. *Investig. Ophthalmol. Vis. Sci.* 2003, *44*, 4682–4688. [CrossRef]

14. Franchi, L.; Munoz-Planillo, R.; Nunez, G. Sensing and reacting to microbes through the inflammasomes. *Nat. Immunol.* 2012, *13*, 325–332. [CrossRef] [PubMed]

15. Shao, B.Z.; Xu, Z.Q.; Han, B.Z.; Su, D.F.; Liu, C. NLRP3 inflammasome and its inhibitors: A review. *Front. Pharmacol.* 2015, *6*, 262. [CrossRef] [PubMed]

16. Negash, A.A.; Ramos, H.J.; Crochet, N.; Lau, D.T.; Doehle, B.; Papic, N.; Delker, D.A.; Jo, J.; Bertoletti, A.; Hagedorn, C.H.; et al. IL-1β production through the NRLP3 inflammasome by hepatic macrophages links hepatitis C virus infection with liver inflammation and disease. *PLoS Pathog.* 2013, *9*, e1003330. [CrossRef] [PubMed]

17. Iannitti, R.G.; Napolioni, V.; Oikonomou, V.; de Luca, A.; Galosi, C.; Pariano, M.; Massi-Benedetti, C.; Borghi, M.; Puccetti, M.; Lucidi, V.; et al. IL-1 receptor antagonist ameliorates inflammasome-dependent inflammation in murine and human cystic fibrosis. *Nat. Commun.* 2016, *7*, 10791. [CrossRef] [PubMed]

18. Netea, M.G.; Joosten, L.A. Inflammasome inhibition: Putting out the fire. *Cell Metab.* 2015, *21*, 513–514. [CrossRef] [PubMed]

19. Youm, Y.H.; Nguyen, K.Y.; Grant, R.W.; Goldberg, E.L.; Bodogai, M.; Kim, D.; D'Agostino, D.; Planavsky, N.; Lupfer, C.; Kanneganti, T.D.; et al. The ketone metabolite β-hydroxybutyrate blocks NLRP3 inflammasome-mediated inflammatory disease. *Nat. Med.* 2015, *21*, 263–269. [CrossRef] [PubMed]

20. Bian, F.; Pelegrino, F.S.; Tukler Henriksson, J.; Pflugfelder, S.C.; Volpe, E.A.; Li, D.Q.; de Paiva, C.S. Differential effects of dexamethasone and doxycycline on inflammation and MMP production in alkali-burned corneas associated with dry eye. *Ocular Surface* 2016, *14*, 242–254. [CrossRef] [PubMed]

21. Chen, G.Y.; Nunez, G. Sterile inflammation: Sensing and reacting to damage. *Nat. Rev. Immunol.* 2010, *10*, 826–837. [CrossRef] [PubMed]

22. Ozaki, E.; Campbell, M.; Doyle, S.L. Targeting the NRLP3 inflammasome in chronic inflammatory diseases: Current perspectives. *J. Inflamm. Res.* 2015, *8*, 15–27. [PubMed]

23. Ghiringhelli, F.; Apetoh, L.; Tesniere, A.; Aymeric, L.; Ma, Y.; Ortiz, C.; Vermaelen, K.; Panaretakis, T.; Mignot, G.; Ullrich, E.; et al. Activation of the NRLP3 inflammasome in dendritic cells induces IL-1β-dependent adaptive immunity against tumors. *Nat. Med.* 2009, *15*, 1170–1178. [CrossRef] [PubMed]

24. Puyang, Z.; Feng, L.; Chen, H.; Liang, P.; Troy, J.B.; Liu, X. Retinal ganglion cell loss is delayed following optic nerve crush in NRLP3 knockout mice. *Sci. Rep.* 2016, *6*, 20998. [CrossRef] [PubMed]

25. Mulay, S.R.; Kulkarni, O.P.; Rupanagudi, K.V.; Migliorini, A.; Darisipudi, M.N.; Vilaysane, A.; Muruve, D.; Shi, Y.; Munro, F.; Liapis, H.; et al. Calcium oxalate crystals induce renal inflammation by NRLP3-mediated IL-1β secretion. *J. Clin. Investig.* 2013, *123*, 236–246. [CrossRef] [PubMed]

26. Zheng, Q.; Ren, Y.; Reinach, P.S.; She, Y.; Xiao, B.; Hua, S.; Qu, J.; Chen, W. Reactive oxygen species activated NRLP3 inflammasomes prime environment-induced murine dry eye. *Exp. Eye Res.* 2014, *125*, 1–8. [CrossRef] [PubMed]

27. Li, J.; Chi, W.; Hua, X.; Bian, F.; Yuan, X.; Deng, R.; Zhang, Z.; de Paiva, C.S.; Pflugfelder, S.C.; Li, D.-Q. Caspase-8 mediated activation of NLRP3 and NLRC4 inflammasomes in experimental dry eye mouse model and human corneal epithelial cells exposed to hyperosmolarity. *Investig. Ophthalmol. Vis. Sci.* 2015, *56*, 4878.

28. McGilligan, V.E.; Gregory-Ksander, M.S.; Li, D.; Moore, J.E.; Hodges, R.R.; Gilmore, M.S.; Moore, T.C.; Dartt, D.A. *Staphylococcus aureus* activates the NRLP3 inflammasome in human and rat conjunctival goblet cells. *PLoS ONE* 2013, *8*, e74010. [CrossRef] [PubMed]

29. Wang, Y.; Hanus, J.W.; Abu-Asab, M.S.; Shen, D.; Ogilvy, A.; Ou, J.; Chu, X.K.; Shi, G.; Li, W.; Wang, S.; et al. NRLP3 upregulation in retinal pigment epithelium in age-related macular degeneration. *Int. J. Mol. Sci.* 2016, *17*, 73. [CrossRef] [PubMed]

30. Corrales, R.M.; Stern, M.E.; de Paiva, C.S.; Welch, J.; Li, D.Q.; Pflugfelder, S.C. Desiccating stress stimulates expression of matrix metalloproteinases by the corneal epithelium. *Investig. Ophthalmol. Vis. Sci.* **2006**, *47*, 3293–3302. [CrossRef] [PubMed]
31. Thermo Fisher Scientific. Available online: www.lifetechnologies.com (accessed on 28 February 2017).
32. De Paiva, C.S.; Chotikavanich, S.; Pangelinan, S.B.; Pitcher, J.D., III; Fang, B.; Zheng, X.; Ma, P.; Pangelinan, W.J.; Siemasko, K.S.; Niederkorn, J.Y.; et al. IL-17 disrupts corneal barrier following desiccating stress. *Mucosal Immunol.* **2009**, *2*, 243–253. [CrossRef] [PubMed]
33. Nakamura, S.; Shibuya, M.; Nakashima, H.; Imagawa, T.; Uehara, M.; Tsubota, K. D-β-Hydroxybutyrate protects against corneal epithelial disorders in a rat dry eye model with jogging board. *Investig. Ophthalmol. Vis. Sci.* **2005**, *46*, 2379–2387. [CrossRef] [PubMed]
34. Yoeruek, E.; Ziemssen, F.; Henke-Fahle, S.; Tatar, O.; Tura, A.; Grisanti, S.; Bartz-Schmidt, K.U.; Szurman, P.; Tübingen Bevacizumab Study Group. Safety, penetration and efficacy of topically applied bevacizumab: Evaluation of eyedrops in corneal neovascularization after chemical burn. *Acta Ophthalmol.* **2008**, *86*, 322–328. [CrossRef] [PubMed]
35. Bian, F.; Pelegrino, F.S.; Pflugfelder, S.C.; Volpe, E.A.; Li, D.Q.; de Paiva, C.S. Desiccating stress-induced MMP production and activity worsens wound healing in alkali-burned corneas. *Investig. Ophthalmol. Vis. Sci.* **2015**, *56*, 4908–4918. [CrossRef] [PubMed]

© 2017 by the authors. Licensee MDPI, Basel, Switzerland. This article is an open access article distributed under the terms and conditions of the Creative Commons Attribution (CC BY) license (http://creativecommons.org/licenses/by/4.0/).

International Journal of
Molecular Sciences

MDPI

Article

Anti-Inflammatory Effect of Titrated Extract of *Centella asiatica* in Phthalic Anhydride-Induced Allergic Dermatitis Animal Model

Ju Ho Park [1], Ji Yeon Choi [1], Dong Ju Son [1], Eun Kyung Park [2], Min Jong Song [2], Mats Hellström [3] and Jin Tae Hong [1,*]

[1]   College of Pharmacy and Medical Research Center, Chungbuk National University, 194-31 Osongsaengmyeong 1-ro, Osong-eup, Heungduk-gu, Cheongju 361-951, Korea; jhp31888@naver.com (J.H.P.); cjy8316@hanmail.net (J.Y.C.); sondj1@chungbuk.ac.kr (D.J.S.)
[2]   Department of Obstetrics & Gynecology, Daejeon St. Mary's Hospital, The Catholic University of Korea, 64 Daeheung-Ro (Daeheung-dong), Jung-gu, Daejeon 301-723, Korea; guevara614@catholic.ac.kr (E.K.P.); bitsugar@catholic.ac.kr (M.J.S.)
[3]   Laboratory for Transplantation and Regenerative Medicine, Sahlgrenska Academy, University of Gothenburg, Gothenburg 411-15, Sweden; mats.hellstrom@gu.se
*   Correspondence: jinthong@chungbuk.ac.kr; Tel.: +82-43-261-2813; Fax: +82-43-268-2732

Academic Editor: Allison Cowin
Received: 22 January 2017; Accepted: 24 March 2017; Published: 30 March 2017

Abstract: *Centella asiatica* has potent antioxidant and anti-inflammatory properties. However, its anti-dermatitic effect has not yet been reported. In this study, we investigated the anti-dermatitic effects of titrated extract of *Centella asiatica* (TECA) in a phthalic anhydride (PA)-induced atopic dermatitis (AD) animal model as well as in vitro model. An AD-like lesion was induced by the topical application of five percent PA to the dorsal skin or ear of Hos:HR-1 mouse. After AD induction, 100 µL of 0.2% and 0.4% of TECA (40 µg or 80 µg/cm^2) was spread on the dorsum of the ear or back skin three times a week for four weeks. We evaluated dermatitis severity, histopathological changes and changes in protein expression by Western blotting for inducible nitric oxide synthase (iNOS), cyclooxygenase-2 (COX-2), and NF-κB activity, which were determined by electromobility shift assay (EMSA). We also measured TNF-α, IL-1β, IL-6, and IgE concentration in the blood of AD mice by enzyme-linked immunosorbent assay (ELISA). TECA treatment attenuated the development of PA-induced atopic dermatitis. Histological analysis showed that TECA inhibited hyperkeratosis, mast cells and infiltration of inflammatory cells. TECA treatment inhibited expression of iNOS and COX-2, and NF-κB activity as well as the release of TNF-α, IL-1β, IL-6, and IgE. In addition, TECA (1, 2, 5 µg/mL) potently inhibited Lipopolysaccharide (LPS) (1 µg/mL)-induced NO production, expression of iNOS and COX-2, and NF-κB DNA binding activities in RAW264.7 macrophage cells. Our data demonstrated that TECA could be a promising agent for AD by inhibition of NF-κB signaling.

Keywords: titrated extract of *Centella asiatica*; skin inflammation; atopic dermatitis; NF-κB; cytokine; IgE

1. Introduction

Atopic dermatitis (AD) is a common chronic inflammatory skin disease inducing intense itching, edema, erythema, thickening, severe pruritus, and eczematous lesions of the skin. Several genetic and environmental factors and immune responses are implicated for the development of AD [1]. Elevated production of serum IgE against many kinds of inhaled allergens and secretion of T helper (Th) 2 cytokines are the main causes of AD [2,3]. Mast cell activation mediated by IgE leads to a release of various chemical mediators which results in infiltration of inflammatory cells such as eosinophils and lymphocytes into the skin lesion [4]. CD4[+] T cells and mast cells in the skin lesions are also involved in the pathogenesis of AD [5–8]. It has been reported that 2,4-dinitrochlorobenzene (DNCB)-induced AD-like skin lesion mouse model showed increased serum IgE and Th2 cytokines such as IL-4, IL-5, and IL-13 [9,10]. These cytokines have direct effects on epidermal keratinocytes, which produce pro-inflammatory cytokines that induce infiltration of immune cells into inflammatory skin lesions [11]. These data indicate that inflammation and activation of immune cells could be significant for the development of AD.

Nuclear factor-κB (NF-κB) is an important transcription factor associated with the allergic inflammatory response in AD. Many studies have shown that NF-κB is an important factor in the regulation of various immune responses in allergic disorders such as AD, asthma, and rheumatoid arthritis [12–14]. Since the activation of NF-κB may exacerbate the allergic inflammation by enhancing the production of inflammatory cytokines and chemokines, various methods have been developed to inhibit NF-κB activation. Moreover, NF-κB inhibitor, IMD-0354 inhibited abnormal proliferation of mast cells, and reduced the allergic response [15]. Potent immunosuppressive drugs such as tacrolimus, corticosteroids, and cyclosporine have been studied as therapeutic agents for AD through inhibition of cytokine production [16]. However, these agents cause severe reverse effects such as tachyphylaxis, recurrence, and exacerbation of AD [17,18]. Thus, other new drugs showing no side effects with strong pharmacological properties could be developed.

Centella asiatica, known by the common name Gotu kola, is a traditional herbal medicine that has been used to exert pharmacological effects in dermatology [19]. The *Centella asiatica* herb is used in the treatment of skin lesions such as burn wounds, excoriations, or eczema as well as in non-dermatological diseases such as diabetic complications [20], and neurodegenerative disorders [21]. *Centella asiatica* has also been effective in chronic venous insufficiency by improvement of microcirculation [22]. The *Centella asiatica* extract was registered in International Nomenclature of Cosmetic Ingredients (INCI) as an ingredient of cosmetics [23]. Although various pharmacological effects of *Centella asiatica* have been reported, its anti-dermatitic effect has not yet been reported. Therefore, we investigated the anti-dermatitic effects of titrated extract of *Centella asiatica* and action mechanism in a phthalic anhydride-induced atopic dermatitis animal model as well as in vitro model.

2. Results

2.1. Effects of TECA Treatment on Ear Thickness and Morphology

Changes in body weight were measured during the experimental period. No significant difference in body weight was detected after any of the treatments (Figure 1A). To investigate whether or not treatment with TECA can suppress the changes in ear phenotype induced by PA treatment, ear thickness and morphology of ear were observed. Ear thickness rapidly increased in PA treated mice compared to control or vehicle treated mice. On the other hand, ear thickness in TECA treated mice was slowly increased in a dose-dependent manner (Figure 1B). Furthermore, symptoms consisting of erythma, edema, and erosion were observed in the PA treated group compared with the control or vehicle treated group. These changes of ear and back morphology and ear thickness were dramatically reversed upon TECA treatment (Figure 1C).

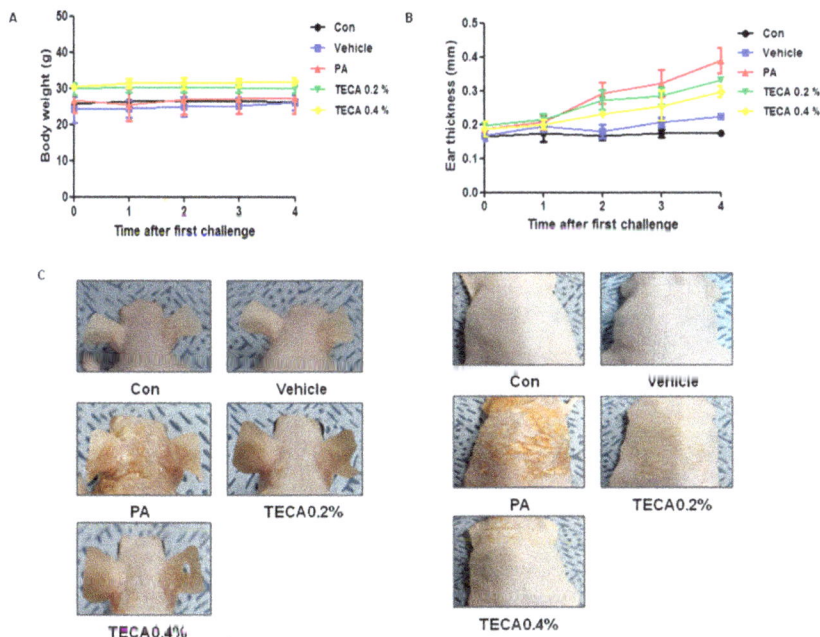

Figure 1. Differences in body weight, ear thickness, ear phenotypes, and back phenotypes. Phthalic anhydride (PA) solution was repeatedly applied to the dorsum of ear and back three times a week during topical application of Titrated extract of Centella asiatica (TECA). After four weeks, body weight (**A**) and ear thickness (**B**) were observed at least three times by following the procedure described in Materials and Methods. Phenotypes (**C**) were randomly selected by one mouse/group. Data shown are the mean ± SD (*n* = 10).

2.2. Effect of TECA Treatment on Lymph Node Weight and IgE Concentration as Well as on Expression of iNOS and COX-2

We investigated whether or not TECA could suppress the increases in lymph node weight and IgE concentration. To accomplish this, we evaluated the auricular lymph node weight and serum IgE concentration. PA treatment induced an increase in lymph node weight compared with control or vehicle treated mice. However, the weight of lymph node was significantly reduced in the TECA treated mice (Figure 2A). In addition, protein expressions of iNOS and COX-2 were significantly upregulated in PA treated AD mice, but significantly suppressed by TECA 0.4% treatment (Figure 2B). It is well known that hyperproduction of IgE is one of the characteristic features of allergic hypersensitivity as well as an indicator of the magnitude of the allergic immune responses in the development of AD [24]. The serum IgE concentration was measured in the blood of mice to determine whether TECA suppressed the allergic responses induced by PA treatment. Repeated topical application of PA solution induced a significant increase in serum IgE concentration. However, a significant decrease of IgE concentration was observed in the TECA treated group (Figure 2C).

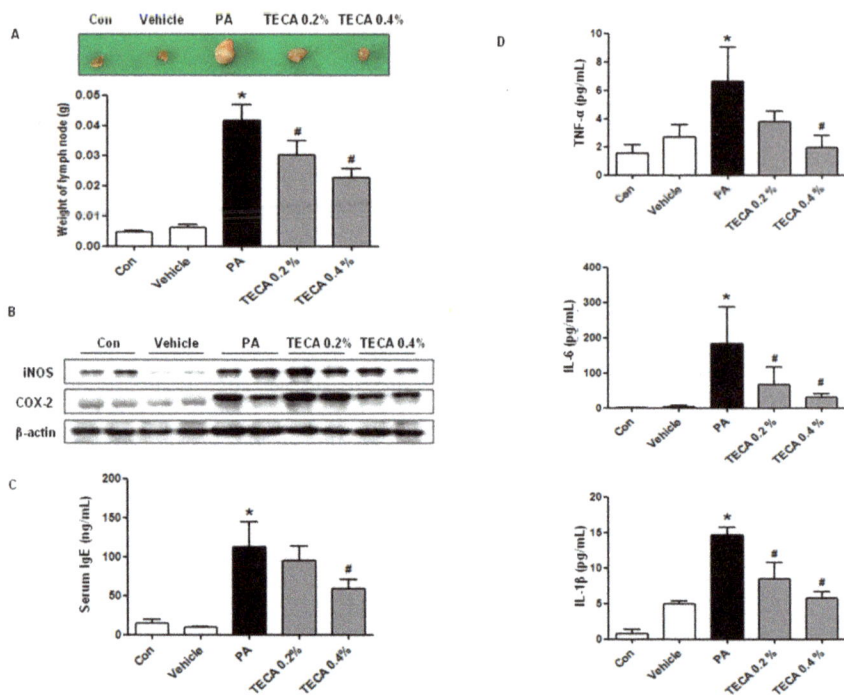

Figure 2. Changes in auricular lymph node weight, expression level of iNOS and COX-2 protein in lymph node, and serum cytokine concentration. After final treatment, mice from each group were sacrificed under anesthesia. The auricular lymph nodes were then harvested from the neck regions of the mice using a microscissor, after which they were weighed (**A**); Alteration of the expression of the two proteins was measured by Western blotting (**B**); Serum used to measure the cytokine concentration was prepared from blood samples collected from the abdominal veins of mice. Serum IgE (**C**), TNF-α, IL-6, and IL-1β (**D**) concentration were quantified by enzyme-linked immunosorbent assay (ELISA). Data shown are gained from the same mice treated shown in Figure 1. Data shown are the mean $\pm$ SD ($n = 10$). * $p < 0.05$ is the significance level compared to the control group. # $p < 0.05$ is the significance level compared to the PA treated group.

2.3. Effect of TECA Treatment on the Release of Inflammatory Cytokines

To determine if TECA treatment could induce alterations in the inflammatory cytokines release in PA-induced skin inflammation, the level of TNF-α, IL-6, and IL-1β was measured in mouse serum of control, vehicle, PA and PA + TECA treated group. The level of TNF-α, IL-6, and IL-1β was generally higher in the PA treated group than the control or vehicle treated group. However, these levels in the TECA treated group were dramatically decreased to the level of the control or vehicle treated group (Figure 2D).

2.4. Effect of TECA Treatment on Inflammatory Responses in Ear and Back

To investigate the suppressive effect of TECA treatment on ear and back histology, histological analysis of the ear and back skin were performed (Figures 3A and 4A). The epidermis and dermis of the ear (Figure 3B), and the epidermis of the back (Figure 4B) were thicker in PA treated group than in the control or vehicle treated group. However, the thickness of them was greatly decreased in the TECA treated group in a dose-dependent manner. In addition, protein expressions of iNOS and COX-2 were significantly upregulated in PA treated AD mice, but significantly suppressed by TECA 0.4% treatment (Figures 3C and 4C).

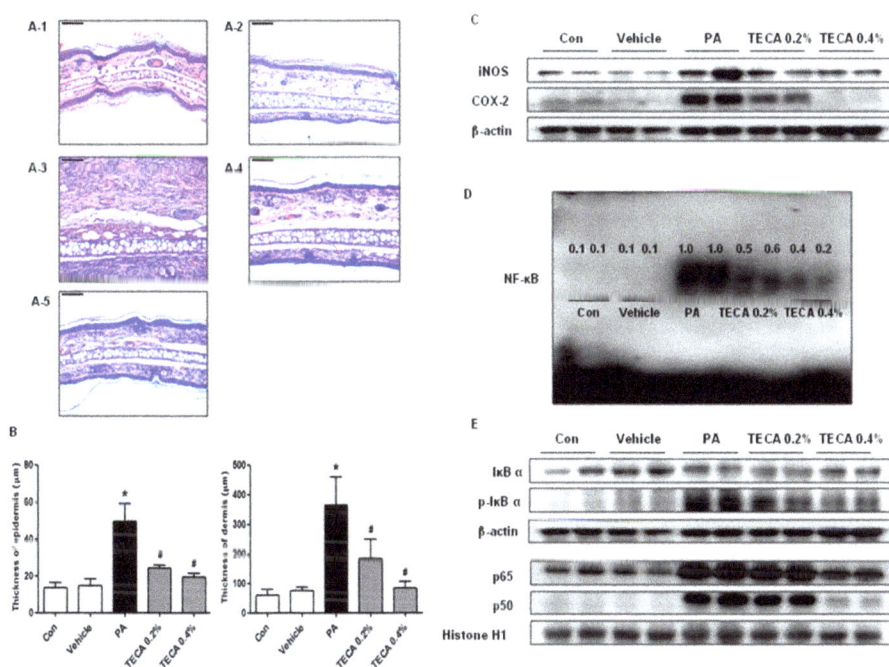

Figure 3. Histopathological analysis of ear tissue and the inhibitions of NF-κB DNA binding activity by topical application of TECA in ear skin. Histopathology of ear skin in control (**A-1**), vehicle (**A-2**), PA (**A-3**), PA + TECA 0.2% (**A-4**), and PA + TECA 0.4% (**A-5**). PA solution was repeatedly applied to the dorsum of ears during topical application of TECA. Histopathological changes in the slide sections of ear tissue were identified by staining with hematoxylin and eosin followed by observation at 200× magnification (Scale bars, 100 μm). (**A**) Histological images and (**B**) thickness of the epidermis and dermis. Alteration of the expression of iNOS and COX-2 proteins were measured by Western blotting (**C**); (**D**) Effect of TECA on NF-κB DNA binding activity in ear skin. The activation of NF-κB was investigated using electromobility shift assay (EMSA) as described in Materials and Methods. Nuclear extracts from homogenized ear skin tissue were incubated in binding reactions of ^{32}P-end-labeled oligonucleotide containing the NF-κB sequence (numbers: relative expression). (**E**) Effect of TECA on translocation of the subunits of NF-κB (p50 and p65) into nucleus, and phosphorylation of IκBα in cytosol in ear skin. Equal amounts of nuclear proteins (20 μg/lane) or total proteins (20 μg/lane) were subjected to 10% SDS-PAGE, and expression of p50, p65, IκBα, and p-IκBα protein were detected by Western blotting using specific antibodies. Histone h1 protein and β-actin protein were used here as an internal control. Data shown are gained from the same mice treated shown in Figure 1. Data shown are the mean ± SD ($n = 10$). * $p < 0.05$ is the significance level compared to the control group. # $p < 0.05$ is the significance level compared to the PA treated group.

2.5. Effect of TECA on NF-κB DNA Binding Activity in PA-Induced AD Mice

NF-κB is implicated for inflammatory responses in AD model. To investigate whether TECA can inhibit NF-κB activation in PA-induced AD model, nuclear extracts from ear and back skin tissue were prepared and assayed with NF-κB DNA binding by EMSA. PA treated mice showed significant NF-κB binding activity when compared to control group in both ear and back skin. On the contrary,

NF-κB binding activity in TECA treated mice was significantly inhibited when compared with PA treated mice (Figures 3D and 4D). In addition, as shown in Figures 3E and 4E, PA treated mice showed significant IκBα degradation in cytosolic fraction when compared to the control group in both ear and back skin. On the contrary, IκBα degradation in TECA treated mice was reduced significantly when compared with PA treated mice. PA treated mice also showed increase in the relocalization of p65 and p50 in nucleus. In contrast, TECA inhibited translocation of p65 and p50 into the nuclear in a dose-dependent manner (Figures 3E and 4E).

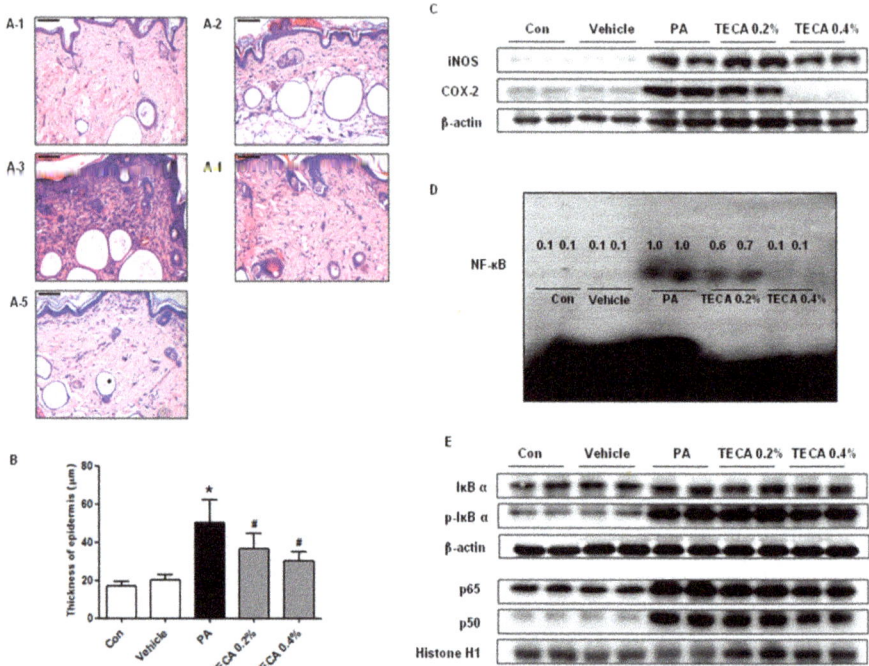

Figure 4. Histopathological analysis of back tissue and the inhibitions of NF-κB DNA binding activity by topical application of TECA in back skin. Histopathology of back skin in control (**A-1**), vehicle (**A-2**), PA (**A-3**), PA + TECA 0.2% (**A-4**), and PA + TECA 0.4% (**A-5**). PA solution was repeatedly applied to the back skin during topical application of TECA. Histopathological changes in the slide sections of back tissue were identified by staining with hematoxylin and eosin followed by observation at $200\times$ magnification (Scale bars, 100 μm). (**A**) Histological images and (**B**) thickness of the epidermis. Alteration of the expression of iNOS and COX-2 proteins were measured by Western blotting (**C**); (**D**) Effect of TECA on NF-κB DNA binding activity in back skin. The activation of NF-κB was investigated using EMSA as described in Materials and Methods. Nuclear extracts from homogenized back skin tissue were incubated in binding reactions of ^{32}P-end-labeled oligonucleotide containing the NF-κB sequence (numbers: relative expression); (**E**) Effect of TECA on translocation of the subunits of NF-κB (p50 and p65) into nucleus, and phosphorylation of IκBα in cytosol in back skin. Equal amounts of nuclear proteins (20 μg/lane) or total proteins (20 μg/lane) were subjected to 10% SDS-PAGE, and expression of p50, p65, IκBα, and p-IκBα protein were detected by Western blotting using specific antibodies. Histone h1 protein and β-actin protein were used here as an internal control. Data shown are gained from the same mice treated shown in Figure 1. Data shown are the mean ± SD (*n* = 10). * $p < 0.05$ is the significance level compared to the control group. $^{\#}$ $p < 0.05$ is the significance level compared to the PA treated group.

2.6. Effect of TECA on LPS-Induced NO Production, and iNOS and COX-2 Expression in RAW264.7 Cells

The effect of TECA on LPS-induced NO production in RAW264.7 cells was investigated by measuring the released nitrite in the culture medium by Griess reaction. After co-treatment with LPS and TECA (1, 2, 5 μg/mL) for 24 h, LPS-induced elevation of nitrite concentration in the medium were decreased in a concentration-dependent manner (Figure 5A). In addition, we determined iNOS and COX-2 expression by Western blot analysis. As shown in Figure 5B, LPS-induced iNOS and COX-2 expression were significantly inhibited by TECA (1, 2, 5 μg/mL) in a concentration-dependent manner.

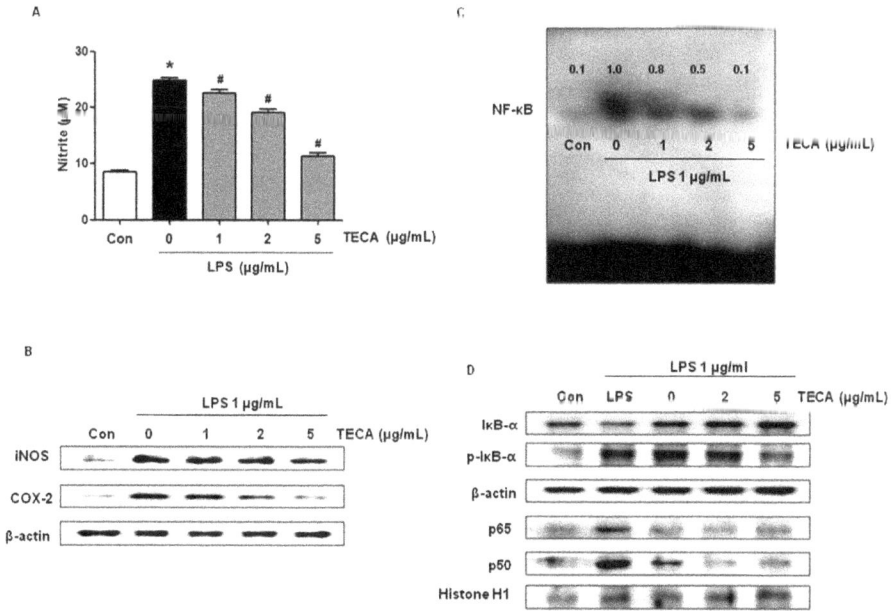

Figure 5. Effects of TECA on LPS-induced NO production, and iNOS and COX-2 expression in RAW264.7 cells. RAW 264.7 cells were pre-treated with different concentration (1, 2, and 5 μg/mL) of TECA for 2 h and then stimulated with LPS (1 μg/mL) for 24 h. Effect of TECA on LPS-induced NO production was measured by the Griess reaction as described in Materials and Methods (**A**); The expression of iNOS and COX-2 after stimulated 24 h was determined by Western blot (**B**); Effect of TECA on LPS-indcued NF-κB DNA binding activity was measured by EMSA as described in Materials and Methods (**C**); Effects of TECA on LPS-induced phosphorylation of IκBα in cytosol, and translocation of the subunits of NF-κB (p50 and p65) into nucleus were measured by Western blot (**D**). Data shown are gained from the same mice treated shown in Figure 1. Data shown are the mean ± SD (*n* = 10). * *p* < 0.05 is the significance level compared to the control group. # *p* < 0.05 is the significance level compared to the PA treated group.

2.7. Effect of TECA on NF-κB DNA Binding Activity in RAW 264.7 Cells

Because activation of NF-κB is critical for induction of both iNOS and COX-2 by LPS or other inflammatory cytokines, we determined whether TECA might suppress NF-κB activation in LPS-activated RAW264.7 cells. RAW264.7 cells were co-treated with LPS and TECA for 1 h, respectively, which is the time to activate NF-κB maximally from its LPS treatment (data are not shown). Nuclear extracts from co-treated cells were prepared and assayed NF-κB DNA binding by EMSA. In RAW264.7 cells, LPS induced a strong NF-κB binding activity, which was markedly inhibited by co-treatment with TECA in a concentration-dependent manner (Figure 5C). We further investigated the inhibitory effect of TECA on the translocation of NF-κB subunit and IκB phosphorylation. Consistent with the inhibitory effect on NF-κB activity, nuclear translocation of p65 and p50 was inhibited in a concentration-dependent manner, and the LPS-induced phosphorylation of IκBα was also inhibited by TECA in a concentration-dependent manner (Figure 5D).

3. Discussion

Topical application of corticosteroids have been used for the treatment of AD because of their great anti-inflammatory and anti-allergic activities [25]. However, they cause irreversible side effects from long-term usage such as common pathogenic infections and immune suppression [26]. For this reason, usage of natural products including various plants, herbs, flowers, yeasts, and fungi is being emphasized as anti-inflammatory and anti-allergic agents. AD is characterized by skin inflammation with eczema-like lesions, itching, and dry skin [27]. In an experimental model, the thickness of ear, epidermis and dermis were important indexes to evaluate the severity of skin inflammation. We found that TECA effectively reduces the skin inflammation and allergic responses induced by PA treatment. In in vitro assay, we also found that TECA inhibited LPS-induced inflammatory responses. It was proven that *Centella asiatica* has an excellent effect on deposition of extracellular matrix proteins. It stimulates proliferation of fibroblasts, increases the synthesis of collagen, decreases metalloproteinases activity and thus increases the deposition of collagen and intracellular free proline levels [28–31]. It also inhibits the inflammatory phase of wound healing [32]. TECA contains asiatic acid (30%), madecassic acid (29–30%), and asiaticoside (40%). The influence of asiatic acid, madecassic acid, and asiaticoside on human skin fibroblast type I collagen synthesis was also found [33]. In a recent study, it was reported that components of *Centella asiatica*, asiaticoside and madecassoside possess wound healing, collagen synthesis, as well as vasodilation activities [34]. These effects are associated with the reduced activation of macrophages and the production of IL-1β [35]. Furthermore, it has been reported that the component of *Centella asiatica*, madecassic acid, plays a role in anti-inflammatory activity through the downregulation of iNOS and COX-2 expression and TNF-α, IL-1β, and IL-6 release in RAW264.7 macrophage cells [36]. *Centella asiatica* applied in the recommended doses is not toxic and possible side effects are rare [19]. These data thus indicate that TECA could be applicable for AD.

It is well known that macrophages play an important role in both acquired and nonspecific immune responses. Activation of macrophage leads to various series of responses including the production of pro-inflammatory cytokines which exert their inflammatory effects by activating a diverse spectrum of signaling cascades in the cells that lead to the induction of inflammatory genes such as iNOS and COX-2 [37]. In this study, PA-induced expression of iNOS and COX-2 was also reduced by TECA in the skin as well as cultured macrophage. IgE-induced activation of mast cells, which resulted in the release of various allergic mediators such as cytokines and histamine [38]. Therefore, the low level of IgE induces lesser allergic responses, and reduces the levels of cytokine. In this regard, TECA potently reduced the level of IgE and release of inflammatory cytokine. These data indicate that TECA could inactivate macrophage in the skin, thus lead to less skin inflammation and atopic responses.

NF-κB is implicated for cytokine release, which is important for anti-inflammatory activity. Pro-inflammatory cytokines, including IL-4, IL-6, IL-1β, and TNF-α, commonly contribute to the regulation of inflammation and immune responses in AD skin lesion [39]. Release of IL-4 primarily regulates hyper-production of IgE [40], and expression of TNF-α and IL-6 stimulate the synthesis of acute phase response protein, which attenuates secretion of IgE and disruption of skin barrier function during allergic reactions [41,42]. In a recent study, *Spirodela polyrhiza* remarkably inhibited expression levels of NF-κB and p-IκBα as well as inflammatory cytokines such as IL-4, IL-6, and TNF-α in AD mice model [43]. Tanaka et al. demonstrated topical application of IMD-0354, an NF-κB inhibitor, is effective in suppressing the activation of NF-κB and in reducing the development of AD in atopic NC/NgaTnd mice [44]. Moreover, it was reported that treatment of NF-κB inhibitor

Xanthii fructus (XF) strongly suppressed IL-4, IL-1β, IFN-γ and TNF-α in AD-like skin lesions [45]. In addition, several natural products inhibited AD development through inhibition of cytokine releases. In TPA-induced skin inflammation, TNF-α and IL-1β in the serum were reduced by 70% ethanol extract from *Asparagus cochinchinensis* [46]. Following treatment with *Liriope platyphylla* (LP) extract, expression of IL-6 and VEGF was significantly reduced in ear tissue of IL-4/Luc/CNS-1 Tg mice treated with PA [5]. In the present study, the levels of two cytokines (TNF-α and IL-6) were elevated in the serum of mice treated with PA, but significantly reduced cytokine release was observed in the TECA treated group. In both PA-induced atopic dermatitis animal model and RAW 264.7 murine macrophage cells, TECA also decreased the degradation of IκBα and nuclear translocation of NF-κB. It has been reported that asiaticoside, a component of *Centella asiatica*, plays a role in the anti-inflammatory effect via downregulation of NF-κB signaling pathway [47]. In our present study, the data demonstrated that TECA attenuates activation of NF-κB, contributing to the reduced TNF-α, IL-6, and IL-1β level and expression of iNOS and COX-2. Therefore, our data suggest that TECA should be considered a candidate agent for AD.

4. Materials and Methods

4.1. Ethical Approval

The experimental protocols were carried out according to the guidelines for animal experiments of the Institutional Animal Care and Use Committee (IACUC) of the Laboratory Animal Research Center at Chungbuk National University, Korea (CBNUA-929-16-01). All efforts were made to minimize animal suffering, and to reduce the number of animals used. All mice were housed in three mice per cage with an automatic temperature control (21–25 °C), relative humidity (45–65%), and 12 h light–dark cycle illuminating from 08:00 a.m. to 08:00 p.m. Food and water were available ad libitum. They were fed a pellet diet consisting of crude protein 20.5%, crude fat 3.5%, crude fiber 8.0%, crude ash 8.0%, calcium 0.5%, and phosphorus 0.5% per 100 g of the diet (collected from Daehan Biolink, Chungcheongbuk-do, Korea). During this study, all mice were particularly observed for normal body posture, piloerection, ataxia, urination, etc., 2 times per day.

4.2. Preparation and Extraction of Centella asiatica

Collected aerial parts of *Centella asiatica* were oven-dried at 50 °C and then powdered using a milling machine. The powdered plant (1 kg) was extracted with 75% (*v/v*) ethanol (3 × 4 L, 3 days each) at room temperature. The extracts were filtrated with a depth-filter coated with active carbon and concentrated at 80 °C under reduced pressure. The concentrate was divided to precipitate and filtrate fraction by filtration. The precipitate fraction was dried at 50 °C to make asiaticoside powder; its yield was 0.12% (*w/w*) of dried plant. Subsequently, the filtrates fraction was hydrolyzed with an alkaline solution containing sodium hydroxide (1% *w/v*) at 80 °C. It was then concentrated, precipitated and dried using the procedure described above. The yield of the genins (asiatic acid and madecassic acid) powder obtained from the filtrate fraction was 0.18% (*w/w*) of dried plant. Both powder extracts were mixed to give the titrated extract of *Centella asiatica*. The components of the titrated extract of *Centella asiatica* were asiaticoside (40%), asiatic acid (30%), and madecassic acid (29–30%) (Table 1). All solvents used were of commercial grade and obtained from Dongkook Pharmaceutical Company, Chungbuk, Korea.

Table 1. Composition of titrated extracts of *Centella asiatica*.

Extract	Composition of Extract
Titrated extract of *Centella asiatica* (TECA)	Asiaticoside (40%), Asiatic acid (30%), Madecassic acid (29–30%)

Titrated extract of *Centella asiatica* includes 40% of asiaticoside, 30% of asiatic acid, and 29–30% of madecassic acid.

4.3. Animal Treatment

The protocols for the animal experiment used in this study were carefully reviewed for ethical and scientific care procedures and approved by the Chungbuk National University-Institutional Animal Care and Use Committee (Approval Number CBNUA-929-16-01). Hos:HR-1 mice (eight-week-old, $n = 40$) were randomly divided into one of four groups. In the first group (Vehicle, $n = 10$), 100 µL of AOO (4:1 acetone: olive oil, v/v: AOO) was spread on the dorsum of the ears and back skin three times a week for four weeks. In the second group (phthalic anhydride (PA), $n = 10$), 100 µL (20 µL/cm^2) of 5% phthalic anhydride solution was applied. The third group (TECA 0.2%, $n = 10$) and fourth group (TECA 0.4%, $n = 10$) were applied with PA, and 3 h later 100 µL of 0.2% and 0.4% titrated extract of *Centella asiatica* (40 µg or 80 µg/cm^2) were applied. Age-matched Hos:HR-1 mice were used as the control group (Control, $n = 10$).

4.4. Measurement of Ear Thickness, and Body and Lymph Node Weight

Body weights of all mice were measured during the experimental period using an electronic balance (Mettler Toledo, Greifensee, Switzerland) once a week for 4 weeks. Additionally, weights of lymph nodes were measured using an electronic balance lymph nodes were collected from sacrificed mice and weighed using an electronic balance (Mettler Toledo, Greifensee, Switzerland). Thickness of ear skin was measured using a thickness gauge (Digimatic Indicator, Matusutoyo Co., Tokyo, Japan).

4.5. Histological Techniques

The ear and back skins were removed from mice, fixed with 10% formalin, embedded in paraffin wax, routinely processed, and then sectioned into 5 µm thick slices. The skin sections were then stained with hematoxylin and eosin (H & E). The thickness of the epidermis and dermis were also measured using the Leica Application Suite (Leica Microsystems, Wetzlar, Germany).

4.6. Mesurement of Serum IgE Concentration

IgE level in the serum was measured by enzyme-linked immunosorbent assay (ELISA) using the mouse IgE kit (Shibayagi, Inc., Gunma, Japan), according to the manufacturer's instructions. The final concentration of IgE was calculated using a linear regression equation obtained from standard absorbance values.

4.7. Cytokine Assay

By the end of the study period, blood specimens were collected. Serum levels of mouse TNF-α, IL-6, and IL-1β were measured by enzyme-linked immunosorbent assay (ELISA) kits provided by Thermoscientific Inc. (Meridian Rd, Rockford, IL, USA) according to the manufacturer's protocol.

4.8. Western Blot Analysis

One hundred milligrams of skin or ear tissues or about 1×10^6 cells were harvested and homogenized with a lysis buffer (50 mM Tris pH 8.0, 150 mM NaCl, 0.2% Sodium dodecyl sulfate (SDS), 1 mM phenyl methylsulfonyl fluoride (PMSF), and 0.5% sodium deoxycholate). After lysis, the lysates were centrifuged at 13,000 rpm for 20 min. Equal amounts of protein (20 µg) were denatured at 95 °C for 5 min after mixing with 5 µL of SDS loading buffer were applied on SDS/10% polyacrylamide gel for electrophoresis and were transferred to nitrocellulose membranes (Hybond ECL, Amersham Pharmacia Biotech Inc., Piscataway, NJ, USA). The membrane was incubated for 4 h at room temperature with specific antibodies: rabbit polyclonal antibodies against iNOS, COX-2, p65 and IκB-α (1:500), and mouse monoclonal antibody against p50 (1:500) (Santa Cruz Biotechnology Inc., Santa Cruz, CA, USA) were used in study. The blot was then incubated with the corresponding conjugated anti-rabbit immunoglobulin G-horseradish peroxidase (Santa Cruz Biotechnology Inc.,

Santa Cruz, CA, USA). Band signals were detected on X-ray film using enhanced chemiluminescence (ECL) detection reagents.

4.9. Gel Electromobility Shift Assay (EMSA)

Gel shift assays were performed according to the manufacturer's recommendations (Promega, Madison, WI, USA). The oligonucleotide sequences for NF-κB were 5'-AGT TGA GGG GAC TTT CCC AGG C-3'. Consensus oligonucleotides were end-labeled using T4 polynucleotide kinase and [^{32}P] ATP for 10 min at 37 °C. Briefly, 10 μg nuclear protein were incubated with the labeled probe for 20 min at room temperature. Subsequently, 1 μL of gel loading buffer was added to each reaction and loaded onto a 5% non-denaturing gel and electrophoresis until the dye was three-fourths of the way down the gel. The gel was dried at 80 °C for 1 h and exposed to film overnight at 70 °C. Quantification and the relative density of the protein bands were performed by phosphorimaging (UVP Inc., Upland, CA, USA).

4.10. Cell Culture

The RAW 264.7 murine macrophage cell line was obtained from the Korea Cell Line Bank (Seoul, Korea). These cells were grown at 37 °C in Dulbecco's modified Eagle's medium (DMEM) medium supplemented with 10% FBS, penicillin (100 units/mL) and streptomycin sulfate (100 μg/mL) in humidified atmosphere of 5% CO_2. Cells were incubated with TECA at various concentrations (1, 2, or 5 μg/mL) or positive chemicals and then stimulated with LPS 1 μg/mL for the indicated time in figure legends. Various concentrations of TECA dissolved in ethanol were added together with LPS. The final concentration of ethanol used was less than 0.05%. Cells were treated with 0.05% ethanol as vehicle control.

4.11. Nitrite Quantification Assay

The NO was determined through the indication of nitrite level in the cell culture media. The RAW264.7 murine macrophages were seeded in 6-well plates (1×10^6 cells/well) with 2 mL of cell culture media and incubated for 24 h. This was followed by discarding the old culture media and replacing them with the new media to maintain the cells. Different concentrations of TECA (1, 2, and 5 μg/mL) were pretreated with the RAW264.7 macrophages. Induction of RAW264.7 macrophages with LPS (1 μg/mL) for all samples was conducted except in control for another 24 h. Then, 100 μL of the collected supernatants was added with 100 μL of Griess reagent (0.1% N-1-napthylethylenediamine dihydrochloride (NED), 1% sulphanilamide, and 2.5% phosphoric acid) and incubated in room temperature for 10 min in dark condition. The absorbance was determined by using a microplate reader at 540 nm wavelength. The NO concentration was determined by comparison to the standard curve.

4.12. Statistical Analysis

The experiments were conducted in triplicate, and all experiments were repeated at least three times with similar results. All statistical analysis was performed with GraphPad Prism 5 software (Version 5.03; GraphPad software, Inc., San Diego, CA, USA). Group differences were analyzed by one-way ANOVA followed by Tukey's multiple comparison test. All values are presented as mean ± SD. Significance was set at $p < 0.05$ for all tests.

5. Conclusions

In our present study, TECA treatment inhibited activation of NF-κB contributing to the reduced pro-inflammatory cytokine and expression of iNOS and COX-2 in PA-induced allergic dermatitis animal model as well as RAW 264.7 murine macrophage. Therefore, our data suggest that TECA could be a promising agent for AD.

Acknowledgments: This work is financially supported by the National Research Foundation of Korea (NRF) Grant funded by the Korea government (MSIP) (No. MRC, 2008-0062275) and by the Ministry of Trade, Industry & Energy (MOTIE, 1415139249) through the fostering project of Osong Academy-Industry Convergence (BAIO).

Author Contributions: Jin Tae Hong and Mats Hellström designed the study and prepared the manuscript. Ju Ho Park and Ji Yeon Choi performed overall experiments. Eun Kyung Park and Min Jong Song discussed the study. All authors have read and approved the final version of this manuscript.

Conflicts of Interest: The authors declare no conflict of interest.

References

1. Sehra, S.; Krishnamurthy, P.; Koh, B.; Zhou, H.M.; Seymour, L.; Akhtar, N.; Travers, J.B.; Turner, M.J.; Kaplan, M.H. Increased Th2 activity and diminished skin barrier function cooperate in allergic skin inflammation. *Eur. J. Immunol.* **2016**, *46*, 2609–2613. [CrossRef] [PubMed]
2. Eichenfield, L.F.M.; Friedlander, S.F.M.; Simpson, E.L.M.M.; Irvine, A.D.M. Assessing the New and Emerging Treatments for Atopic Dermatitis. *Semin. Cutan. Med. Surg.* **2016**, *35* (Suppl **65**), S92–S96. [CrossRef] [PubMed]
3. Choi, J.K.; Kim, S.H. Inhibitory effect of galangin on atopic dermatitis-like skin lesions. *Food Chem. Toxicol.* **2014**, *68*, 135–141. [CrossRef] [PubMed]
4. Kuramoto, T.; Yokoe, M.; Tanaka, D.; Yuri, A.; Nishitani, A.; Higuchi, Y.; Yoshimi, K.; Tanaka, M.; Kuwamura, M.; Hiai, H.; et al. Atopic dermatitis-like skin lesions with IgE hyperproduction and pruritus in KFRS4/Kyo rats. *J. Dermatol. Sci.* **2015**, *80*, 116–123. [CrossRef] [PubMed]
5. Kwak, M.H.; Kim, J.E.; Hwang, I.S.; Lee, Y.J.; An, B.S.; Hong, J.T.; Lee, S.H.; Hwang, D.Y. Quantitative evaluation of therapeutic effect of Liriope platyphylla on phthalic anhydride-induced atopic dermatitis in IL-4/Luc/CNS-1 Tg mice. *J. Ethnopharmacol.* **2013**, *148*, 880–889. [CrossRef] [PubMed]
6. Sung, J.E.; Lee, H.A.; Kim, J.E.; Go, J.; Seo, E.J.; Yun, W.B.; Kim, D.S.; Son, H.J.; Lee, C.Y.; Lee, H.S.; et al. Therapeutic effect of ethyl acetate extract from *Asparagus cochinchinensis* on phthalic anhydride-induced skin inflammation. *Lab. Anim. Res.* **2016**, *32*, 34–45. [CrossRef] [PubMed]
7. Park, H.J.; Jang, Y.J.; Yim, J.H.; Lee, H.K.; Pyo, S. Ramalin Isolated from Ramalina Terebrata Attenuates Atopic Dermatitis-like Skin Lesions in Balb/c Mice and Cutaneous Immune Responses in Keratinocytes and Mast Cells. *Phytother. Res.* **2016**, *30*, 1978–1987. [CrossRef] [PubMed]
8. Zhang, Y.Y.; Wang, A.X.; Xu, L.; Shen, N.; Zhu, J.; Tu, C.X. Characteristics of peripheral blood CD4$^+$CD25$^+$ regulatory T cells and related cytokines in severe atopic dermatitis. *Eur. J. Dermatol.* **2016**, *26*, 240–246. [PubMed]
9. Klewicka, E.; Cukrowska, B.; Libudzisz, Z.; Slizewska, K.; Motyl, I. Changes in gut microbiota in children with atopic dermatitis administered the bacteria *Lactobacillus casei* DN-114001. *Pol. J. Microbiol.* **2011**, *60*, 329–333. [PubMed]
10. Harada, D.; Takada, C.; Tsukumo, Y.; Takaba, K.; Manabe, H. Analyses of a mouse model of the dermatitis caused by 2,4,6-trinitro-1-chlorobenzene (TNCB)-repeated application. *J. Dermatol. Sci.* **2005**, *37*, 159–167. [CrossRef] [PubMed]
11. Homey, B.; Steinhoff, M.; Ruzicka, T.; Leung, D.Y. Cytokines and chemokines orchestrate atopic skin inflammation. *J. Allergy Clin. Immunol.* **2006**, *118*, 178–189. [CrossRef] [PubMed]
12. Wullaert, A.; Bonnet, M.C.; Pasparakis, M. NF-kappaB in the regulation of epithelial homeostasis and inflammation. *Cell Res.* **2011**, *21*, 146–158. [CrossRef] [PubMed]
13. Barnes, P.J.; Karin, M. Nuclear factor-κB: A pivotal transcription factor in chronic inflammatory diseases. *N. Engl. J. Med.* **1997**, *336*, 1066–1071. [CrossRef]
14. Nakamura, H.; Aoki, M.; Tamai, K.; Oishi, M.; Ogihara, T.; Kaneda, Y.; Morishita, R. Prevention and regression of atopic dermatitis by ointment containing NF-κB decoy oligodeoxynucleotides in NC/Nga atopic mouse model. *Gene Ther.* **2002**, *9*, 1221–1229. [CrossRef] [PubMed]
15. Tanaka, A.; Konno, M.; Muto, S.; Kambe, N.; Morii, E.; Nakahata, T.; Itai, A.; Matsuda, H. A novel NF-κB inhibitor, IMD-0354, suppresses neoplastic proliferation of human mast cells with constitutively activated C-kit receptors. *Blood* **2005**, *105*, 2324–2331. [CrossRef] [PubMed]
16. Leung, D.Y. Atopic dermatitis: New insights and opportunities for therapeutic intervention. *J. Allergy Clin. Immunol.* **2000**, *105*, 860–876. [CrossRef] [PubMed]

17. Simpson, E.L. Atopic dermatitis: A review of topical treatment options. *Curr. Med. Res. Opin.* **2010**, *26*, 633–640. [CrossRef] [PubMed]
18. Gonzales, F.; Ramdane, N.; Delebarre-Sauvage, C.; Modiano, P.; Duhamel, A.; Lasek, A. Monitoring of topical corticosteroid phobia in a population of parents with children with atopic dermatitis using the TOPICOP(R) scale: Prevalence, risk factors and the impact of therapeutic patient education. *J. Eur. Acad. Dermatol. Venereol.* **2016**, *31*, e172–e174. [CrossRef] [PubMed]
19. Bylka, W.; Znajdek-Awizen, P.; Studzinska-Sroka, E.; Danczak-Pazdrowska, A.; Brzezinska, M. *Centella asiatica* in dermatology: An overview. *Phytother. Res.* **2014**, *28*, 1117–1124. [CrossRef] [PubMed]
20. Incandela, L.; Cesarone, M.R.; DeSanctis, M.T.; Belcaro, G.; Dugall, M.; Acerbi, G. Treatment of diabetic microangiopathy and edema with HR (Paroven, Venoruton; O-(β-hydroxyethyl)-rutosides): A prospective, placebo-controlled, randomized study. *J. Cardiovasc. Pharmacol. Ther.* **2002**, *7* (Suppl. S1), S11–S15. [CrossRef] [PubMed]
21. Subathra, M.; Shila, S.; Devi, M.A.; Panneerselvam, C. Emerging role of *Centella asiatica* in improving age-related neurological antioxidant status. *Exp. Gerontol.* **2005**, *40*, 707–715. [CrossRef] [PubMed]
22. Chong, N.J.; Aziz, Z. A Systematic Review of the Efficacy of *Centella asiatica* for Improvement of the Signs and Symptoms of Chronic Venous Insufficiency. *Evid. Based Complement. Altern. Med.* **2013**, *2013*, 627182. [CrossRef] [PubMed]
23. Bylka, W.; Znajdek-Awizen, P.; Studzinska-Sroka, E.; Brzezinska, M. *Centella asiatica* in cosmetology. *Postepy Dermatol. Alergol.* **2013**, *30*, 46–49. [CrossRef] [PubMed]
24. Suzuki, H.; Makino, Y.; Nagata, M.; Furuta, J.; Enomoto, H.; Hirota, T.; Tamari, M.; Noguchi, E. A rare variant in CYP27A1 and its association with atopic dermatitis with high serum total IgE. *Allergy* **2016**, *71*, 1486–1489. [CrossRef] [PubMed]
25. Takahashi-Ando, N.; Jones, M.A.; Fujisawa, S.; Hama, R. Patient-reported outcomes after discontinuation of long-term topical corticosteroid treatment for atopic dermatitis: A targeted cross-sectional survey. *Drug Healthc. Patient Saf.* **2015**, *7*, 57–62. [CrossRef] [PubMed]
26. Bebawy, J.F. Perioperative steroids for peritumoral intracranial edema: A review of mechanisms, efficacy, and side effects. *J. Neurosurg. Anesthesiol.* **2012**, *24*, 173–177. [CrossRef] [PubMed]
27. Park, S.J.; Lee, Y.H.; Lee, K.H.; Kim, T.J. Effect of eriodictyol on the development of atopic dermatitis-like lesions in ICR mice. *Biol. Pharm. Bull.* **2013**, *36*, 1375–1379. [CrossRef] [PubMed]
28. Hashim, P.; Sidek, H.; Helan, M.H.; Sabery, A.; Palanisamy, U.D.; Ilham, M. Triterpene composition and bioactivities of *Centella asiatica*. *Molecules* **2011**, *16*, 1310–1322. [CrossRef] [PubMed]
29. Tang, B.; Zhu, B.; Liang, Y.; Bi, L.; Hu, Z.; Chen, B.; Zhang, K.; Zhu, J. Asiaticoside suppresses collagen expression and TGF-β/Smad signaling through inducing Smad7 and inhibiting TGF-βRI and TGF-βRII in keloid fibroblasts. *Arch. Dermatol. Res.* **2011**, *303*, 563–572. [CrossRef] [PubMed]
30. Nowwarote, N.; Osathanon, T.; Jitjaturunt, P.; Manopattanasoontorn, S.; Pavasant, P. Asiaticoside induces type I collagen synthesis and osteogenic differentiation in human periodontal ligament cells. *Phytother. Res.* **2013**, *27*, 457–462. [CrossRef] [PubMed]
31. Maquart, F.X.; Chastang, F.; Simeon, A.; Birembaut, P.; Gillery, P.; Wegrowski, Y. Triterpenes from *Centella asiatica* stimulate extracellular matrix accumulation in rat experimental wounds. *Eur. J. Dermatol.* **1999**, *9*, 289–296. [PubMed]
32. Nhiem, N.X.; Tai, B.H.; Quang, T.H.; Kiem, P.V.; Minh, C.V.; Nam, N.H.; Kim, J.H.; Im, L.R.; Lee, Y.M.; Kim, Y.H. A new ursane-type triterpenoid glycoside from *Centella asiatica* leaves modulates the production of nitric oxide and secretion of TNF-α in activated RAW264.7 cells. *Bioorg. Med. Chem. Lett.* **2011**, *21*, 1777–1781. [CrossRef] [PubMed]
33. Bonte, F.; Dumas, M.; Chaudagne, C.; Meybeck, A. Influence of asiatic acid, madecassic acid, and asiaticoside on human collagen I synthesis. *Planta Med.* **1994**, *60*, 133–135. [CrossRef] [PubMed]
34. Hou, Q.; Li, M.; Lu, Y.H.; Liu, D.H.; Li, C.C. Burn wound healing properties of asiaticoside and madecassoside. *Exp. Ther. Med.* **2016**, *12*, 1269–1274. [CrossRef] [PubMed]
35. Kimura, Y.; Sumiyoshi, M.; Samukawa, K.; Satake, N.; Sakanaka, M. Facilitating action of asiaticoside at low doses on burn wound repair and its mechanism. *Eur. J. Pharmacol.* **2008**, *584*, 415–423. [CrossRef] [PubMed]
36. Won, J.H.; Shin, J.S.; Park, H.J.; Jung, H.J.; Koh, D.J.; Jo, B.G.; Lee, J.Y.; Yun, K.; Lee, K.T. Anti-inflammatory effects of madecassic acid via the suppression of NF-κB pathway in LPS-induced RAW264.7 macrophage cells. *Planta Med.* **2010**, *76*, 251–257. [CrossRef] [PubMed]

37. Murakami, Y.; Shoji, M.; Hirata, A.; Tanaka, S.; Yokoe, I.; Fujisawa, S. Dehydrodiisoeugenol, an isoeugenol dimer, inhibits lipopolysaccharide-stimulated nuclear factor kappa B activation and cyclooxygenase-2 expression in macrophages. *Arch. Biochem. Biophys.* **2005**, *434*, 326–332. [CrossRef] [PubMed]
38. Liu, F.T.; Goodarzi, H.; Chen, H.Y. IgE, mast cells, and eosinophils in atopic dermatitis. *Clin. Rev. Allergy Immunol.* **2011**, *41*, 298–310. [CrossRef] [PubMed]
39. Schreiber, S.; Kilgus, O.; Payer, E.; Kutil, R.; Elbe, A.; Mueller, C.; Stingl, G. Cytokine pattern of Langerhans cells isolated from murine epidermal cell cultures. *J. Immunol.* **1992**, *149*, 3524–3534. [PubMed]
40. Friedmann, P.S. Contact sensitisation and allergic contact dermatitis: Immunobiological mechanisms. *Toxicol. Lett.* **2006**, *162*, 49–54. [CrossRef] [PubMed]
41. Kim, G.D.; Lee, S.E.; Park, Y.S.; Shin, D.H.; Park, G.G.; Park, C.S. Immunosuppressive effects of fisetin against dinitrofluorobenzene-induced atopic dermatitis-like symptoms in NC/Nga mice. *Food Chem. Toxicol.* **2014**, *66*, 341–349. [CrossRef] [PubMed]
42. Yang, G.; Choi, C.H.; Lee, K.; Lee, M.; Ham, I.; Choi, H.Y. Effects of Catalpa ovata stem bark on atopic dermatitis-like skin lesions in NC/Nga mice. *J. Ethnopharmacol.* **2013**, *145*, 416–423. [CrossRef] [PubMed]
43. Lee, H.J.; Kim, M.H.; Choi, Y.Y.; Kim, E.H.; Hong, J.; Kim, K.; Yang, W.M. Improvement of atopic dermatitis with topical application of *Spirodela polyrhiza*. *J. Ethnopharmacol.* **2016**, *180*, 12–17. [CrossRef] [PubMed]
44. Tanaka, A.; Muto, S.; Jung, K.; Itai, A.; Matsuda, H. Topical application with a new NF-κB inhibitor improves atopic dermatitis in NC/NgaTnd mice. *J. Investig. Dermatol.* **2007**, *127*, 855–863. [CrossRef] [PubMed]
45. Park, J.H.; Kim, M.S.; Jeong, G.S.; Yoon, J. Xanthii fructus extract inhibits TNF-α/IFN-γ-induced Th2-chemokines production via blockade of NF-κB, STAT1 and p38-MAPK activation in human epidermal keratinocytes. *J. Ethnopharmacol.* **2015**, *171*, 85–93. [CrossRef] [PubMed]
46. Kim, H.; Lee, E.; Lim, T.; Jung, J.; Lyu, Y. Inhibitory effect of Asparagus cochinchinensis on tumor necrosis factor-α secretion from astrocytes. *Int. J. Immunopharmacol.* **1998**, *20*, 153–162. [CrossRef]
47. Qiu, J.; Yu, L.; Zhang, X.; Wu, Q.; Wang, D.; Wang, X.; Xia, C.; Feng, H. Asiaticoside attenuates lipopolysaccharide-induced acute lung injury via down-regulation of NF-κB signaling pathway. *Int. Immunopharmacol.* **2015**, *26*, 181–187. [CrossRef] [PubMed]

© 2017 by the authors. Licensee MDPI, Basel, Switzerland. This article is an open access article distributed under the terms and conditions of the Creative Commons Attribution (CC BY) license (http://creativecommons.org/licenses/by/4.0/).

International Journal of
Molecular Sciences

MDPI

Article

One Year Follow-Up Risk Assessment in SKH-1 Mice and Wounds Treated with an Argon Plasma Jet

Anke Schmidt [1,*], Thomas von Woedtke [1,2], Jan Stenzel [3], Tobias Lindner [3], Stefan Polei [3], Brigitte Vollmar [4] and Sander Bekeschus [1]

[1] Leibniz-Institute for Plasma Science and Technology (INP Greifswald), Departments of Plasma Life Science and ZIK Plasmatis, Felix-Hausdorff-Str. 2, 17489 Greifswald, Germany; woedtke@inp-greifswald.de (T.v.W.); sander.bekeschus@inp-greifswald.de (S.B.)

[2] Department of Hygiene and Environmental Medicine, University Medicine Greifswald, 17475 Greifswald, Germany

[3] Core Facility Multimodal Small Animal Imaging, 18057 Rostock, Germany; jan.stenzel@med.uni-rostock.de (J.S.); tobias.lindner@med.uni-rostock.de (T.L.); stefan.polei@med.uni-rostock.de (S.P.)

[4] Institute for Experimental Surgery, Rostock University Medical Center, Schillingallee 69a, 18057 Rostock, Germany; brigitte.vollmar@med.uni-rostock.de

* Correspondence: anke.schmidt@inp-greifswald.de; Tel.: +49-3834-5543958; Fax: +49-3834-554301

Academic Editor: Allison Cowin
Received: 20 March 2017; Accepted: 12 April 2017; Published: 19 April 2017

Abstract: Multiple evidence in animal models and in humans suggest a beneficial role of cold physical plasma in wound treatment. Yet, risk assessment studies are important to further foster therapeutic advancement and acceptance of cold plasma in clinics. Accordingly, we investigated the long term side effects of repetitive plasma treatment over 14 consecutive days in a rodent full-thickness ear wound model. Subsequently, animals were housed for 350 days and sacrificed thereafter. In blood, systemic changes of the pro-inflammatory cytokines interleukin 1β and tumor necrosis factor α were absent. Similarly, tumor marker levels of α-fetoprotein and calcitonin remained unchanged. Using quantitative PCR, the expression levels of several cytokines and tumor markers in liver, lung, and skin were found to be similar in the control and treatment group as well. Likewise, histological and immunohistochemical analysis failed to detect abnormal morphological changes and the presence of tumor markers such as carcinoembryonic antigen, α-fetoprotein, or the neighbor of *Punc 11*. Absence of neoplastic lesions was confirmed by non-invasive imaging methods such as anatomical magnetic resonance imaging and positron emission tomography-computed tomography. Our results suggest that the beneficial effects of cold plasma in wound healing come without apparent side effects including tumor formation or chronic inflammation.

Keywords: dermal full-thickness wounds; *kINPen* plasma jet; plasma medicine; reactive oxygen and nitrogen species; risk evaluation; SKH1 mouse model

---

1. Introduction

Cold atmospheric pressure plasma has emerged as a promising tool for biomedical and clinical applications [1]. In this field called plasma medicine, encouraging results have been achieved for disinfection purposes [2], in vitro [3–5] and in patients [6–8]. Notably, a key feature determining healing is the state of wound oxygenation [9], and evidence suggests that the scavenging of active oxygen species impairs wound healing [10]. Accordingly, strategies influencing redox signaling may be used as an accessory therapy in chronic wound management [11–13], establishing a link to plasma medicine [14–16].

Cold physical plasmas are partially ionized gases that mediate biological responses especially via generation of reactive oxygen (ROS) and nitrogen species (RNS) [17–19]. Crucially, such species are translated in cells via redox enzymes [20] and therefore actively participate in intracellular signaling events [21–23]. Most mammalian cells maintain and benefit from a residual concentration of ROS and possess a complex system to sense a relay of ROS-related signals [24,25]. In contrast to low ROS and RNS quantities, higher concentrations are known to be responsible for apoptotic signaling [26–28] or finally for DNA damage [29]. Previous studies assessed the mutagenic risks of plasma in vitro [30–32]. The data provided in these studies demonstrated the absence of mutagenic or genotoxic effects in plasma-treated cells or in a hen's egg test model for micronuclei induction (HET-MN), suggesting that a clinical application of the argon plasma jet does not pose mutagenic risks. This is confirmed by a clinical long-term observation of laser skin lesions which were treated by cold atmospheric plasma [33,34]. However, systematic in vivo studies investigating any malignant side effects of plasma have not carried out to date.

To detect tumorigenicity, the usage of rodent models is commonly proposed. With regard to potential plasma applications in human wound healing, we utilized a full-thickness immunocompetent mouse model subjected to plasma treatment to monitor long-term effects. Non-invasive methods such as magnetic resonance imaging (MRI) and positron-emission tomography/computed tomography (PET/CT) are able to detect neoplastic lesions throughout the body. Using both technologies, tumorigenic effects were investigated in animals one year after plasma treatment. Using quantitative PCR, ELISA measurements, and immunohistochemical analysis of several tumor markers, we also investigated primary tumor formation or metastasis of malignant tumors.

2. Results

2.1. Evaluation of Histological Architecture and Inflammation Status after One Year

The aim of this study was the risk evaluation of cold plasma 350 days after wound treatment in a total of 84 hairless mice. In our long-term observation, plasma-treated mice showed typical health state, nutrition, and behavior. Moreover, we did not identify any toxic side effects or chronic wound infections. Two untreated animals developed a hepatocellular carcinoma (HCC, male) or skin abnormalities (female), which served as positive controls for ex vivo analyses. One of forty-two animals in the plasma group showed an enlargement of the organ spleen (data not shown). Structurally, wound areas were similar between experimental groups (arrowhead in Figure 1A–C). Yet, hematoxylin and eosin (H&E) staining showed some histologic changes in the dermal wound region. We found a separation and disconnection of dermal layers (I) without inflammatory cell infiltration into the dermis (I'). This effect was observed in most animals independent from treatment regime or gender (controls, A; males, B; females, C), indicating a normal healing output without excessive scar formation on day 350. Macroscopically, the organs of plasma-treated animals did not exhibit morphological changes, signs of tumor formation or metastatic processes, or differences in size or weight, such as in lung, liver, brain, thyroid gland, kidney, spleen, and heart (II, clockwise rotation).

Next, H&E staining was performed for the heart, kidney, brain, and thyroid tissues (Figure 2). Normal architecture of the heart with cardiac myocytes and centrally placed nuclei (I) was observed. Similarly, kidney tissues showed typical glomeruli and convoluted tubule structures without signs of necrosis (II). Brain sections from mice treated with plasma did not reveal either neuronal injury, focal inflammatory cell infiltration, hyperchromatic cells, or cellular shrinkage or swelling (III). In thyroid histology, follicles were surrounded with thyrocytes without cellular infiltrations (IV).

Similar to the controls (A), H&E-stained liver sections from mice treated with plasma showed normal hepatic architecture with the central vein (cv), radially surrounding hepatocytes (h), sinusoids (s), and nuclei (n) in males (B) and females (C). Additionally, we could not identify swollen or multinucleated macrophages in H&E staining, or scattered inflammatory cell aggregation (Figure 3A–C). Immunostaining with F4/80 antibody was performed to show the morphological architecture associated with a non-activated state. In an untreated control mouse we found a hepatocellular carcinoma (HCC) with a remarkable cellular infiltration in the liver (inlet in 0), which served as the "positive control" (+ve ctrl). The major histologic finding in the liver of the HCC-bearing mouse was a broad F4/80 positive cell staining, suggesting a strong inflammation with inflammatory foci (arrows, 0I). Contrary to that, we found in all mice studied a normal ramified structure of Kupffer cells. In addition, we observed no differences in the quality and amount of F4/80 positive cells, which were not swollen or amoeboid (AI–CI). One remarkable pathological characteristic of the liver is increased production of collagen, which is the main component of the extracellular matrix in fibrotic tissue [35]. Depositions of collagen fibers were examined in liver sections showing no hyperplasia of the fibrous tissue in plasma-treated mice (II). In contrast, arrows indicate a high level of collagen deposition stained by picrosirius red (PSR) in the fibrotic septa between nodules of HCC-bearing liver tissue (0II). Moreover, mRNA for the pro-inflammatory cytokine tumor-necrosis factor α (*TNFα*) was not elevated in the liver of plasma-treated mice relative to controls after one year (D). This finding was confirmed systemically in blood plasma, with a small but significant down-regulation of *TNFα* in females (E) 15 days after wounding (Figure 3).

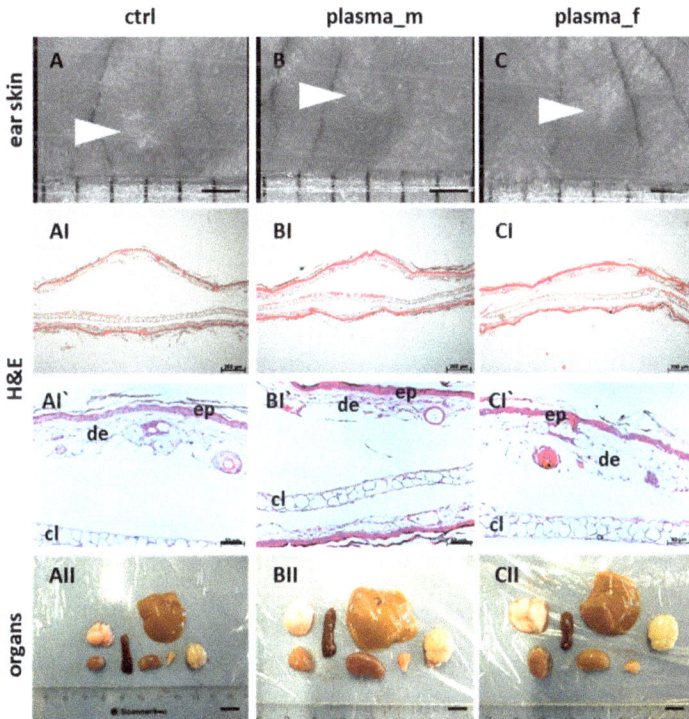

Figure 1. Macro- and microscopic skin wounds and organs 350 days after plasma treatment. Similar to untreated control animals (**A**), stereo microscopy of wound region revealed no differences of morphology (arrowheads) in ear tissue 350 days after injury in plasma- and untreated males (**B**) and females (**C**). Hematoxylin and eosin (H&E)-stained skin sections of dermal layers in control and plasma-treated mice were similar showing normal dermal architecture with a disconnection between the cartilage layer (cl) and dermis (de, ep, epidermis) at the wound site without inflammatory cell infiltration into the dermis (**I–I'**). Macroscopic evaluation of different organs (lung, spleen, liver, heart, kidney, thyroid glands, and brain) lacking visible tumor formation (**II**). Representative images are shown. Scale bar 1 cm (**II**), 1 mm (**A–C**), 100 μm (**I**), and 50 μM (**I'**).

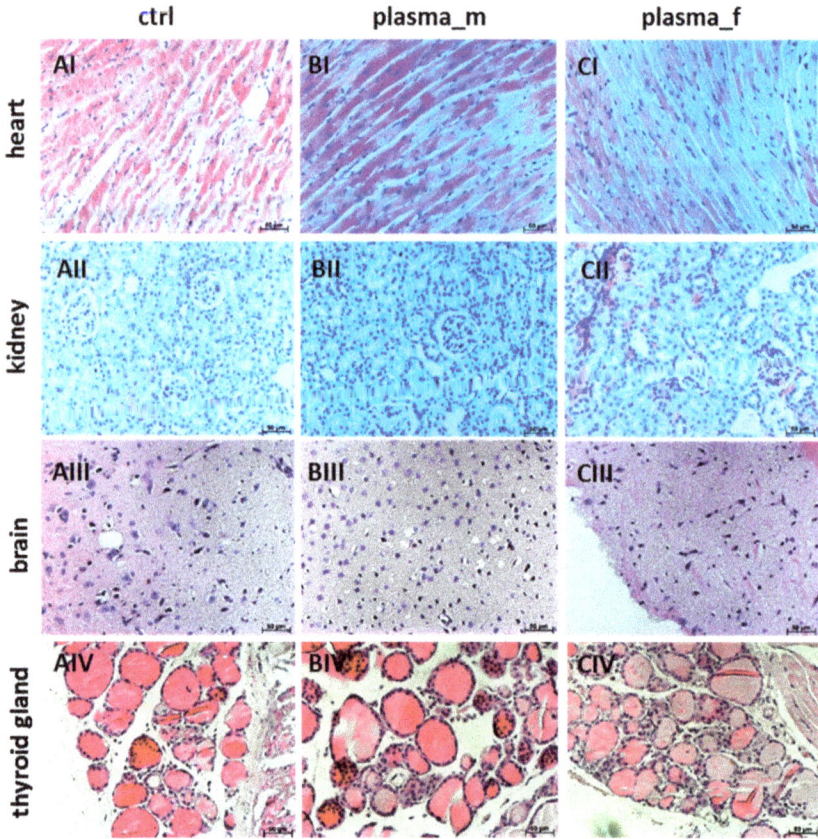

Figure 2. Histology in different organs of mice treated with plasma. H&E-stained heart sections showing normal architecture with cardiac myocytes and centrally placed nuclei (**I**) in controls (**A**), and plasma-treated males (**B**) and females (**C**). Kidney sections showing normal histological structure with glomerulus and convoluted tubules without necrosis (**II**). Brain sections from mice treated with plasma showing no signs of either neuronal injury or focal inflammatory cell infiltration nor hyperchromatic cells or cellular shrinkage or swelling (**III**). Normal thyroid histology showing follicles surrounded with thyrocytes without cellular infiltrations (**IV**). One representative picture of H&E staining is shown for selected organs. Scale bar 50 μm.

Histological examination of the lung sections did not show changes consistent with lung pathology. Macroscopically, lungs were not less aerated or covered with white fibrin patches. Moreover, in H&E stained sections we did not find scattered inflammatory cell aggregations with interstitial neutrophilic infiltration, septal thickening of the alveolar capillary membrane, or focal hemorrhage of the mesenchyme in plasma-treated males (B) or females (C) similar to the controls (A). Likewise, *TNFα* mRNA expression was not altered in the lungs of plasma-treated mice relative to controls (Figure 4D).

Next, no morphological abnormalities of the spleen or excessive neutrophil infiltration in the red splenic pulp were found in plasma-treated mice (B,C) similar to untreated controls (A). Immunostaining with F4/80 antibody was performed to visualize macrophages (II) in the spleen and immunostaining with Ly6G showed the granulocytes distribution (III). Taken together, experimental groups exhibited no differences in staining of immune cell types (Figure 5).

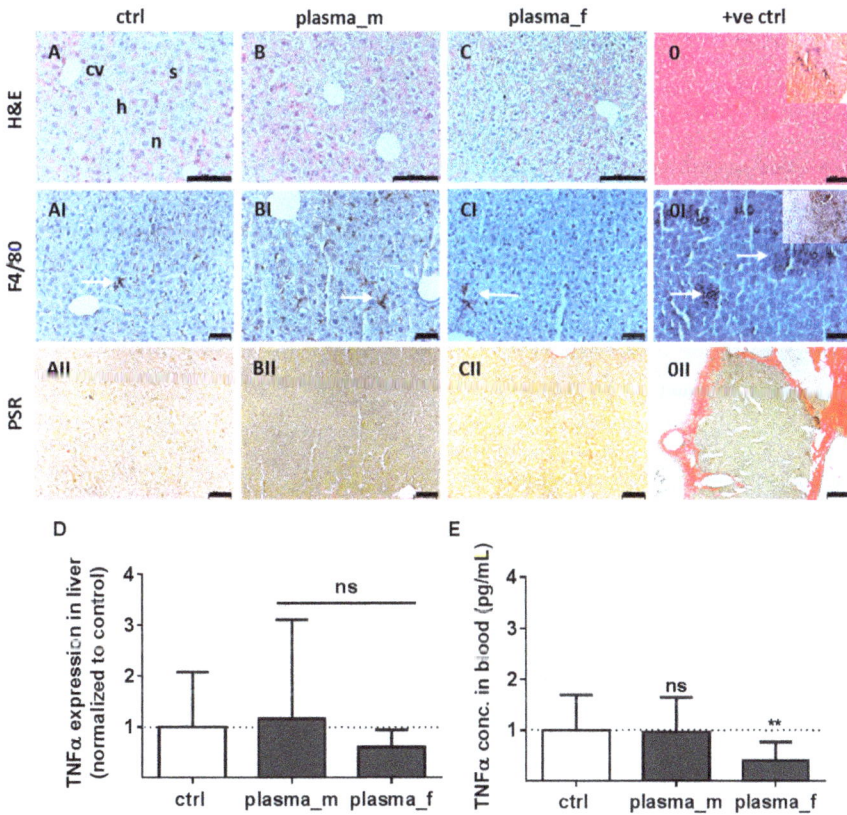

Figure 3. Liver histology, Kupffer cells, local and systemic tumor necrosis factor α (*TNFα*) levels. Similar to controls (**A**), H&E-stained liver sections from mice treated with plasma showing normal hepatic architecture with central vein (cv) and surrounding hepatocytes (h), sinusoids (s), and nuclei (n) in males (**B**) and females (**C**). Scale bar 100 μm. Immunostaining with F4/80 antibody was performed to show scattered inflammatory cell aggregations and to visualize ramified Kupffer cells in liver sections, which were not swollen or amoeboid (**AI–CI**). A hepatocellular carcinoma in liver (from the untreated control group) showed a strong cell infiltration (inlet, (**0**)) hyperplasia of liver tissue and a broad F4/80 positive cell-staining (inlet in (**0I**)) with inflammatory foci (arrows in (**0I**), +ve ctrl). Picrosirius red (PSR) staining of tissue slices from liver showing collagen fibers in cancer tissue of a hepatocellular carcinoma (HCC)-bearing mouse (arrow, (**0II**)) in contrast to non-tumor sections of plasma- and untreated animals (**A–CII**). Representative images are shown with scale bars of 50 μm (**I–II**). Using quantitative polymerase chain reaction (qPCR), *TNFα* mRNA expression was quantified in the liver of plasma- and untreated mice (**D**). Analysis of blood samples has shown a significant down-regulation of *TNFα* in females 15 days after wounding (**E**). Results are means + S.D. for $n > 4$ (**D**; ns, not significant) or nine mice (**E**, ** $p < 0.01$).

Cervical lymph nodes are normally subject to a number of different pathological conditions including tumor development and inflammation [36]. Therefore, cervical lymph nodes were removed from plasma-treated animals and compared to untreated animals. The size of lymph nodes were similar in both experimental groups indicating the absence of lymphomas (data not shown). H&E staining showed neither follicular hyperplasia nor dilated parafollicular zones (Figure 6A–C). In the lymph nodes of plasma-treated mice, interleukin 1β (*IL-1β*) mRNA was in the range of control levels

(Figure 6D). Moreover, blood plasma levels of *IL-1β* were similar in all treatment groups 15 days after wounding (Figure 6E). Additionally, parafollicular hyperplasia, plasmocytosis, and increased collagen density in the lymph nodes of several tumor models are reported [37]. We clearly obtained a follicular hyperplasia in the HCC-bearing mouse (Figure 6FI) in contrast to the plasma-treated mice (Figure 6GI). The collagen fiber density in cervical and mesenteric lymph nodes from this mouse was significantly increased compared to plasma and untreated mice, indicating no transformational changes in the latter groups. Collagen fibers are shown in red in a PSR staining using light microscopy (Figure 6FII) or by fluorescence microscopy (Figure 6GII) [38]. No histologic abnormalities of bones and other unmentioned organs were observed (data not shown).

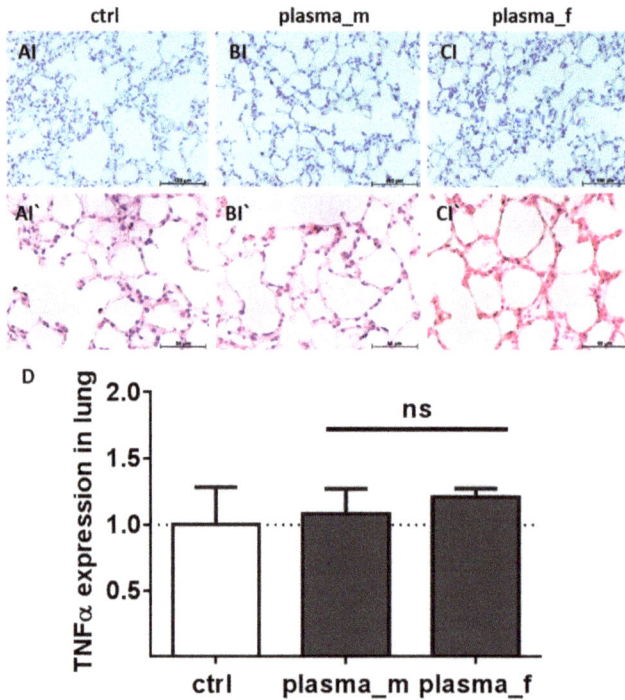

Figure 4. Lung histology and TNFα expression. Similar to controls (**A**), H&E staining did not show scattered inflammatory cell aggregations with interstitial lymphocytic and neutrophilic infiltration, thickening of the alveolar capillary membrane, or focal hemorrhage in the mesenchyme of lungs of plasma-treated males (**B**) or females (**C**). Using qPCR, *TNFα* mRNA expression was quantified in the lungs of plasma- and untreated mice (**D**). Representative images with 100 μm (**I**) or 50 μm (**I′**). Results are means + S.D. for *n* > 4 (**D**; ns, not significant).

2.2. Evaluation of Tumor Marker (TM) Expression and Secretion

To explore the safety aspects of cold plasma as a precondition for a clinical application, we next investigated the tumor marker (TM) expression in several organs, skin tissue, and blood plasma using immunohistochemistry, quantitative polymerase chain reaction (qPCR), and ELISA 350 d after plasma treatment. Representative images of immunohistochemical staining revealed no enhanced tumor marker expression in liver, brain, lung, and skin tissue (Figure 7A). The immunohistochemical staining of α-fetoprotein (*AFP*, I), a traditional tumor marker for hepatocellular carcinoma (HCC), β2 microglobulin (*β2M*, II), a tumor marker for some blood cell cancers such as lymphoma and multiple myelomas [39], as well as the carcinoembryonic antigen (*CEA*, III), the most widely used tumor marker in patients with non-small cell lung cancer [40], were significantly enhanced in organs of the HCC-bearing mouse (+ve ctrl, left panel) compared to plasma-treated mice (right panel). Additionally, the neuron specific enolase (*NSE*), a TM of neuroendocrine tumors such as Merkel-cell carcinoma of the skin [41], was obviously increased in skin sections of the positive control mouse (+ve ctrl), which has developed a skin abnormality (IV, left). Nevertheless, no positive *NSE* staining was found in plasma-treated mice (IV, right). Blood plasma analysis of tumor markers revealed no differences between groups in *AFP* (I) or calcitonin level (*CT*, II), a TM of medullary thyroid carcinoma [42] (Figure 7B).

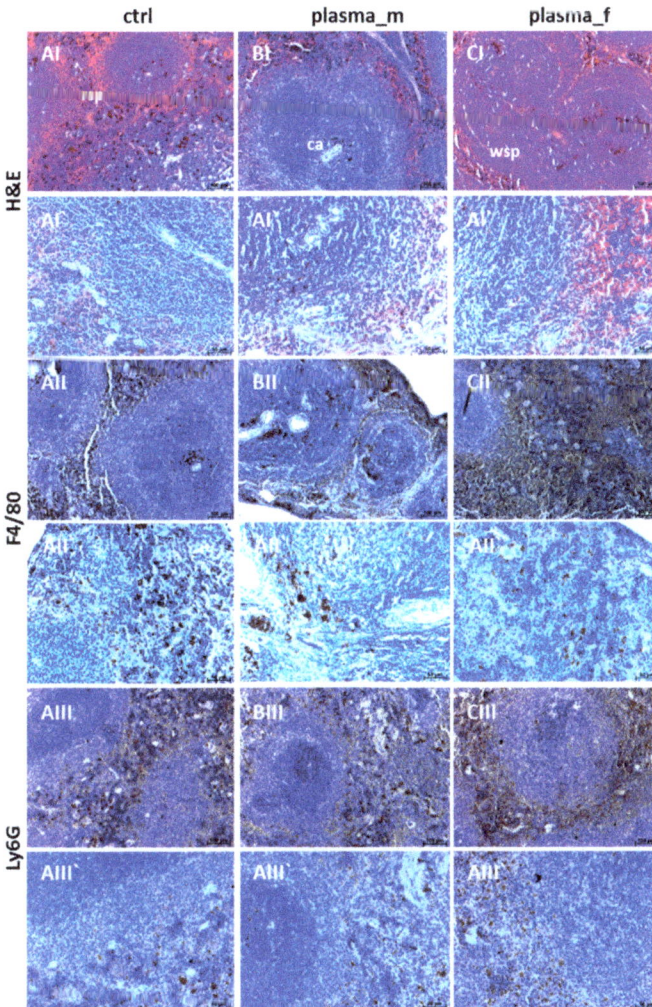

Figure 5. Macrophage and granulocyte distribution in spleens. Similar to controls (**A**), H&E staining did not show abnormalities in spleens in plasma-treated males (**B**) or females (**C**); rsp, red splenic pulp; ca, central artery; wsp, white splenic pulp. Intra-individual differences appeared in relation to the size of white and red pulp without a clear tendency. Immunostaining with F4/80 and Ly6G antibodies were performed to visualize macrophages (**II**) as well as normally dispersed granulocyte distribution (**III**). Representative images with 100 μm (**I–III**) or 50 μm (**I'–III'**).

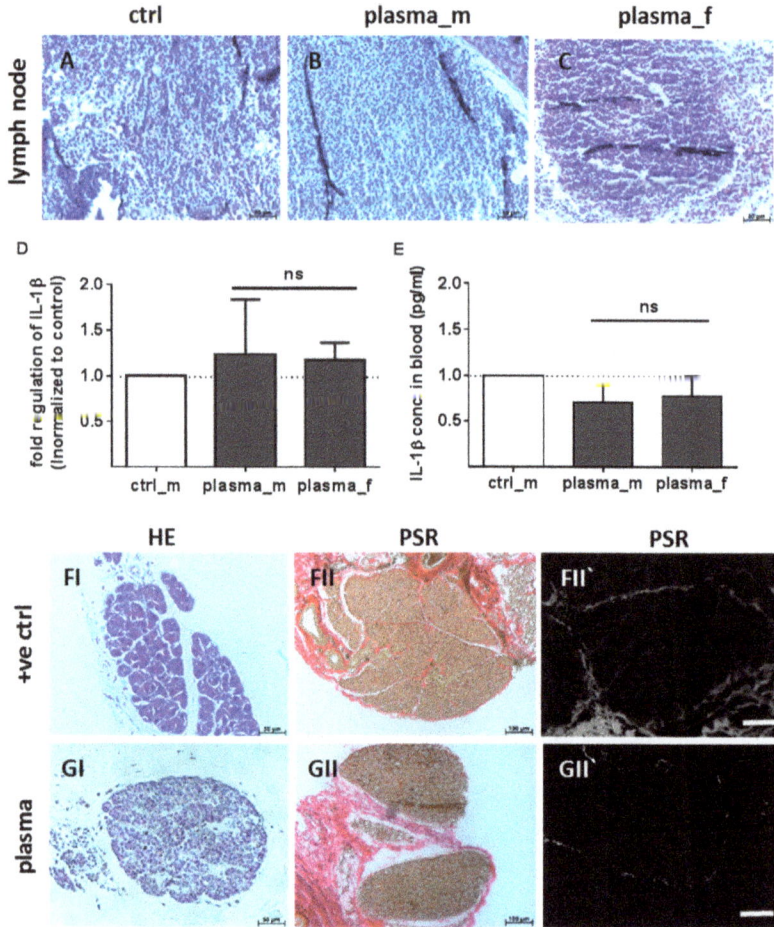

Figure 6. Lymph node histology, *IL-1β* expression, and comparison to tumor samples. H&E staining did not show hypertrophy, hyperplasia, or dilated parafollicular zones in lymph nodes of plasma-treated animals in comparison to controls (**A**) in males (**B**) and females (**C**). *IL-1β* mRNA expression was quantified by qPCR in cervical lymph nodes ((**D**), *n* > 4) and cytokine measurement of *IL-1β* concentration in blood samples 15 days after injury ((**E**), *n* > 9). Parafollicular hyperplasia, plasmocytosis, and increased collagen fiber density in lymph nodes of mice carrying a primary hepatocellular carcinoma (**F**) in contrast to plasma-treated mice (**G**) was visualized by H&E (**I**) and PSR staining (**II**) using light (**I–II**) or fluorescence microscopy (**II`**). Scale bar 50 μm (**A–C**), 100 μm (**F,G**).

Next, we performed quantitative PCR analysis of five tumor markers such as *AFP*, *NOPE* (neighbor of *Punc 11*), a novel TM of the liver [43], β2 microglobulin (*β2M*), carcinoembryonic antigen (*CEA*), or neuron-specific enolase (*NSE*). No changes were observed in, e.g., liver, brain, lung, or ear skin tissue relative to the controls and in contrast to organs of the HCC-bearing mouse (+ve ctrl). Actin-containing elements of the cytoskeleton are changed in pathophysiological conditions such as malignant tumors [44]. We investigated alteration of either β-actin protein expression in skin tissue sections using IHC as well as gene expression using qPCR. Structural changes of the cytoskeleton and changes in mRNA levels were absent in plasma-treated mice (Figure 7C).

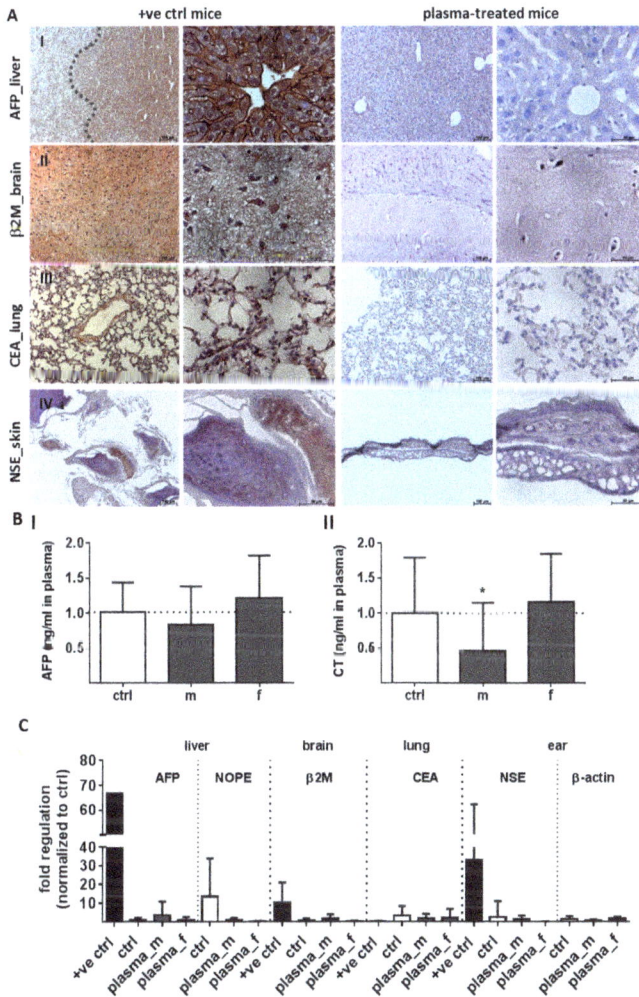

Figure 7. Unchanged tumor marker levels in liver, lung, brain, and skin of plasma-treated mice. Representative images of immunohistochemistry of different tumor markers (TM): *AFP* (**I**) in liver, *β2M* in brain (**II**), *CEA* in lung (**III**), as well as *NSE* staining in skin tissue (**IV**) after one year in positive control (+ve ctrl, left) and plasma-treated (right) animals (**A**). Scale bar 50 μm (right columns) or 100 μm (left columns). Using ELISA, we analyzed *AFP* (**I**), and calcitonin (*CT*) (**II**), a TM of medullary thyroid carcinoma, in blood serum (**B**; * $p < 0.05$; ($n > 9$). The mRNA expression levels of five TM and β-actin in liver, brain, lung, and ear skin tissues from plasma- and untreated (ctrl) mice were compared with organs from a HCC-bearing mouse (+ve ctrl). At least three independent experiments were performed and summarized in the indicated experimental groups (m, male; f, female). *AFP*, α-fetoprotein; *NOPE*, neighbor of *Punc 11*; *β2M*, β2-microglobulin; *NSE*, neuron specific enolase; *CEA*, carcinoembryonic antigen (**C**).

2.3. Multimodal Imaging of Mice 350 Days after Plasma Treatment

Positron-emission tomography/computed tomography (PET/CT) and anatomical magnetic resonance imaging (MRI) were employed to detect any long-term side effects in the plasma group.

In PET/CT imaging with 18F-FDG, indicating tissue metabolic activity corresponding to glucose uptake, the plasma-treated male as well as female mice showed no significant differences in tracer uptake in the heart (III), brain (IV), liver (V), or kidney (VI). Throughout the body (VII), pathological differences in tracer uptake were absent in the organs (e.g., thymus) of plasma-treated and control mice (Figure 8). Similarly, representative three-dimensional MRI scans derived from T2-weighted rapid acquisition with relaxation enhancement (RARE)-sequences (at an in-plane resolution of 180 μm) did not show any indication of tumor formation in control (A) or plasma-treated mice (B) one year after plasma treatment (Figure 9). Longitudinal and axial whole body view did not show pathophysiological glucose uptake in the kidneys (I) or stomach (II). This was confirmed by lateral scans of the abdomen with liver (III), kidney (IV), and head with brain (V). By contrast, glucose uptake is shown in two representative images of a liver tumor in non-plasma-treated animals (CI–II).

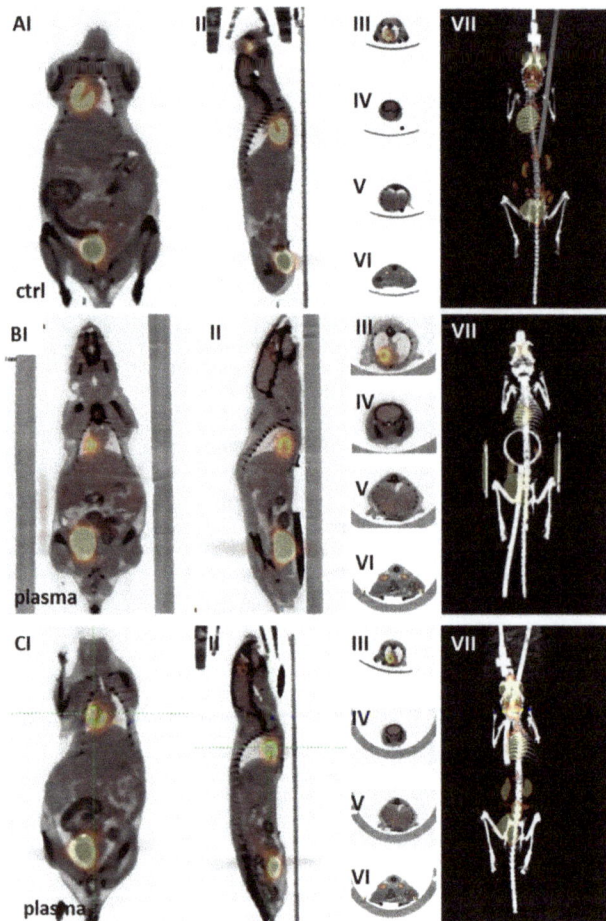

Figure 8. Positron emission tomography-computed tomography (PET/CT) imaging with 18F-fluorodeoxyglucose (18F-FDG) tracer for risk estimation after plasma treatment. (PET/CT with 18F-FDG tracer indicates tissue metabolic activity corresponding to glucose uptake. The spatial resolution was 1.5 mm. Representative scans of control (**A**), plasma-treated male (**B**), and female (**C**) mice after one year: whole body (**I,II**), heart (**III**), brain (**IV**), liver (**V**), and kidney (**VI**) scans and complete three-dimensional body PET/CT fusion image (**VII**).

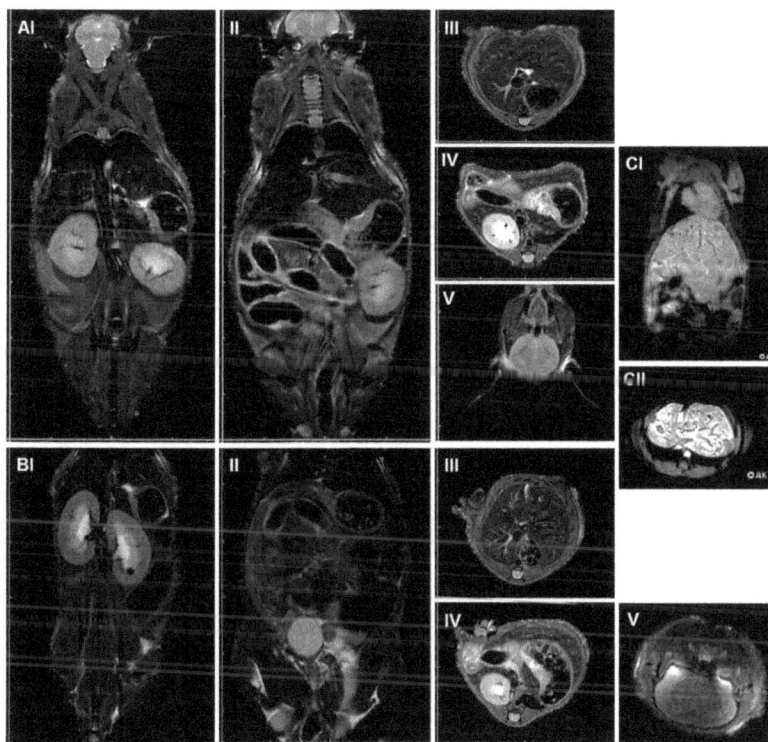

Figure 9. Risks evaluation in plasma-treated mice using anatomical magnetic-resonance imaging (MRI). Representative MRI scans of control (**A**) and plasma-treated mice one year after plasma treatment (**B**). Images derived from T2-weighted rapid acquisition with relaxation enhancement (RARE)-sequences with a deliverable spatial resolution of 0.18 mm in-plane, echo time (TE/TR 29/4100 ms): coronal slices with obvious visualization of kidneys (**I**) or stomach (**II**), as well as axial scans of abdomen with liver (**III**), kidney (**IV**), and head with the brain (**V**). Two representative images of a liver tumor are shown in a control (not plasma-treated) mouse (**CI–II**).

3. Discussion

Chemical and physico-chemical clinical interventions require safety testing with regard to cytotoxicity and genotoxicity. Although cell culture experiments are commonly used to investigate the extent of damage of substances, animal experiments are more meaningful studies with regard to long-term effects. Plasma application was shown to lack genotoxic effects in vitro [30,32] and using an egg HET-MN model [31], which takes an intermediate position between cell culture and rodent models. Here, we have further quantitatively characterized the long-term side effects using an immunocompetent mouse model. In this model, inflammation and immune responses in wound repair are not attenuated, allowing us to follow not only wound healing in a physiological setting [45] but also the long-term impact on organs and skin tissue.

Cold plasmas are a significant source of reactive oxygen species (ROS, e.g., HO, O_2^-, O_3, and H_2O_2) and reactive nitrogen species (RNS, e.g., NO, $ONOO^-$, and NO_2^-) [46]. At low concentration, some of these reactive components are beneficial—leading to cell proliferation and survival pathways—and are considered second messengers in the field of redox biology. Nevertheless, these species can be detoxified and metabolically controlled in cells and organisms via anti-oxidant protection mechanisms. We have previously demonstrated in vitro that plasma significantly altered anti-oxidant and phase II detoxification enzymes and proteins such as glutathione peroxidases, catalase, superoxide dismutase, heme oxygenase 1, NAD(P)H dehydrogenase 1, mitogen-activated protein kinases, molecules of the Jun pathway, and growth factors to protect skin cells from oxidative stress [47–50]. At higher quantities, reactive species may be tumorigenic [51] and mutagenic [52]. Regardless of the plasma device or the treatment time, no genotoxic effects of the kINPen plasma were found using a HPRT1 mutation assay [30] or the HET-MN model [31]. In the latter study, the global antioxidant defense was also not significantly challenged in fertilized plasma-treated chicken eggs. Furthermore, a long exposure time of test agents is recommended to increase the possibility of tumor development and to receive resilient statements about tumorigenicity. The regular clinical protocol for plasma treatment of chronic wounds aims for two to three treatments per week and a treatment time of 30 to 60 s per cm^2 [53]. Regarding these observations and preclinical plasma studies in humans [54–57], we decided on a daily treatment procedure of 20 s over 14 consecutive days, which seems advisable for toxicological investigations. Because of the brief exposure during the plasma treatment, acute health risks via ozone formation are not to be expected. The kINPen expels maximum ozone concentrations of 0.10–0.13 ppm at lower distances (<10 cm), which is in the range of the maximum working concentration value (MAK) of 0.1 ppm in the air [53].

Using whole body imaging as well as histological and biochemical analysis, our data suggest the absence of long-term side effects in the skin tissue and other organs of plasma-treated mice. This is in agreement with the literature where no histological damage was observed in human skin after 1 min [58] or 10 min of plasma treatment [59]. The absence of expression of a panel of tumor markers in liver, lung, brain, and ear tissue using qPCR suggests that plasma can support wound healing [45] without alterations in the tissue architecture and/or function. Additionally, serum levels were measured for AFP, a screening substance for a subset of abnormalities such as hepatocellular carcinoma [42] and calcitonin, which is produced in humans and rodents primarily by the parafollicular cells in the thyroid. A malignancy of thyroid glands (e.g., medullary thyroid carcinoma, C-cell hyperplasia, non-thyroidal carcinoma) typically produces elevated serum calcitonin levels of >5 ng/L in females, and 12 ng/L in males [60] and can therefore be used diagnostically as a tumor marker. Calcitonin and AFP serum levels were in the physiological range in plasma-treated and untreated animals of both genders.

Moreover, cervical and mesenteric lymph node and liver sections showed a strong inflammatory pathology in the HCC-bearing mouse as a response to a primary tumor formation in contrast to our plasma-treated animals. In the tumor-bearing mouse of the control group, alterations in the macrophage population indicated an activated state in the liver [36]. In all other animals in the plasma and control group, F4/80 positive cells within the liver confirmed the classic morphology of Kupffer cells consistent with an unaltered state of the liver. We also characterized the presence, morphology, and localization of granulocytes and macrophages in spleens suggesting that the mice exhibited dominant healthy features. Ly6G has also been implicated in the development of antitumor responses: It has been demonstrated that lymphoid cancers derive by chronic stimulation or dysfunction of B cells in association with a sustained inflammatory reaction [61]. Here, multisystem inflammatory reactions with cell infiltrations in several organs such as thyroid glands, spleen, and lymph nodes, or secretion of pro-inflammatory cytokines were not observed 350 days after plasma treatment.

Our study has limitations. First, we did not include positive controls in the animal experiments. Yet, in our view, it is not only experimentally too demanding but also unethical to include several dozens of different (possible) tumor groups with many animals each only for the sake of "generating" positive controls for downstream assays. Second, it was beyond the scope of this study to investigate every single tissue and organ in the animal body for possible side effects. Yet, using MRI and PET/CT scans, we believe to have sufficiently searched for any tumor formation in a large number of animals. Altogether, the findings of this study show the absence of long-term side effects after one year in mice receiving 14 consecutive plasma treatments. Hence, and given a profound efficacy, clinical plasma applications continue to be a promising therapy especially in dermatology.

4. Materials and Methods

4.1. Animals

A total of 84 SKH1-hr hairless immunocompetent mice (Charles River Laboratories, Sulzfeld, Germany) were housed under standard conditions in an animal facility (Rostock University Medical Center, Rostock, Germany). All experiments were approved by the local ethics committee according to the guidelines for care and use of laboratory animals (number 7221.3-1-013/14).

4.2. Wounding and Exposure to Cold Physical Plasma

Wounding was performed on the left and right ears of 10–12 weeks old animals as previously described [45]. Full-thickness circular dermal wounds of approximately 3 mm^2 size were created by removing the epidermis and dermis but not the cartilage using a micro-scissor. For reproducibility, the procedure was carried out by a single operator. Mice were assigned randomly into four groups ($n = 21$, males and females), and either received daily treatment of plasma (20 s) or were left untreated over 14 consecutive days. The argon plasma jet *kINPen* 11 (neoplas tools, Greifswald, Germany) ionizes a flow (5 standard liters per minute) of argon gas at a frequency of 1 MHz [53]. Organs and tissues were collected 350 days post intervention.

4.3. Wound Closure Observations and Histological Evaluation

Wound tissues was fixed in 4% paraformaldehyde, and 5 μm-sections were stained with hematoxylin/eosin (H&E) to visualize morphological changes by means of light microscopy. The collagen content of lymph nodes was qualitatively determined within picrosirius red (PSR) stained paraffin-embedded tissue sections using fluorescence microscopy [38]. For immunohistostaining, paraffin sections were deparaffinized, rehydrated, and boiled for 5 min for antigen retrieval. Prior to imaging by light microscopy, sections were incubated with α-fetoprotein (MAB1368, R&D Systems, Germany), β-2-microglobulin (sc-15366, Santa Cruz, Heidelberg, Germany), neuron-specific enolase (#R30242, NSJ Bioreagents, Hamburg, Germany), and carcinoembryonic antigen (AF6480, R&D Systems, Wiesbaden, Germany) overnight at 4 °C. Detection of Kupffer cells and macrophages in paraffin-embedded liver and spleen sections was performed using anti-mouse F4/80 antibody (#14-4801; Affymetrix, Frankfurt, Germany). For the identification of monocytes and macrophages in the spleen, Ly6G was used (#14-5931; Affymetrix, Frankfurt, Germany). After washing, sections were incubated with an immuno-peroxidase polymer (N-Histofine staining reagent; Medac, Wedel, Germany). Stained sections were mounted onto glass microscope slides using mounting medium containing DAPI (VectaShield; Biozol, Eching, Germany) prior to analysis using an Axio Observer Z.1 (Zeiss, Jena, Germany).

4.4. Homogenisation of Tissues and Gene Expression Analysis

For expression analysis, tissue and organ samples were collected at day 350 ($n = 84$). Fresh tissues from ear and several organs (e.g., lung, liver, brain, thyroid, and lymph nodes) were snap-frozen in liquid nitrogen, and stored at −80 °C. For gene expression analysis, homogenization was performed in RNA lysis buffer (Bio&Sell, Nürnberg, Germany) using a FastPrep-24 5G homogenisator (Biomedicals, Heidelberg, Germany). Total RNA was isolated (Bio&Sell) and quantitative PCR (qPCR) was performed as described previously [49]. Briefly, 1 μg of RNA was transcribed into cDNA, and qPCR was conducted in triplicate using SYBR Green I Master in a 96-well LightCycler 480 qPCR system (Roche Diagnostics, Mannheim, Germany). Gene primers were used from Biomol (Hamburg, Germany) and Biotez (Berlin, Germany; Table A1). The housekeeping gene *GAPDH* expression was unaffected by plasma and was used as an internal control for normalization. Genes expression was analyzed using the ΔΔCt method [62]. The final value for gene expression was determined by calculating the ratio of expression in the respective sample related to the control. Band intensities were quantified using

ImageQuantTL Software (GE Healthcare, Freiburg, Germany), and expressed as fold change compared to the corresponding control. Experiments were performed with six to ten mice for each group.

4.5. ELISA Measurements and Bead-Based Cytokine Analysis

Blood serum were collected retrobulbar in EDTA-tubes (d0, d15 or d350), and centrifuged. Samples were stored until use at −80 °C. TNFα and IL-1β secretion was measured using bead-based multiplex cytokine analysis (BioLegend,Uithoorn, The Netherlands). Enzyme-linked immunosorbent assay (ELISA) for a liver specific tumor marker "α fetoprotein" (AFP, BioLegend, Uithoorn, The Netherlands) and a medullary thyroid carcinoma specific tumor marker "calcitonin" (CT, BioLegend, Uithoorn, The Netherlands) was performed.

4.6. MRI, PET/CT Scans, and Interpretation of Data

For positron emission tomography-computed tomography (PET/CT) and magnetic resonance imaging (MRI) sessions 350 days post intervention, animals were anaesthetized with isoflurane (1.5–2.5%) supplemented with oxygen. During both scans, respiration of the mice was monitored and adjusted to a breathing rate of 35–50 breaths/min. All MRI scans were additionally respiration triggered. Temperature of mice were also controlled during the entire imaging period of all scans. Anatomical MRI T2-weighted TurboRARE sequences with coronal and axial slices over the whole body were acquired with the following parameters: TE/TR: 29/4100 ms, FoV: cor. 94.4 mm × 35 mm, axial 33.7 mm × 37 mm, in-plane resolution 0.18 mm, slice thickness 1 mm. For small animal PET/CT imaging, mice were injected intravenously with 15 MBq ^{18}F-FDG via a microcatheter placed in the tail vein. Static PET scans in prone position (duration: 15 min) were performed starting 60 min after tracer application using the Inveon dedicated PET/CT scanner (Siemens Preclinical Solutions, Knoxville, TN, USA). Whole body CT scans were acquired for attenuation correction and anatomical reference. CT images were reconstructed with a Feldkamp algorithm. PET data were first Fourier rebinned into a 2D dataset from which real-space images were reconstructed with an ordered subset expectation maximization (OSEM) algorithm with 16 subsets and 4 iterations. Attenuation correction was carried out using the CT data.

4.7. Statistical Analysis

Means of at least three independent experiments were calculated using *prism* software (GraphPad, La Jolla, CA, USA). Data were subjected to statistical analysis using the unpaired Student's *t* test or one-way analysis of variance (ANOVA) followed by the Dunnett's post hoc test. Significant differences were indicated as mean + S.D. with *p*-values from * $p < 0.05$, ** $p < 0.01$, *** $p < 0.001$.

5. Conclusions

One year after fourteen consecutive treatments of murine full-thickness ear wounds with the argon plasma jet kINPen, we were able to conclude key aspects with regard to efficacy and safety of that device. First, skin tissue healed physiologically without excessive scar formation similar to control animals. Second, the application of plasma was safe as demonstrated by the absence of tumor formation at the treatment site. Third, no malignant tissue was found in any other major organ including liver, lymph node, spleen, heart, lung, and brain. Fourth, the lack of neoplastic markers in blood plasma of plasma-treated animals suggested an absence of tumor tissue at sites in the murine body not investigated in this study. Fifth, the analysis of cytokines and immune cells indicated a physiological immune regulation without pathological or excessive inflammation. These data suggest that cold plasma applied topically is a safe adjuvant strategy in dermatology with regard to the absence of neoplastic side effects.

Acknowledgments: The authors thank Caner Özer, Liane Kantz, Juliane Moritz (all INP Greifswald, Germany), and Dorothea Frenz (IEC, Rostock University Medical Center, Germany) for technical support as well as

Änne Glass (IBIMA, Rostock University Medical Center, Germany) for biometrical planning. This work was founded by the Ministry of Education, Science, and Culture of the State of Mecklenburg-Western Pomerania (Germany), the European Social Fund (grant numbers AU 11 038; ESF/IV-BM-B35-0010/13 and AU 15 001), and the German Federal Ministry of Education and Research (grant number 03Z22DN11).

Author Contributions: All authors contributed to this study: Anke Schmidt, Thomas von Woedtke, and Brigitte Vollmar designed the study. Anke Schmidt performed animal and ex vivo experiments; Jan Stenzel performed PET/CT. Tobias Lindner and Stefan Polei performed MRI. Sander Bekeschus conducted ELISA and bead-based cytokine measurements. All authors co-wrote the manuscript.

Conflicts of Interest: The authors declare no conflict of interest.

Abbreviations

FoV	field of view
MRI	magnetic resonance imaging
PET/CT	positron emission tomography/computed tomography
ROS	reactive oxygen species
RNS	reactive nitrogen species
TE	echo time
TR	repetition time

Appendix A

Table A1. Murine primer sequences or catalog numbers (Biomol, Germany) for use in quantitative real-time PCR.

Name	Gene	Sequences/# (Biomol)
mGAPDH_s	House keeping gene	CATGGCCTCCAAGGAGTAAG
mGAPDH_as		TGTGAGGGAGATGCTCAGTG
mTNFα_s	Tumor necrosis factor α	TCTCATGCACCACCATCAAGGACT
mTNFα_as		ACCACTCTCCCTTTGCAGAACTCA
mIL-1β_s	Interleukin 1β	GCAACTGTTCCTGAACTCAACT
mIL-1β_as		ATCTTTTGGGGTCCGTCAACT
mAFP_s	α fetoprotein	CCAGGAAGTCTGTTTCACAGAAG
mAFP_as		CAAAAGGCTCACACCAAAGAG
mβ2M	β2 microglobulin	VMPS-563
mCEA1	Carcinoembryonic antigen	VMPS-1054
mCT	Calcitonin	VMPS-785
mNSE	Neuron specific enolase	VMPS-1942

References

1. Weltmann, K.-D.; von Woedtke, T. Plasma medicine-current state of research and medical application. *Plasma Phys. Control. Fusion* **2017**, *59*, 014031. [CrossRef]
2. Kramer, A.; Bekeschus, S.; Matthes, R.; Bender, C.; Stope, M.B.; Napp, M.; Lademann, O.; Lademann, J.; Weltmann, K.-D.; Schauer, F. Cold Physical Plasmas in the Field of Hygiene-Relevance, Significance, and Future Applications. *Plasma Process. Polym.* **2015**, *12*, 1410–1422. [CrossRef]
3. Matthes, R.; Hubner, N.O.; Bender, C.; Koban, I.; Horn, S.; Bekeschus, S.; Weltmann, K.-D.; Kocher, T.; Kramer, A.; Assadian, O. Efficacy of different carrier gases for barrier discharge plasma generation compared to chlorhexidine on the survival of Pseudomonas aeruginosa embedded in biofilm in vitro. *Skin Pharmacol. Physiol.* **2014**, *27*, 148–157. [CrossRef] [PubMed]
4. Daeschlein, G.; Napp, M.; von Podewils, S.; Lutze, S.; Emmert, S.; Lange, A.; Klare, I.; Haase, H.; Gumbel, D.; von Woedtke, T.; Junger, M. In Vitro Susceptibility of Multidrug Resistant Skin and Wound Pathogens Against Low Temperature Atmospheric Pressure Plasma Jet (APPJ) and Dielectric Barrier Discharge Plasma (DBD). *Plasma Process. Polym.* **2014**, *11*, 175–183. [CrossRef]
5. Mai-Prochnow, A.; Clauson, M.; Hong, J.; Murphy, A.B. Gram positive and Gram negative bacteria differ in their sensitivity to cold plasma. *Sci. Rep.* **2016**, *6*, 38610. [CrossRef] [PubMed]

6. Isbary, G.; Heinlin, J.; Shimizu, T.; Zimmermann, J.L.; Morfill, G.; Schmidt, H.U.; Monetti, R.; Steffes, B.; Bunk, W.; Li, Y.; et al. Successful and safe use of 2 min cold atmospheric argon plasma in chronic wounds: Results of a randomized controlled trial. *Br. J. Dermatol.* **2012**, *167*, 404–410. [CrossRef] [PubMed]

7. Isbary, G.; Stolz, W.; Shimizu, T.; Monetti, R.; Bunk, W.; Schmidt, H.-U.; Morfill, G.; Klämpfl, T.; Steffes, B.; Thomas, H. Cold atmospheric argon plasma treatment may accelerate wound healing in chronic wounds: Results of an open retrospective randomized controlled study in vivo. *Clin. Plasma Med.* **2013**, *1*, 25–30. [CrossRef]

8. Klebes, M.; Ulrich, C.; Kluschke, F.; Patzelt, A.; Vandersee, S.; Richter, H.; Bob, A.; von Hutten, J.; Krediet, J.T.; Kramer, A.; et al. Combined antibacterial effects of tissue-tolerable plasma and a modern conventional liquid antiseptic on chronic wound treatment. *J. Biophotonics* **2015**, *8*, 382–391. [CrossRef] [PubMed]

9. Sen, C.K. Wound healing essentials: Let there be oxygen. *Wound Repair Regen.* **2009**, *17*, 1–18. [CrossRef] [PubMed]

10. Roy, S.; Khanna, S.; Nallu, K.; Hunt, T.K.; Sen, C.K. Dermal wound healing is subject to redox control. *Mol. Ther.* **2006**, *13*, 211–220. [CrossRef] [PubMed]

11. Kranke, P.; Bennett, M.H.; James, M.M.; Schnabel, A.; Debus, S.E. Hyperbaric Oxygen Therapy for Chronic Wounds. *Cochrane Database Syst Rev.* **2012**, *18*, CD004123.

12. Sen, C.K. The general case for redox control of wound repair. In *Wound Repair and Regeneration: Official Publication of the Wound Healing Society [and] the European Tissue Repair Society*; European Tissue Repair Society: St. Louis, MO, USA, 2003; Volume 11, pp. 431–438.

13. Sunkari, V.G.; Lind, F.; Botusan, I.R.; Kashif, A.; Liu, Z.J.; Ylä-Herttuala, S.; Brismar, K.; Velazquez, O.; Catrina, S.B. Hyperbaric oxygen therapy activates hypoxia-inducible factor 1 (HIF-1), which contributes to improved wound healing in diabetic mice. *Wound Repair Regen.* **2015**, *23*, 98–103. [CrossRef] [PubMed]

14. Park, J.E.; Barbul, A. Understanding the role of immune regulation in wound healing. *Am. J. Surg.* **2004**, *187*, 11S–16S. [CrossRef]

15. Isbary, G.; Shimizu, T.; Zimmermann, J.L.; Thomas, H.M.; Morfill, G.E.; Stolz, W. Cold atmospheric plasma for local infection control and subsequent pain reduction in a patient with chronic post-operative ear infection. *New Microbes New Infect.* **2013**, *1*, 41–43. [CrossRef] [PubMed]

16. Vandersee, S.; Terhorst, D.; Humme, D.; Beyer, M. Treatment of indolent primary cutaneous B-cell lymphomas with subcutaneous interferon-alfa. *J. Am. Acad. Dermatol.* **2014**, *70*, 709–715. [CrossRef] [PubMed]

17. Jablonowski, H.; von Woedtke, T. Research on plasma medicine-relevant plasma–liquid interaction: What happened in the past five years? *Clin. Plasma Med.* **2015**, *3*, 42–52. [CrossRef]

18. Reuter, S.; Tresp, H.; Wende, K.; Hammer, M.U.; Winter, J.; Masur, K.; Schmidt-Bleker, A.; Weltmann, K.-D. From RONS to ROS: Tailoring Plasma Jet Treatment of Skin Cells. IEEE Trans. *Plasma Sci.* **2012**, *40*, 2986–2993. [CrossRef]

19. Graves, D.B. The emerging role of reactive oxygen and nitrogen species in redox biology and some implications for plasma applications to medicine and biology. *J. Phys. D Appl. Phys.* **2012**, *45*, 263001. [CrossRef]

20. Hanschmann, E.M.; Godoy, J.R.; Berndt, C.; Hudemann, C.; Lillig, C.H. Thioredoxins, glutaredoxins, and peroxiredoxins—Molecular mechanisms and health significance: From cofactors to antioxidants to redox signaling. *Antioxid. Redox Signal.* **2013**, *19*, 1539–1605. [CrossRef] [PubMed]

21. Castro Fernandes, D.; Bonatto, D.; Laurindo, F.M. The Evolving Concept of Oxidative Stress. In *Studies on Cardiovascular Disorders*; Sauer, H., Shah, A.M., Laurindo, F.R.M., Eds.; Humana Press: New York, NY, USA, 2010; pp. 1–41.

22. Schmidt, A.; Rödder, K.; Hasse, S.; Masur, K.; Toups, L.; Lillig, C.H.; von Woedtke, T.; Wende, K.; Bekeschus, S. Redox-regulation of activator protein 1 family members in blood cancer cell lines exposed to cold physical plasma-treated medium. *Plasma Process. Polym.* **2016**, *13*, 1179–1188. [CrossRef]

23. Ray, P.D.; Huang, B.-W.; Tsuji, Y. Reactive oxygen species (ROS) homeostasis and redox regulation in cellular signaling. *Cell. Signal.* **2012**, *24*, 981–990. [CrossRef] [PubMed]

24. Poljsak, B.; Milisav, I. The neglected significance of "antioxidative stress". *Oxid. Med. Cell. Longev.* **2012**, *2012*, 480895. [CrossRef] [PubMed]

25. Sena, L.A.; Chandel, N.S. Physiological roles of mitochondrial reactive oxygen species. *Mol. Cell* **2012**, *48*, 158–167. [CrossRef] [PubMed]

26. Ahn, H.J.; Kim, K.I.; Kim, G.; Moon, E.; Yang, S.S.; Lee, J.-S. Atmospheric-Pressure Plasma Jet Induces Apoptosis Involving Mitochondria via Generation of Free Radicals. *PLoS ONE* **2011**, *6*, e28154. [CrossRef] [PubMed]
27. Bekeschus, S.; Rödder, K.; Schmidt, A.; Stope, M.B.; von Woedtke, T.; Miller, V.; Fridman, A.; Weltmann, K.-D.; Masur, K.; Metelmann, H.-R.; et al. Cold physical plasma selects for specific T helper cell subsets with distinct cells surface markers in a caspase-dependent and NF-κB-independent manner. *Plasma Process. Polym.* **2016**, *13*, 1144–1150. [CrossRef]
28. Weiss, M.; Gumbel, D.; Hanschmann, E.M.; Mandelkow, R.; Gelbrich, N.; Zimmermann, U.; Walther, R.; Ekkernkamp, A.; Sckell, A.; Kramer, A.; et al. Cold Atmospheric Plasma Treatment Induces Anti Proliferative Effects in Prostate Cancer Cells by Redox and Apoptotic Signaling Pathways. *PLoS ONE* **2015**, *10*, e0130350. [CrossRef] [PubMed]
29. Dickinson, B.C.; Chang, C.J. Chemistry and biology of reactive oxygen species in signaling or stress responses. *Nat. Chem. Biol.* **2011**, *7*, 504–511. [CrossRef] [PubMed]
30. Wende, K.; Bekeschus, S.; Schmidt, A.; Jatsch, L.; Hasse, S.; Weltmann, K.-D.; Masur, K.; von Woedtke, T. Risk assessment of a cold argon plasma jet in respect to its mutagenicity. *Mutat. Res. Genet. Toxicol. Environ. Mutagen.* **2016**, *798–799*, 48–54. [CrossRef] [PubMed]
31. Kluge, S.; Bekeschus, S.; Bender, C.; Benkhai, H.; Sckell, A.; Below, H.; Stope, M.B.; Kramer, A. Investigating the Mutagenicity of a Cold Argon-Plasma Jet in an HET-MN Model. *PLoS ONE* **2016**, *11*, e0160667. [CrossRef] [PubMed]
32. Boxhammer, V.; Li, Y.F.; Koritzer, J.; Shimizu, T.; Maisch, T.; Thomas, H.M.; Schlegel, J.; Morfill, G.E.; Zimmermann, J.L. Investigation of the mutagenic potential of cold atmospheric plasma at bactericidal dosages. *Mutat. Res.* **2013**, *753*, 23–28. [CrossRef] [PubMed]
33. Metelmann, H.-R.; Vu, T.T.; Do, H.T.; Le, T.N.B.; Hoang, T.H.A.; Phi, T.T.T.; Luong, T.M.L.; Doan, V.T.; Nguyen, T.T.H.; Nguyen, T.H.M.; et al. Podmelle. Scar formation of laser skin lesions after cold atmospheric pressure plasma (CAP) treatment: A clinical long term observation. *Clin. Plasma Med.* **2013**, *1*, 30–35. [CrossRef]
34. Isbary, G.; Zimmermann, J.L.; Shimizu, T.; Li, Y.-F.; Morfill, G.E.; Thomas, H.M.; Steffes, B.; Heinlin, J.; Karrer, S.; Stolz, W.; et al. Non-thermal plasma—More than five years of clinical experience. *Clin. Plasma Med.* **2013**, *1*, 19–23. [CrossRef]
35. Fahmy, S.R. Anti-fibrotic effect of Holothuria arenicola extract against bile duct ligation in rats. *BMC Complement. Altern. Med.* **2015**, *15*, 14. [CrossRef] [PubMed]
36. Mizukami, H.; Mi, Y.; Wada, R.; Kono, M.; Yamashita, T.; Liu, Y.; Werth, N.; Sandhoff, R.; Sandhoff, K.; Proia, R.L. Systemic inflammation in glucocerebrosidase-deficient mice with minimal glucosylceramide storage. *J. Clin. Investig.* **2002**, *109*, 1215–1221. [CrossRef] [PubMed]
37. Rizwan, A.; Bulte, C.; Kalaichelvan, A.; Cheng, M.; Krishnamachary, B.; Bhujwalla, Z.M.; Jiang, L.; Glunde, K. Metastatic breast cancer cells in lymph nodes increase nodal collagen density. *Sci. Rep.* **2015**, *5*, 10002. [CrossRef] [PubMed]
38. Vogel, B.; Siebert, H.; Hofmann, U.; Frantz, S. Determination of collagen content within picrosirius red stained paraffin-embedded tissue sections using fluorescence microscopy. *MethodsX* **2015**, *2*, 124–134. [CrossRef] [PubMed]
39. Miyata, T.; Oda, O.; Inagi, R.; Iida, Y.; Araki, N.; Yamada, N.; Horiuchi, S.; Taniguchi, N.; Maeda, K.; Kinoshita, T. beta 2-Microglobulin modified with advanced glycation end products is a major component of hemodialysis-associated amyloidosis. *J. Clin. Investig.* **1993**, *92*, 1243–1252. [CrossRef] [PubMed]
40. Aarons, C.B.; Bajenova, O.; Andrews, C.; Heydrick, S.; Bushell, K.N.; Reed, K.L.; Thomas, P.; Becker, J.M.; Stucchi, A.F. Carcinoembryonic antigen-stimulated THP-1 macrophages activate endothelial cells and increase cell-cell adhesion of colorectal cancer cells. *Clin. Exp. Metastasis* **2007**, *24*, 201–209. [CrossRef] [PubMed]
41. Baudin, E.; Gigliotti, A.; Ducreux, M.; Ropers, J.; Comoy, E.; Sabourin, J.C.; Bidart, J.M.; Cailleux, A.F.; Bonacci, R.; Ruffie, P.; et al. Neuron-specific enolase and chromogranin A as markers of neuroendocrine tumours. *Br. J. Cancer* **1998**, *78*, 1102–1107. [CrossRef] [PubMed]
42. Fingas, C.D.; Altinbas, A.; Schlattjan, M.; Beilfuss, A.; Sowa, J.P.; Sydor, S.; Bechmann, L.P.; Ertle, J.; Akkiz, H.; Herzer, K.; et al. Expression of apoptosis- and vitamin D pathway-related genes in hepatocellular carcinoma. *Digestion* **2013**, *87*, 176–181. [CrossRef] [PubMed]

43. Marquardt, J.U.; Quasdorff, M.; Varnholt, H.; Curth, H.M.; Mesghenna, S.; Protzer, U.; Goeser, T.; Nierhoff, D. Neighbor of Punc E11, a novel oncofetal marker for hepatocellular carcinoma. *Int. J. Cancer.* **2011**, *128*, 2353–2363. [CrossRef] [PubMed]

44. Rifkin, D.B.; Crowe, R.M.; Pollack, R. Tumor promoters induce changes in the chick embryo fibroblast cytoskeleton. *Cell* **1979**, *18*, 361–368. [CrossRef]

45. Schmidt, A.; Bekeschus, S.; Wende, K.; Vollmar, B.; von Woedtke, T. A cold plasma jet accelerates wound healing in a murine model of full-thickness skin wounds. *Exp. Dermatol.* **2017**, *26*, 156–162. [CrossRef] [PubMed]

46. Weltmann, K.-D.; Kindel, E.; Brandenburg, R.; Meyer, C.; Bussiahn, R.; Wilke, C.; von Woedtke, T. Atmospheric Pressure Plasma Jet for Medical Therapy: Plasma Parameters and Risk Estimation. *Contrib. Plasma Phys.* **2009**, *49*, 631–640. [CrossRef]

47. Schmidt, A.; von Woedtke, T.; Bekeschus, S. Periodic exposure of keratinocytes to cold physical plasma—An in vitro model for redox-related diseases of the skin. *Oxid. Med. Cell. Longev.* **2016**, *2016*, 9816072. [CrossRef] [PubMed]

48. Schmidt, A.; Wende, K.; Bekeschus, S.; Bundscherer, L.; Barton, A.; Ottmuller, K.; Weltmann, K.D.; Masur, K. Non-thermal plasma treatment is associated with changes in transcriptome of human epithelial skin cells. *Free Radic. Res.* **2013**, *47*, 577–592. [CrossRef] [PubMed]

49. Schmidt, A.; Dietrich, S.; Steuer, A.; Weltmann, K.-D.; von Woedtke, T.; Masur, K.; Wende, K. Non-thermal plasma activates human keratinocytes by stimulation of antioxidant and phase II pathways. *J. Biol. Chem.* **2015**, *290*, 6731–6750. [CrossRef] [PubMed]

50. Bundscherer, L.; Wende, K.; Ottmuller, K.; Barton, A.; Schmidt, A.; Bekeschus, S.; Hasse, S.; Weltmann, K.-D.; Masur, K.; Lindequist, U. Impact of non-thermal plasma treatment on MAPK signaling pathways of human immune cell lines. *Immunobiology* **2013**, *218*, 1248–1255. [CrossRef] [PubMed]

51. Halliwell, B. Oxidative stress and cancer: Have we moved forward? *Biochem. J.* **2007**, *401*, 1–11. [CrossRef] [PubMed]

52. Cairns, R.A.; Harris, I.S.; Mak, T.W. Regulation of cancer cell metabolism. *Nat. Rev. Cancer* **2011**, *11*, 85–95. [CrossRef] [PubMed]

53. Bekeschus, S.; Schmidt, A.; Weltmann, K.-D.; von Woedtke, T. The Plasma Jet kINPen—A Powerful Tool for Wound Healing. *Clin. Plasma Med.* **2016**, *4*, 19–28. [CrossRef]

54. Lademann, J.; Richter, H.; Alborova, A.; Humme, D.; Patzelt, A.; Kramer, A.; Weltmann, K.D.; Hartmann, B.; Ottomann, C.; Fluhr, J.W.; et al. Risk assessment of the application of a plasma jet in dermatology. *J. Biomed. Opt.* **2009**, *14*, 054025. [CrossRef] [PubMed]

55. Kramer, A.; Lademann, J.; Bender, C.; Sckell, A.; Hartmann, B.; Münch, S.; Hinz, P.; Ekkernkamp, A.; Matthes, R.; Koban, I.; et al. Suitability of tissue tolerable plasmas (TTP) for the management of chronic wounds. *Clin. Plasma Med.* **2013**, *1*, 11–18. [CrossRef]

56. Metelmann, H.-R.; von Woedtke, T.; Bussiahn, R.; Weltmann, K.-D.; Rieck, M.; Khalili, R.; Podmelle, F.; Waite, P.D. Experimental recovery of CO_2-laser skin lesions by plasma stimulation. *Am. J. Cosmet. Surg.* **2012**, *29*, 52–56. [CrossRef]

57. Von Woedtke, T.; Metelmann, H.-R.; Weltmann, K.-D. Clinical Plasma Medicine: State and Perspectives of in Vivo Application of Cold Atmospheric Plasma. *Contrib. Plasma Phys.* **2014**, *54*, 104–117. [CrossRef]

58. Hasse, S.; Tran, T.D.; Hahn, O.; Kindler, S.; Metelmann, H.R.; von Woedtke, T.; Masur, K. Induction of proliferation of basal epidermal keratinocytes by cold atmospheric-pressure plasma. *Clin. Exp. Dermatol.* **2016**, *41*, 202–209. [CrossRef] [PubMed]

59. Von Woedtke, T.; Metelmann, H.-R.; Weltmann, K.-D. Editorial. *Clin. Plasma Med.* **2013**, *1*, 1–2. [CrossRef]

60. Basuyau, J.P.; Mallet, E.; Leroy, M.; Brunelle, P. Reference intervals for serum calcitonin in men, women, and children. *Clin. Chem.* **2004**, *50*, 1828–1830. [CrossRef] [PubMed]

61. Shoenfeld, Y.; Gallant, L.A.; Shaklai, M.; Livni, E.; Djaldetti, M.; Pinkhas, J. Gaucher's disease: A disease with chronic stimulation of the immune system. *Arch. Pathol. Lab. Med.* **1982**, *106*, 388–391. [PubMed]

62. Livak, K.J.; Schmittgen, T.D. Analysis of Relative Gene Expression Data Using Real-Time Quantitative PCR and the 2CT Method. *Methods* **2001**, *25*, 402–408. [CrossRef] [PubMed]

© 2017 by the authors. Licensee MDPI, Basel, Switzerland. This article is an open access article distributed under the terms and conditions of the Creative Commons Attribution (CC BY) license (http://creativecommons.org/licenses/by/4.0/).

International Journal of
Molecular Sciences

MDPI

Communication

First Insights into Human Fingertip Regeneration by Echo-Doppler Imaging and Wound Microenvironment Assessment

Paris Jafari [†], Camillo Muller [†], Anthony Grognuz, Lee Ann Applegate, Wassim Raffoul, Pietro G. di Summa [‡] and Sébastien Durand [*,‡]

Plastic and Hand Surgery Department, Lausanne University Hospital, 1011 Lausanne, Switzerland; Paris.Jafari@chuv.ch (P.J.); Camillo.Muller@chuv.ch (C.M.); Anthony.Grognuz@chuv.ch (A.G.); Lee.Laurent-Applegate@chuv.ch (L.A.A.); Wassim.Raffoul@chuv.ch (W.R.); Pietro.Di-Summa@chuv.ch (P.G.d.S.)
* Correspondence: Sebastien.Durand@chuv.ch; Tel.: +41-79-556-7893
† These authors contributed equally to this work.
‡ These authors contributed equally to this work.

Academic Editor: Allison Cowin
Received: 9 April 2017; Accepted: 6 May 2017; Published: 13 May 2017

Abstract: Fingertip response to trauma represents a fascinating example of tissue regeneration. Regeneration derives from proliferative mesenchymal cells (blastema) that subsequently differentiate into soft and skeletal tissues. Clinically, conservative treatment of the amputated fingertip under occlusive dressing can shift the response to tissue loss from a wound repair process towards regeneration. When analyzing by Immunoassay the wound exudate from occlusive dressings, the concentrations of brain-derived neurotrophic factor (BDNF) and leukemia inhibitory factor (LIF) were higher in fingertip exudates than in burn wounds (used as controls for wound repair versus regeneration). Vascular endothelial growth factor A (VEGF-A) and platelet-derived growth factor (PDGF) were highly expressed in both samples in comparable levels. In our study, pro-inflammatory cytokines were relatively higher expressed in regenerative fingertips than in the burn wound exudates while chemokines were present in lower levels. Functional, vascular and mechanical properties of the regenerated fingertips were analyzed three months after trauma and the data were compared to the corresponding fingertip on the collateral uninjured side. While sensory recovery and morphology (pulp thickness and texture) were similar to uninjured sides, mechanical parameters (elasticity, vascularization) were increased in the regenerated fingertips. Further studies should be done to clarify the importance of inflammatory cells, immunity and growth factors in determining the outcome of the regenerative process and its influence on the clinical outcome.

Keywords: fingertip regeneration; clinical assessment; Doppler imaging; angiogenesis

---

1. Introduction

Humans maintain regenerative capability of fingertips [1,2], replacing the lost tissue following substantial trauma. This regeneration occurs in a level dependent manner as long as the proximal nail matrix remains intact [3]. Regenerative mechanisms in humans are not elucidated but the involvement of Wnt activation in nail stem cells and fibroblast growth factor 2 (FGF-2) signaling have been described in mice and amphibian models for limb regeneration [3–6]. Fingertip regeneration derives from proliferative mesenchymal cells (blastema) [7] that subsequently differentiate into soft and skeletal tissues [8,9]. In humans the conservative treatment of the amputated fingertip under occlusive dressing is essential to shift the response of tissue loss from a wound repair process towards regeneration [10]. Therefore, the regenerative microenvironment created under the occlusive dressing should play an important role in providing signals that initiate regeneration and control proliferation and patterning [11]. Following skin injury in humans, immediate release of various growth factors, cytokines and chemokines triggers and controls sequential steps of wound repair. The role and importance of these factors in the repairing microenvironment of wounds has been well established [12]. Among these growth factors, platelet-derived growth factor (PDGF) is known to have a central role in all different steps of wound repair by triggering cell migration into the wound, enhancing the proliferation of fibroblasts and extracellular matrix production and also contraction [13]. Epidermal growth factor (EGF) family of growth factors is also important for reepithelialization and vascularization during wound repair [12,14]. Vascular endothelial growth factor (VEGF) and other growth factors that increase expression such as hepatocyte growth factor (HGF) [15] have crucial roles in wound angiogenesis [16] which then affects overall wound repair processes [17]. Apart from angiogenesis, nerve in-growth is also important in wound healing [18] and the growth factors that have an effect on innervation such as nerve growth factor (NGF) have been shown to be essential to normal wound repair [19]. Along with several growth factors, chemokines such as macrophage inflammatory protein 1-α (MIP 1-α) play an important role in the recruitment of inflammatory cells to the wound site [20] and initiation of wound repair processes [21] within the inflammatory phase that implicates several proinflammatory cytokines such as IL1-α, IL1-β, IL-6 and TNF-α [22]. There might be similarities between the implication of the above mentioned factors between wound repair and wound regeneration. However, knowledge on the role of these factors during regeneration is lacking and limited to animal models. PDGF is shown to promote the expansion of the blastema mesenchymal precursor population which is differentiated to bone and dermis [23] while VEGF is considered to be inhibitory for regeneration [24]. The implication of immune signaling in regeneration is shown in axolotl model of regeneration with an increase in the expression of several proinflammatory cytokines early after limb amputation [25]. However, cellular and molecular mechanisms in human fingertip regeneration remain poorly understood. In this preliminary study, we investigated the early pro-regenerative microenvironment of human fingertip amputations treated with occlusive dressing. Our primary purpose was to determine factors that favor regeneration over repair in these wounds. In three consecutive patients, growth factor, chemokine and cytokine contents of the regenerating wound fluid were analyzed. Since burn wounds are considered to be repairing rather than regenerating wounds [26], we compared the fingertip wound fluid cytokine and growth factor content, with burn wound fluid from three burn patients. Our secondary goal was to assess the quality of the regenerated tissue. We quantified the functional, vascular and mechanical properties of the regenerated fingertips three months after trauma and compared the data to the corresponding fingertip on the collateral uninjured side. Clinical (sensory recovery), morphological (pulp thickness and texture) and mechanical parameters (elasticity, vascularization) were recorded and analyzed.

2. Results

2.1. Clinical Evaluation of Regenerated Fingertips

Five healthy male patients (mean age 50 ± 15 years) were included in this preliminary study (Table 1). The occlusive dressing was applied upon admission (Figure 1a) and after primary debridement and changed once a week without rinsing the wound for a mean duration of 5 weeks (4–7). Patients had 9 (7–12) consultations within the first 6 months. Mean follow-up was 6 months (5–7). Clinical and morphologic evaluation was done at three months post-trauma (Figure 1b).

Sensitivity of the regenerated fingertips was assessed by two points discrimination that was measured less than 4 mm and overlapped with the uninjured side for each finger. We assessed eventual pain using the Verbal Descriptor Scale representing different growing intensities of pain varying from none to very severe pain. Our patients had no or mild pain in the regenerated fingertips. Thumb-finger Pinch was recorded as an index of functional force that did not reveal substantial difference with collateral control fingers.

Table 1. Patient and amputated fingertip characteristics.

Patient	Age	Amputated Finger/Hand	Two Points Discrimination	Pinch Test	Pain (0–5)
Patient 1	30	Middle/left	≤4 mm on both fingers	2 kg/2 kg	0/0
Patient 2	45	Middle/right	≤4 mm on both fingers	2.5 kg/2 kg	0/0
Patient 3	76	Index/right	≤4 mm on both fingers	Not obtained	0/0
Patient 4	47	Middle/left	≤4 mm on both fingers	Not obtained	0/0
Patient 5	52	Middle/right	≤4 mm on both fingers	1.5 kg/2.5 kg	1/0

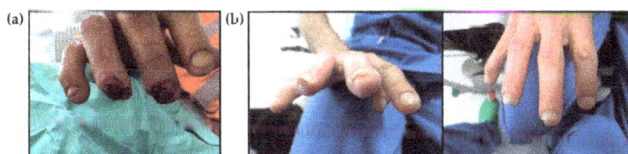

Figure 1. Representative images of amputated fingers. (**a**) At admission and before the application of occlusive dressing; (**b**) Three months post-trauma and at clinical and morphological evaluation.

2.2. Evaluation of Morphological, Mechanical and Vascular Properties of Regenerated Fingertips

We used different modalities of Ultrasound imaging (aixplorer®, Aix-en-Provence, France) to measure the above parameters. Pulp thickness assessment was based on B-mode echography, pulp vascularization was measured by Doppler ultrasound test and the elasticity of the regenerative tissue was assessed by the elasticity modulus from Shear Wave Elastography (SWE).

We observed no significant differences in pulp thickness between regenerated fingertips and the collateral control finger (Figures 2a and 3). The average soft tissue coverage was 7.2 ± 0.6 mm (6.7–8.2) compared to 7.7 ± 0.6 mm (6.9–8.3) in control collateral finger with no significant difference observed. Interestingly, elasticity (Figures 2b and 3) was measured 4.8 ± 1.3 m/s (3.5–6.7) compared to 3.6 ± 0.6 m/s (3.1–4.4) in control collateral fingers and this represented a significant increase of $132.3 \pm 23\%$ in regenerated fingertips. Likewise, vascularization (Figures 2c and 3) was significantly higher ($162.4 \pm 53.8\%$) in the regenerated fingertip 58.8 ± 8.6 % (50.7–72.1) compared to controls with a vascularization percentage of $38.6 \pm 10\%$ (26.3–51.4).

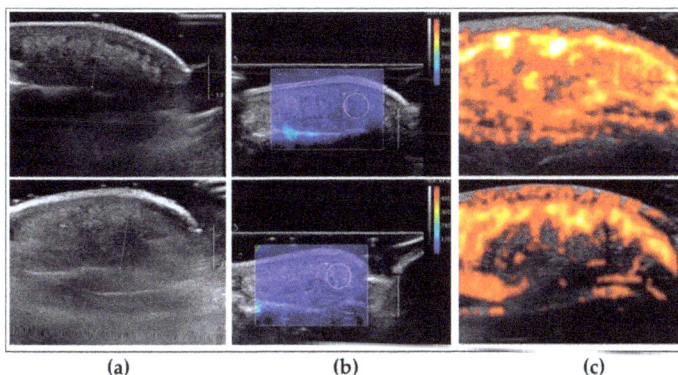

Figure 2. Representative echo-doppler images of regenerated (upper row) and un-injured collateral (lower row) fingertips. (**a**) Ultrasound b-mode imaging for the measurement of the pulp thickness; (**b**) b-mode imaging with superimposed share wave mapping for the measurement of the pulp elasticity. The white circle has a diameter of 5 mm; (**c**) b-mode imaging with superimposed power Doppler for the measurement of vascularity.

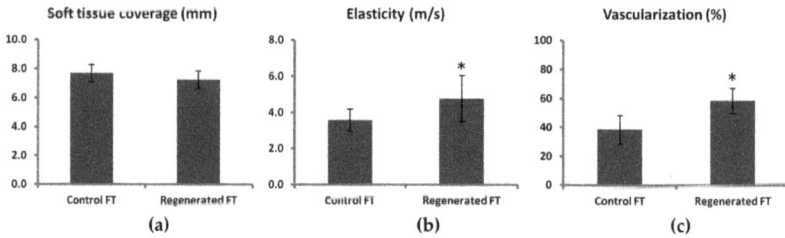

Figure 3. Morphologic, mechanical and vascular characteristics of regenerated fingertips compared to control collateral healthy finger. (a) Soft tissue coverage, (b) Elasticity, (c) vascularization. Data are presented as mean ± SD for five patients (* $p < 0.05$).

2.3. Regenerative Wound Microenvironment

We measured the concentrations of different inflammatory cytokines, chemokines and growth factors in the wound exudate from three amputated fingertips and three burn patients using Luminex technology. Despite a large variation in wound cytokine levels among different patients (Figure 4), the concentrations of brain-derived neurotrophic factor (BDNF), EGF and leukemia inhibitory factor (LIF) were higher in fingertip exudates than in burn wounds. VEGF-A, PDGF and HGF were highly expressed in both samples in comparable levels (Figure 4a). We observed a general trend towards higher values of inflammatory cytokines in fingertip exudates whereas chemokine levels of burn wound exudates were globally higher (Figure 4b). Bone Morphogenetic Protein signaling has been shown to be important during fingertip regeneration in mice [27] but we did not observe detectable levels of BMP in the fingertip exudate samples.

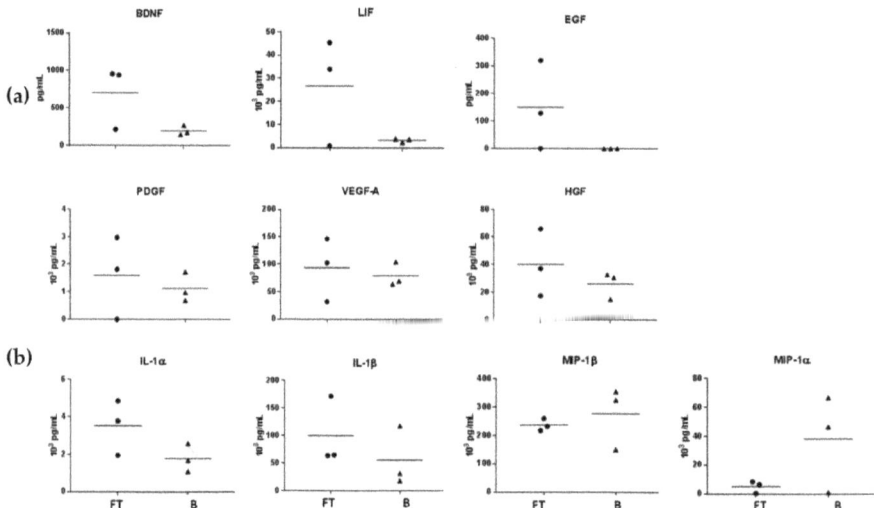

Figure 4. Measured levels of (a) growth factors, (b) cytokines and chemokines in fingertip exudate samples (FT) and burn wound exudate samples (B), seven days post-trauma.

3. Discussion

General belief is that the ability to regenerate fingertips is lost or decreased in human by age [10]. Here we show that in all our adult patients (mean age 50 years) fingertips underwent regeneration with satisfactory clinical outcome. By definition, re-growth of cells or tissues during regeneration replace both form and function in damaged organs [28]. Indeed, clinical assessment of the regenerated fingertips in our patients revealed that they have comparable morphological and functional characteristics with contralateral uninjured fingertips. However, we observed key mechanical modifications in the regenerated fingertips. Increase in elasticity paralleled high vascularization in regenerated fingertips suggesting that the enhanced vascularity during regeneration affects the elasticity of these tissues. Nevertheless, long-term studies are necessary to confirm whether these events are associated and if they persist in time after the regeneration process is completed. Increased vascularization could be transient and a result of increased expression of pro-angiogenic factors in the regenerative microenvironment of the fingertips as we observed in this study. We analyzed the wound fluid that was accumulated under the occlusive dressing over amputated fingertips over the primary phase of regeneration and most probably blastema formation [29]. We measured high levels of VEGF in the wound fluid that explains increased vascularity which seems to be essential for the complete regeneration in human fingertips. Interestingly, in mouse models of regeneration, an inverse phenomena is observed where the expression of VEGF is down regulated in blastema [11] and even induction of angiogenesis by VEGF inhibited the regeneration by altering the transition of blastema to the differentiation phase [24]. Thus, for the first time, our results reveal discrepancies between mice and human fingertip regeneration. However, in a recent study, regenerated fingertips in fast-healing mice showed 70% increased vascularization compared to non-healing mice with impaired regeneration [30]. These data along with our observations, suggest an important role for higher blood flow in the fingertip regeneration in mammalians. In surgical wounds and chronic vascular ulcers, high expression of b-FGF, which is another pro-angiogenic factor, is observed at very early phases of wound healing and the levels further decline over time during healing [31,32]. We did not detect this growth factor in the accumulative exudates collected at day 7, which is most probably too late for b-FGF expression. A more comprehensive analysis of pro-angiogenic factors over time in fingertip exudate is necessary and will help to determine the factors that control angiogenesis during regeneration.

Deficiency in inflammatory response has been suggested to be an explanation for the absence of fibrotic response during fingertip regeneration [33]. However, our preliminary data showed that, at least during the first seven days post-amputation in human fingertips, a strong inflammatory response is ongoing in the regenerative microenvironment compared to healing microenvironment of burn wounds as shown in Figure 4. The central role of inflammation and innate immune response are known in wound healing but still poorly understood in regeneration models [34], where a prominent expression of inflammatory and immune-related genes was revealed [35,36]. In these models, both pro- and anti-inflammatory cytokines are highly expressed during the first two weeks of regeneration with pro-inflammatory molecules being prominent during the first week and anti-inflammatory cytokines at the second week and after [34]. Classical models of regeneration such as salamander [25], Xenopus [35] or mice digit tip [37] also show that protein factors triggering the immediate events following traumatic injury are similar in both wound healing and organ regeneration [25,35], but the immune factors that favor one pathway over the other remain unknown. In our study, pro-inflammatory cytokines were relatively higher expressed in regenerative fingertips than in the burn wound exudates while chemokines where present in lower levels. Further studies should be done to clarify the importance of inflammatory cells and immunity in determining the outcome of the regenerative process.

In amphibians and mice, complete regeneration is dependent to nerve-derived growth factors that are secreted by nerve-associated Schwann cell precursors and promote the expansion of blastema and digit regeneration [38,39]. Transcriptome analysis these cells revealed that PDGF, LIF and could be potential paracrine factors that might be important for digit tip regeneration [23]. We observed an increased expression of LIF and BDNF in the regenerating fingertip exudates compared to burn wound exudates. The role of enervation as a regeneration triggering factor in humans is not known but our clinical observations with restoration of normal sensitivity seems to reveal an important nerve regeneration process (supported by increased expression of growth factors such as BDNF) to be active in human fingertips. Application of occlusive dressing leads to the creation of local permissive niche that provides necessary signals to initiate regeneration which is influenced by local microenvironment rather than systemic factors [40].

We detected the expression of other growth factors that promote wound reepithelialization, in fingertip exudates such as HGF and EGF that should have a role in the proliferation and further redifferentiation of blastema cell mass. Further work with higher number of patients is required to establish eventual difference between growth factor profiles and signaling differentiation pathways in regenerating or healing wounds. Moreover, in this study we were focused only on the first week post-amputation that we considered being crucial for the initiation of regeneration and blastema formation in regenerative fingertips. In further studies, it would be important to perform sample collection at different time points of regeneration in order to investigate which signaling molecules are involved in different phases of regeneration and how their expression might change over time.

4. Materials and Methods

4.1. Ethics

Written consent was obtained from all patients, and the procedures were performed in line with the Helsinki Declaration of 1975 (Recommendations guiding medical doctors in biomedical research involving human subjects). Exudate collection were approved by the State Ethics Committee (Protocol 488/13) and regulated by the DAL (Department of Musculoskeletal Medicine) and Institutional Biobank Procedures.

4.2. Fingertip Exudate Collection and Analysis

Fingertip exudates were collected from occlusive dressing at the first dressing change seven days after trauma. Samples were diluted in assay buffer, centrifuged and the supernatants were stored at −80 °C until further analysis.

4.3. Burn Wound Exudate Collection

Burn wound exudate was collected as described in [41] and with ethical approval. Briefly, a negative pressure dressing was applied over second-degree burn wounds that collected the accumulated fluid into a reservoir bottle. Samples were collected by changing the reservoir bottle and stored at −80 °C until further analysis.

4.4. Wound Exudate Immunoassay

Wound exudate levels of different cytokines, chemokines and growth factors were quantified using Luminex analysis on samples collected from three patients with regenerating fingertips under occlusive dressing and three burn patients according to the instructions of the manufacturer (R&D systems, Minneapolis, MN, USA). Absorbance was read on a Luminex 200 IS device and absorbance data were converted to protein concentrations with Luminex 100 IS Software (version 2.3, Luminex, Oosterhout, The Netherlands). Patient samples were tested in triplicate, and the mean value was used for analysis.

4.5. Clinical Assessment of the Regenerated Fingertips (Two Point Discrimination, Force and Pain)

Two-point discrimination was performed with the 2pt Disk-criminator from Alimed®. Control finger was tested first and regenerated fingertip afterwards. Fingertips were tested in 1 and 2 point discrimination. The patient, with closed eyes, stated whether he felt one or two points. We began the test at 15 mm distance and then moved the two points closer until the patient could not discriminate anymore (until 4 mm).

Two fingers round pinch (tip to tip pinch) was measured using 50 lb hydraulic pinch gauge (FE-120601) Jamar®. The gauge was place between the tip of the thumb and the tip of the concerned finger. The patient applied a spark of maximal power. Pain assessment was performed using visual descriptor scale representing different growing intensities of pain varying from zero to five (0 corresponding to no pain; 1 mild pain; 2 moderate pain; 3 severe pain; 4 very severe pain and 5 worst pain imaginable).

4.6. Power-Doppler and Elastography Assessment of Regenerated Fingertips

Digital pulp thickness assessment was based on B-mode echography (aixplorer®, Aix-en-Provence, France). Pulp vascularization was measured by power doppler ultrasound test and the elasticity of the regenerative tissue was assessed by the elasticity modulus from Shear Wave Elastography (SWE). Power Doppler sonography is a technique that displays the strength of the Doppler signal in color, rather than the speed and direction information. It has three times the sensitivity of conventional color Doppler for detection of flow and is particularly useful for small capillaries with low-velocity flow. SWE on the other hand is a technique measures the elasticity of tissue by measuring the propagation speed of shear waves, resulting in mechanical disturbances of ultrasonic waves applied to the tissue by the ultrasound probe [42].

The SWE measurements were taken using equipment (Aixplorer®, Aix-en-Provence, France), with a high-frequency SHL 15-4 probe (medium frequency 12 kHz). We first used normal ultrasound imaging in B-mode for topographic orientation and then superimposed and further shear wave elastography mapping in penetration mode. A mapping of the shear wave velocity, from blue (soft tissue, low speed) to red (hard tissue, high speed), produced the first qualitative information. A 3 mm^2 Q-Box (quantitative box) focus area was established in the middle of the digital pulp, and the quantitative values were obtained in m/s (shear wave velocity) and in kPa (elasticity modulus). To asses microvascularisation we performed echo-doppler in the same manner. After positioning of the ultrasound probe in B-mode, and then superimposed the ultrasound Doppler mapping. All images were saved in dicom format.

4.7. Image and Statistical Analysis

The pulp vascularization of regenerated fingertips and of collateral control fingers was observed by Doppler ultrasound (aixplorer®, Aix-en-Provence, France) and images were recorded as DICOM files. In absence of vascularization, the pixels remained gray or black. Regions that were vascularized appeared in color varying from red to white depending on the intensity. Images were processed within ImageJ software (NIH, Bethesda, MD, USA) to extract vascularization pixels which were selected with a color threshold based on Hue Saturation and Brightness (HSB) color mode. Pixels with Hue between 0 and 60, saturation between 5 and 255 and brightness between 15 and 255 were selected and considered as vascularization. The percentage of vascularization was calculated as a ratio between vascularization pixels divided by total number of pixels in the evaluation zone.

Independent 2-sided *t* test was used to evaluate statistical difference between the morphological and mechanical characteristics of regenerated fingertips and control healthy fingers. $p < 0.05$ was considered as significant.

Acknowledgments: This research was supported by the SwissTransmed CRUS grant number 14/2013 to the B5 platform. Authors would like to acknowledge Sandra Monnier and Jessie Lamouille for recruiting the patients to the study and their help for sample collection.

Author Contributions: Lee Ann Applegate and Wassim Raffoul conceived and designed the experiments; Paris Jafari and Camillo Muller performed the experiments; Paris Jafari, Pietro G. di Summa, Anthony Gronguz and Sébastien Durand analyzed the data; Paris Jafari, Pietro G. di Summa and Sébastien Durand wrote the paper and all authors made final corrections.

Conflicts of Interest: The authors declare no conflict of interest.

Abbreviations

FGF-2	Fibroblast Growth Factor 2
SWE	Sear Wave Elastography
BDNF	Brain-derived Neurotrophic factor
PDGF	Platelet-Derived Growth Factor
VEGF-A	Vascular Endothelial Growth Factor A
NGF	Nerve Growth Factor
b-FGF	Basic Fibroblast Growth Factor
MIP	Macrophage Inflammatory Protein
LIF	Leukemia Inhibitory Factor
CNTF	Ciliary Neurotrophic Factor
HGF	Hepatocyte Growth Factor
EGF	Epidermal Growth Factor

References

1. Douglas, B.S. Conservative management of guillotine amputation of the finger in children. *Aust. Paediatr. J.* **1972**, *8*, 86–89. [CrossRef] [PubMed]
2. Illingworth, C.M. Trapped fingers and amputated finger tips in children. *J. Pediatr. Surg.* **1974**, *9*, 853–858. [CrossRef]
3. Takeo, M.; Chou, W.C.; Sun, Q.; Lee, W.; Rabbani, P.; Loomis, C.; Taketo, M.M.; Ito, M. Wnt activation in nail epithelium couples nail growth to digit regeneration. *Nature* **2013**, *499*, 228–232. [CrossRef] [PubMed]
4. Yokoyama, H.; Ogino, H.; Stoick-Cooper, C.L.; Grainger, R.M.; Moon, R.T. Wnt/β-catenin signaling has an essential role in the initiation of limb regeneration. *Dev. Biol.* **2007**, *306*, 170–178. [CrossRef] [PubMed]
5. Yamada, Y.; Yokoyama, S.; Fukuda, N.; Kidoya, H.; Huang, X.Y.; Naitoh, H.; Satoh, N.; Takakura, N. A novel approach for myocardial regeneration with educated cord blood cells cocultured with cells from brown adipose tissue. *Biochem. Biophys. Res. Commun.* **2007**, *353*, 182–188. [CrossRef] [PubMed]
6. Mullen, L.M.; Bryant, S.V.; Torok, M.A.; Blumberg, B.; Gardiner, D.M. Nerve dependency of regeneration: The role of distal-less and fgf signaling in amphibian limb regeneration. *Development* **1996**, *122*, 3487–3497. [PubMed]
7. Poss, K.D. Advances in understanding tissue regenerative capacity and mechanisms in animals. *Nat. Rev. Genet.* **2010**, *11*, 710–722. [CrossRef] [PubMed]
8. Vidal, P.; Dickson, M.G. Regeneration of the distal phalanx. A case report. *J. Hand Surg.* **1993**, *18*, 230–233. [CrossRef]
9. Lee, L.P.; Lau, P.Y.; Chan, C.W. A simple and efficient treatment for fingertip injuries. *J. Hand Surg.* **1995**, *20*, 63–71. [CrossRef]
10. Shieh, S.J.; Cheng, T.C. Regeneration and repair of human digits and limbs: Fact and fiction. *Regeneration* **2015**, *2*, 149–168. [CrossRef] [PubMed]
11. Muneoka, K.; Allan, C.H.; Yang, X.; Lee, J.; Han, M. Mammalian regeneration and regenerative medicine. *Birth Defects Res. C: Embryo Today* **2008**, *84*, 265–280. [CrossRef] [PubMed]
12. Werner, S.; Grose, R. Regulation of wound healing by growth factors and cytokines. *Physiol. Rev.* **2003**, *83*, 835–870. [PubMed]
13. Heldin, C.H.; Westermark, B. Mechanism of action and in vivo role of platelet-derived growth factor. *Physiol. Rev.* **1999**, *79*, 1283–1316. [PubMed]
14. Broadley, K.N.; Aquino, A.M.; Woodward, S.C.; Buckley-Sturrock, A.; Sato, Y.; Rifkin, D.B.; Davidson, J.M. Monospecific antibodies implicate basic fibroblast growth factor in normal wound repair. *Lab. Investig.* **1989**, *61*, 571–575. [PubMed]
15. Xin, X.; Yang, S.; Ingle, G.; Zlot, C.; Rangell, L.; Kowalski, J.; Schwall, R.; Ferrara, N.; Gerritsen, M.E. Hepatocyte growth factor enhances vascular endothelial growth factor-induced angiogenesis in vitro and in vivo. *Am. J. Pathol.* **2001**, *158*, 1111–1120. [CrossRef]
16. Tsou, R.; Fathke, C.; Wilson, L.; Wallace, K.; Gibran, N.; Isik, F. Retroviral delivery of dominant-negative vascular endothelial growth factor receptor type 2 to murine wounds inhibits wound angiogenesis. *Wound Repair Regen.* **2002**, *10*, 222–229. [CrossRef] [PubMed]

17. Howdieshell, T.R.; Callaway, D.; Webb, W.L.; Gaines, M.D.; Procter, C.D., Jr.; Sathyanarayana; Pollock, J.S.; Brock, T.L.; McNeil, P.L. Antibody neutralization of vascular endothelial growth factor inhibits wound granulation tissue formation. *J. Surg. Res.* **2001**, *96*, 173–182. [CrossRef] [PubMed]

18. Harsum, S.; Clarke, J.D.; Martin, P. A reciprocal relationship between cutaneous nerves and repairing skin wounds in the developing chick embryo. *Dev. Biol.* **2001**, *238*, 27–39. [CrossRef] [PubMed]

19. Apfel, S.C.; Arezzo, J.C.; Brownlee, M.; Federoff, H.; Kessler, J.A. Nerve growth factor administration protects against experimental diabetic sensory neuropathy. *Brain Res.* **1994**, *634*, 7–12. [CrossRef]

20. DiPietro, L.A.; Burdick, M.; Low, Q.E.; Kunkel, S.L.; Strieter, R.M. Mip-1alpha as a critical macrophage chemoattractant in murine wound repair. *J. Clin. Investig.* **1998**, *101*, 1693–1698. [CrossRef] [PubMed]

21. Gillitzer, R.; Goebeler, M. Chemokines in cutaneous wound healing. *J. Leukoc. Biol.* **2001**, *69*, 513–521. [PubMed]

22. Hubner, G.; Brauchle, M.; Smola, H.; Madlener, M.; Fassler, R.; Werner, S. Differential regulation of pro-inflammatory cytokines during wound healing in normal and glucocorticoid-treated mice. *Cytokine* **1996**, *8*, 548–556. [CrossRef] [PubMed]

23. Johnston, A.P.; Yuzwa, S.A.; Carr, M.J.; Mahmud, N.; Storer, M.A.; Krause, M.P.; Jones, K.; Paul, S.; Kaplan, D.R.; Miller, F.D. Dedifferentiated schwann cell precursors secreting paracrine factors are required for regeneration of the mammalian digit tip. *Cell Stem Cell* **2016**, *19*, 433–448. [CrossRef] [PubMed]

24. Yu, L.; Yan, M.; Simkin, J.; Ketcham, P.D.; Leininger, E.; Han, M.; Muneoka, K. Angiogenesis is inhibitory for mammalian digit regeneration. *Regeneration* **2014**, *1*, 33–46. [CrossRef] [PubMed]

25. Godwin, J.W.; Pinto, A.R.; Rosenthal, N.A. Macrophages are required for adult salamander limb regeneration. *Proc. Natl. Acad. Sci. USA* **2013**, *110*, 9415–9420. [CrossRef] [PubMed]

26. Rowan, M.P.; Cancio, L.C.; Elster, E.A.; Burmeister, D.M.; Rose, L.F.; Natesan, S.; Chan, R.K.; Christy, R.J.; Chung, K.K. Burn wound healing and treatment: Review and advancements. *Crit. Care* **2015**, *19*, 243. [CrossRef] [PubMed]

27. Yu, L.; Han, M.; Yan, M.; Lee, E.C.; Lee, J.; Muneoka, K. Bmp signaling induces digit regeneration in neonatal mice. *Development* **2010**, *137*, 551–559. [CrossRef] [PubMed]

28. Goss, R.J. Regeneration versus repair. In *Wound Healing: Biochemical and Clinical Aspects*; Cohen, I.K., Diegelman, R.F., Lindblad, W.J., Eds.; W.B. Saunders Co.: Philadelphia, PA, USA, 1992; pp. 20–39.

29. Simkin, J.; Sammarco, M.C.; Dawson, L.A.; Schanes, P.P.; Yu, L.; Muneoka, K. The mammalian blastema: Regeneration at our fingertips. *Regeneration* **2015**, *2*, 93–105. [CrossRef] [PubMed]

30. Kwiatkowski, A.; Piatkowski, M.; Chen, M.; Kan, L.; Meng, Q.; Fan, H.; Osman, A.H.; Liu, Z.; Ledford, B.; He, J.Q. Superior angiogenesis facilitates digit regrowth in mrl/mpj mice compared to c57bl/6 mice. *Biochem. Biophys. Res. Commun.* **2016**, *473*, 907–912. [CrossRef] [PubMed]

31. Gohel, M.S.; Windhaber, R.A.; Tarlton, J.F.; Whyman, M.R.; Poskitt, K.R. The relationship between cytokine concentrations and wound healing in chronic venous ulceration. *J. Vasc. Surg.* **2008**, *48*, 1272–1277. [CrossRef] [PubMed]

32. Nissen, N.N.; Polverini, P.J.; Koch, A.E.; Volin, M.V.; Gamelli, R.L.; DiPietro, L.A. Vascular endothelial growth factor mediates angiogenic activity during the proliferative phase of wound healing. *Am. J. Pathol.* **1998**, *152*, 1445–1452. [PubMed]

33. Lehoczky, J.A. Are fingernails are a key to unlocking the puzzle of mammalian limb regeneration? *Exp. Dermatol.* **2016**. [CrossRef] [PubMed]

34. Mescher, A.L.; Neff, A.W.; King, M.W. Inflammation and immunity in organ regeneration. *Dev. Comp. Immunol.* **2017**, *66*, 98–110. [CrossRef] [PubMed]

35. Grow, M.; Neff, A.W.; Mescher, A.L.; King, M.W. Global analysis of gene expression in xenopus hindlimbs during stage-dependent complete and incomplete regeneration. *Dev. Dyn.* **2006**, *235*, 2667–2685. [CrossRef] [PubMed]

36. King, M.W.; Neff, A.W.; Mescher, A.L. The developing xenopus limb as a model for studies on the balance between inflammation and regeneration. *Anat. Rec. (Hoboken)* **2012**, *295*, 1552–1561. [CrossRef] [PubMed]

37. Fernando, W.A.; Leininger, E.; Simkin, J.; Li, N.; Malcom, C.A.; Sathyamoorthi, S.; Han, M.; Muneoka, K. Wound healing and blastema formation in regenerating digit tips of adult mice. *Dev. Biol.* **2011**, *350*, 301–310. [CrossRef] [PubMed]

38. Kumar, A.; Brockes, J.P. Nerve dependence in tissue, organ, and appendage regeneration. *Trends Neurosci.* **2012**, *35*, 691–699. [CrossRef] [PubMed]

39. Rinkevich, Y.; Montoro, D.T.; Muhonen, E.; Walmsley, G.G.; Lo, D.; Hasegawa, M.; Januszyk, M.; Connolly, A.J.; Weissman, I.L.; Longaker, M.T. Clonal analysis reveals nerve-dependent and independent roles on mammalian hind limb tissue maintenance and regeneration. *Proc. Natl. Acad. Sci. USA* **2014**, *111*, 9846–9851. [CrossRef] [PubMed]

40. Rinkevich, Y.; Lindau, P.; Ueno, H.; Longaker, M.T.; Weissman, I.L. Germ-layer and lineage-restricted stem/progenitors regenerate the mouse digit tip. *Nature* **2011**, *476*, 409–413. [CrossRef] [PubMed]

41. Baudoin, J.; Jafari, P.; Meuli, J.; Applegate, L.A.; Raffoul, W. Topical negative pressure on burns: An innovative method for wound exudate collection. *Plast. Reconstr. Surg. Glob. Open* **2016**, *4*, e1117. [CrossRef] [PubMed]

42. Bercoff, J.; Tanter, M.; Fink, M. Supersonic shear imaging: A new technique for soft tissue elasticity mapping. *IEEE Trans. Ultrason. Ferroelectr. Freq. Control* **2004**, *51*, 396–409. [CrossRef] [PubMed]

© 2017 by the authors. Licensee MDPI, Basel, Switzerland. This article is an open access article distributed under the terms and conditions of the Creative Commons Attribution (CC BY) license (http://creativecommons.org/licenses/by/4.0/).

International Journal of
Molecular Sciences

MDPI

Review

Diabetes and Wound Angiogenesis

Uzoagu A. Okonkwo [1,2] and Luisa A. DiPietro [2,*]

[1] Department of Microbiology and Immunology, University of Illinois at Chicago College of Medicine, Chicago, IL 60612, USA; uokonk2@uic.edu
[2] Center for Wound Healing and Tissue Regeneration, University of Illinois at Chicago College of Dentistry, Chicago, IL 60612, USA
* Correspondence: Ldipiet@uic.edu; Tel.: +1-312-355-0432

Received: 30 May 2017; Accepted: 22 June 2017; Published: 3 July 2017

Abstract: Diabetes Mellitus Type II (DM2) is a growing international health concern with no end in sight. Complications of DM2 involve a myriad of comorbidities including the serious complications of poor wound healing, chronic ulceration, and resultant limb amputation. In skin wound healing, which has definite, orderly phases, diabetes leads to improper function at all stages. While the etiology of chronic, non-healing diabetic wounds is multi-faceted, the progression to a non-healing phenotype is closely linked to poor vascular networks. This review focuses on diabetic wound healing, paying special attention to the aberrations that have been described in the proliferative, remodeling, and maturation phases of wound angiogenesis. Additionally, this review considers therapeutics that may offer promise to better wound healing outcomes.

Keywords: diabetes; wound healing; angiogenesis

---

1. Introduction

Diabetes mellitus type II (DM2) is a metabolic disorder defined by hyperglycemia due to insulin resistance. DM2 has become a major global health epidemic with a particularly large incidence in the United States. As of 2010 the global prevalence of DM2 was estimated at 280 million people, with recent measures by the Centers for Disease Control and Prevention (CDC) in 2014 indicating that 86 million Americans are in the pre-diabetic state and 29.1 million have been diagnosed with the disease [1]. DM2 is associated with numerous co-morbidities, including, but not limited to, cardiovascular disease, stroke, chronic renal failure, peripheral neuropathy, and diabetic skin wounds or ulcerations [2]. Diabetic skin ulcerations present as painful sores with disintegration of dermal tissue including the epidermis, dermis, and in many cases, subcutaneous tissue [3]. In diabetes, chronic skin ulcerations are common on the lower extremities, particularly the foot. Diabetic foot ulcers (DFU) affect 15% of diabetic patients. Of those patients with DFUs, 14–24% subsequently experience a lower extremity amputation, with the mortality rate from amputation approaching 50–59% five-year post-amputation [3–6]. Studies of the pathology of diabetic foot ulceration have focused on microbial invasion, epithelial breakdown, and impaired immune function as some of the causative factors for the non-healing phenotype [7]. One underlying factor that accompanies all diabetic ulcerations is poor vascular flow, a circumstance that impedes proper wound healing. Numerous studies have highlighted the importance of adequate vascular sufficiency and vessel proliferation in tissue repair and the lack thereof in diabetic wound healing [8]. More studies, albeit limited, have looked at whether disarrayed capillary remodeling and maturation of vessels might play a role in impaired diabetic wound healing. This review will synthesize the current findings in the literature about the role of appropriate capillary growth, function, and maturation in the context of diabetic wound healing.

2. The Anatomy and Maintenance of Blood Vessels

The human body is composed of a vast network of blood vessels that receive nutrient-rich, oxygenated blood from the heart starting with the largest in size called arteries, then subsequently decreasing in diameter to the arteriole, and finally to the capillary, where actual gas, nutrient, and waste exchange between vessels and tissue occurs. Waste is then carried back to the great veins of the venous system through venules. Capillaries, the cornerstone of nutrient diffusion, are so abundant in the body that no cell is more than 100–200 microns from one [9]. The anatomy of a capillary can be described as a microvessel that is composed of a one-cell layer thick lumen of endothelium with a diameter of 8–20 microns, just wide enough for the flow of erythrocytes and leukocytes [10]. The endothelial cells (EC) that constitute capillaries are dynamic cell types that have only recently been appreciated for their many activities. Endothelial cells line the lumen of every blood vessel in the body and are involved in filtration, hemostasis, barrier function, inflammation, and angiogenesis [11]. In the unwounded state, endothelial cells lining vessels are in a setting of quiescence demonstrated by the mostly basal expression of pro or anti-angiogenic stimuli in the vascular bed [12].

At the level of the capillaries, endothelial cells are in an intimate relationship with an enigmatic cell type called the pericyte. The pericyte is a type of mural cell embedded in the vascular basement membrane that wraps around endothelial cells [13,14]. Much remains to be discovered about this cell. Studies suggest that pericytes participate in regulation of capillary blood flow, clearance of cellular debris, stabilization, and paracrine communication of endothelial cells [15–17]. Loss of pericytes perturbs capillary function and results in extravascular leakage and edema [18]. The clear-cut identification of pericytes has long been a source of consternation. Due to their morphological similarities to vascular smooth muscle, the detection of distinct and ubiquitous markers that identify the unique pericyte population in vessels has proved to be an arduous task. Markers used with some success include α-smooth muscle actin (α-SMA) neural/glial antigen-2 (NG-2), desmin, Regulator of G-protein signaling 5 (RGS5), and platelet-derived growth factor receptor (PDGFR)-β [15,19]. Unfortunately, none of these markers are truly specific for the pericyte, and each marker is known to be expressed in other cell types. Nonetheless, several studies have linked some of these markers with pericyte function. Of note, PDGFR-β has been shown to be responsible for the recruitment of pericytes, and inhibition of PDGFR-β by imantinib results in decreased pericyte recruitment and migration in vitro [20]. In a mouse model, imantinib treatment caused decreased levels of NG-2$^+$ pericytes in dermal wound tissue [21]. The role of PDGFR-β derives from interaction with its ligand, PDGF-β. PDGF-β deficient embryos show insufficient mural cell recruitment, resulting in uncontrolled EC development, vessel enlargement, permeability, and impaired perfusion [22,23]. Other factors that have been implicated in the maintenance of pericytes include the endothelial cell receptor, angiopoietin-2 receptor (Tie2) and its ligand, angiopoietin-1 (Ang1). Many studies showcase the importance of the Tie2/Ang1 complex to vessel maturity and integrity, including down regulation of the pro-angiogenic factor vascular endothelial growth factor (VEGF), and in turn, prevention of vascular leakage [24,25].

3. Vasculogenesis and Angiogenesis

Vasculogenesis is defined as the formation of new vessels from precursor cells that coalesce into primitive vascular networks. This process, which is prominent during development, contrasts with angiogenesis, which is the formation of new capillaries from established vasculature [26]. Angiogenesis occurs primarily during embryonic and post-natal development and is quite rare in the healthy adult state except in the female reproductive system [27]. Several female reproductive organs, including the ovary, endometrial lining of the uterus, and placenta undergo angiogenesis as part of their normal physiologic responses [28]. Both physiologic and pathologic angiogenesis are regulated by a balance of proangiogenic factors and anti-angiogenic factors. Vessels produced during non-pathologic angiogenesis are characterized by their refinement, integrity, and ability to deliver nutrients to tissues in a controlled, timely manner [29]. Disarrayed, runaway angiogenesis is a distinguishing feature of many pathologic processes including solid tumor growth and fibrosis [29,30]. In such pathological disease states, angiogenesis can be persistent, disorganized, and never reaches attenuation [31,32]. While basal levels of pro-angiogenic factors and the expression of their corresponding receptors are maintained during the healthy state of homeostasis, the diseased state exhibits upregulated expression of pro-angiogenic factors such as VEGF, fibroblast growth factor (FGF), and others to promote unfettered growth of vessel beds and continual proliferation and migration of endothelial cells [33]. Pathological angiogenesis is most often associated with malignancy, where excessive angiogenesis can continue without order or end as tumor growth [34]. Dysfunctional angiogenesis is also seen in Crohn's disease, psoriasis, endometriosis, and rheumatoid arthritis [35]. Dysfunctional angiogenesis has been implicated in the pathophysiology of atherosclerosis and diabetic retinopathy as well as many other comorbidities that affect diabetic patients [36].

4. Angiogenesis in Wound Healing

Wound healing is a complex process that can be divided into a series of stages that include hemostasis, inflammation, proliferation, and remodeling [37]. Prior to injury, the vasculature is in a state of quiescence in which blood vessels are adequately perfused to deliver sufficient nutrients, and oxygen to the tissue. Basal levels of pro-angiogenic factors such as VEGF and FGF in addition to anti-angiogenic factors such as Ang-1 and pigment epithelium derived factor (PEDF) are expressed to maintain a functional vascular network that is neither proliferating nor diminishing [38,39]. When an assault to tissue occurs that produces injury, this homeostasis is interrupted, leading to a hypoxic state. Hypoxia is an important activator of the endothelial cells in the injured and adjacent vasculature [40]. In this hypoxic environment, the innate immune system recruits leukocytes to the site of injury, with neutrophils being the first responders in the acute phase of inflammation [41]. Following the influx of neutrophils comes the later arrival of macrophages, and shortly thereafter the tissue reaches the zenith of the proliferative phase of wound healing. Here macrophages, emerging new capillaries, and loose connective tissue, characterized by edema and immaturity, form granulation tissue [42]. One hallmark of the proliferative phase of wound healing is robust angiogenesis. Following the oxygen gradient that was established by injury, numerous proangiogenic factors are produced in wounds. These factors, the most notable of which is VEGF, stimulate capillaries to form nascent immature loops and branches (Figure 1). VEGF has been shown to be one of the most important angiogenic factors in wounds, and its production lies downstream of hypoxia. Hypoxia following injury activates hypoxia-inducible factor-1 (HIF-1), a transcriptional activator that promotes angiogenesis by upregulating target genes such VEGF-A [43]. VEGF-A, the main isoform in the wound, binds to its receptors on endothelial cells, directing vessel growth [44]. VEGF and other pro-angiogenic factors guide vascular growth to areas of low oxygen starting from the wound periphery into the wound bed [40,45].

One of the defining attributes of wound angiogenesis is the creation of a disorganized and poorly perfused vasculature, characterized by a malformed capillary bed with blind-ended sprouts and torturous loops [29]. Although malformed, the amount of capillaries in wounds reaches numbers much higher than normal skin and peaks at approximately day 7–10 post wounding [46,47]. Following the apex of angiogenesis, a switch from pro- to anti-angiogenic factors is ushered in, and vessels begin to regress with the help of programmed cell death, termed apoptosis (Figure 1) [48,49]. The proliferative phase, which can be characterized as chaotic, robust, and abundant, is contrasted by maturation, a process that refines and selects for competent nascent vessels to become durable mature vessels similar to the pre-injured state [50]. As capillary refinement occurs, vessel stabilization is mediated by smooth muscle cell recruitment in the form of pericytes [51]. Pericytes are important to the stabilization and maturation of newly formed vascular bed. In wounds, pericytes are actively recruited in response to several factors, with the best described being PDGF [52,53].

Once pericytes arrive in the healing wound, they interact with endothelial cells and the basement membrane (Figure 1). Pericyte-covered capillaries in the wound bed are resistant to the anti-angiogenic factors which are produced during the remodeling phase of repair. Capillary pruning in wounds is mediated by the active production of several anti-angiogenic, vascular maturation factors, with the best described including pigment epithelium derived factor (PEDF) and sprouty-2 (SPRY2) [54,55]. PEDF is a member of the serine protease inhibitor (SERPIN) family, and is one the most potent anti-angiogenic factors in the vasculature [56]. It is constitutively expressed in unwounded skin and serves a powerful homeostatic factor. PEDF has been shown to induce EC apoptosis and to reduce the permeability of leakage-prone neovasculature [57]. A role for PEDF has been described in several other tissue pathologies with dysregulated angiogenesis, with one of the most widely studied being cancer. In a study by Chen et al., a popular drug used for the treatment of diabetes, metformin, was shown to decrease prostate cancer cell proliferation, migration and tumor growth through a mechanism that involved the upregulation of PEDF [58]. PEDF anti-angiogenic activity has been shown to reduce tumor growth and vascularity, and PEDF is currently under investigation as an anti-tumor and anti-angiogenic therapy [59]. In the context of skin wound healing, studies by our laboratory have shown that the production of PEDF is essential to vascular remodeling and maturation [60,61]. Sprouty-2 (SPRY2) is a second factor that has been demonstrated to assist in capillary remodeling in wounds. SPRY2 is an intracellular protein that inhibits mitogen activated protein kinase (MAPK) signaling, ultimately downregulating the effect of VEGF on EC proliferation in wounds [55,62].

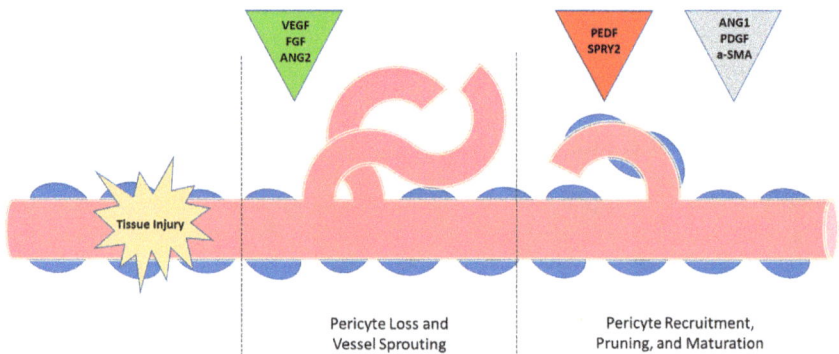

Figure 1. Events in wound angiogenesis. In the normal quiescent state, capillaries (pink) are surrounded by pericytes (blue). Following injury, the hypoxia that is created by the disruption of the vasculature stimulates the production of pro-angiogenic factors (green triangle), resulting in the sprouting of immature and disorganized new capillaries. In the remodeling phase, anti-angiogenic factors (red triangle) cause most of the newly formed capillaries to undergo apoptosis, and the capillary bed is pruned. Maturation factors (gray triangle) support the recruitment of stabilizing pericytes and the maturation of the basement membrane on the new capillaries. The result is a stable, well perfused capillary bed with a vessel density similar to normal uninjured tissue.

In addition to PEDF and SPRY2, another group of factors known to influence the normal progression of wound capillary growth and remodeling are the angiopoietins 1 and 2, which work in concert with the Tie2 receptor [63]. Ang1 is a potent maturation factor that stabilizes pericytes and ECs in capillaries. Angiopoietin-2 (Ang2) has an antagonistic effect and destabilizes vessels [64]. Both angiopoietins compete for binding to Tie2 tyrosine kinase receptor [65]. Consequently, during the normal angiogenic process, high levels of Ang2 are seen during the proliferative and pro-angiogenic phase of wound healing, while increased levels of Ang1 are seen during the maturation phase [66].

As the remodeling phase of wound healing ends, the tissue achieves normal vascular permeability, blood flow, and shows normal vascular branching [32]. Additionally, the high oxygen demand of the early stages of wound healing, a situation that activates many immune mediators and pro-angiogenic factors, returns to pre-injury levels [67]. The return to normoxia and normal oxygen demands signals a maintenance stage and return to quiescence. The quiescent environment includes anti-angiogenic mediators that ensure that vessel integrity is maintained and that endothelial cell migration, vessel sprouting, and branching are kept in check [68] (Figure 1).

In healing wounds, angiogenesis supports and intersects with the other ongoing proliferative activities and with the remodeling phase of repair [69]. During the proliferative phase, the new capillary growth in wounds is interwoven with multiple components of dermal repair. Fibroblast migration, proliferation, and collagen synthesis all occur during the same period as the angiogenic response. Similarly, epithelial proliferation and closure of the wound is also ongoing during the time of capillary growth. These many proliferative processes occur nearly simultaneously and support one another. For example, while new capillaries respond to the oxygen and nutrient needs of the proliferating tissues, stimulated epithelial cells produce VEGF to spur the capillary growth. During the remodeling phase, a phase triggered as oxygen levels return to normal, capillaries are pruned, the hyperproliferative epithelium thins to normal thickness, and the collagen in the wound bed undergoes maturation and cross-linking to achieve greater tissue strength. Recent studies suggest that capillary pruning and collagen maturation may interact in the remodeling phase, as the pruning of capillaries has been suggested to influence the final extracellular matrix (ECM) structure [70]. Like the proliferative phase, then, the remodeling phase exhibits a confluence of remodeling in capillaries, connective tissue, and epithelium.

5. Diabetes: An Altered Angiogenic State

As described above, angiogenesis in normal wound healing relies on a delicate balance between the promotion of vessel growth and proliferation and the promotion of vessel maturation and quiescence. The diabetic disease state can significantly perturb this balance, disrupting proper wound healing, tissue regeneration, and the restoration of a healthy vascular system. Perturbations in vascular integrity are also a feature of diabetes. Diabetic hyperglycemia, particularly in DM2, has been implicated in the progression of vascular disease in a multitude of both animal and clinical studies. The elevated systemic glucose levels seen in diabetic patients are the root cause of many micro and macrovascular complications that ultimately can affect angiogenesis [71]. ECs exposed to elevated blood glucose for extended periods of time have been shown to become dysfunctional, leading to integrity loss and increased susceptibility to apoptosis, detachment, and circulation into the bloodstream [72,73]. Free flowing, detached ECs have been shown to be a predictive marker of coronary heart disease and other pro-atherosclerotic processes in diabetes [74].

Insufficient angiogenesis plays a significant role in the pathogenesis of diabetic wound healing and micro and microvascular disease. Interestingly, though, while diabetic wounds have an angiogenic deficit, diabetes can lead to either increased or decreased angiogenesis depending upon the pathologic process. Numerous studies have shown that the diabetes-related changes in the angiogenic response can be tissue and/or organ dependent. For example, in diabetic retinopathy (DR) excessive angiogenesis occurs, leading to a pathology that is characterized by microaneurysms, hemorrhages, and vascular edema [75]. Along with increased capillary growth, a hallmark vascular modification seen in diabetic retinopathy is the loss of pericyte coverage in the retinal capillary network. This loss of pericytes is conducive to vascular edema and leakage. In response, the damaged capillaries experience hypoxia, leading to increased, abnormal expression of HIF-1 transcription factor and subsequent upregulation of VEGF-A, a known factor for increasing vascular permeability. This situation further contributes to the formation of an excessive neovasculature that is leaky and unrefined [75]. Importantly, an increase in the vitreous plasma levels of VEGF-A correlates with the severity of DR in diabetic patients [76]. Diabetic nephropathy (DN) is also characterized by excessive angiogenesis and resulting damage of the glomerular filtration system. The diabetic kidney may secrete excessive levels of VEGF-A in the initial stages of DN, which can lead to hyperpermeability of vessels, accelerated EC proliferation, and inhibition of apoptosis [77]. Studies by Cooper et al. showed in the early pathogenesis of diabetic nephropathy, mRNA levels of VEGF-A and its receptor VEGFR-2 are increased; in later disease progression levels of VEGF-A remain increased [78]. Disarrayed VEGF signaling is closely tied to impaired VEGF receptor activation, which is responsible for EC cell activation and proliferation, in addition to monocyte and endothelial progenitor cell (EPC) recruitment [75]. Abnormal receptor activation leads to increased circulating VEGF-A, due to decreased VEGF-A sensing in the vasculature, leading to such phenomena as plaque destabilization, and as mentioned previously, retinopathy [79].

6. Diabetes and Wound Angiogenesis

In contrast to diabetic nephropathy and retinopathy, diabetes leads to a decrease in angiogenesis in healing wounds. Diabetic wounds, impacted by insufficient angiogenesis, show decreased vascularity and capillary density [80]. Wound closure is greatly delayed in diabetes, and chronic non-healing wounds are common. An association of impaired angiogenesis to the pathologic wound repair seen in diabetic patients has been suggested by many studies. Below we review many of the described alterations in wound angiogenic response that are seen in the context of diabetes.

Macrophages, an important cell type of the innate immune system that are required for wound repair, have been shown to have altered functions in diabetic wounds [81]. In normal wounds, macrophages switch from a proinflammatory to pro-reparative phenotype, with the latter supporting tissue regrowth. In diabetic wounds, macrophage deficits include altered phenotypes that fail to stimulate tissue repair. One animal model of diabetes that has been widely used for wound healing studies is the db/db mouse. The db/db mouse is a genetic model of obesity, diabetes, and dyslipidemia that results from a mutation in the leptin receptor gene, and db/db mice are well documented to have significantly delayed healing [82]. In regard to macrophage function, Khanna et al. showed that macrophages at the wound site of db/db mice showed decreased efferocytosis, leading to increased apoptotic burden and inflammatory profile in the wound [83]. Since macrophages are an important source of VEGF and other pro-angiogenic mediators in wounds, the macrophage deficit may be linked to the documented decrease in wound angiogenesis that is seen in diabetic wounds. In one of the first studies to investigate the mechanisms that influence angiogenesis in diabetic skin wounds, Seitz et al. showed that VEGF-A protein and mRNA levels in wounds of db/db mice were significantly decreased compared to normal healthy controls [84]. Subsequent studies by Galiano et al. also identified this deficit, and went on to show that wounds of db/db mice treated with VEGF-A exhibited accelerated wound closure compared to untreated mice [85]. Of note, this study also showed that VEGF-A treated mice exhibited more of an early leaky, malformed vasculature and more edema until VEGF therapy was ceased [85].

In addition to the decrease in pro-angiogenic stimulus, several studies now demonstrate diabetes-associated changes in anti-angiogenic factors and capillary maturation factors in wounds. The production of the anti-angiogenic factor PEDF has been examined in the context of diabetic wound healing, although these studies investigated systemic serum levels rather than tissue expression levels. In one such study, PEDF was shown to occur at higher circulating levels in patients with diabetic foot ulcers compared to both non-diabetic and diabetic patients without DFU [86]. While this study might suggest that elevated levels of PEDF could negatively impact wound healing outcomes, the PEDF level was quantified only systemically and not within wound tissue.

One vascular maturation pathway that has been implicated in the deficits seen in diabetic wound angiogenesis is the Ang1/Ang2/Tie2 complex. In diabetic wounds it has been shown that the ratio of Ang1 to Ang2 is decreased, meaning that the ability of diabetic wound vasculature to progress to a mature phenotype is likely disturbed [87,88]. A role for Ang 1 as a maturation factor has also been described in streptozotocin (STZ)-induced diabetic mice. These studies have shown that topical application of neutralizing antibodies to Ang1 on wounded skin reduced the maturation of nascently formed blood vessels [89] Another study has shown that treatment of wounds of STZ-induced diabetic mice with transplanted bone marrow treated with Ang1 led to increased endothelial progenitor cells (EPCs) and increased neovascularization at day 7 post wounding [90].

MicroRNAs (miRNAs) are another class of molecule that can regulate angiogenesis and other aspects of wound repair, and miRNAs are known to be differentially expressed in the diabetic wound milieu. miRNAs are small, non-coding RNAs that are involved in post-translational modifications or gene silencing. Many miRNAs have been shown to be perturbed in diabetic wound healing, and specific miRNAs have been demonstrated to have modified expression in diabetic wound healing [91]. miR26-b is one miRNA that is highly expressed in diabetic ECs, and neutralization of this miRNA in diabetic wound models leads to increased wound closure and granulation tissue production [92]. Downregulation of miR-200b, shown to enhance TNF-α expression, leads to increased angiogenesis in diabetic wound skin [93]. In vivo and in vitro studies using local miR27-b, believed to affect levels of the anti-angiogenic molecule thrombospondin 1 (TSP1) in the wound bed, showed that restoration of miR27-b regulates angiogenesis in diabetic mouse models [94].

In the resolution and maturation phase of wound angiogenesis, platelet derived growth factor (PDGF) is one maturation factor that appears to be perturbed in the diabetic state. As mentioned above, PDGF encourages capillary maturation by nurturing and recruiting pericytes and retarding vessel regression [95]. PDGF has been extensively studied in diabetic skin wound healing, and db/db mice express lower levels of PDGF and its receptor in wounds [96]. Moreover, the topical application of PDGF has been shown to accelerate closure rates in diabetic wound healing [97]. PDGF is likely to have effects beyond capillary stabilization, as this factor is also a mitogen for fibroblasts. As further discussed below, recombinant PDGF is one of the only currently available growth factor therapies for non-healing diabetic ulcers.

In addition to changes in the production of pro-angiogenic and vascular maturation factors, the diabetic state leads to an inherently decreased population of endothelial progenitor cells (EPCs) from the bone marrow [98]. This deficit in turn, reduces the baseline vascularity in diabetic tissues and likely affects wound angiogenesis [99]. The EPCs of diabetics have a reduced capacity to produce functional angiogenic sprouts and tubes in ischemic models [100]. Without the appropriate function of these vital cells, processes such as granulation tissue formation, capillary growth, and collagen deposition are impaired in the wound bed [37,101]. Several studies suggest that EPC therapy might benefit diabetic wounds. For example, studies by Asai et al. showed that topical introduction of the sonic hedgehog (*Shh*) gene induced EPC proliferation, adhesion, and tube formation in vitro and increased wound vascularity in vivo in diabetic mice [102].

Given the many changes in pro-angiogenic and vascular maturation factors in diabetes, it is perhaps not surprising that the vascular architecture is known to be perturbed both in the normal skin and in the wounds of diabetics. Corrosion casting and scanning electron microscopy (SEM) studies have shown that diabetic patients suffering from diabetic microangiopathy in the toe region exhibit damaged capillary architecture and evidence of vascular leakage [103]. Deficiencies in the vascular architecture in diabetic wounds have also been noted. Recent studies in our laboratory have found that wounds of mice subjected to a high fat diet (HFD), a diet induced obesity model (DIO) that closely approximates DM2, present with more tortuous and aberrant architecture [104]. Overall, the diabetic state creates a large array of angiogenic deficits that occur during both the early and late stages of wound healing and affect both the proliferation and maturation of vessels. Many of the known anti-angiogenic and maturation factors in wounds have not yet been studied in the context of diabetes. In recent preliminary studies, though, we found that excisional wounds from db/db mice express lower levels of multiple anti-angiogenic and maturation factors during wound healing resolution as compared to control [105]. Conceptually, this suggests that vascular pruning and maturation may be delayed in diabetic wounds, a situation that might lead to chronic wounds or recurring wounds due to lack of a well-perfused and durable vascular bed. Table 1 gives a brief overview of the angiogenic events that are altered in diabetes. Further investigations of anti-angiogenic and maturation factor levels in the diabetic wound will need to be conducted to further evaluate the changes that occur in the resolution phase of healing.

Table 1. Diabetes-associated changes in wound healing.

Event	Diabetes-Associated Changes	References
Normal Quiescent Capillary Bed	Microangiopathies, loss of pericytes	[80]
Proangiogenic Stimulus in Wounds	Decreased response to hypoxia, decreased production of pro-angiogenic factors, impaired receptor function, miRNA misregulation, macrophage dysfunction	[43,81,83,85,90,92]
Angiogenic Response in Wounds	Blunted, miRNA misregulation, decreased in endothelial progenitor cells	[91,93,94,101,102]
Capillary Pruning and Maturation during Wound Resolution	Not yet well studied, but altered production of anti-angiogenic factors reported	[17,86,89]

7. Current Therapies to Ameliorate Diabetic Wound Healing

The hunt for effective therapies to treat the sizable patient population affected by the chronic, non healing wounds brought on by diabetes has been elusive. While there has been research that has progressed from laboratory, to clinical trials, and finally to clinical practice, these treatments have failed to be the silver bullet that will heal chronic diabetic wounds. Here we discuss those available therapies that involve mechanisms that improve angiogenesis and vascular perfusion.

One therapy that has been heavily relied upon in the clinic has been hyperbaric oxygen therapy (HBOT). HBOT requires that a patient inhale 100 per cent oxygen in an enclosed chamber where the pressure has been increased to above that found at sea level. The treatment has been shown to improve tissue hypoxia, vessel perfusion, reduce inflammation and edema, and increase angiogenesis [106]. Numerous studies in patients with DFUs have shown that patients undergoing HBOT show increased healing rates and reduced risk of a major limb amputation [107,108]. Unfortunately, HBOT is cost-prohibitive to the average patient, and while it has an over 20-year track record in the clinic, is still not a complete answer for treating non-healing diabetic foot wounds. Another therapy that has been attempted is the use of growth factors such as VEGF and PDGF. As mentioned before, these factors are important in the proliferative and maturation phase of wound healing, respectively. Numerous animal studies have shown that topical application of VEGF and its isoforms improve wound healing in diabetic mice [85,109,110]. However, topical VEGF therapy in human DFUs with recombinant VEGF (rh-VEGF) (Telbermin) has been met with limited success. Although Phase I trials suggested that patients who received topical VEGF compared to placebo exhibited improved healing, the drug was abandoned after Phase II clinical trials demonstrated no significant effect [111]. PDGF treatment has also been investigated in both mouse model and clinical trials. Recombinant PDGF became available as becaplermin (Regranex) in 1997 for the treatment of DFUs [112–114]. Topical treatment with becaplermin in clinical trial showed a 43 per cent increase in wound closure versus placebo in patients, in addition to reduced time to wound close of 32 per cent and complete healing of ulcers in 57.5% of patients [112–114]. Unfortunately, becaplermin has been met with many issues. Firstly, it is an expensive treatment that may not be readily accessible or feasible for many patients. Additionally, adverse side effects of rash and burning sensation at the site of application, as well as increased risks of osteomyelitis and cellulitis have been reported [113]. The most worrisome adverse side effect from the drug is the possible increased risk of malignancy in users who undergo more than 3 tubes of topical treatment, warranting the FDA to release a black box warning for the drug in 2008 [111]. Consequently, topical growth factors, while having promising results in animal models, have not yet translated well to the clinic. While single growth factors have met with limited success in the treatment of wounds, the use of platelet derived therapies has been suggested as a possible improvement as it provides a myriad of factors [115]. Platelets themselves are a rich source of many growth factors, including PDGF, transforming growth factor β (TGFβ), FGF-2, epidermal growth factor (EGF), and VEGF. Platelet derivatives such as platelet-rich plasma, platelet gel, and platelet-rich fibrin therefore have been explored for repair and regenerative strategies for both hard and soft tissues. Some of the advantages of platelet derivatives include the polyfactor approach as well as the ability to prepare the derivatives from the patient's own platelets, thus limiting patient exposure to exogenous agents. Beyond growth factors, multiple promising new therapies for diabetic wounds are currently under investigation. These include the application of cells such as stem cells and macrophages, and the use of sophisticated bioengineering approaches to provoke tissue repair responses.

8. Limitations of Current Knowledge and Future Directions

Our current knowledge of the pathology of DFUs and of the angiogenic response in diabetic wounds is incomplete. One major stumbling block to our understanding has been the lack of readily available animal models that sufficiently approximate human diabetic wound healing, especially chronic wounds. Moreover, many of the preclinical animal studies to date have focused on monotherapies, an approach which seems unlikely to be adequate for the treatment of the multifactorial problem of diabetic chronic skin wounds. Furthermore, work in diabetic wound healing has until recently focused only on the initial stages of wound healing and not the deficits that might prevent diabetic wound resolution. Adequate healing requires vascular maturation, including a return to quiescence and a normal vascular network. Much more work is needed to understand the differences during all stages of wound healing that occur in the diabetic state, with the goal of finding therapies and interventions to help drive these non-healing wounds to a state of health and integrity. Considering the enormous health care costs, decreased quality of life, and numerous comorbidities that are associated with diabetes, it is imperative that a holistic and multi-factorial approach to treating diabetes and to advancing wound care is sought.

Acknowledgments: Research in the authors' laboratory is currently supported by NIH R01-GM50875 and NIH R21-DE025926.

Author Contributions: Uzoagu A. Okonkwo and Luisa A. DiPietro planned the content. Uzoagu A. Okonkwo drafted the review. Luisa A. DiPietro provided guidance and edited the manuscript.

Conflicts of Interest: The authors declare no conflict of interest.

Abbreviations

α-SMA	α-smooth muscle actin
Ang	angiopoietin
CDC	Centers for Disease Control and Prevention
db/db	genetically diabetic mouse model
DIO	diet-induced obesity
DN	diabetic nephropathy
DFU	diabetic foot ulcer
DM2	diabetes mellitus type II
EC	endothelial cell
ECM	extracellular matrix
EGF	epidermal growth factor
EPC	endothelial progenitor cell
FGF	fibroblast growth factor
HBOT	hyperbaric oxygen therapy
HFD	high fat diet
HIF	hypoxia inducible factor
MAPK	mitogen-activated protein kinase
NG2	neural/glial antigen-2
PDGF	platelet derived growth factor
PDGFR	platelet derived growth factor receptor
PEDF	pigment epithelium derived factor
RGS5	regulator of G-protein signaling 5
SERPIN	serine protease inhibitor
Shh	sonic hedgehog
SPRY2	sprouty homolog 2
STZ	streptozotocin
Tie2	angiopoietin-2 receptor
TGFβ	transforming growth factor β
TNF-α	tumor necrosis factor-α
VEGF	vascular endothelial growth factor

References

1. Centers for Disease Control and Prevention. *National Diabetes Statistical Report: Estimates of Diabetes and Its Burden in the United States, 2014*; US Department of Health and Human Services: Atlanta, GA, USA, 2014.
2. Nathan, D.M. Long-Term Complications of Diabetes Mellitus. *N. Engl. J. Med.* **1993**, *328*, 1676–1685. [CrossRef] [PubMed]
3. Alavi, A.; Sibbald, R.G.; Mayer, D.; Goodman, L.; Botros, M.; Armstrong, D.G.; Woo, K.; Boeni, T.; Ayello, E.A.; Kirsner, R.S. Diabetic foot ulcers: Part I. pathophysiology and prevention. *J. Am. Acad. Dermatol.* **2014**, *70*, 1.e1–1.e18. [CrossRef] [PubMed]
4. Jeffcoate, W.J.; Harding, K.G. Diabetic foot ulcers. *Lancet* **2003**, *361*, 1545–1551. [CrossRef]
5. Martins-Mendes, D.; Monteiro-Soares, M.; Boyko, E.J.; Ribeiro, M.; Barata, P.; Lima, J.; Soares, R. The independent contribution of diabetic foot ulcer on lower extremity amputation and mortality risk. *J. Diabetes Complicat.* **2013**, *28*, 632–638. [CrossRef] [PubMed]
6. Wu, S.C.; Driver, V.R.; Wrobel, J.S.; Armstrong, D.G. Foot ulcers in the diabetic patient, prevention and treatment. *Vasc. Health Risk Manag.* **2007**, *3*, 65–76. [PubMed]
7. Noor, S.; Zubair, M.; Ahmad, J. Diabetic foot ulcer—A review on pathophysiology, classification and microbial etiology. *Diabetes Metab. Syndr. Clin. Res. Rev.* **2015**, *9*, 192–199. [CrossRef] [PubMed]
8. Prompers, L.; Schaper, N.; Apelqvist, J.; Edmonds, M.; Jude, E.; Mauricio, D.; Uccioli, L.; Urbancic, V.; Bakker, K.; Holstein, P.; et al. Prediction of outcome in individuals with diabetic foot ulcers: Focus on the differences between individuals with and without peripheral arterial disease. The EURODIALE Study. *Diabetologia* **2008**, *51*, 747–755. [CrossRef] [PubMed]
9. Alberts, B.; Johnson, A.; Lewis, J.; Raff, M.; Roberts, K.; Walter, P. *Molecular Biology of the Cell*, 4th ed.; Garland Science: New York, NY, USA, 2002.
10. Pittman, R.N. Regulation of Tissue Oxygenation. *Morgan Claypool Life Sci.* **2011**, *3*, 100. [CrossRef]
11. Galley, H.F.; Webster, N.R. Physiology of the endothelium. *Br. J. Anaesth.* **2004**, *93*, 105–113. [CrossRef] [PubMed]
12. Aird, W.C. Endothelium in health and disease. *Pharmacol. Rep.* **2008**, *60*, 139–143. [PubMed]
13. Caporali, A.; Martello, A.; Miscianinov, V.; Maselli, D.; Vono, R.; Spinetti, G. Contribution of pericyte paracrine regulation of the endothelium to angiogenesis. *Pharmacol. Ther.* **2017**, *171*, 56–64. [CrossRef] [PubMed]
14. Shepro, D.; Morel, N.M. Pericyte physiology. *FASEB J.* **1993**, *7*, 1031–1038. [PubMed]
15. Armulik, A.; Genové, G.; Betsholtz, C. Pericytes: Developmental, Physiological, and Pathological Perspectives, Problems, and Promises. *Dev. Cell* **2011**, *21*, 193–215. [CrossRef] [PubMed]
16. Armulik, A.; Abramsson, A.; Betsholtz, C. Endothelial/pericyte interactions. *Circ. Res.* **2005**, *97*, 512–523. [CrossRef] [PubMed]
17. Geevarghese, A.; Herman, I.M. Pericyte-endothelial crosstalk: Implications and opportunities for advanced cellular therapies. *Transl. Res.* **2014**, *163*, 296–306. [CrossRef] [PubMed]
18. Dulmovits, B.M.; Herman, I.M. Microvascular remodeling and wound healing: A role for pericytes. *Int. J. Biochem. Cell Biol.* **2012**, *44*, 1800–1812. [CrossRef] [PubMed]
19. Mills, S.J.; Cowin, A.J.; Kaur, P. Pericytes, mesenchymal stem cells and the wound healing process. *Cells* **2013**, *2*, 621–634. [CrossRef] [PubMed]
20. Van Dijk, C.G.M.; Nieuweboer, F.E.; Pei, J.Y.; Xu, Y.J.; Burgisser, P.; Van Mulligen, E.; El Azzouzi, H.; Duncker, D.J.; Verhaar, M.C.; Cheng, C. The complex mural cell: Pericyte function in health and disease. *Int. J. Cardiol.* **2015**, *190*, 75–89. [CrossRef] [PubMed]
21. Rajkumar, V.S.; Shiwen, X.; Bostrom, M.; Leoni, P.; Muddle, J.; Ivarsson, M.; Gerdin, B.; Denton, C.P.; Bou-Gharios, G.; Black, C.M.; et al. Platelet-derived growth factor-β receptor activation is essential for fibroblast and pericyte recruitment during cutaneous wound healing. *Am. J. Pathol.* **2006**, *169*, 2254–2265. [CrossRef] [PubMed]
22. Hellström, M.; Gerhardt, H.; Kalén, M.; Li, X.; Eriksson, U.; Wolburg, H.; Betsholtz, C. Lack of pericytes leads to endothelial hyperplasia and abnormal vascular morphogenesis. *J. Cell Biol.* **2001**, *152*, 543–553. [CrossRef]
23. Abramsson, A.; Berlin, Ö.; Papayan, H.; Paulin, D.; Shani, M.; Betsholtz, C. Analysis of mural cell recruitment to tumor vessels. *Circulation* **2002**, *105*, 112–117. [CrossRef] [PubMed]

24. Sato, T.N.; Tozawa, Y.; Deutsch, U.; Wolburg Buchholz, K.; Fujiwara, Y.; Gendron Maguire, M.; Gridley, T.; Wolburg, H.; Risau, W.; Qin, Y. Distinct roles of the receptor tyrosine kinases Tie-1 and Tie-2 in blood vessel formation. *Nature* **1995**, *376*, 70–74. [CrossRef] [PubMed]
25. Sundberg, C.; Kowanetz, M.; Brown, L.F.; Detmar, M.; Dvorak, H.F. Stable expression of angiopoietin-1 and other markers by cultured pericytes: Phenotypic similarities to a subpopulation of cells in maturing vessels during later stages of angiogenesis in vivo. *Lab. Investig.* **2002**, *82*, 387–401. [CrossRef] [PubMed]
26. Kolte, D.; McClung, J.A.; Aronow, W.S. Vasculogenesis and Angiogenesis. In *Translational Research in Coronary Artery Disease: Pathophysiology to Treatment*; Aronow, W.S., McClung, J.A., Eds.; Academic Press: Cambridge, MA, USA, 2015; pp. 49–65.
27. Geudens, I.; Gerhardt, H. Coordinating cell behaviour during blood vessel formation. *Development* **2011**, *138*, 4569–4583. [CrossRef] [PubMed]
28. Reynolds, L.P.; Grazul-Bilska, A.T.; Redmer, D.A. Angiogenesis in the female reproductive organs: Pathological implications. *Int. J. Exp. Pathol.* **2002**, *83*, 151–163. [CrossRef] [PubMed]
29. Fukumura, D.A.I.; Jain, R.K. Imaging angiogenesis and the microenvironment. *APMIS* **2008**, *116*, 695–715. [CrossRef] [PubMed]
30. Folkman, J. Angiogenesis in cancer, vascular, rheumatoid and other disease. *Nat. Med.* **1995**, *1*, 27–31. [CrossRef] [PubMed]
31. Carmeliet, P.; Jain, R.K. Angiogenesis in cancer and other diseases. *Nature* **2000**, *407*, 249–257. [CrossRef] [PubMed]
32. Carmeliet, P. Angiogenesis in life, disease and medicine. *Nature* **2005**, *438*, 932–936. [CrossRef] [PubMed]
33. Papetti, M.; Herman, I.M. Mechanisms of normal and tumor-derived angiogenesis. *AJP Cell Physiol.* **2002**, *282*, C947–C970. [CrossRef] [PubMed]
34. Kerbel, R.S. Tumor angiogenesis. *N. Engl. J. Med.* **2008**, *358*, 2039–2049. [CrossRef] [PubMed]
35. Nissen, N.N.; Polverini, P.J.; Koch, A.E.; Volin, M.V.; Gamelli, R.L.; DiPietro, L.A. Vascular endothelial growth factor mediates angiogenic activity during the proliferative phase of wound healing. *Am. J. Pathol.* **1998**, *152*, 1445–1452. [PubMed]
36. Fowler, M.J. Microvascular and macrovascular complications of diabetes. *Clin. Diabetes* **2011**, *29*, 116–122. [CrossRef]
37. Guo, S.; DiPietro, L.A. Factors affecting wound healing. *J. Dent. Res.* **2010**, *89*, 219–229. [CrossRef] [PubMed]
38. Demidova-Rice, T.N.; Durham, J.T.; Herman, I.M. Wound healing angiogenesis: Innovations and challenges in acute and chronic wound healing. *Adv. Wound Care* **2012**, *1*, 17–22. [CrossRef] [PubMed]
39. Tonnesen, M.G.; Feng, X.; Clark, R.A.F. Angiogenesis in wound healing. *J. Investig. Dermatol. Symp. Proc.* **2000**, *5*, 40–46. [CrossRef] [PubMed]
40. Tandara, A.A.; Mustoe, T.A. Oxygen in wound healing—More than a nutrient. *World J. Surg.* **2004**, *28*, 294–300. [CrossRef] [PubMed]
41. Turabelidze, A.; Dipietro, L.A. Inflammation and Wound Healing. In *Oral Wound Healing: Cell Biology and Clinical Management*; Larjava, H., Ed.; John Wiley and Sons: Hoboken, NJ, USA, 2013; pp. 39–56.
42. Koh, T.J.; DiPietro, L.A. Inflammation and wound healing: The role of the macrophage. *Expert Rev. Mol. Med.* **2011**, *13*, e23. [CrossRef] [PubMed]
43. Liu, L.; Marti, G.P.; Wei, X.; Zhang, X.; Zhang, H.; Liu, Y.V.; Semenza, G.L.; Harmon, J.W. Age-dependent Impairment of HIF-1α expression in diabetic mice: Correction with electroporation-facilitated gene therapy increases wound healing, angiogenesis, and circulating angiogenic cells. *J. Cell Physiol.* **2008**, *217*, 319–327. [CrossRef] [PubMed]
44. Eming, S.A.; Brachvogel, B.; Odorisio, T.; Koch, M. Regulation of angiogenesis: Wound healing as a model. *Prog. Histochem. Cytochem.* **2007**, *42*, 115–170. [CrossRef] [PubMed]
45. Broughton, G.; Janis, J.E.; Attinger, C.E. The basic science of wound healing. *Plast. Reconstr. Surg.* **2006**, *117*, 12S–34S. [CrossRef] [PubMed]
46. Johnson, K.E.; Wilgus, T.A. Vascular endothelial growth factor and angiogenesis in the regulation of cutaneous wound repair. *Adv. Wound Care* **2014**, *3*, 647–661. [CrossRef] [PubMed]
47. Szpaderska, A.M.; Zuckerman, J.D.; DiPietro, L.A. Differential injury responses in oral mucosal and cutaneous wounds. *J. Dent. Res.* **2003**, *82*, 621–626. [CrossRef] [PubMed]
48. Greenhalgh, D.G. The role of apoptosis in wound healing. *Int. J. Biochem. Cell Biol.* **1998**, *30*, 1019–1030. [CrossRef]

49. Dimmeler, S.; Zeiher, A.M. Endothelial cell apoptosis in angiogenesis and vessel regression. *Circ. Res.* **2000**, *87*, 434–439. [CrossRef] [PubMed]
50. Chen, R.R.; Silva, E.A.; Yuen, W.W.; Mooney, D.J. Spatio-temporal VEGF and PDGF delivery patterns blood vessel formation and maturation. *Pharm. Res.* **2007**, *24*, 258–264. [CrossRef] [PubMed]
51. Hirschi, K.K.; D'Amore, P.A. Pericytes in the microvasculature. *Cardiovasc. Res.* **1996**, *32*, 687–698. [CrossRef]
52. Allt, G.; Lawrenson, J.G. Pericytes: Cell biology and pathology. *Cells Tissues Organs* **2001**, *169*, 1–11. [CrossRef] [PubMed]
53. Lindblom, P.; Gerhardt, H.; Liebner, S.; Abramsson, A.; Enge, M.; Hellström, M.; Bäckström, G.; Fredriksson, S.; Landegren, U.; Nyström, H.C.; et al. Endothelial PDGF-B retention is required for proper investment of pericytes in the microvessel wall. *Genes Dev.* **2003**, *17*, 1835–1840. [CrossRef] [PubMed]
54. Pries, A.R.; Secomb, T.W. Making Microvascular Networks Work: Angiogenesis, Remodeling, and Pruning. *Physiology* **2014**, *29*, 446–455. [CrossRef] [PubMed]
55. Wietecha, M.S.; Cerny, W.L.; Dipietro, L.A. Mechanisms of vessel regression: Toward an understanding of the resolution of angiogenesis. *Curr. Top. Microbiol. Immunol.* **2013**, *367*, 3–32. [PubMed]
56. Filleur, S.; Nelius, T.; De Riese, W.; Kennedy, R.C. Characterization of PEDF: A multi-functional serpin family protein. *J. Cell. Biochem.* **2009**, *106*, 769–775. [CrossRef] [PubMed]
57. Broadhead, M.L.; Becerra, S.P.; Choong, P.F.M.; Dass, C.R. The applied biochemistry of PEDF and implications for tissue homeostasis. *Growth Factors* **2010**, *28*, 280–285. [CrossRef] [PubMed]
58. Chen, X.; Li, C.; He, T.; Mao, J.; Li, C.; Lyu, J.; Meng, Q.H. Metformin inhibits prostate cancer cell proliferation, migration, and tumor growth through upregulation of PEDF expression. *Cancer Biol. Ther.* **2016**, *17*, 507–514. [CrossRef] [PubMed]
59. Becerra, S.P.; Notario, V. The effects of PEDF on cancer biology: Mechanisms of action and therapeutic potential. *Nat. Rev. Cancer* **2013**, *13*, 258–271. [CrossRef] [PubMed]
60. Wietecha, M.S.; Krol, M.J.; DiPietro, L.A. PEDF functions as an endogenous anti-angiogenic during dermal wound repair. *Wound Repair Regen.* **2012**, *20*, A45.
61. Wietecha, M.S.; Król, M.J.; Michalczyk, E.R.; Chen, L.; Gettins, P.G.; DiPietro, L.A. Pigment epithelium-derived factor as a multifunctional regulator of wound healing. *Am. J. Physiol. Heart Circ. Physiol.* **2015**, *309*, H812–H826. [CrossRef] [PubMed]
62. Wietecha, M.S.; Chen, L.; Ranzer, M.J.; Anderson, K.; Ying, C.; Patel, T.B.; DiPietro, L.A. Sprouty2 downregulates angiogenesis during mouse skin wound healing. *Am. J. Physiol. Heart Circ. Physiol.* **2011**, *300*, H459–H467. [CrossRef] [PubMed]
63. Fagiani, E.; Christofori, G. Angiopoietins in angiogenesis. *Cancer Lett.* **2013**, *328*, 18–26. [CrossRef] [PubMed]
64. Staton, C.A.; Valluru, M.; Hoh, L.; Reed, M.W.R.; Brown, N.J. Angiopoietin-1, angiopoietin-2 and Tie-2 receptor expression in human dermal wound repair and scarring. *Br. J. Dermatol.* **2010**, *163*, 920–927. [CrossRef] [PubMed]
65. Maisonpierre, P.C. Angiopoietin-2, a natural antagonist for Tie2 that disrupts in vivo angiogenesis. *Science* **1997**, *277*, 55–60. [CrossRef] [PubMed]
66. Brudno, Y.; Ennett-Shepard, A.B.; Chen, R.R.; Aizenberg, M.; Mooney, D.J. Enhancing microvascular formation and vessel maturation through temporal control over multiple pro-angiogenic and pro-maturation factors. *Biomaterials* **2013**, *34*, 9201–9209. [CrossRef] [PubMed]
67. Hickey, M.M.; Simon, M.C. Regulation of Angiogenesis by hypoxia and hypoxia-inducible factors. *Curr. Top. Dev. Biol.* **2006**, *76*, 217–257. [PubMed]
68. Jain, R.K. Molecular regulation of vessel maturation. *Nat. Med.* **2003**, *9*, 685–693. [CrossRef] [PubMed]
69. Gurtner, G.C.; Werner, S.; Barrandon, Y.; Longaker, M.T. Wound repair and regeneration. *Nature* **2008**, *453*, 314–321. [CrossRef] [PubMed]
70. DiPietro, L.A. Angiogenesis and wound repair: When enough is enough. *J. Leukoc. Biol.* **2016**, *100*, 1–6. [CrossRef] [PubMed]
71. Altabas, V. Diabetes, endothelial dysfunction, and vascular repair: What should a diabetologist keep his eye on? *Int. J. Endocrinol.* **2015**, *2015*, 1–14. [CrossRef] [PubMed]
72. Piconi, L.; Quagliaro, L.; Assaloni, R.; Da Ros, R.; Maier, A.; Zuodar, G.; Ceriello, A. Constant and intermittent high glucose enhances endothelial cell apoptosis through mitochondrial superoxide overproduction. *Diabetes Metab. Res. Rev.* **2006**, *22*, 198–203. [CrossRef] [PubMed]

73. Yu, J.Q.; Liu, X.F.; Chin, L.K.; Liu, A.Q.; Luo, K.Q. Study of endothelial cell apoptosis using fluorescence resonance energy transfer (FRET) biosensor cell line with hemodynamic microfluidic chip system. *Lab Chip* **2013**, *13*, 2693–2700. [CrossRef] [PubMed]

74. McClung, J.A.; Naseer, N.; Saleem, M.; Rossi, G.P.; Weiss, M.B.; Abraham, N.G.; Kappas, A. Circulating endothelial cells are elevated in patients with type 2 diabetes mellitus independently of HbA1c. *Diabetologia* **2005**, *48*, 345–350. [CrossRef] [PubMed]

75. Kota, S.; Meher, L.; Jammula, S.; Kota, S.; Krishna, S.V.S.; Modi, K. Aberrant angiogenesis: The gateway to diabetic complications. *Indian J. Endocrinol. Metab.* **2012**, *16*, 918. [CrossRef] [PubMed]

76. Okamoto, T.; Yamagishi, S.I.; Inagaki, Y.; Amano, S.; Takeuchi, M.; Kikuchi, S.; Ohno, S.; Yoshimura, A. Incadronate disodium inhibits advanced glycation end products-induced angiogenesis in vitro. *Biochem. Biophys. Res. Commun.* **2002**, *297*, 419–424. [CrossRef]

77. McGinn, S.; Saad, S.; Poronnik, P.; Pollock, C.A. High glucose-mediated effects on endothelial cell proliferation occur via p38 MAP kinase. *Am. J. Physiol. Endocrinol. Metab.* **2003**, *285*, E708–E717. [CrossRef] [PubMed]

78. Cooper, M.E.; Vranes, D.; Youssef, S.; Stacker, S.A.; Cox, A.J.; Rizkalla, B.; Casley, D.J.; Bach, L.A.; Kelly, D.J.; Gilbert, R.E. Increased renal expression of vascular endothelial growth factor (VEGF) and its receptor VEGFR-2 in experimental diabetes. *Diabetes* **1999**, *48*, 2229–2239. [CrossRef] [PubMed]

79. Simons, M. Angiogenesis, arteriogenesis, and diabetes: Paradigm reassessed? *J. Am. Coll. Cardiol.* **2005**, *46*, 835–837. [CrossRef] [PubMed]

80. Dinh, T.; Veves, A. Microcirculation of the Diabetic Foot. *Curr. Pharm. Des.* **2005**, *11*, 2301–2309. [CrossRef] [PubMed]

81. Mirza, R.; Koh, T.J. Dysregulation of monocyte/macrophage phenotype in wounds of diabetic mice. *Cytokine* **2011**, *56*, 256–264. [CrossRef] [PubMed]

82. Michaels, J.; Churgin, S.S.; Blechman, K.M.; Greives, M.R.; Aarabi, S.; Galiano, R.D.; Gurtner, G.C. db/db mice exhibit severe wound-healing impairments compared with other murine diabetic strains in a silicone-splinted excisional wound model. *Wound Repair Regen.* **2007**, *15*, 665–670. [CrossRef] [PubMed]

83. Khanna, S.; Biswas, S.; Shang, Y.; Collard, E.; Azad, A.; Kauh, C.; Bhasker, V.; Gordillo, G.M.; Sen, C.K.; Roy, S. Macrophage dysfunction impairs resolution of inflammation in the wounds of diabetic mice. *PLoS ONE* **2010**, *5*, e9539. [CrossRef] [PubMed]

84. Seitz, O.; Schürmann, C.; Hermes, N.; Müller, E.; Pfeilschifter, J.; Frank, S.; Goren, I. Wound healing in mice with high-fat diet- or ob gene-induced diabetes-obesity syndromes: A comparative study. *Exp. Diabetes Res.* **2010**, *2010*. [CrossRef] [PubMed]

85. Galiano, R.D.; Tepper, O.M.; Pelo, C.R.; Bhatt, K.A.; Callaghan, M.; Bastidas, N.; Bunting, S.; Steinmetz, H.G.; Gurtner, G.C. Topical vascular endothelial growth factor accelerates diabetic wound healing through increased angiogenesis and by mobilizing and recruiting bone marrow-derived cells. *Am. J. Pathol.* **2004**, *164*, 1935–1947. [CrossRef]

86. Qi, W.; Yang, C.; Dai, Z.; Che, D.; Feng, J.; Mao, Y.; Cheng, R.; Wang, Z.; He, X.; Zhou, T.; et al. High levels of pigment epithelium-derived factor in diabetes impair wound healing through suppression of Wnt signaling. *Diabetes* **2015**, *64*, 1407–1419. [CrossRef] [PubMed]

87. Isidori, A.M.; Venneri, M.A.; Fiore, D. Angiopoietin-1 and Angiopoietin-2 in metabolic disorders: Therapeutic strategies to restore the highs and lows of angiogenesis in diabetes. *J. Endocrinol. Investig.* **2016**, *39*, 1235–1246. [CrossRef] [PubMed]

88. Kampfer, H.; Pfeilschifter, J.; Frank, S. Expressional Regulation of Angiopoietin-1 and -2 and the Tie-1 and -2 receptor tyrosine kinases during cutaneous wound healing: A comparative study of normal and impaired repair. *Lab. Investig.* **2001**, *81*, 361–373. [CrossRef] [PubMed]

89. Li, C.; Yu, T.; Liu, Y.; Chen, X.; Zhang, X. Topical Application of insulin accelerates vessel maturation of wounds by regulating angiopoietin-1 in diabetic mice. *Int. J. Low. Extrem. Wounds* **2015**, *14*, 353–364. [CrossRef] [PubMed]

90. Balaji, S.; Han, N.; Moles, C.; Shaaban, A.F.; Bollyky, P.L.; Crombleholme, T.M.; Keswani, S.G. Angiopoietin-1 improves endothelial progenitor cell-dependent neovascularization in diabetic wounds. *Surgery* **2015**, *158*, 846–856. [CrossRef] [PubMed]

91. Xu, J.; Zgheib, C.; Hu, J.; Wu, W.; Zhang, L.; Liechty, K.W. The role of microRNA-15b in the impaired angiogenesis in diabetic wounds. *Wound Repair Regen.* **2014**, *22*, 671–677. [CrossRef] [PubMed]

92. Icli, B.; Nabzdyk, C.S.; Lujan-Hernandez, J.; Cahill, M.; Auster, M.E.; Wara, A.K.M.; Sun, X.; Ozdemir, D.; Giatsidis, G.; Orgill, D.P.; et al. Regulation of impaired angiogenesis in diabetic dermal wound healing by microRNA-26a. *J. Mol. Cell. Cardiol.* **2016**, *91*, 151–159. [CrossRef] [PubMed]

93. Chan, Y.C.; Roy, S.; Khanna, S.; Sen, C.K. Downregulation of endothelial MicroRNA-200b supports cutaneous wound angiogenesis by desilencing GATA binding protein 2 and vascular endothelial growth factor receptor 2. *Arterioscler. Thromb. Vasc. Biol.* **2012**, *32*, 1372–1382. [CrossRef] [PubMed]

94. Wang, J.M.; Tao, J.; Chen, D.D.; Cai, J.J.; Irani, K.; Wang, Q.; Yuan, H.; Chen, A.F. MicroRNA miR-27b rescues bone marrow derived angiogenic cell function and accelerates wound healing in type 2 diabetes mellitus. *Arterioscler. Thromb. Vasc. Biol.* **2014**, *34*, 99–109. [CrossRef] [PubMed]

95. Hellberg, C.; Östman, A.; Heldin, C.-H. PDGF and Vessel Maturation. *Recent Results Cancer Res.* **2010**, *180*, 103–114. [PubMed]

96. Beer, H.D.; Longaker, M.T.; Werner, S. Reduced expression of PDGF and PDGF receptors during impaired wound healing. *J.Investig. Dermatol.* **1997**, *109*, 132–138. [CrossRef] [PubMed]

97. Brown, R.L.; Breeden, M.P.; Greenhalgh, D.G. PDGF and TGF-α act synergistically to improve wound healing in the genetically diabetic mouse. *J. Surg. Res.* **1994**, *56*, 562–570. [CrossRef] [PubMed]

98. Drela, E.; Stankowska, K.; Kulwas, A.; Rość, D. Endothelial progenitor cells in diabetic foot syndrome. *Adv. Clin. Exp. Med.* **2012**, *21*, 249–254. [PubMed]

99. Kolluru, G.K.; Bir, S.C.; Kevil, C.G. Endothelial dysfunction and diabetes: Effects on angiogenesis, vascular remodeling, and wound healing. *Int. J. Vasc. Med.* **2012**, *2012*. [CrossRef] [PubMed]

100. Tamarat, R.; Silvestre, J.-S.; Le Ricousse-Roussanne, S.; Barateau, V.; Lecomte-Raclet, L.; Clergue, M.; Duriez, M.; Tobelem, G.; Lévy, B.I. Impairment in ischemia-induced neovascularization in diabetes: Bone marrow mononuclear cell dysfunction and therapeutic potential of placenta growth factor treatment. *Am. J. Pathol.* **2004**, *164*, 457–466. [CrossRef]

101. Fiorina, P.; Pietramaggiori, G.; Scherer, S.S.; Jurewicz, M.; Mathews, J.C.; Vergani, A.; Thomas, G.; Orsenigo, E.; Staudacher, C.; La Rosa, S.; et al. The mobilization and effect of endogenous bone marrow progenitor cells in diabetic wound healing. *Cell Transplant.* **2010**, *19*, 1369–1381. [CrossRef] [PubMed]

102. Asai, J.; Takenaka, H.; Kusano, K.F.; Ii, M.; Luedemann, C.; Curry, C.; Eaton, E.; Iwakura, A.; Tsutsumi, Y.; Hamada, H.; et al. Topical sonic hedgehog gene therapy accelerates wound healing in diabetes by enhancing endothelial progenitor cell-mediated microvascular remodeling. *Circulation* **2006**, *113*, 2413–2424. [CrossRef] [PubMed]

103. Sangiorgi, S.; Manelli, A.; Reguzzoni, M.; Ronga, M.; Protasoni, M.; Dell'Orbo, C. The cutaneous microvascular architecture of human diabetic toe studied by corrosion casting and scanning electron microscopy analysis. *Anat. Rec.* **2010**, *293*, 1639–1645. [CrossRef] [PubMed]

104. Urao, N.; Okonkwo, U.A.; Fang, M.M.; Zhuang, Z.W.; Koh, T.J.; DiPietro, L.A. MicroCT angiography detects vascular formation and regression in skin wound healing. *Microvasc. Res.* **2016**, *106*, 57–66. [CrossRef] [PubMed]

105. Okonkwo, O.U.; Chen, L.; Modilevsky, B.; Zhao, Y.; DiPietro, L.A. Vascular Maturity and Integrity in Diabetic skin wounds. In *Wound Repair and Regeneration*; Davidson, J.M., Ed.; Wound Healing Society: San Diego, CA, USA, 2017; pp. w7–w8.

106. Tiaka, E.K.; Papanas, N.; Manolakis, A.C.; Maltezos, E. The role of hyperbaric oxygen in the treatment of diabetic foot ulcers. *Angiology* **2011**, *63*, 302–314. [CrossRef] [PubMed]

107. Kaya, A.; Aydin, F.; Altay, T.; Karapinar, L.; Ozturk, H.; Karakuzu, C. Can major amputation rates be decreased in diabetic foot ulcers with hyperbaric oxygen therapy? *Int. Orthop.* **2009**, *33*, 441–446. [CrossRef] [PubMed]

108. Stoekenbroek, R.M.; Santema, T.B.; Legemate, D.A.; Ubbink, D.T.; Van Den Brink, A.; Koelemay, M.J.W. Hyperbaric oxygen for the treatment of diabetic foot ulcers: A systematic review. *Eur. J. Vasc. Endovasc. Surg.* **2014**, *47*, 647–655. [CrossRef] [PubMed]

109. Kwon, M.J.; An, S.; Choi, S.; Nam, K.; Jung, H.S.; Yoon, C.S.; Ko, J.H.; Jun, H.J.; Kim, T.K.; Jung, S.J.; et al. Effective healing of diabetic skin wounds by using nonviral gene therapy based on minicircle vascular endothelial growth factor DNA and a cationic dendrimer. *J. Gene Med.* **2012**, *14*, 272–278. [CrossRef] [PubMed]

110. Bao, P.; Kodra, A.; Tomic-canic, M.; Golinko, M.S.; Ehrlich, H.P.; Brem, H. The role of vascular endothelial groth factor in wound healing. *J. Surg. Res.* **2010**, *153*, 347–358. [CrossRef] [PubMed]

111. Barrientos, S.; Brem, H.; Stojadinovic, O.; Tomic-Canic, M. Clinical application of growth factors and cytokines in wound healing. *Wound Repair Regen.* **2014**, *22*, 569–578. [CrossRef] [PubMed]
112. Steed, D.L. Clinical evaluation of recombinant human platelet-derived growth factor for the treatment of lower extremity diabetic ulcers. *J. Vasc. Surg.* **1995**, *21*, 71–81. [CrossRef]
113. Smiell, J.M.; Wieman, T.J.; Steed, D.L.; Perry, B.H.; Sampson, A.R.; Schwab, B.H. Efficacy and safety of becaplermin (recombinant human platelet-derived growth factor-BB)in patients with nonhealing, lower extremity diabetic ulcers: A combined analysis of four randomized studies. *Wound Repair Regen.* **1999**, *7*, 335–346. [CrossRef] [PubMed]
114. Embil, J.M.; Papp, K.; Sibbald, G.; Tousignant, J.; Smiell, J.M., Wong, B.; Lau, C.Y. Recombinant human platelet-derived growth factor BB (becaplermin) for healing chronic lower extremity diabetic ulcers: An open-label clinical evaluation of efficacy. *Wound Repair Regen.* **2000**, *8*, 162–168. [CrossRef] [PubMed]
115. De Pascale, M.R.; Sommese, L.; Casamassimi, A.; Napoli, C. Platelet derivatives in regenerative medicine: An update. *Transfus. Med. Rev.* **2017**, *29*, 52–61. [CrossRef] [PubMed]

© 2017 by the authors. Licensee MDPI, Basel, Switzerland. This article is an open access article distributed under the terms and conditions of the Creative Commons Attribution (CC BY) license (http://creativecommons.org/licenses/by/4.0/).

International Journal of
Molecular Sciences

MDPI

Review

The Importance of Pericytes in Healing: Wounds and other Pathologies

Hannah M. Thomas [1,2,3], Allison J. Cowin [1,2,3] and Stuart J. Mills [1,2,3,*]

[1]   Centre for Regenerative Medicine, Future Industries Institute, University of South Australia,
     Mawson Lakes 5095, Australia; hannah.thomas@mymail.unisa.edu.au (H.M.T.);
     allison.cowin@unisa.edu.au (A.J.C.)
[2]   School of Pharmacy and Medical Sciences, University of South Australia, Adelaide 5000, Australia
[3]   Cooperative Research Centre for Cell Therapy Manufacturing, North Terrace, Adelaide 5000, Australia
*   Correspondence: stuart.mills@unisa.edu.au; Tel.: +61-08-8302-3896

Academic Editor: Terrence Piva
Received: 10 April 2017; Accepted: 15 May 2017; Published: 24 May 2017

Abstract: Much of current research investigates the beneficial properties of mesenchymal stem cells (MSCs) as a treatment for wounds and other forms of injury. In this review, we bring attention to and discuss the role of the pericyte, a cell type which shares much of the differentiation potential and regenerative properties of the MSC as well as specific roles in the regulation of angiogenesis, inflammation and fibrosis. Pericytes have been identified as dysfunctional or depleted in many disease states, and observing the outcomes of pericyte perturbation in models of disease and wound healing informs our understanding of overall pericyte function and identifies these cells as an important target in the development of therapies to encourage healing.

Keywords: pericyte; MSC; wound healing; cell therapy

1. An Introduction to Pericytes

1.1. Morphology, Location and Function

Pericytes are cells found on the outside of blood vessels [1]. Their long processes encircle the abluminal surface of those vessels and attribute structural integrity to the vessel wall. Pericyte morphology is characterised by minimal cytoplasm, a prominent nucleus and projecting processes which wrap around associated capillaries (Figure 1) [2].

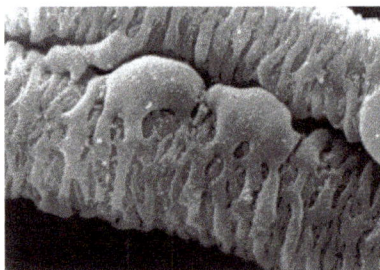

Figure 1. SEM showing pericyte processes spanning a venous capillary of the rete mirabile from an eel swimbladder. Image was kindly supplied by Professor Roger C. Wagner, University of Delaware.

From their position on the outer surface of the blood vessel, pericytes interact with endothelial cells (ECs), which reside on the other side of the basement membrane, through adhesion plaques which provide adherence between the ECs and the pericytes. Peg-and-socket contacts facilitate the diffusion of molecules between the two cell types [3]. Pericytes and ECs together create and maintain the shared basement membrane, the acellular component of the vessel wall [4]. This relationship allows pericytes to regulate the blood flow within vessels by virtue of high levels of α-smooth muscle actin (αSMA) and myosin expression, which can bring about vessel constriction [5]. Pericyte density and the EC to pericyte ratio is found to differ between organs, with ratios estimated to range from 1:1 in the central nervous system (CNS) to 10:1 in tissues such as skeletal muscle [6]. Consequently, in any given organ the proportion of the abluminal vessel surface which is covered by pericytes can be anywhere from 10–70% [7]. At the interface between the endothelial tube and the surrounding tissue, pericytes are ideally located to regulate processes associated with the vasculature, including the control of angiogenesis, which is well documented for these cells both in the context of general homeostasis and in response to trauma. Pericytes mechanically regulate vessel wall integrity, and serve as signalling mediators of EC behaviour. Paracrine pericyte signalling directs EC proliferation and migration to form new vessel sprouts when appropriate and inhibits aberrant pro-angiogenic behaviour in ECs when vessel sprouting is not required [8]. Pericytes have also recently been found to regulate the diffusion of cells and proteins from the vessel to the surrounding tissue, influencing the infiltration of neutrophils [9,10] and macrophages [11], which suggests an additional role for pericytes as mediators of the inflammatory process.

1.2. Pericyte Origin and MSC Properties

The developmental origins of the pericyte across all tissues are still not fully understood. For the most part, pericytes develop from the mesoderm during embryogenesis, with the origins of pericytes in the gut [12], lungs [13] and liver [14] having been tracked to the mesothelium. Similarly, cardiac pericytes have been shown to stem from the epicardial mesothelium [15,16]. In the central nervous system, however, chick-chimera studies show that while pericytes in the spinal cord and brain stem have developed from the mesoderm, pericytes of the forebrain are more likely to be derived from neural crest cells of the neuroectoderm [17].

The diverse origins of tissue-specific pericytes are reflected in the antigenic heterogeneity of pericytes observed between tissues. Currently, there are no markers identified as being expressed exclusively by pericytes, nor any constitutively expressed across pericytes of all locations. Within varying tissues, pericytes are found to display morphological changes and differential expression of markers dependent on their differentiation state and specific function within that tissue [18]. Changes in expression are also noted between different developmental stages and disease states [7]. Further, some markers are only expressed when pericytes are actively involved in remodelling of the vasculature, such as RGS5, which is expressed on activated pericytes in tumour development and vascular remodelling but is absent at other times [2,19]. While the list of recognised pericyte markers is growing, there remains a distinct absence within the field of a method by which pericytes can be identified indiscriminately of tissue, disease or developmental factors. As such, pericyte identification still relies on the concurrent identification of perivascular location, morphology and expression of multiple markers. For example, pericytes express many of the same markers as fibroblasts and exhibit similar morphology, so colocalisation between blood vessels and pericytes is often necessary to distinguish between the two cell types. The current struggles in pericyte identification and therefore isolation have been comprehensively reviewed in recent years [7,20,21]. Ansell and Izita discuss the difficulties encountered when comparing previous studies of pericyte function, particularly in vitro, and identify the potential for confounding results due to the unintentional selection of different pericyte subtypes for inclusion in experimental studies [22]. Much of the discussion of pericytes in the literature today addresses the current limitations in our ability to accurately define and isolate pericyte subtypes for experimental purposes.

There is significant overlap between markers expressed by pericytes and mesenchymal stem cells (MSCs), which is perhaps unsurprising given the predominately mesenchymal origin of many pericyte populations. Expression of CD105, CD73, CD90 and CD44 is observed in both pericytes and MSCs [23], and the observation that some subsets of pericytes express αSMA while others do not leads some researchers to postulate that αSMA$^+$ pericytes are more likely to carry out a structural support role at the blood vessel wall while αSMA$^-$ pericytes possess a more regenerative MSC-like phenotype [24].

Much of the current research in regenerative medicine is invested in evaluating the potential of cells with MSC-like properties as treatments to improve wound healing. We now understand that pericytes not only express MSC markers but also possess MSC-like properties, and display the ability to differentiate in vitro into an array of mesenchymal cell types. These include adipocytes [25], osteoblasts, chondrocytes [26], phagocytes and granulocytes [2]. The potential for pericytes to differentiate into beneficial cell types during the proliferative and regenerative stages of wound healing is an exciting prospect, and in the context of wound healing a cell type with the potential to positively contribute to direct regeneration of lost tissue represents a possible target for therapeutics or a source for the development of a cell-based therapy. Prior to the consideration of these cells for application in cell therapies however, the differentiation potential of all pericyte populations must be comprehensively understood.

Pericyte differentiation potential is extensive and highly dependent on lineage and the surrounding environment [27]. In fact there is a body of observations surrounding pericyte plasticity which, in conjunction with a shared perivascular location, suggests that MSCs and pericytes are, in fact, the same cell type [28]. Recently however, only CD146$^+$ bone marrow MSCs (BM-MSCs) and pericytes (also CD146$^+$) were found to maintain endothelial tube networks and improve angiogenic sprouting in vitro, while CD146$^-$ subtypes of the BM-MSC population did not, suggesting that pericytes are perhaps a subset of MSCs with vascular biology functions which not all MSCs possess [29]. One school of thought with regards to the difficulty of defining the difference between a pericyte and an MSC suggests that a pericyte which is in direct contact through gap junctions with ECs should be termed a pericyte, but that upon liberation of a pericyte from the vessel wall, that same cell should then be termed an MSC [30].

2. Pericytes in Wound Healing

Wound healing is a complex process made up of a series of overlapping events that include inflammation, matrix formation and remodelling. Further to our initial understanding of their involvement in the stabilisation of blood vessels and the control of blood pressure, the number of recognised functions of pericytes has broadened drastically, with many of these functions involved in wound healing (Table 1).

Table 1. Pericyte functions and their contributions to wound healing.

Wound Healing Process	Pericyte Functions
Angiogenesis	Structural support of existing blood vessels [19] Regulation of EC proliferation and migration to form new vessels [8] Prevention of capillary tube regression by TIMP-3 expression [31] Stabilisation of newly formed capillaries [32,33]
Inflammation	Regulation of vessel permeability [34–36] Regulation of neutrophil extravasion [9,10] Regulation of macrophage extravasion [11,37] Control of leukocyte trafficking [38] Control of T cell activation [39,40] Response to inflammatory signals [41,42]
Re-epithelialisation	Regulation of keratinocyte migration [43]
Fibrosis	Production of collagen [44,45] Differentiation into myofibroblasts [46]
Tissue regeneration	MSC-like properties: differentiation potential includes adipocytes, osteoblasts, chondrocytes, phagocytes and granulocytes [2,25,26]

2.1. Pericytes and Inflammation

Inflammation is one of the key phases during wound healing, and any perturbation in this carefully controlled process can lead to delayed healing, fibrosis or the incomplete healing seen in chronic wounds. It is triggered within 30–40 min of wounding and begins with the influx of neutrophils from the blood vessels to the wound site. Once in the wound, neutrophils act to phagocytose invading pathogens and cellular debris to clear the wound of infection. Early studies of pericytes in inflammation have noted that these cells form umbrella-like covers over gaps between ECs following histamine treatment, which prevents cells and proteins from leaving the vessel. Interestingly, the opposite is seen after IL-2 treatment, which causes pericytes to realign at EC junctions and results in leaky microvessels [34,35]. Neng et al. showed that pericytes also regulate tight junctions in a paracrine manner in EC monolayers in the mouse ear. This was shown to in turn regulate EC monolayer permeability and to control the diffusion of proteins and cells out of the blood [36]. These studies suggest that pericytes play a significant role in controlling inflammation. More recently pericytes have been shown to be directly involved in the extravasation of neutrophils from the vessels. Direct contact between neutrophils and pericytes induces a relaxation of the pericyte cytoskeleton via the inhibition of RhoA/ROCK signalling, which allows neutrophils to leave the vessel at regions displaying low expression of matrix proteins, termed "low expression regions" (LERs) [10]. This process is facilitated by the expression of Intercellular Adhesion Molecule-1 (ICAM-1), Macrophage antigen-1 (Mac-1) and Leukocyte function associated antigen-1 (LFA-1) [9]. ICAM-1 expression by pericytes, in conjunction with the expression of chemoattractant MIF, has also been shown not only to attract and activate neutrophils and macrophages, but also to facilitate efficient trafficking of these cells to areas of infection [38]. Pericytes have also been shown to influence T cell activity: brain pericytes are able to present antigens to T cells in order to induce lymphocyte activation [39], whereas retinal pericytes inhibit T cell action [40]. Whether these inherent differences in action are due to differences in the local population of pericytes or a result of microenvironmental cues is unclear. Brain pericytes also respond to inflammatory signals, such as lipopolysaccharides (LPS), resulting in activated NF-κB and expression of Interferon gamma-induced protein 10 (IP-10) and Monocyte Chemoattractant Protein-1 (MCP-1) [41]. Blockade of pericyte recruitment to vessels and therefore pericyte-EC intractions induces inflammatory reactions in ECs and results in increased extravasation of macrophages in an adult mouse model of diabetic retinopathy, again illustrating the importance of pericyte influence in the correct regulation of inflammatory infiltration [37]. Hung et al. suggest a role for pericytes not only in the recruitment of immune cells but also in the direct detection of proinflammatory molecules during infection, and show that decreasing the presence of pericyte-like cells in a model of lung injury leads to decreased inflammatory response to infection, leading the authors to propose that pericytes be considered "interstitial immune sentinel cells" [42]. Together, these studies intimately link pericyte action with regulation of the inflammatory response.

2.2. Pericytes and Re-Epithelialisation

Another key process during wound healing is the reformation of the epithelial barrier post-wounding. This helps to prevent wound infection and begins to restore some of the vital functions of the skin, such as the prevention of excess water loss and the regulation of temperature. Pericytes have been implicated in this process, and this action is quite distinct from pericyte action at the surface of the vessels. Paquet-Fifield et al. isolated pericytes from skin by means of FACS sorting with a pericyte specific antibody, and created organotypic cultures (OCs), with or without pericytes, which also contained fibroblasts and keratinocytes. When pericytes were present in the OCs there was a drastic improvement in the epidermal layer formed: the epidermis in these OCs was multilayered and sustainable for much greater periods of time when compared to the epidermis of OCs which did not contain pericytes [43].

2.3. Pericytes and Angiogenesis

Pericytes have been known to play a role in blood vessel formation from some of the earliest studies of their function, which identified these cells as being distinct from endothelial cells and originally labelled them Rouget cells [47]. More recently, they have been shown to respond to platelet derived growth factor β (PDGFβ) and transforming growth factor β (TGF-β) released by platelets upon injury [2]. The chemotactic response of pericytes to PDGFβ causes these cells to leave the outer layers of blood vessels and migrate into the wound site. This was established by Rajkumar et al. in studies using the PDGFβ inhibitor imatinib [48]. This migration allows ECs to proliferate into the wound site in response to vascular endothelial growth factor (VEGF), which is also released upon platelet activation [49]. This process is aided by the production of fibronectin, vitronectin and laminins, which provide a structure to support EC migration and capillary tube formation [50,51]. The provisional matrix, formed by these ECM components, is frequently remodelled during healing by proteases released by macrophages [52]. This can expose matricryptic sites such as Arg-Gly-Asp (RGD), which act as adhesion sites for EC receptors and can, therefore, regulate EC migration, proliferation and survival [52]. Interestingly, pericytes may regulate protease action via their expression of tissue inhibitor of metalloproteinase-3 (TIMP-3), which prevents capillary tube regression normally caused by matrix metalloproteinase-1 (MMP-1) and -10 (MMP-10) [31]. Pericytes also act to stabilise newly formed capillaries by the expression of TGF-β [32] and by Rho GTPase regulated alterations in pericyte contractility, which inhibit EC proliferation [33]. PDGFβ also appears to be essential in this process, as PDGFβ and PDGFRβ knockout mice exhibit endothelial hyperplasia with a distinct lack of pericytes present on blood vessels [53]. Interestingly, control of PDGFβ expression appears to be via Tie2 and Ang1/Ang2 interactions, where Tie2/Ang1 interaction leads to PDGFβ expression and pericyte recruitment, while Tie2/Ang2 interaction leads to the opposite [54]. In addition to PDGFβ and PDGFRβ, the PDGFβ retention motif is also crucial for pericyte-EC signalling. This sequence of amino acids acts to hold PDGF in close proximity to the EC for it to be recognised by the PDGFRβ expressing pericytes. This allows direct pericyte-EC interaction, as well as creating a PDGF gradient which enables pericytes to migrate to ECs [55]. This motif has been studied using a *pdgf-b^{ret/ret}* mouse knockout model, and has been shown to be crucial for maintaining vascular function in the retina, brain and liver [56–58].

2.4. Pericytes and Matrix Deposition/Fibrosis

Under normal conditions, matrix deposition is initiated once the wound has been cleared of infection and cellular debris. The main cell type responsible for this is the fibroblast. These cells initially deposit fibronectin and collagen III, but in later phases replace these proteins with collagen I and elastin. Fibroblasts, like pericytes, are attracted to the wound site by the expression of PDGF by resident cells and platelets [48]. Once in the wound, fibroblasts may become activated to differentiate into myofibroblasts, expressing α-SMA to physically contract the wound [59]. Interestingly, pericytes are also able to produce collagen [44,45]. The pericytes in these studies appear to remain as collagen secreting cells and don't express αSMA, suggesting that they do not convert to myofibroblasts unlike the resident fibroblasts within the wound. In an interesting study, Dulauroy et al. were able to use a Cre-transgenic mouse to label ADAM12, which is induced only during embryogenesis and fibrosis. They showed that the majority of collagen producing cells were positive for αSMA and thus were myofibroblasts. These perivascular cells were also shown to be positive for PDGFRβ and NG2, and were presumed to be pericytes [60]. In other studies, pericytes have been shown to differentiate into myofibroblasts to promote fibrosis, particularly in the kidneys where the pericytes present are called mesangial cells [46]. Interestingly, deletion of pericytes does not alter the recruitment of myofibroblasts or alter kidney fibrosis, which suggests that resident MSCs may also play a role in promoting fibrosis, and lends credence to the theory that pericytes are derived from MSC populations rather than the reverse [61]. Birbrair et al. suggest that pericytes could be split into two subsets dependent on their expression of Nestin (type-1: Nestin$^-$NG2$^+$ and type-2: Nestin$^+$NG2$^+$). They find that type-1 pericytes accumulate near sites of fibrosis but are not solely responsible for the resultant fibrosis, whereas type-2 pericytes appear to play a role in angiogenesis [62,63]. Pericytes have also been show to play a significant role in fibrosis in the liver as hepatic stellate cells. Mederacke et al. use a Cre-transgenic mouse that marks all stellate cells to show that 82–96% of myofibroblasts in a model of toxic, cholestatic and fatty liver disease are of stellate origin [64]. These studies illustrate that pericytes have a major role in important matrix deposition, but under negative circumstances may promote fibrosis.

Clearly, pericytes can influence each phase of the wound healing process (Table 1), and as such should be considered a major cell type that can regulate healing. With increasing evidence that pericytes can promote fibrosis, these cells may not only be a potential target for therapies to accelerate healing but also to prevent fibrosis. Many of the beneficial aspects of pericytes are due to their plasticity and ability to act in a stem cell-like manner to regulate the microenvironment, resulting in improved healing.

3. Pericytes in Other Pathologies

Pericytes mediate both angiogenesis and vessel permeability, consequently they are important in the development of solid tumours, which rely on sufficient vascularization and therefore blood supply to grow.

Pericyte stabilization of the vessel wall supports vascularization within a tumour and can prevent the passing of cancer cell-targeting drugs such as chemotherapeutic agents from the blood stream to the tumour tissue [65]. Consequently, there has been some anti-angiogenic targeting of pericytes within tumours, with a view to destabilizing the vessels that feed tumours and increasing the permeability of cancer drugs into the tumour. Under normal circumstances, however, pericyte signaling represents a fine balance between pro- and anti-angiogenic activities, as pericyte presence not only stabilizes the function of preexisting vessels but also prevents the aberrant proliferation of ECs to form new vasculature. As such, insufficient pericyte coverage in tumours can also be detrimental, resulting in excessive vascular sprouting and increased vascularization of tumours. This suggests that pro-angiogenic targeting of pericytes in tumours may also be beneficial. Additionally, pericyte-EC cross-talk and the resultant regulation of ECs in tumours limits the metastasis of cancer cells, and depleted pericyte coverage of vessels in PDGFβ-deficient mice leads to increased metastasis of solid tumours [66]. Research into the targeting of pericytes in cancer aims to balance pro- and anti-angiogenic signaling to achieve 'vascular normalisation' within the tumour microenvironment. Interestingly the neuron-glial antigen 2 (NG2) proteoglycan, which is often used as a pericyte marker in conjunction with the expression of other proteins, appears to play an important role in pericyte biology within the context of tumour angiogenesis. Altered vascularity, including vessel leakiness and decreased pericyte coverage, is noted in models of brain and breast cancer in NG2 null mice [67,68], and this is thought to be the result of decreased pericyte-EC crosstalk. NG2 knockout in a mouse brain melanoma model also appears to decrease tumour blood supply and increase hypoxia, hinting at a possible therapeutic pathway for the treatment of this disease [69,70]. The result of pericyte perturbation in the context of tumour growth however is complex and multifaceted, directly reflecting the delicate nature of normal angiogenic control, and as such the development of pericyte-targeted therapeutics for cancer is difficult.

Pericyte numbers decline in the dermal and muscle capillary networks of diabetic patients, where they also exhibit an altered morphology with hypertrophy, abnormal cytoplasmic branching and gaps in the basement membrane [71]. These pericytes also appear to promote a fibrotic or sclerotic vessel [72,73]. A study which investigated pericyte changes in patients with chronic venous insufficiency found that 31 out of 42 patients displayed an altered pericyte phenotype [74]. Significantly diminished pericyte coverage is also observed on blood vessels in the retinas of diabetic patients experiencing retinopathy, and in fact this is identified as one of the main mechanisms of disease progression [75]. Hyperglycemia in these patients has been shown both in vivo and in vitro to cause pericyte apoptosis leading to increased production of acellular capillaries in the retina [76]. Mechanistic studies identify activation of NF-κB, PKCδ and SHP-1 as effectors in this outcome [76,77].

Given that one of the most common diabetic pathologies is angiogenic dysregulation, pericyte dysfunction is not a surprising observation in diabetic patients and models, and indicates the normalisation of pericyte number and function as a promising therapeutic approach for the treatment of diabetic complications.

In humans, a two-fold increase in the pericyte number observed on pulmonary arteries is noted in the lungs of patients with pulmonary arterial hypertension (PAH) when compared with the vessels of healthy control samples. Ricard et al. [78] show that these finding are recapitulated in vivo in models of PAH, and show that these additional pericytes serve as a source of smooth muscle-like cells leading to endothelial dysfunction and excessive remodelling of the pulmonary vasculature which is associated with PAH.

Table 2. Pathologies exhibiting pericyte perturbation and likely outcomes of altered pericyte number or function.

Disorder	Pericyte Aberrance Observed	Pericyte Functions Likely to Impact Disease
Diabetic chronic healing	Decreased pericyte numbers in dermis, pericytes exhibit altered morphology [72–75]	Angiogenesis-decreased vascularisation
		Vessel permeability-leaky vessels lead to prolonged and uncontrolled inflammation
		Fibrosis-pericytes promote fibrotic vessels
		Stem cell properties-replacement of lost cell/tissue types
Diabetic retinopathy	Decreased pericyte numbers, increased pericyte apoptosis [76]	Angiogenesis-decreased control of endothelial proliferation
		Vessel permeability-leakiness of vessels
Solid tumour	Unknown, however control of angiogenesis has long been recognised as an important target in treatment of solid tumours	Angiogenesis-tumour relies on new vasculature for blood supply
		Endothelial control-metastasis of cancer
		Vessel permeability-ability of chemotherapeutic agents to pass from bloodstream to tumour tissue
Pulmonary arterial hypertension (PAH)	Increased pericyte coverage on pulmonary arteries [78]	Angiogenesis-excessive remodelling of pulmonary vasculature and endothelial dysfunction
Alzheimers (AD)	Degeneration at blood brain barrier (BBB) [79]	Angiogenesis-break down of vessels causes decreased cereberal bloodflow leading to neurodegeneration
		Vessel permeability-accumulation of damaging molecules in the brain
Chronic kidney disease (CKD)	Differentiation of pericytes into myofibroblasts [80]	Fibrosis-pericytes thought to be source of myofibroblasts contributing to excessive fibrotic activity
		Angiogenesis-differentiation of pericytes into myofibroblasts leaves less pericytes to stabilise vasculature

Degeneration of pericytes is also observed at the blood brain barrier (BBB) in patients with Alzheimers disease (AD) [79]. This leads to neurodegeneration, caused by vascular breakdown and decreased cerebral blood flow. It is also suggested that pericyte loss further contributes to neurodegeneration by allowing increased permeability of blood vessels, leading to the buildup of damaging molecules such as plasmin, fibrin and thrombin in the brain [81]. Here, in yet another model of disease, the importance of vascular stability and permeabilisation and how these parameters are regulated by pericytes is illustrated once again.

Pericytes have also been implicated in fibrosis of the kidney. Mouse models of kidney fibrosis indicate that collagen-producing myofibroblasts appear to originate from a perivascular location [83], and genetic tagging of pericytes illustrates these cells can gain αSMA expression and differentiate into myofibroblasts [80]. It is suggested that the activation of pericytes in the kidney to differentiate into myofibroblasts not only leads to fibrosis but also leaves the endothelium of capillaries unstable, leading to the decreased renal vascularization which is documented in chronic kidney disease [83]. With investigations into pericyte action, uncovering the greater role that pericytes play in many divergent pathologies (Table 2) increases the possibilities for future therapeutic treatments targeting pericytes.

4. Pericytes as a Therapeutic Agent

With an ever-growing understanding of pericyte potential and function comes new possibilities in the form of pericyte-based therapies. In comparison to the significant momentum that MSC application has gained in recent years, the therapeutic potential of pericytes is under-investigated, however early preclinical studies in mice indicate that pericytes can contribute positively to healing in several different tissues.

The myogenic potential of in vitro pericytes is recapitulated by in vivo studies of pericyte function in mouse models of muscle damage. Human pericytes isolated by expression of CD146, NG2 and PDGFRβ from both skeletal muscle and nonmuscular tissues produce myofibres and contribute to muscle regeneration when injected into a cardiotoxin-induced mouse model of muscle damage [84]. Similarly, cells positive for pericyte markers CD146, NG2 and PDGFRβ isolated from the culture of human pluripotent stem cells (hPSCs) promote muscle regeneration and increased vascularization when applied to a mouse model of limb ischemia. At 21 days, these transplanted cells are found incorporated into both the vasculature and the muscle tissue of the recovering limb [85]. In another study, human placental cells isolated with a CD146$^+$CD34$^-$CD45$^-$CD56$^-$ expression profile are shown to produce myofibres and promote increased angiogenesis in the muscles of SCID/*mdx* mice [86]. With their myogenic abilities confirmed in vivo, the development of a pericyte-based therapy holds great potential to enhance the healing of muscles.

Chen et al. have also investigated the therapeutic potential of CD146$^+$CD34$^-$CD45$^-$CD56$^-$ pericytes on the ischemic heart. In their study, pericyte application to a mouse model of myocardial infarction (MI) resulted in improved cardiac recovery and contractility, as well as decreased fibrosis and decreased infiltration of inflammatory cell types [11]. The authors report superior recovery following pericyte transplantation when compared to transplantation of CD56$^+$ myogenic progenitor cells, which suggests that the positive outcome was not solely due to the myogenic capabilities of pericytes but rather a cumulative effect of pericyte function which may also include regulation of angiogenesis and inflammation. This study also suggests an anti-fibrotic function for pericytes, in contrast to other studies which identify pericytes as significant contributors to fibrosis by way of differentiation into myofibroblasts [87–89]. It is possible that in this model the tissue signalling environment did not induce pericyte differentiation into myofibroblasts, and this highlights the fact that pericyte function and heterogeneity are heavily influenced not only by pericyte origin but also by the surrounding tissue environment.

Pericytes (CD146$^+$NG2$^+$CD45$^-$) isolated from mouse fat tissue display osteogenic potential in vitro which is mirrored in vivo by contributing to the regeneration of mouse bone injury when applied in a seeded scaffold [26]. Similarly, human adipose derived pericytes (CD146$^+$CD34$^-$CD45$^-$CD31$^-$) enhance bone healing and encourage bone-union in mouse bone fractures with equal efficacy to BM-MSCs [90]. Given this, the authors suggest that adipose derived pericytes present a more preferable option for transplantation than BM-MSCs as they can be isolated abundantly from fat tissue such as that resulting from liposuction and represent a more defined and homogenous population that BM-MSCs.

In the diabetic retina, there is a loss of pericytes which leads to collapse of the vasculature and ultimately blindness. Human cells derived from adipose stem cells and expressing αSMA, PDGFRβ and NG2 can protect against diabetic retinopathy in a mouse model, causing revascularisation of the retina after injection which was not achieved by injection of human BM-MSCs [91]. This effect of pericyte injection was enhanced when the cells were pre-treated with TGF-β1. These results indicate that replacement of lost endogenous pericytes in a diabetic setting can encourage angiogenesis and vascular support, as is seen in acute models of injury, and are particularly promising when considering the possibility that pericyte therapies may hold for treating diabetic pathologies associated with pericyte loss.

Human umbilical cord perivascular cells (HUCPV) isolated by expression pattern $CD45^-CD34^-SH2^+SH3^+Thy-1^+CD44^+$ and also expressing the pericyte marker 3G5 cause enhanced healing at 3 and 7 days when applied in a fibrin gel to full thickness skin defects in Balb/C mice, as assessed by dermal thickness and re-epithelialisation [92]. In another study, the application of human adipose derived stem cells expressing pericyte markers αSMA, PDGFRβ, NG2 and Ang1 in a PEG-fibrin gel to a rat model of excisional wounding resulted in earlier collagen deposition and remodelling, as well as increased angiogenesis [93]. Contrastingly, the application of human neonatal foreskin dermal pericytes ($CD45^-VLA-1/\alpha1/CD49a^{bright}$) to a mouse model of excisional wound healing did not enhance re-epithelialisation and in fact resulted in a decrease in dermal wound closure at day 7 when compared to wounds treated with $CD45^-VLA-1^{dim}$ fibroblasts [94]. The authors suggest that as this model assessed the capability of applied human pericytes to influence the migration of endogenous keratinocytes in wounded mouse skin, signalling between the human and murine cells in question may not have been successful. In addition, this study applied cells isolated based on a different expression pattern when compared to other studies investigating the application of pericytes to wounded skin, and serves to further illustrate the heterogeneity of pericytes and how important standardisation of identification and isolation will be before these cells can be considered a realistic source for the development of cell therapies.

Overall, early studies of pericyte application to mouse models of muscle, heart, bone and skin injury show promising signs that these regenerative and plastic cells can positively contribute to healing, but also raise questions as to how the origin of cell isolation and method of delivery can affect the ability of these cells to carry out beneficial functions. Little is known about the effect of pericyte delivery to chronic tissues, including chronic wounds, but the demonstration in other tissues that applying pericytes can encourage enhanced angiogenesis and decreased inflammatory infiltration as well as regenerating lost tissue is promising when considering the treatment of non-healing wounds. The fibrotic activity of the pericyte and how this is influenced by tissue-specific environments remains incompletely understood, and this is an area which would require significant investigation before the use of pericytes as a clinical cell therapy for wound healing could be considered.

5. Conclusions

The functions and capabilities of pericytes are impressive and, as yet, incompletely understood. These cells regulate the vasculature and the inflammatory response, and in addition possess MSC-like regenerative qualities. As such, the pericyte is well placed to significantly influence healing outcomes. A decrease in pericytes associated with the vasculature is well documented in the retinas of diabetic patients, and this results in the onset of diabetic retinopathy. Loss of pericytes is also documented in other disease states, and aberrant pericyte function is identified as an important target in the development of cancer therapies. With each observation of pericyte function or dysfunction in the context of new disease environments, the body of knowledge illustrating the importance of pericytes in the regulation of homeostatic and healing processes grows. There has been a lot of interest in the idea of MSC application as a wound therapy, and it is possible that pericytes, which possess both MSC-like behaviours and distinct regulatory roles in angiogenesis and inflammation, may represent another promising cell population for the development of treatments. In fact, recent in vivo studies show that the transplantation of isolated pericytes can positively influence the healing of bone, muscle and skin and can support revascularisation in a mouse model of diabetic retinopathy. It seems that pericytes have an important part to play in chronic and acute healing processes, and must be considered a crucial cell type as we continue to work towards a comprehensive understanding of healing processes to better advise the development of effective therapies.

Acknowledgments: The authors would like to acknowledge the support of the Cooperative Research Centre for Cell Therapy Manufacturing and the Australian Government's Cooperative Research Centres Program.

Conflicts of Interest: The authors declare no conflict of interest.

References

1. Sims, D.E. The pericyte—A review. *Tissue Cell* **1986**, *18*, 153–174. [CrossRef]
2. Mills, S.J.; Cowin, A.J.; Kaur, P. Pericytes, mesenchymal stem cells and the wound healing process. *Cells* **2013**, *2*, 621–634. [CrossRef] [PubMed]
3. Rucker, H.K.; Wynder, H.J.; Thomas, W.E. Cellular mechanisms of cns pericytes. *Brain Res. Bull.* **2000**, *51*, 363–369. [CrossRef]
4. Stratman, A.N.; Malotte, K.M.; Mahan, R.D.; Davis, M.J.; Davis, G.E. Pericyte recruitment during vasculogenic tube assembly stimulates endothelial basement membrane matrix formation. *Blood* **2009**, *114*, 5091–5101. [CrossRef] [PubMed]
5. Tilton, R.G.; Kilo, C.; Williamson, J.R. Pericyte-Endothelial relationships in cardiac and skeletal-muscle capillaries. *Microvasc. Res.* **1979**, *18*, 325–335 [CrossRef]
6. Armulik, A.; Genové, G.; Betsholtz, C. Pericytes: Developmental, physiological, and pathological perspectives, problems, and promises. *Dev. Cell* **2011**, *21*, 193–215. [CrossRef] [PubMed]
7. Geevarghese, A.; Herman, I.M. Pericyte-endothelial cross-talk: Implications and opportunities for advanced cellular therapies. *Transl. Res.* **2014**, *163*, 296–306. [CrossRef] [PubMed]
8. Gaengel, K.; Genové, G.; Armulik, A.; Betsholtz, C. Endothelial-Mural cell signaling in vascular development and angiogenesis. *Arterioscler. Thromb. Vasc. Biol.* **2009**, *29*, 630. [CrossRef] [PubMed]
9. Proebstl, D.; Voisin, M.-B.; Woodfin, A.; Whiteford, J.; D'Acquisto, F.; Jones, G.E.; Rowe, D.; Nourshargh, S. Pericytes support neutrophil subendothelial cell crawling and breaching of venular walls in vivo. *J. Exp. Med.* **2012**, *209*, 1219–1234. [CrossRef] [PubMed]
10. Wang, S.; Cao, C.; Chen, Z.; Bankaitis, V.; Tzima, E.; Sheibani, N.; Burridge, K. Pericytes regulate vascular basement membrane remodeling and govern neutrophil extravasation during inflammation. *PLoS ONE* **2012**, *7*, e45499. [CrossRef] [PubMed]
11. Chen, C.-W.; Okada, M.; Proto, J.D.; Gao, X.; Sekiya, N.; Beckman, S.A.; Corselli, M.; Crisan, M.; Saparov, A.; Tobita, K.; et al. Human pericytes for ischemic heart repair. *Stem Cells* **2013**, *31*, 305–316. [CrossRef] [PubMed]
12. Wilm, B.; Ipenberg, A.; Hastie, N.D.; Burch, J.B.E.; Bader, D.M. The serosal mesothelium is a major source of smooth muscle cells of the gut vasculature. *Development* **2005**, *132*, 5317. [CrossRef] [PubMed]
13. Que, J.; Wilm, B.; Hasegawa, H.; Wang, F.; Bader, D.; Hogan, B.L.M. Mesothelium contributes to vascular smooth muscle and mesenchyme during lung development. *Proc. Natl. Acad. Sci. USA* **2008**, *105*, 16626–16630. [CrossRef] [PubMed]
14. Asahina, K.; Zhou, B.; Pu, W.T.; Tsukamoto, H. Septum transversum-derived mesothelium gives rise to hepatic stellate cells and perivascular mesenchymal cells in developing mouse liver. *Hepatology* **2011**, *53*, 983–995. [CrossRef] [PubMed]
15. Cai, C.L.; Martin, J.C.; Sun, Y.; Cui, L.; Wang, L.; Ouyang, K.; Yang, L.; Bu, L.; Liang, X.; Zhang, X.; et al A myocardial lineage derives from tbx18 epicardial cells. *Nature* **2008**, *454*, 104–108. [CrossRef] [PubMed]
16. Wessels, A.; Pérez-Pomares, J.M. The epicardium and epicardially derived cells (EPDCs) as cardiac stem cells. *Anat. Rec. A Discov. Mol. Cell Evol. Biol.* **2004**, *276A*, 43–57. [CrossRef] [PubMed]
17. Etchevers, H.C.; Vincent, C.; Le Douarin, N.M.; Couly, G.F. The cephalic neural crest provides pericytes and smooth muscle cells to all blood vessels of the face and forebrain. *Development* **2001**, *128*, 1059. [PubMed]

18. Díaz-Flores, L.; Gutiérrez, R.; Madrid, J.F.; Varela, H.; Valladares, F.; Acosta, E.; Martín-Vasallo, P.; Díaz-Flores, L. Pericytes. Morphofunction, interactions and pathology in a quiescent and activated mesenchymal cell niche. *Histol. Histopathol.* **2009**, *24*, 909–969. [PubMed]

19. Bergers, G.; Song, S. The role of pericytes in blood-vessel formation and maintenance. *Neuro Oncol.* **2005**, *7*, 452–464. [CrossRef] [PubMed]

20. Van Dijk, C.G.M.; Nieuweboer, F.E.; Pei, J.Y.; Xu, Y.J.; Burgisser, P.; van Mulligen, E.; el Azzouzi, H.; Duncker, D.J.; Verhaar, M.C.; Cheng, C. The complex mural cell: Pericyte function in health and disease. *Int. J. Cardiol.* **2015**, *190*, 75–89. [CrossRef] [PubMed]

21. Wong, S.-P.; Rowley, J.E.; Redpath, A.N.; Tilman, J.D.; Fellous, T.G.; Johnson, J.R. Pericytes, mesenchymal stem cells and their contributions to tissue repair. *Pharmacol. Ther.* **2015**, *151*, 107–120. [CrossRef] [PubMed]

22. Ansell, D.M.; Izeta, A. Pericytes in wound healing: Friend or foe? *Exp. Dermatol.* **2015**, *24*, 833–834. [CrossRef] [PubMed]

23. Corselli, M.; Chen, C.-W.; Crisan, M.; Lazzari, L.; Péault, B. Perivascular ancestors of adult multipotent stem cells. *Arterioscler. Thromb. Vasc. Biol.* **2010**, *30*, 1104. [CrossRef] [PubMed]

24. V Nehls, D.D. Heterogeneity of microvascular pericytes for smooth muscle type α-actin. *J. Cell Biol.* **1991**, *113*, 147–154. [CrossRef]

25. Chen, W.C.W.; Baily, J.E.; Corselli, M.; Diaz, M.; Sun, B.; Xiang, G.; Gray, G.A.; Huard, J.; Péault, B. Human myocardial pericytes: Multipotent mesodermal precursors exhibiting cardiac specificity. *Stem Cells* **2015**, *33*, 557–573. [CrossRef] [PubMed]

26. König, M.A.; Canepa, D.D.; Cadosch, D.; Casanova, E.; Heinzelmann, M.; Rittirsch, D.; Plecko, M.; Hemmi, S.; Simmen, H.-P.; Cinelli, P.; et al. Direct transplantation of native pericytes from adipose tissue: A new perspective to stimulate healing in critical size bone defects. *Cytotherapy* **2016**, *18*, 41–52.

27. Herrmann, M.; Bara, J.J.; Sprecher, C.M.; Menzel, U.; Jalowiec, J.M.; Osinga, R.; Scherberich, A.; Alini, M.; Verrier, S. Pericyte plasticity—Comparative investigation of the angiogenic and multilineage potential of pericytes from different human tissues. *Cells Mater.* **2016**, *31*, 236–241.

28. Caplan, A.I. All mscs are pericytes? *Cell Stem Cell* **2008**, *3*, 229–230. [CrossRef] [PubMed]

29. Blocki, A.; Wang, Y.; Koch, M.; Peh, P.; Beyer, S.; Law, P.; Hui, J.; Raghunath, M. Not all MSCs can act as pericytes: Functional in vitro assays to distinguish pericytes from other mesenchymal stem cells in angiogenesis. *Stem Cells Dev.* **2013**, *22*, 2347–2355. [CrossRef] [PubMed]

30. Da Silva Meirelles, L.; Caplan, A.I.; Nardi, N.B. In search of the in vivo identity of mesenchymal stem cells. *Stem Cells* **2008**, *26*, 2287–2299. [CrossRef] [PubMed]

31. Davis, G.E.; Saunders, W.B. Molecular balance of capillary tube formation versus regression in wound repair: Role of matrix metalloproteinases and their inhibitors. *J. Investig. Dermatol. Symp. Proc.* **2006**, *11*, 44–56. [CrossRef] [PubMed]

32. Antonelli-Orlidge, A.; Saunders, K.B.; Smith, S.R.; D'Amore, P.A. An activated form of transforming growth factor β is produced by cocultures of endothelial cells and pericytes. *Proc. Natl. Acad. Sci. USA* **1989**, *86*, 4544–4548. [CrossRef] [PubMed]

33. Kutcher, M.E.; Herman, I.M. The pericyte: Cellular regulator of microvascular blood flow. *Microvasc. Res.* **2009**, *77*, 235–246. [CrossRef] [PubMed]

34. Miller, F.N.; Sims, D.E.; Schuschke, D.A.; Abney, D.L. Differentiation of light-dye effects in the microcirculation. *Microvasc. Res.* **1992**, *44*, 166–184. [CrossRef]

35. Sims, D.E.; Miller, F.N.; Horne, M.M.; Edwards, M.J. Interleukin-2 alters the positions of capillary and venule pericytes in rat cremaster muscle. *J Submicrosc. Cytol. Pathol.* **1994**, *26*, 507–513. [PubMed]

36. Neng, L.; Zhang, F.; Kachelmeier, A.; Shi, X. Endothelial cell, pericyte, and perivascular resident macrophage-type melanocyte interactions regulate cochlear intrastrial fluid–blood barrier permeability. *J. Assoc. Res. Otolaryngol.* **2013**, *14*, 175–185. [CrossRef] [PubMed]

37. Ogura, S.; Kurata, K.; Hattori, Y.; Takase, H.; Ishiguro-Oonuma, T.; Hwang, Y.; Ahn, S.; Park, I.; Ikeda, W.; Kusuhara, S.; et al. Sustained inflammation after pericyte depletion induces irreversible blood-retina barrier breakdown. *JCI Insight* **2017**, *2*, e90905. [CrossRef] [PubMed]

38. Stark, K.; Eckart, A.; Haidari, S.; Tirniceriu, A.; Lorenz, M.; von Bruhl, M.-L.; Gartner, F.; Khandoga, A.G.; Legate, K.R.; Pless, R.; et al. Capillary and arteriolar pericytes attract innate leukocytes exiting through venules and 'instruct' them with pattern-recognition and motility programs. *Nat. Immunol.* **2013**, *14*, 41–51. [CrossRef] [PubMed]

39. Balabanov, R.; Dore-Duffy, P. Role of the cns microvascular pericyte in the blood-brain barrier. *J. Neurosci. Res.* **1998**, *53*, 637–644. [CrossRef]

40. Tu, Z.; Li, Y.; Smith, D.S.; Sheibani, N.; Huang, S.; Kern, T.; Lin, F. Retinal pericytes inhibit activated T cell proliferation. *Invest. Ophthalmol. Vis. Sci.* **2011**, *52*, 9005–9010. [CrossRef] [PubMed]

41. Jansson, D.; Rustenhoven, J.; Feng, S.; Hurley, D.; Oldfield, R.L.; Bergin, P.S.; Mee, E.W.; Faull, R.L.; Dragunow, M. A role for human brain pericytes in neuroinflammation. *J. Neuroinflam.* **2014**, *11*, 104. [CrossRef] [PubMed]

42. Hung, C.F.; Mittelsteadt, K.L.; Brauer, R.; McKinney, B.L.; Hallstrand, T.S.; Parks, W.C.; Chen, P.; Schnapp, L.M.; Liles, W.C.; Duffield, J.S.; et al. Lung pericyte-like cells are functional interstitial immune sentinel cells. *Am. J. Physiol. Lung Cell. Mol. Physiol.* **2017**, *312*, L556. [CrossRef] [PubMed]

43. Paquet-Fifield, S.; Schlüter, H.; Li, A.; Aitken, T.; Gangatirkar, P.; Blashki, D.; Koelmeyer, R.; Pouliot, N.; Palatsides, M.; Ellis, S.; et al. A role for pericytes as microenvironmental regulators of human skin tissue regeneration. *J. Clin. Investig.* **2009**, *119*, 2795–2806. [CrossRef] [PubMed]

44. Popescu, F.C.; Busuioc, C.J.; Mogosanu, G.D.; Pop, O.T.; Parvanescu, H.; Lascar, I.; Nicolae, C.I.; Mogoanta, L. Pericytes and myofibroblasts reaction in experimental thermal third degree skin burns. *Rom. J. Morphol. Embryol.* **2011**, *52*, 1011–1017. [PubMed]

45. Sundberg, C.; Ljungstrom, M.; Lindmark, G.; Gerdin, B.; Rubin, K. Microvascular pericytes express platelet-derived growth factor-beta receptors in human healing wounds and colorectal adenocarcinoma. *Am. J. Pathol.* **1993**, *143*, 1377–1388. [PubMed]

46. Lin, S.L.; Kisseleva, T.; Brenner, D.A.; Duffield, J.S. Pericytes and perivascular fibroblasts are the primary source of collagen-producing cells in obstructive fibrosis of the kidney. *Am. J. Pathol.* **2008**, *173*, 1617–1627. [CrossRef] [PubMed]

47. Clark, E.R.; Clark, E.L. The relation of 'rouget' cells to capillary contractility. *Dev. Dyn.* **1925**, *35*, 265–282. [CrossRef]

48. Rajkumar, V.S.; Shiwen, X.; Bostrom, M.; Leoni, P.; Muddle, J.; Ivarsson, M.; Gerdin, B.; Denton, C.P.; Bou-Gharios, G.; Black, C.M.; et al. Platelet-derived growth factor-beta receptor activation is essential for fibroblast and pericyte recruitment during cutaneous wound healing. *Am. J. Pathol.* **2006**, *169*, 2254–2265. [CrossRef] [PubMed]

49. Mohle, R.; Green, D.; Moore, M.A.; Nachman, R.L.; Rafii, S. Constitutive production and thrombin-induced release of vascular endothelial growth factor by human megakaryocytes and platelets. *Proc. Natl. Acad. Sci. USA* **1997**, *94*, 663–668. [CrossRef] [PubMed]

50. Li, J.; Zhang, Y.P.; Kirsner, R.S. Angiogenesis in wound repair: Angiogenic growth factors and the extracellular matrix. *Microsc. Res. Tech.* **2003**, *60*, 107–114. [CrossRef] [PubMed]

51. Tonnesen, M.G.; Feng, X.; Clark, R.A. Angiogenesis in wound healing. *J. Investig. Dermatol. Symp. Proc.* **2000**, *5*, 40–46. [CrossRef] [PubMed]

52. Arroyo, A.G.; Iruela-Arispe, M.L. Extracellular matrix, inflammation, and the angiogenic response. *Cardiovasc. Res.* **2010**, *86*, 226–235. [CrossRef] [PubMed]

53. Hellstrom, M.; Gerhardt, H.; Kalen, M.; Li, X.; Eriksson, U.; Wolburg, H.; Betsholtz, C. Lack of pericytes leads to endothelial hyperplasia and abnormal vascular morphogenesis. *J. Cell Biol.* **2001**, *153*, 543–553. [CrossRef] [PubMed]

54. Maisonpierre, P.C.; Suri, C.; Jones, P.F.; Bartunkova, S.; Wiegand, S.J.; Radziejewski, C.; Compton, D.; McClain, J.; Aldrich, T.H.; Papadopoulos, N.; et al. Angiopoietin-2, a natural antagonist for tie2 that disrupts in vivo angiogenesis. *Science* **1997**, *277*, 55. [CrossRef] [PubMed]

55. Lindblom, P.; Gerhardt, H.; Liebner, S.; Abramsson, A.; Enge, M.; Hellstrom, M.; Backstrom, G.; Fredriksson, S.; Landegren, U.; Nystrom, H.C.; et al. Endothelial PDGF-β retention is required for proper investment of pericytes in the microvessel wall. *Genes Dev.* **2003**, *17*, 1835–1840. [CrossRef] [PubMed]

56. Genove, G.; Mollick, T.; Johansson, K. Photoreceptor degeneration, structural remodeling and glial activation: A morphological study on a genetic mouse model for pericyte deficiency. *Neuroscience* **2014**, *279*, 269–284. [CrossRef] [PubMed]

57. Raines, S.M.; Richards, O.C.; Schneider, L.R.; Schueler, K.L.; Rabaglia, M.E.; Oler, A.T.; Stapleton, D.S.; Genove, G.; Dawson, J.A.; Betsholtz, C.; et al. Loss of PDGF-β activity increases hepatic vascular permeability and enhances insulin sensitivity. *Am. J. Physiol. Endocrinol. Metab.* **2011**, *301*, E517–E526. [CrossRef] [PubMed]

58. Villasenor, R.; Kuennecke, B.; Ozmen, L.; Ammann, M.; Kugler, C.; Gruninger, F.; Loetscher, H.; Freskgard, P.O.; Collin, L. Region-specific permeability of the blood-brain barrier upon pericyte loss. *J. Cereb. Blood Flow. Metab.* **2017**. [CrossRef] [PubMed]

59. Darby, I.; Skalli, O.; Gabbiani, G. Alpha-smooth muscle actin is transiently expressed by myofibroblasts during experimental wound healing. *Lab. Investig.* **1990**, *63*, 21–29. [PubMed]

60. Dulauroy, S.; Di Carlo, S.E.; Langa, F.; Eberl, G.; Peduto, L. Lineage tracing and genetic ablation of adam12+ perivascular cells identify a major source of profibrotic cells during acute tissue injury. *Nat. Med.* **2012**, *18*, 1262–1270. [CrossRef] [PubMed]

61. LeBleu, V.S.; Taduri, G.; O'Connell, J.; Teng, Y.; Cooke, V.G.; Woda, C.; Sugimoto, H.; Kalluri, R. Origin and function of myofibroblasts in kidney fibrosis. *Nat. Med.* **2013**, *19*, 1047–1053. [CrossRef] [PubMed]

62. Birbrair, A.; Zhang, T.; Files, D.C.; Mannava, S.; Smith, T.; Wang, Z.M.; Messi, M.L.; Mintz, A.; Delbono, O. Type-1 pericytes accumulate after tissue injury and produce collagen in an organ-dependent manner. *Stem Cell Res. Ther.* **2014**, *5*, 122. [CrossRef] [PubMed]

63. Birbrair, A.; Zhang, T.; Wang, Z.-M.; Messi, M.L.; Olson, J.D.; Mintz, A.; Delbono, O. Type-2 pericytes participate in normal and tumoral angiogenesis. *Am. J. Physiol. Cell Physiol.* **2014**, *307*, C25–C38. [CrossRef] [PubMed]

64. Mederacke, I.; Hsu, C.C.; Troeger, J.S.; Huebener, P.; Mu, X.; Dapito, D.H.; Pradere, J.P.; Schwabe, R.F. Fate tracing reveals hepatic stellate cells as dominant contributors to liver fibrosis independent of its aetiology. *Nat. Commun.* **2013**, *4*, 2823. [CrossRef] [PubMed]

65. Meng, M.-B.; Zaorsky, N.G.; Deng, L.; Wang, H.-H.; Chao, J.; Zhao, L.-J.; Yuan, Z.-Y.; Ping, W. Pericytes: A double-edged sword in cancer therapy. *Future Oncol.* **2014**, *11*, 169–179. [CrossRef] [PubMed]

66. Xian, X.; Håkansson, J.; Ståhlberg, A.; Lindblom, P.; Betsholtz, C.; Gerhardt, H.; Semb, H. Pericytes limit tumor cell metastasis. *J. Clin. Investig.* **2006**, *116*, 642–651. [CrossRef] [PubMed]

67. Huang, F.-J.; You, W.-K.; Bonaldo, P.; Seyfried, T.N.; Pasquale, E.B.; Stallcup, W.B. Pericyte deficiencies lead to aberrant tumor vascularization in the brain of the ng2 null mouse. *Dev. Biol.* **2010**, *344*, 1035–1046. [CrossRef] [PubMed]

68. Gibby, K.; You, W.-K.; Kadoya, K.; Helgadottir, H.; Young, L.J.T.; Ellies, L.G.; Chang, Y.; Cardiff, R.D.; Stallcup, W.B. Early vascular deficits are correlated with delayed mammary tumorigenesis in the mmtv-pymt transgenic mouse following genetic ablation of the ng2 proteoglycan. *Breast Cancer Res.* **2012**, *14*, R67. [CrossRef] [PubMed]

69. Stallcup, W.B.; You, W.K.; Kucharova, K.; Cejudo-Martin, P.; Yotsumoto, F. Ng2 proteoglycan-dependent contributions of pericytes and macrophages to brain tumor vascularization and progression. *Microcirculation* **2016**, *23*, 122–133. [CrossRef] [PubMed]

70. You, W.K.; Yotsumoto, F.; Sakimura, K.; Adams, R.H.; Stallcup, W.B. Ng2 proteoglycan promotes tumor vascularization via integrin-dependent effects on pericyte function. *Angiogenesis* **2014**, *17*, 61–76. [CrossRef] [PubMed]

71. Braverman, I.M.; Sibley, J.; Keh, A. Ultrastructural analysis of the endothelial-pericyte relationship in diabetic cutaneous vessels. *J. Investig. Dermatol.* **1990**, *95*, 147–153. [CrossRef] [PubMed]

72. Braverman, I.M.; Keh-Yen, A. Ultrastructural abnormalities of the microvasculature and elastic fibers in the skin of juvenile diabetics. *J. Investig. Dermatol.* **1984**, *82*, 270–274. [CrossRef] [PubMed]

73. Shakya, S.; Wang, Y.; Mack, J.A.; Maytin, E.V. Hyperglycemia-induced changes in hyaluronan contribute to impaired skin wound healing in diabetes: Review and perspective. *Int. J. Cell Biol.* **2015**, *2015*, 701738. [CrossRef] [PubMed]

74. Laaff, H.; Vandscheidt, W.; Weiss, J.M.; Schaefer, H.E.; Schoepf, E. Immunohistochemical investigation of pericytes in chronic venous insufficiency. *Vasa* **1991**, *20*, 323–328. [PubMed]

75. Safi, S.Z.; Qvist, R.; Kumar, S.; Batumalaie, K.; Ismail, I.S.B. Molecular mechanisms of diabetic retinopathy, general preventive strategies, and novel therapeutic targets. *BioMed Res. Int.* **2014**, *2014*, 801269. [CrossRef] [PubMed]

76. Geraldes, P.; Hiraoka-Yamamoto, J.; Matsumoto, M.; Clermont, A.; Leitges, M.; Marette, A.; Aiello, L.P.; Kern, T.S.; King, G.L. Activation of pkcδ and shp1 by hyperglycemia causes vascular cell apoptosis and diabetic retinopathy. *Nat. Med.* **2009**, *15*, 1298–1306. [CrossRef] [PubMed]

77. Ejaz, S.; Chekarova, I.; Ejaz, A.; Sohail, A.; Lim, C.W. Importance of pericytes and mechanisms of pericyte loss during diabetic retinopathy. *Diabetes Obes. Metab.* **2008**, *10*, 53–63. [PubMed]

78. Ricard, N.; Tu, L.; Le Hiress, M.; Huertas, A.; Phan, C.; Thuillet, R.; Sattler, C.; Fadel, E.; Seferian, A.; Montani, D.; et al. Increased pericyte coverage mediated by endothelial-derived fibroblast growth factor-2 and interleukin-6 is a source of smooth muscle–like cells in pulmonary hypertension. *Circulation* **2014**, *129*, 1586. [CrossRef] [PubMed]

79. Sengillo, J.D.; Winkler, E.A.; Walker, C.T.; Sullivan, J.S.; Johnson, M.; Zlokovic, B.V. Deficiency in mural vascular cells coincides with blood-brain barrier disruption in alzheimer's disease. *Brain Pathol.* **2013**, *23*, 303–310. [CrossRef] [PubMed]

80. Humphreys, B.D.; Lin, S.-L.; Kobayashi, A.; Hudson, T.E.; Nowlin, B.T.; Bonventre, J.V.; Valerius, M.T.; McMahon, A.P.; Duffield, J.S. Fate tracing reveals the pericyte and not epithelial origin of myofibroblasts in kidney fibrosis. *Am. J. Pathol.* **2010**, *176*, 85–97. [CrossRef] [PubMed]

81. Winkler, E.A.; Sagare, A.P.; Zlokovic, B.V. The pericyte: A forgotten cell type with important implications for Alzheimer's disease? *Brain Pathol.* **2014**, *24*, 371–386. [CrossRef] [PubMed]

82. Faulkner, J.L.; Szczykalski, L.M.; Springer, F.; Barnes, J.L. Origin of interstitial fibroblasts in an accelerated model of angiotensin ii-induced renal fibrosis. *Am. J. Pathol.* **2005**, *167*, 1190–1205. [CrossRef]

83. Kramann, R.; Humphreys, B.D. Kidney pericytes: Roles in regeneration and fibrosis. *Semin. Nephrol.* **2014**, *34*, 374–383. [CrossRef] [PubMed]

84. Crisan, M.; Yap, S.; Casteilla, L.; Chen, C.-W.; Corselli, M.; Park, T.S.; Andriolo, G.; Sun, B.; Zheng, B.; Zhang, L.; et al. A perivascular origin for mesenchymal stem cells in multiple human organs. *Cell Stem Cell* **2008**, *3*, 301–313. [CrossRef] [PubMed]

85. Dar, A.; Domev, H.; Ben-Yosef, O.; Tzukerman, M.; Zeevi-Levin, N.; Novak, A.; Germanguz, I.; Amit, M.; Itskovitz-Eldor, J. Multipotent vasculogenic pericytes from human pluripotent stem cells promote recovery of murine ischemic limb. *Circulation* **2012**, *125*, 87. [CrossRef] [PubMed]

86. Park, T.S.; Gavina, M.; Chen, C.-W.; Sun, B.; Teng, P.-N.; Huard, J.; Deasy, B.M.; Zimmerlin, L.; Péault, B. Placental perivascular cells for human muscle regeneration. *Stem Cells Dev.* **2011**, *20*, 451–463. [CrossRef] [PubMed]

87. Birbrair, A.; Zhang, T.; Wang, Z.-M.; Messi, M.L.; Mintz, A.; Delbono, O. Type-1 pericytes participate in fibrous tissue deposition in aged skeletal muscle. *Am. J. Pathol.* **2013**, *305*, C1098–C1113. [CrossRef] [PubMed]

88. Greenhalgh, S.N.; Conroy, K.P.; Henderson, N.C. Healing scars: Targeting pericytes to treat fibrosis. *Int. J. Med.* **2015**, *108*, 3–7. [CrossRef] [PubMed]

89. Kida, Y.; Duffield, J.S. Frontiers in research: Chronic kidney diseases: The pivotal role of pericytes in kidney fibrosis. *Clin. Exp. Pharmacol. Physiol.* **2011**, *38*, 417–423. [CrossRef] [PubMed]

90. Tawonsawatruk, T.; West, C.C.; Murray, I.R.; Soo, C.; Péault, B.; Simpson, A.H.R.W. Adipose derived pericytes rescue fractures from a failure of healing—Non-union. *Sci. Rep.* **2016**, *6*, 22779. [CrossRef] [PubMed]

91. Mendel, T.A.; Clabough, E.B.D.; Kao, D.S.; Demidova-Rice, T.N.; Durham, J.T.; Zotter, B.C.; Seaman, S.A.; Cronk, S.M.; Rakoczy, E.P.; Katz, A.J.; et al. Pericytes derived from adipose-derived stem cells protect against retinal vasculopathy. *PLoS ONE* **2013**, *8*, e65691. [CrossRef]

92. Zebardast, N.; Lickorish, D.; Davies, J.E. Human umbilical cord perivascular cells (HUCPVC): A mesenchymal cell source for dermal wound healing. *Organogenesis* **2010**, *6*, 197–203. [CrossRef] [PubMed]

93. Zamora, D.O.; Natesan, S.; Becerra, S.; Wrice, N.; Chung, E.; Suggs, L.J.; Christy, R.J. Enhanced wound vascularization using a dsASCs seeded FPEG scaffold. *Angiogenesis* **2013**, *16*, 745–757. [CrossRef] [PubMed]

94. Mills, S.J.; Zhuang, L.; Arandjelovic, P.; Cowin, A.J.; Kaur, P. Effects of human pericytes in a murine excision model of wound healing. *Exp. Dermatol.* **2015**, *24*, 881–882. [CrossRef] [PubMed]

© 2017 by the authors. Licensee MDPI, Basel, Switzerland. This article is an open access article distributed under the terms and conditions of the Creative Commons Attribution (CC BY) license (http://creativecommons.org/licenses/by/4.0/).

International Journal of
Molecular Sciences

MDPI

Review

Wound-Healing Studies in Cornea and Skin: Parallels, Differences and Opportunities

Anne Bukowiecki [1,2], Deniz Hos [1,3], Claus Cursiefen [1,3] and Sabine A. Eming [2,3,4,*]

1   Department of Ophthalmology, University Hospital of Cologne, 50937 Cologne, Germany;
     anne.bukowiecki@uk-koeln.de (A.B.); deniz.hos@uk-koeln.de (D.H.); claus.cursiefen@uk-koeln.de (C.C.)
2   Department of Dermatology, University of Cologne, Kerpener Strasse 62, 50937 Cologne, Germany
3   Center for Molecular Medicine Cologne (CMMC), University of Cologne, 50931 Cologne, Germany
4   Excellence Cluster: Cellular Stress Responses in Aging-associated Diseases, CECAD, University of Cologne,
     50931 Cologne, Germany
*   Correspondence: sabine.eming@uni-koeln.de; Tel.: +49-221-478 3196, Fax: +49-221-478-5949

Academic Editor: Allison Cowin
Received: 13 April 2017; Accepted: 31 May 2017; Published: 12 June 2017

Abstract: The cornea and the skin are both organs that provide the outer barrier of the body. Both tissues have developed intrinsic mechanisms that protect the organism from a wide range of external threats, but at the same time also enable rapid restoration of tissue integrity and organ-specific function. The easy accessibility makes the skin an attractive model system to study tissue damage and repair. Findings from skin research have contributed to unravelling novel fundamental principles in regenerative biology and the repair of other epithelial-mesenchymal tissues, such as the cornea. Following barrier disruption, the influx of inflammatory cells, myofibroblast differentiation, extracellular matrix synthesis and scar formation present parallel repair mechanisms in cornea and skin wound healing. Yet, capillary sprouting, while pivotal in proper skin wound healing, is a process that is rather associated with pathological repair of the cornea. Understanding the parallels and differences of the cellular and molecular networks that coordinate the wound healing response in skin and cornea are likely of mutual importance for both organs with regard to the development of regenerative therapies and understanding of the disease pathologies that affect epithelial-mesenchymal interactions. Here, we review the principal events in corneal wound healing and the mechanisms to restore corneal transparency and barrier function. We also refer to skin repair mechanisms and their potential implications for regenerative processes in the cornea.

Keywords: cornea; skin; wound healing; inflammation; regeneration; repair

1. Introduction

The cornea is the most anterior part of the eye and accounts for about 70% of its refractive power. The transparent nature and defined curvature of the cornea assure that light is focused and transmitted without scatter through the lens and onto the retina. Therefore, the tissue integrity of the cornea is of particular importance for clear vision. Eye injuries such as physical or chemical trauma or severe infections may result in permanent corneal damage leading to opacification and loss of visual acuity. Thus, rapid restoration of corneal integrity after injury is key to prevent intraocular inflammation that can cause permanent loss of vision or even loss of the eye itself.

Like in skin wound healing, inflammation is a fundamental process in corneal wound healing [1]. While a prolonged inflammatory response may exacerbate tissue damage, therapeutic suppression of the inflammatory response (e.g., by glucocorticosteroids) may also impair healing and might lead to a delay in epithelial wound closure [2,3] or promote infections [4,5]. This review will focus on the corneal immune system as an important factor in corneal wound healing and regeneration. We will first give an overview of corneal (microscopic) anatomy and physiology. Second, we will focus on corneal angiogenic and immune privileges, which are essential for corneal function and homeostasis. Finally, we will provide deeper insight into the cellular events in corneal inflammation, their consequences for non-immune cell function and the outcome of the healing response.

2. Corneal Anatomy

The cornea consists of five layers, with different regenerative capacities: (1) a stratified non-keratinizing squamous epithelial layer, which together with the tear film forms the outermost barrier of the eye; (2) the Bowman layer, an acellular collagenous layer beneath the epithelial basement membrane; (3) a collagen-rich stromal layer, which accounts for 80–90% of the corneal thickness in humans and 60–70% in mice; (4) Descemet's membrane, which forms the basement membrane for the (5) endothelial cell layer on the posterior side of the cornea. The endothelial cell layer is a monolayer of polygonal cells that is in contact with the aqueous humor (Figure 1).

The corneal epithelium is developmentally derived from surface ectoderm and consists of stratified squamous epithelial cells. The epithelial surface is protected against damage and pathogens by the tear film, which also contains growth factors like epidermal growth factor (EGF) to promote epithelial regeneration [6]. Consisting of a lipid top layer to prevent evaporation, an aqueous middle layer and a mucin layer, the tear film protects the epithelial cells, washes away foreign particles and creates a smooth surface for clear vision. The tear film is uniformly spread across the cornea through the interaction of the mucin layer with the glycocalyx of the surface epithelial cells. Tight junctions connect the superficial cells to create an impenetrable barrier [7]. Beneath the superficial cell layer, the corneal epithelium consists of 2–3 layers of suprabasal wing-shaped cells, which rest on a single layer of basal cells attached to a basement membrane via hemidesmosomes. The basal cells are capable of proliferation and constantly replenish wing cells and superficial cells in the upper layers. In this process, proliferating basal cells move upwards and take on a more flattened shape. Superficial cells are continuously shed from the epithelial surface supported by eyelid blinking [7]. Similar to the skin, the epithelial basement membrane consists of various collagen types (IV, VII, XII, XV, XVII, XVIII), heparin sulfate proteoglycans, fibronectin, laminins and nidogens. Interestingly, the composition of the membrane has been shown to undergo changes during postnatal development and is not homogeneous, but considerably varies between the limbal region and the central regions above the Bowman layer [8–10]. The regions differ in collagen type IV and laminin isoforms [11], as well as fibrillin and tenascin expression [8]. The Bowman layer is located between the epithelial basement membrane and the anterior stroma. This layer is acellular and consists of randomly-arranged fibrils of collagen types I, III, V and VI [12,13]. It is well developed in humans, birds and higher mammals, such as cattle, whereas the Bowman layer is considerably thinner in rodents, such as mouse, rat, rabbit and guinea pig, while the collagen fiber diameters do not seem to vary among species [14].

While the epithelium provides a smooth surface to avoid light scatter, the stroma provides the corneas' spheroid shape and accounts for its transparent nature and refractive power. The unique shape of the cornea generates only small spherical aberration, ensures that light is transmitted without scatter and is focused on the lens and retina. The stroma consists of tightly-packed collagen fibers (mainly consisting of fibrillary collagen types I and V), which form lamellae with parallel distributions. These collagen complexes are surrounded by small leucine-rich proteoglycans [15]. In the central cornea, the collagen lamellae are arranged in right angles relative to the fibers in neighboring layers, while towards the corneal periphery lamellae, they are orientated circumferentially [7]. This uniquely arranged network of collagen fibers is assumed to reduce light scatter and provide corneal clarity. The corneal stroma is devoid of blood and lymphatic vessels, but contains a dense network of autonomic and sensory nerve fibers. To ensure transparency, the central corneal axons lack myelin sheaths. As the primary cell type, the stroma contains sparsely-spread resting keratocytes, which continuously secrete collagen and proteoglycans to assure turnover of the extracellular matrix (ECM), although corneal collagen turnover is much slower than in skin [16]. Keratocytes are the major cell type involved in corneal repair. After injury, these mesenchymal-derived cells become activated and are able to transform into two phenotypes: fibroblasts and myofibroblasts.

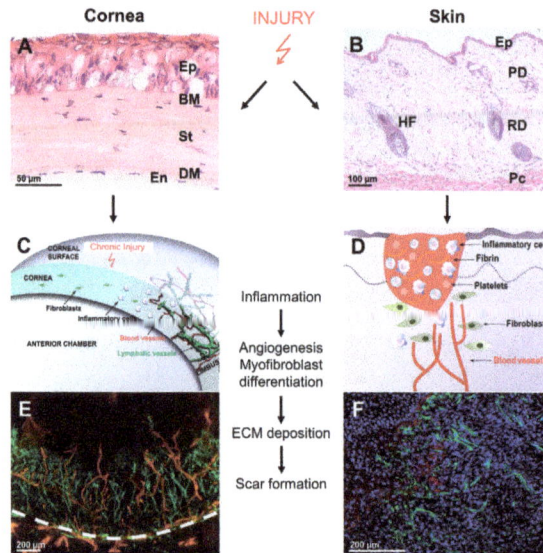

Figure 1. Parallels and differences in corneal and skin wound healing. Injury and wound healing in cornea and skin involves a similar sequence of events: inflammation, myofibroblast differentiation, extracellular matrix (ECM) deposition and eventually development of fibrosis. Yet, specific cellular responses, such as immediate keratocyte apoptosis after wounding, are unique events in corneal wound healing and presumably serve to avoid excessive corneal inflammatory reactions and opacification after tissue injury. Due to the physiological avascular nature of the cornea, hemostasis and activation of platelets are absent in corneal healing; however, both processes provide an important source of growth factors and cytokines in skin wound healing. Instead, in cornea, several of these factors (e.g., platelet-derived growth factor (PDGF)) are provided by epithelial cells [17]. Furthermore, to maintain its transparency, the cornea does not respond with the induction of capillary sprouts to minor injuries, which is essential to maintain good vision; however, chronic injury leads to an angiogenic response. (**A**) Histological view of the unwounded mouse cornea (HE staining); Ep: epithelium; BM: Bowman layer; St: stroma; DM: Descemet's membrane; En: endothelium; (**B**) Histological view of the unwounded mouse skin (HE staining); EP: epidermis; PD: papillary dermis; RD: reticular dermis; Pc: panniculus carnosus; HF: hair follicle; (**C**) Infiltration of inflammatory cells into the physiologically avascular cornea and sprouting of blood and lymphatic vessels from the limbus after severe corneal injury (modified from Hos et al. [18]). Of note, corneal angiogenesis is a rare event and usually does not occur during wound healing. However, in severe and eye-threatening conditions, proangiogenic stimuli might overcome the corneal antiangiogenic mechanisms and lead to the secondary ingrowth of pathological vessels from the limbus into the corneal center. This is in contrast to skin wound healing, where capillary sprouts are important for an adequate healing response; (**D**) In skin wound healing, hemostasis, platelet activation and angiogenesis are key repair mechanisms. Similar to corneal wound healing, skin repair involves immune cell recruitment, activation and transdifferentiation of fibroblasts, which synthesize and remodel extracellular matrix and mediate wound contraction; (**E**) Murine corneal whole mount after experimentally-induced corneal injury and neovascularization (green: blood vessels, CD31 stain; red: lymphatic vessels, lymphatic vessel endothelial hyaluronan receptor 1 (LYVE-1) stain); the dashed white line indicates the limbal border. (**F**) Skin wound three days post-injury; immunohistochemical double staining for fibrin and blood vessels (blue: DAPI stain; red: fibrin stain; green: blood vessels, CD31 stain).

The corneal endothelium like the stroma is derived from the neural crest and consists of a single layer of hexagonal-shaped cells resting on a basal membrane, named Descemet's membrane. Descemet's membrane is mainly composed of collagen type VIII, but also contains collagen type IV. The function of endothelial cells is of particular importance for corneal homeostasis, as endothelial cells feature a high number of Na^+-K^+-ATPases on their lateral membranes, creating an osmotic gradient that ensures that the cornea remains relatively dehydrated. The cornea ideally remains at a 78% hydration level, for which a sufficient number of endothelial cells is required to counteract corneal swelling and opacification. At birth, endothelial cell density amounts to 6000 cells/mm^2 and decreases at a rate of about 0.6% a year. To preserve a sufficient dehydration rate, a minimum number of about 500–800 cells/mm^2 is needed [7,19,20].

3. Distinct Regenerative Capacities of the Different Cellular Corneal Layers

The epithelium as the outer barrier is constantly self-renewing and has the highest regenerative capacity, as epithelial cells are replenished every 7–10 days. Epithelial stem cells reside in the limbal palisades and migrate towards the corneal center, where they differentiate to transient amplifying cells and basal cells [21,22]. Thus, renewal of epithelial cells not only involves a vertical movement of differentiating cells from deep to superficial layers, like in skin, but also a centripetal migration of stem cells from the limbus to the central cornea as they undergo differentiation [22]. The limbal stem cell niche is essential for corneal epithelial homeostasis, and patients suffering from limbal stem cell deficiency, e.g., due to severe chemical trauma to the limbal region or inherited diseases affecting limbal stem cells, often show severe and vision-limiting corneal alterations [23]. In these patients, overgrowth of the adjacent conjunctiva on the cornea (conjunctivalization) can frequently be observed, indicating that limbal stem cells are not only important to provide a cellular source for corneal epithelial turnover, but also to maintain the identity and barrier between various cellular compartments in the eye.

While stromal cell populations are quickly replenished, the challenge in stromal wound healing is to restore the exceptionally regular collagen organization, which is initially lost at the site of injury. This remodeling process and restoration of normal transparency can take several years [24]. Among all corneal layers, the endothelium has the lowest mitotic activity and lowest regenerative capacity. Therefore, to close small endothelial ruptures, the remaining endothelial cells will migrate and enlarge to remodel the monolayer. Consequently, cells will re-establish barrier function and resume their pumping activity [25]. As the endothelium is essential for the constant dehydration and thereby the transparency of the cornea, diseases or injuries leading to extensive loss of endothelial cells generally lead to corneal opacification due to excessive fluid accumulation and swelling. To date, the only treatment available to treat patients suffering from endothelial dysfunction or endothelial cell loss is corneal transplantation (keratoplasty) by procedures like DSAEK (Descemet's stripping automated endothelial keratoplasty) or DMEK (Descemet's membrane endothelial keratoplasty) where the endothelial cell layer and Descemet's membrane (with or without stromal parts, respectively) are replaced by donor tissue.

4. The Corneal Angiogenic and Immune Privilege

In almost all organs, blood and lymphatic vessels form extensive networks that are vital for multiple functions, including the transport of nutrients, liquids, signaling molecules and cells. The cornea is one of the few tissues of the body that in its healthy state is entirely devoid of blood and lymphatic vessels. However, corneal homeostasis also depends on oxygen and nutrients, which the cornea receives via the tear film from the outside, the aqueous humor from the anterior chamber at the inner side and the limbal vasculature from the corneal periphery. Furthermore, in contrast to other tissues including the skin, the cornea usually does not respond with the induction of blood and/or lymphatic vessels to minor injuries and angiogenic stimuli. Such reactions would interfere with its transparency and result in vision loss. Therefore, the avascular state of the cornea, termed corneal angiogenic privilege, is essential for its function and is actively maintained.

In this regard, it was demonstrated that the corneal epithelium expresses soluble forms of the three major vascular endothelial growth factor (VEGF) receptors (sVEGFR-1, sVEGFR-2, sVEGFR-3), which are all assumed to act as decoy receptors to trap the key angiogenic growth factors VEGF-A, VEGF-C and VEGF-D, which might contribute to maintaining corneal avascularity [26–28]. Additional potent antiangiogenic molecules that are expressed in the cornea are angiostatin, endostatin, thrombospondin-1, thrombospondin-2 and pigment epithelium-derived factor [29–32]. These factors exert their antiangiogenic functions by blockade of vascular endothelial cell migration and/or proliferation or interfere with growth factor bioavailability. In addition to these molecules, the cornea also expresses inhibitory PAS domain protein (IPAS), a negative regulator of hypoxia inducible factor (HIF). It has been shown that IPAS inhibits HIF-mediated upregulation of VEGF in the cornea and supports to maintain corneal angiogenic privilege even under hypoxic conditions [33].

Although the corneal angiogenic privilege is essential for good vision and is therefore strictly and (in part) redundantly regulated, this condition is not invulnerable. In fact, severe inflammatory and potentially eye-threatening conditions can result in a massive upregulation of proangiogenic growth factors, which might overwhelm the antiangiogenic mechanisms of the cornea and result in a secondary ingrowth of both blood and lymphatic vessels from the limbal area into the corneal center (Figure 1). This angiogenic response immolates the transparency of the cornea and its functionality for the safety of the whole eye. Therefore, after the immune response is complete and barrier function has been achieved, corneal blood and lymphatic vessels need to resolve promptly to restore corneal transparency and functionality. However, in certain pathological conditions, corneal blood and lymphatic vessel regress very slowly, and may even persist and contribute to detrimental corneal diseases [34–36].

In addition to its angiogenic privilege, the cornea also belongs to the so-called immune privileged sites of the body [37]. In this regard, it has been shown that antigens (e.g., alloantigens) introduced into the anterior chamber result in the generation of antigen-specific regulatory T cells (Tregs), leading to tolerance and suppression of immune responses against these antigens. The immune privilege of the cornea has been attributed to several components: The healthy cornea contains only a few major histocompatibility complex (MHC) II-positive antigen-presenting cells (APCs), and corneal expression of MHC I molecules is reduced compared to other tissues. Furthermore, immunomodulatory factors, such as Fas ligand (CD95L) and programmed death ligand 1 (PD-L1), are present in the cornea at high levels and are able to suppress excessive immune responses by inhibiting T cell proliferation and inducing T cell apoptosis [38,39]. One additional and important factor responsible for corneal immune privilege is the so-called anterior chamber-associated immune deviation (ACAID). ACAID causes immune tolerance by antigen-specific downregulation of delayed-type hypersensitivity (DTH) reactions. This mechanism suppresses cell-mediated immune responses that are potentially caused by tissue damage. Antigens introduced into the anterior chamber are captured by APCs migrating to the marginal zone of the spleen, where they are involved in the activation of regulatory T cells, which subsequently suppress DTH immune reactions [40–43]. The spleen is indispensable for the development of ACAID, as it has been shown that removal of the spleen during introduction of various antigens into the anterior chamber prevents the development of ACAID and leads to rapid antigen-mediated rejection [44]. The immune and angiogenic privilege of the cornea are closely linked, as corneas with pathological corneal neovascularization often show a loss of their immune privilege in parallel [45,46]. In this context, it is generally accepted that the presence of pathological corneal blood and lymphatic vessels are the main risk factor for immunological graft rejection episodes after corneal transplantation [47,48].

5. Immune Cells in the Resting Cornea

Initially, it was assumed that antigen-presenting cells (APCs) were virtually absent from corneal tissue, which was believed to be one of the underlying mechanisms of the corneal immune privilege [49,50]. Since then, the view of the corneal cellular immune system has dramatically changed, and virtually all types of immune cells have been shown to be residing in the epithelium and stroma of the naive cornea. Immunohistological analyses in mice carried out with the pan-leukocyte marker CD45 could identify leukocytes located mostly in the corneal periphery, but also in the epicentral and central cornea [50,51]. CD45$^+$ cells are distributed through the entire depth of the stroma, although most cells are either located in the anterior or posterior third of the stroma [51]. In bone marrow transplantation experiments after corneal irradiation, it was demonstrated that these myeloid-derived CD45$^+$ cells are constantly reconstituted from the bone marrow, with 75% of cells being replenished within eight weeks [52]. However, one has to keep in mind that irradiation of the cornea might have substantially affected the turnover rate of these cells.

Macrophages stained positive for CD11b were shown to make up about 50% of the resident corneal leukocytes [51]. They are localized primarily in the posterior stroma and are uniformly distributed in the periphery and center of the cornea [50]. Recently, the origin of these corneal macrophages has been investigated in detail, considering their CCR2 expression profiles. The current concept divides macrophages and monocytes into a proinflammatory CX3CR1^{low} CCR2^{positive} type and a tissue resident, anti-inflammatory CX3CR1^{high} CCR2^{negative} type [53]. Liu and colleagues reported that CCR2^{negative} macrophages possibly originating from the yolk sac or fetal liver can be detected in the stroma starting from E12.5, while CCR2^{positive} macrophages in the cornea are presumably fetal liver- or bone marrow-derived and appear at later stages in embryonic development starting from E17.5. Additionally, this work could demonstrate that CCR2^{negative} macrophages in the cornea are, for the most part, replenished by resident macrophage proliferation, whereas a large proportion of the CCR2^{positive} population is continuously reconstituted from the bone marrow [54].

Langerhans-cells (LCs), an epithelial type of dendritic cell, are found in the corneal epithelium in the limbus region, periphery and central cornea. Notably, central cells have been described to be MHC II negative, while peripheral cells are mainly MHC II positive [55]. In rat cornea, slow-cycling MHC II$^+$ cells have been identified in the limbal basal epithelium, which are putative LC precursor cells [56]. Similar to the LC distribution, stromal dendritic cells (DCs) are found in the periphery and center of the anterior stroma, with the central cells lacking MHC II expression [50,57]. These DC populations have been shown to be replenished by bone marrow-derived cells [58]. In this regard, it has also been shown that the fractalkine receptor CX3CR1 is involved in homeostatic homing of corneal MHC II$^+$ cells to the cornea [59].

LysM-positive neutrophils are located around the limbal vessels in the periphery, but are not found in the central cornea [60]. Similarly, mast cells are found in the corneal limbus and conjunctival parenchyma, but not in the central cornea of adult mice. They are present in the developing cornea starting from E12.5 and accumulate around the corneal vasculature, during embryonic development. Interestingly, the distribution of mast cells is influenced by eye-opening, as cells in the central cornea have been observed to disappear after eye-opening and thus after the first contact of the cornea with external factors [61]. Although in earlier reports, CD3$^+$ T cells have been reported to be absent from the cornea [51], by now, resident CD4$^+$ and CD8$^+$ T cells have been identified in the central and peripheral region of naive corneas [62].

6. Corneal Wound Healing

Corneal wound healing involves a complex cascade of cellular events of which we can only highlight some aspects in this article. Depending on the extent of the injury, cells of the epithelium, stroma and endothelium react differently to restore corneal integrity. Corneal repair is orchestrated by a complex network of growth factors that largely overlaps with those factors reported in skin repair (reviewed in: [63–66]).

6.1. Epithelial Healing

Epithelial healing in the cornea has been studied extensively. As the outermost layer, the epithelium is prone to injuries and needs to repair quickly to prevent infection of deeper layers. Small epithelial defects are usually covered within 24 hours (h). Epithelial injury and apoptosis of the injured cells leads to disruption of the attachment to the underlying basement membrane. Subsequently, cells from the wound margin rapidly respond with flattening and centripetal migration. As cell-to-cell adhesions are partially maintained, the defect is slowly covered by a sliding cellular sheet. This process is followed by proliferation and differentiation of basal cells to restore the cell layer. During epithelial wound healing, a temporary ECM facilitates epithelial migration to cover the wound. Fibrin, fibronectin and hyaluronic acid are some of the molecules identified in this matrix [67]. During the terminal phase of epithelial healing, the cells generate new hemidesmosomes for anchorage to the underlying layer [67,68]. Multiple growth factors, also identified as key mediators of epidermal repair in skin, are involved in epithelial healing, including EGF, transforming growth factor β (TGFβ), hepatocyte growth factor (HGF) and keratinocyte growth factor (KGF) [69].

Epithelial regeneration is also supported by corneal nerves, which regulate the blinking reflex and thereby the turnover of corneal epithelial cells. Furthermore, corneal nerves have been shown to provide epitheliotropic factors, such as substance P, calcitonin gene related peptide (CGRP) and nerve growth factor (NGF), among others [70–75]. Patients suffering from degeneration of corneal nerves (called neurotrophic keratopathy), e.g., due to herpetic disease or systemic diseases such as diabetic polyneuropathy, often show delayed epithelial wound healing and might even suffer from persistent and potentially sight-threatening epithelial wounds that might lead to corneal perforation or serve as entry for, e.g., bacteria into the eye, with devastating consequences. Topical treatment with recombinant NGF was shown to promote epithelial healing [75,76] and is currently in clinical trials for the treatment of several diseases of the ocular surface, like dry eye disease and neurotrophic keratitis, and for the promotion of corneal nerve regeneration after refractive and cataract surgery. Besides NGF, several other growth factors have already entered clinical trials, including EGF, fibroblast growth factor (FGF) and insulin-like growth factor (IGF), with promising results [77–79].

6.2. Stromal Healing

The first response observed in the corneal stroma after epithelial damage is keratocyte apoptosis beneath the wound, which is assumed to be initiated by cytokines released from damaged epithelial cells, including IL-1α, IL-1β, Fas ligand and TNFα [80–84]. Keratocyte death results in an area beneath the wound site virtually devoid of cells. The remaining keratocytes become activated, differentiate into fibroblasts and begin to migrate to the wound site within 24 h after injury [67]. Using a rabbit model of transcorneal freeze injury, it was recently observed that during this migration process, the fibroblasts form connected streams aligned in parallel to the collagen lamellae, suggesting that lamellae provide "contact guidance" for fibroblast migration [85]. Subsequently, fibroblasts proliferate to repopulate the wound site. During stromal healing, their migration and activation are assumed to be mediated by several growth factors, including TGFβ, platelet-derived growth factor (PDGF), FGF-2 and EGF [85–88]. TGFβ and PDGF have also been identified as the main growth factors involved in the transdifferentiation from fibroblasts to myofibroblasts, which occurs in stromal wounds [88–94].

Myofibroblasts are identified by their expression of contractile α-smooth muscle actin (αSMA) stress fibers, which enable them to mediate wound closure and contraction [95,96]. Additionally, myofibroblasts are characterized by the expression of vimentin and desmin and reduced transparency, compared to quiescent keratocytes [97,98]. The number of myofibroblasts at the wound site seems to vary between different types of wounds: In incisional wounds, where wound contraction is fundamental for proper healing, myofibroblasts seem to be most abundant [99,100]. Studies indicate that myofibroblast are not solely generated by local fibroblast proliferation and transformation, but by a large proportion are derived from bone-marrow cells recruited to the cornea as observed to some extent in skin [101–103]. As soon as the myofibroblast/fibroblast-mediated stromal wound closure is complete, the number of myofibroblasts in the stroma slowly decreases. The following remodeling of the stroma to restore transparency can take several years. In this process, the disorganized repair matrix is slowly substituted by regular corneal ECM.

6.3. Endothelial Damage Response

Endothelial wound healing is in most situations limited to the reorganization and enlargement of the remaining endothelial cells, as in humans their mitotic activity is negligible. As soon as the cells close the damaged areas, usually after a few days, they resume their pumping activity and begin to secrete new basement membrane. After corneal injury, endothelial cells may also undergo a process of endothelial mesenchymal transition (EnMT), similar to epithelial-mesenchymal transition (EMT) observed in skin. During EnMT, the usually quiescent endothelial cells attain fibroblast-like characteristics and begin to proliferate. EnMT is characterized by loss of cell-junctions, loss of apical-basal polarity and actin-skeleton reorganization (reviewed in [104]). Apart from numerous other factors possibly involved, this process has been shown to rely on TGFβ, FGF-2, IL-1β and involves NFκB activation [105–108]. EnMT may lead to fibrous ECM deposition (so-called retrocorneal fibrous membrane) posterior to Descemet's membrane. The consequences of EnMT are therefore considered undesirable, as it might result in loss of endothelial cells and corneal opacification due to ECM deposition [109]. Moreover, EnMT is also observed in ex vivo cultured endothelial cells, leading to a loss of endothelial cell characteristics during attempts of bioengineering endothelial grafts. Efforts to expand endothelial cells in vitro aim to inhibit the process of EnMT and preserve cell morphology for instance by using inhibitors of TGF-β signaling or the Rho-kinase/ROCK pathway [108,110–112].

7. The Fibrotic Response in Corneal Repair

Corneal injury extending into the corneal stroma causes fluid influx and fibrin deposition, leading to swelling and subsequent local loss of transparency. During corneal stromal repair, fibroblasts and myofibroblasts deposit multiple elements of the ECM including type III collagen, fibronectin, tenascin and glycosaminoglycans, facilitating the migration of fibroblasts [113]. Prolonged myofibroblast activation and ongoing deposition of repair matrix can cause long-lasting corneal opacification. In addition, the collagens and glycosaminoglycans produced in the early phase of the repair response show irregularities in their fibril size, arrangement and composition, which are likely to contribute to permanent opacification and scarring [114,115]. The contraction of repair tissue may also alter the curvature of the cornea and its optic properties. However, corneal stromal injuries do not always result in corneal scaring and permanent opacification. Smaller and superficial wounds often heal with restitutio ad integrum. The exact mechanisms leading to full restoration or scarring of the injured cornea are so far poorly understood.

It has been shown that epithelial abrasion alone does not activate the fibrotic response. Fibrosis is only observed in corneal injuries that include rupture of the epithelial basement membrane and wounding of the stroma [116]. Similar to dermal repair in the skin, PDGF and TGF-β are pivotal mediators of the corneal fibrotic response [117]. Due to the angiogenic privilege, platelets as critical source of PDGF and TGF-β are virtually absent, although their accumulation around the limbal vessels after wounding has been reported [118]. In the cornea, PDGF is produced by epithelial cells and (like TGFβ) has been detected in tears [119–121]. For controlling the impact of these factors on stromal keratocytes, the integrity of the epithelial basement membrane appears important, by creating a barrier for the growth factors produced by epithelial cells and derived from tears. In the case of basement membrane rupture, PDGF gains access to the stroma and subsequently acts on keratocytes, stimulating their migration and proliferation [122]. Likewise, it was shown that upon loss of the basement membrane, TGF-β2 produced by the epithelial cells can get in contact with the stroma and induces keratocyte activation [116]. Accordingly, Han and colleagues recently reported that epithelial cells produce exosomes that after rupture of the basement membrane gain access to the stroma. These exosomes subsequently fuse with stromal keratocytes and are able to promote their differentiation into myofibroblasts in wounds affecting epithelium and stroma [123].

Interestingly, similar to scarless healing in embryonic skin, also corneal wounds in the embryo heal without a scar. In the chick embryo, incisional wounds affecting epithelium, basement membrane and anterior stroma do not involve fibrotic response and scar formation. Intriguingly, the stromal response in the embryo does not involve enhanced keratocyte apoptosis or proliferation and re-epithelialization in these wounds is observed to be very slow compared to adult [124]. The underlying cellular mechanisms involved in this scarless type of healing and whether this alternative pathway can still be activated in adulthood remain to be elucidated.

An exaggerated fibrotic response after corneal injury may rarely result in hypertrophic scar or keloid formation, characterized by excessive fibrous tissue. Compared to skin, the development of keloids is a very uncommon event in the cornea following corneal trauma or disease. As opposed to hypertrophic scars, which are restricted to the initial site of injury, keloids may overgrow the initial lesion covering large parts of the corneal surface and are often observed to be recurrent. Keloids clinically appear as defined, white, elevated tissue and are characterized by a hyperplastic epithelium, disruption of the Bowman layer and contain disorganized stromal collagen fibers, as well as activated fibroblasts and may be paralleled by neovascularization [125–128]. Interestingly, also spontaneous keloid formations without previous corneal injury have been reported [129–131]. It is unresolved whether a predisposition for dermal keloid formation, as observed in patients from Asian or African American ethnicity, is associated with a higher risk to develop corneal keloid after corneal injury or surgery [131]. However, corneal keloids have been observed to occur in genetic syndromes such as Lowe's syndrome [132,133] and Rubinstein–Taybi syndrome [131,134], the latter of which has been shown to be accompanied by dermal keloid formation, which may occur spontaneously or after preceding trauma in up to 24% of the patients affected [135–137].

8. Cellular Events in Corneal Inflammation

Knowledge acquired about wound healing and cellular infiltration into the cornea has mostly been provided by corneal injury models in rabbit and mice, including epithelial abrasion [138], intrastromal suture placement [139], chemical burn injuries [140] or corneal incisions (for a review, see [141]). As each of these injury models is characterized by a different wound healing response, kinetics and quality of inflammation, combining the results into an overall picture of cellular events after injury is a difficult task. The results obtained may vary among the nature of injury, species and strains [142]. In general, immune cell recruitment after corneal injury is mediated by proinflammatory cytokines released from epithelial cells and keratocytes at the injured site. Il-1, Il-6 and TNFα have been shown to be important mediators [143–147]. Being attracted by these and several other cytokines, recruited leukocytes from the limbal blood vessels enter the stroma and migrate towards the wound site [148].

Neutrophils are the first cells infiltrating the cornea after injury: they can be detected as soon as 2 h after injury and have been observed to enter the cornea in two major waves at 18 and 30 h after epithelial abrasion; as soon as 48 h after injury, their numbers normalize again [149]. Neutrophils seem to be involved in several processes in corneal wound healing, as in neutropenic mice, re-epithelialization after abrasion was observed to be delayed [149,150]. Moreover, the absence of neutrophils was reported to affect the recovery of the corneal nerves, which was attributed to their secretion of VEGF-A as a neurotrophic factor [151,152]. Interestingly, the recruitment of neutrophils into the cornea seems to be dependent on platelet accumulation and vice versa, as it was shown that depletion of platelets decreased neutrophil influx, and in turn, the depletion of neutrophils led to reduced numbers of platelets accumulating around the limbal vessels [117].

Comparable to myeloid cell recruitment in skin injury, shortly after neutrophils have entered the cornea, macrophages extravasate from the limbal vessels, infiltrate the stroma from superficial to deeper layers and migrate towards the corneal center [148]. Macrophages remove debris and apoptotic cells at the wound site, but have also been shown to be essential mediators of angiogenesis after severe and prolonged corneal injury. The newly-formed blood and lymphatic vessels presumably serve to supply oxygen, growth factors and immune cells and in turn mediate their clearance. During corneal inflammation, macrophages express high levels of VEGF-A, VEGF-C and VEGF-D, thereby inducing the proliferation of vascular endothelial cells [153,154]. In addition, studies indicate that macrophages are able to form lymphatic vessel-like structures de novo [155]. It has recently been reconfirmed that macrophages are capable of forming tubular structures, by demonstrating that they form non-endothelial vascular channels in a subcutaneous tumor model [156]. Macrophages also take part in corneal wound closure by secreting TGF-β to promote the differentiation of fibroblasts to myofibroblasts. The experimental depletion of macrophages in both cutaneous and corneal injury in mice leads to a delay in wound healing and a reduction of αSMA-positive fibroblasts [157], as well as a delay in epithelial closure after abrasion [53,158].

Corneal injury also leads to an increase in Langerhans cell numbers in the corneal epithelium. LC recruitment into the cornea upon central corneal cautery was shown to be mediated by IL-1, TNF and CCR5 signaling in mice [159]. In rats, centripetal migration of MHCII⁺ LCs from the limbus to the central corneal epithelium was observed 4 h after central cauterization. The majority of these cells were shown to be recruited from the limbal basal epithelium [160]. The rise in LC numbers at the corneal surface increases the antigen presenting capacity of the cornea, which is important to adequately respond to foreign antigens, which might have been introduced through the wound. This might also apply to stromal DCs: while most DCs in the corneal center are MHC II negative, in the case of inflammation, their MHC II and costimulatory molecule expression are induced [160,161]. Yet, little is known about the role of stromal DCs in corneal wound healing, although it has been shown that corneal DCs migrate together with the epithelial sheet to cover the wound after epithelial abrasion in mice. Consistently, the experimental depletion of DCs delays epithelial closure and alters epithelial gene expression [162,163].

Natural killer (NK) cells, which are rare in the naive, intact cornea, accumulate around the limbal vessels and infiltrate the corneal stroma from the periphery to the center. Their influx peaks at 24 h after epithelial abrasion [164]. Experimental depletion of NK cells in mice results in delayed epithelial wound closure and impaired regeneration of corneal nerves. Moreover, in the absence of NK cells, neutrophil numbers were observed to be elevated [164], while DC numbers were reduced [162], indicating that NK cells possibly play a role in orchestrating the corneal inflammatory response [164]. However, the mechanisms involved are not yet resolved. Similar to findings in skin wound healing, a possible role in modulating the corneal immune response is also proposed for γδ T cells. This type of T cell, recruited via ICAM1 and CD18, was shown to promote wound healing and influence neutrophil and platelet numbers in corneal inflammation [165–167].

To date, it is still an open question by which mechanisms immune cells are recruited to the injured cornea during the inflammatory phase. Several chemokines and their receptors have been identified in the inflamed cornea, yet their functional role in immune cell recruitment to the cornea is poorly characterized. For instance, the CXC chemokines CXCL1 and CXCL5 [149], as well as CXCL10, CCL7 and macrophage chemoattractant protein 1 (MCP-1) expression [164] were found to be elevated after epithelial abrasion in mice. Correspondingly, CXCL1, CXCL8 and MCP-1 mRNA levels were found to be elevated in human inflamed corneas [168]. CCL5 and MCP-1 have also been shown to be secreted by corneal keratocytes after stimulation with IL-1 or TNF [143]. Additionally, CCR7 and its ligand CCL21 were found to be upregulated in inflamed corneas, mediating MHC II⁺ cell recruitment [169].

For the recruitment of neutrophils, it has been postulated that recruitment is mostly mediated by P- and E-selectins, but also involves ICAM-1 and CD18 interaction [149], whereas myeloid cell/macrophage recruitment to the cornea is assumed to be (partially) mediated by the chemokine receptor CCR2 and its main ligand MCP-1 [170,171]. In the cornea, MCP-1 is expressed by epithelial cells, as well as keratocytes [171] and was shown to be upregulated after epithelial scrape injury [144]. Furthermore, MCP-1 introduced into the rabbit cornea in a corneal-micropocket-assay led to increased macrophage infiltration and angiogenesis [172]. Consistently, in CCR2- or MCP-1-deficient mice, macrophage recruitment was demonstrated to be impaired, while neutrophil numbers were not affected [170]. The latter is unexpected given the critical role of macrophages in neutrophil phagocytosis and clearance. The number of corneal CD11b⁺ cells was also reduced in a model of topical CCR2 antagonist treatment [173]. Correspondingly, in skin wound healing, it was shown that CCR2-mediated recruitment of VEGF-A-expressing macrophages is critical for the initiation of capillary sprouting during the wound healing response [174].

Although macrophage polarization has been extensively studied in cutaneous wound healing [1,174], macrophage polarization phenotypes have hardly been investigated during corneal repair. It has been shown that CCR2-positive, as well as CCR2-negative macrophages are required for proper corneal epithelial healing [53]. In addition, alterations in gene expression levels of inflammatory factors such as inducible nitric oxide synthase (iNOS), but also an impaired transition of macrophages to the M2-like phenotype are associated with impaired wound healing following epithelial injury [158]. An experimental study by Uchiyama and colleagues in a rat alkali burn model has shown that treatment with a peroxisome proliferator-activated receptor gamma agonist prevents excessive corneal inflammation and results in a decreased profibrotic response with less corneal opacification. Interestingly, this study also demonstrated that although the overall number of macrophages was decreased in the treatment group, M2-like macrophage numbers were significantly higher, suggesting a superior role of this macrophage subpopulation in corneal wound healing, after chemical trauma [175]. Although the literature suggests that alterations in macrophage numbers and polarization phenotypes are associated with impaired corneal wound healing, definitive evidence for a functional role is missing.

Apart from MCP-1, it was shown that stromal fibroblasts upregulate a variety of additional chemokines after stimulation with proinflammatory molecules such as IL-1α or TNFα. Among those are granulocyte colony-stimulating factor (G-CSF), monocyte chemotactic and activating factor (MCAF), neutrophil-activating peptide (NAP) and monocyte-derived neutrophil chemotactic factor (MDNCF) [144], which all might be involved in monocyte/macrophage and granulocyte recruitment and activation after corneal damage. Collectively, current findings propose that homing and inflammatory influx of leucocytes into the cornea are mediated by multiple recruitment mechanisms. As yet, the predominant mechanisms are not identified. Furthermore, it remains unclear how tissue resident immune cells are involved in the inflammatory response. Due to the corneal immune privilege, these mechanisms may possibly vary from those observed in other tissues. After the repair response is complete and corneal integrity is reestablished, inflammation also has to resolve rapidly. In this regard, it is still unclear how inflammatory cells sense that tissue integrity is finally restored and how these cells are removed from the cornea. The mechanisms involved in immune cell clearance from the cornea still remain to be investigated.

9. Stem Cells in Corneal Regeneration

The limbus region provides the primary niche for corneal stem cells [176]. This area contains both the corneal epithelial and mesenchymal (stromal) stem cells [176–178]; although the term limbal stem cells (LSCs) is exclusively used for the epithelial progenitor subpopulation. LSC biology has been extensively studied over the past decades, and already several years ago, LSC transplantation has been transferred to the clinic for the treatment of various corneal epithelial deficiencies (reviewed in [179,180]). Recent experimental studies showed that some cells (3–4%) in the adult corneal stroma subjacent to the epithelial basement membrane also express markers of ocular progenitor cells that are not present in differentiated keratocytes [178,181]. These cells, termed corneal stromal stem cells (CSSCs), have the ability to divide extensively, generate adult keratocytes and have the potential to give rise to several non-corneal cell types, such as cartilage or neural cells [181]. CSSCs express genes indicative for mesenchymal stem cells (such as BMI1, CXCR4, ABCG2), as well as genes indicative for pluripotent cells (such as SOX2, KLF4, OCT4, NANOG). Furthermore, CSSCs express genes that are present in early corneal development (such as PAX6 and Six2) and genes associated with neural development (CDH2, NESTIN, NGFR) [182].

Several studies have analyzed the function of CSSCs in the corneal steady state and after injury. In vitro, CSSCs can produce connective tissue including collagen fibrils when cultured under low-mitogenic conditions [183]. Furthermore, when cultured on parallel aligned nanofibers, silk or polycarbonate substrates, CSSCs produce layers of highly parallel collagen fibrils with uniform diameter and regular interfibrillar spacing similar to that of the normal corneal stroma [184–186]. An important function of CSSCs is their support of the LSC population in the limbal stem cell niche [187,188]. Furthermore, experimental studies in mice have shown that CSSCs have the ability to support stromal regeneration and prevent scarring [189,190]: injection of CSSCs into the corneal stoma of Lumican-deficient mice, which have opaque corneas due to increased corneal thickness and altered collagen organization, results in restoration of the normal corneal architecture and reestablishment of corneal transparency [190]. Moreover, the application of CSSCs to surgical wounds comprising the epithelium and the anterior stroma (which usually leads to the deposition of disorganized and opaque scar tissue) results in deposition of extracellular matrix with normal collagen organization and clear corneas [189]. In addition, CSSCs have been reported to modulate immune responses in the cornea [190].

Collectively, CSSCs have been shown to facilitate corneal stromal regeneration without scarring and excessive inflammation. Although the therapeutic use of CSSCs is still far behind the use of LSCs, novel CSSCs-based therapeutic strategies might open up novel and effective treatment strategies for corneal stromal wound healing in the future.

10. Concluding Remarks and Future Directions

Corneal transparency is essential for proper vision, and a detailed understanding of the mechanisms involved in corneal wound healing and regeneration is of great importance to treat patients suffering from corneal diseases. At present, epithelial regeneration can be supported relatively satisfactorily, and it is possible to treat superficial corneal scars, e.g., with laser therapy. However, for deep corneal scars or endothelial diseases, which are frequent, corneal transplantation is still the only option to restore clear vision. Nonetheless, due to the scarcity of corneal donor tissue or limited access to corneal surgery, corneal opacification is still one of the major causes of blindness worldwide [191–193].

Ongoing research aims to support corneal regeneration by stem cell-based therapies. To support epithelial regeneration, research focuses on the ex vivo expansion and subsequent transplantation of LSCs derived from limbal biopsies, as already performed to treat patients suffering from limbal stem cell deficiency. As mentioned earlier, studies using LSCs and CSSCs have demonstrated beneficial effects in several disease models, promoting epithelial and stromal repair [189,190,194]. Due to the limited availability of corneal stem cells and the potential risks related to biopsy excision from donor eyes, several non-ocular tissues have been investigated as potential autologous sources of epithelial-like cells [195]. Apart from potentially simplifying the process of cell harvesting and culture conditions, these approaches would avoid immune reaction to donor tissue in patients with bilateral corneal disease. Here, oral mucosa epithelial cells [196,197] and conjunctival epithelial cells [198–200], already in clinical use, have been shown to successfully support epithelial regeneration in limbal stem cell deficiency, though data on long-term outcomes are limited. Interestingly, alternative approaches demonstrate that corneal epithelial-like cells may also be transdifferentiated from dermal sources: skin-derived precursor cells, epidermal adult stem cells and hair follicle stem cells were shown to attain corneal epithelial-cell like properties [201,202] and improve corneal healing [203]. For stromal regeneration, it has been demonstrated that adult stem cells isolated from dental pulp may be differentiated to attain a keratocyte-like phenotype, secreting collagen fibrils and corneal proteoglycans [204]

Innovative approaches show that corneal epithelial cell-like or keratocyte-like cells can also be differentiated from induced pluripotent stem cells (iPSCs) generated from corneal and non-ocular cells that might be used in future treatment options [205–209]. In an experimental therapeutic approach, iPSCs generated from human corneal keratocytes enhanced corneal healing in rats, when topically applied to cornea after epithelial abrasion [210]. Moreover, a broad area of research focusses on mesenchymal stem cells (MSCs) derived from non-ocular tissues to promote corneal and skin regeneration. MSCs derived from bone marrow, adipose tissue, umbilical cord or dental pulp [211,212] were demonstrated to acquire both epithelial cell-like [213,214] and keratocyte-like characteristics [215–217] (see [195] for a systematic review). The therapeutic use of bone-marrow or adipose tissue-derived MSCs, by local injection or topical application in animal models of chemical burns, was shown to improve corneal healing and reduce corneal inflammation and neovascularization [218–221].

Ongoing efforts in the field of endothelial regeneration focus on the establishment of methods for the isolation, cultivation, expansion and transplantation of corneal endothelial cells, with promising results [222–224]. These techniques are novel approaches to substitute for corneal transplantation procedures like DMEK in patients suffering from endothelial cell loss or dysfunction. Conclusively, cell-based therapies have the potential to accelerate corneal wound healing, compensate for epithelial cell, keratocyte and endothelial cell loss in certain injuries and might promote the substitution of opaque repair tissue with regular transparent corneal ECM. Further research will reveal whether these novel approaches will be successful in supporting corneal regeneration.

Acknowledgments: German Research Foundation (DFG) (FOR2240 to Deniz Hos, Claus Cursiefen and Sabine A. Eming; SFB829 to Sabine A. Eming), Center for Molecular Medicine Cologne (CMMC), University of Cologne (to Deniz Hos, Claus Cursiefen and Sabine A. Eming); Excellence Cluster: Cellular Stress Responses in Aging-associated Diseases (CECAD), University of Cologne (to Sabine A. Eming).

Conflicts of Interest: The authors declare no conflict of interest.

References

1. Lucas, T.; Waisman, A.; Ranjan, R.; Roes, J.; Krieg, T.; Müller, W.; Roers, A.; Eming, S.A. Differential roles of macrophages in diverse phases of skin repair. *J. Immunol.* **2010**, *184*, 3964–3977. [CrossRef] [PubMed]
2. Chung, J.H.; Kang, Y.G.; Kim, H.J. Effect of 0.1% dexamethasone on epithelial healing in experimental corneal alkali wounds: Morphological changes during the repair process. *Graefe's Arch. Clin. Exp. Ophthalmol.* **1998**, *236*, 537–545. [CrossRef]
3. Eming, S.A.; Martin, P.; Tomic-Canic, M. Wound repair and regeneration: Mechanisms, signaling, and translation. *Sci. Transl. Med.* **2014**, *6*, 265sr6. [CrossRef] [PubMed]
4. Gritz, D.C.; Lee, T.Y.; Kwitko, S.; McDonnell, P.J. Topical anti-inflammatory agents in an animal model of microbial keratitis. *Arch. Ophthalmol.* **1990**, *108*, 1001–1005. [CrossRef] [PubMed]
5. Tuli, S.S. Topical Corticosteroids in the Management of Bacterial Keratitis. *Curr. Ophthalmol. Rep.* **2013**, *1*, 190–193. [CrossRef] [PubMed]
6. Klenkler, B.; Sheardown, H.; Jones, L. Growth factors in the tear film: Role in tissue maintenance, wound healing, and ocular pathology. *Ocul. Surf.* **2007**, *5*, 228–239. [CrossRef]
7. DelMonte, D.W.; Kim, T. Anatomy and physiology of the cornea. *J. Cataract Refract. Surg.* **2011**, *37*, 588–598. [CrossRef] [PubMed]
8. Kabosova, A.; Azar, D.T.; Bannikov, G.A.; Campbell, K.P.; Durbeej, M.; Ghohestani, R.F.; Jones, J.C.R.; Kenney, M.C.; Koch, M.; Ninomiya, Y.; et al. Compositional differences between infant and adult human corneal basement membranes. *Investig. Ophthalmol. Vis. Sci.* **2007**, *48*, 4989–4999. [CrossRef] [PubMed]
9. Herwig, M.C.; Müller, A.M.; Holz, F.G.; Loeffler, K.U. Immunolocalization of different collagens in the cornea of human fetal eyes: A developmental approach. *Curr. Eye Res.* **2013**, *38*, 60–69. [CrossRef] [PubMed]
10. Torricelli, A.A.M.; Singh, V.; Santhiago, M.R.; Wilson, S.E. The corneal epithelial basement membrane: Structure, function, and disease. *Investig. Ophthalmol. Vis. Sci.* **2013**, *54*, 6390–6400. [CrossRef] [PubMed]
11. Ljubimov, A.V.; Burgeson, R.E.; Butkowski, R.J.; Michael, A.F.; Sun, T.T.; Kenney, M.C. Human corneal basement membrane heterogeneity: Topographical differences in the expression of type IV collagen and laminin isoforms. *Lab. Investig.* **1995**, *72*, 461–473. [PubMed]
12. Marshall, G.E.; Konstas, A.G.; Lee, W.R. Immunogold fine structural localization of extracellular matrix components in aged human cornea. II. Collagen types V and VI. *Graefe's Arch. Clin. Exp. Ophthalmol.* **1991**, *229*, 164–171. [CrossRef]
13. Newsome, D.A.; Foidart, J.M.; Hassell, J.R.; Krachmer, J.H.; Rodrigues, M.M.; Katz, S.I. Detection of specific collagen types in normal and keratoconus corneas. *Investig. Ophthalmol. Vis. Sci.* **1981**, *20*, 738–750.

14. Hayashi, S.; Osawa, T.; Tohyama, K. Comparative observations on corneas, with special reference to Bowman's layer and Descemet's membrane in mammals and amphibians. *J. Morphol.* **2002**, *254*, 247–258. [CrossRef] [PubMed]

15. Hassell, J.R.; Birk, D.E. The molecular basis of corneal transparency. *Exp. Eye Res.* **2010**, *91*, 326–335. [CrossRef] [PubMed]

16. Davison, P.F.; Galbavy, E.J. Connective tissue remodeling in corneal and scleral wounds. *Investig. Ophthalmol. Vis. Sci.* **1986**, *27*, 1478–1484.

17. Fini, M.E.; Stramer, B.M. How the cornea heals: Cornea-specific repair mechanisms affecting surgical outcomes. *Cornea* **2005**, *24*, S2–S11. [CrossRef] [PubMed]

18. Hos, D.; Schlereth, S.L.; Bock, F.; Heindl, L.M.; Cursiefen, C. Antilymphangiogenic therapy to promote transplant survival and to reduce cancer metastasis: What can we learn from the eye? *Semin. Cell Dev. Biol.* **2015**, *38*, 117–130. [CrossRef] [PubMed]

19. Bourne, W.M. Biology of the corneal endothelium in health and disease. *Eye* **2003**, *17*, 912–918. [CrossRef] [PubMed]

20. Joyce, N.C. Proliferative capacity of corneal endothelial cells. *Exp. Eye Res.* **2012**, *95*, 16–23. [CrossRef] [PubMed]

21. Davanger, M.; Evensen, A. Role of the pericorneal papillary structure in renewal of corneal epithelium. *Nature* **1971**, *229*, 560–561. [CrossRef] [PubMed]

22. Nowell, C.S.; Radtke, F. Corneal epithelial stem cells and their niche at a glance. *J. Cell Sci.* **2017**, *130*, 1021–1025. [PubMed]

23. Nowell, C.S.; Odermatt, P.D.; Azzolin, L.; Hohnel, S.; Wagner, E.F.; Fantner, G.E.; Lutolf, M.P.; Barrandon, Y.; Piccolo, S.; Radtke, F. Chronic inflammation imposes aberrant cell fate in regenerating epithelia through mechanotransduction. *Nat. Cell Biol.* **2016**, *18*, 168–180. [CrossRef] [PubMed]

24. Cintron, C.; Covington, H.I.; Kublin, C.L. Morphologic analyses of proteoglycans in rabbit corneal scars. *Investig. Ophthalmol. Vis. Sci.* **1990**, *31*, 1789–1798.

25. Jumblatt, M.M.; Willer, S.S. Corneal endothelial repair. Regulation of prostaglandin E2 synthesis. *Investig. Ophthalmol. Vis. Sci.* **1996**, *37*, 1294–1301.

26. Albuquerque, R.J.C.; Hayashi, T.; Cho, W.G.; Kleinman, M.E.; Dridi, S.; Takeda, A.; Baffi, J.Z.; Yamada, K.; Kaneko, H.; Green, M.G.; et al. Alternatively spliced vascular endothelial growth factor receptor-2 is an essential endogenous inhibitor of lymphatic vessel growth. *Nat. Med.* **2009**, *15*, 1023–1030. [CrossRef] [PubMed]

27. Ambati, B.K.; Nozaki, M.; Singh, N.; Takeda, A.; Jani, P.D.; Suthar, T.; Albuquerque, R.J.C.; Richter, E.; Sakurai, E.; Newcomb, M.T.; et al. Corneal avascularity is due to soluble VEGF receptor-1. *Nature* **2006**, *443*, 993–997. [CrossRef] [PubMed]

28. Singh, N.; Tiem, M.; Watkins, R.; Cho, Y.K.; Wang, Y.; Olsen, T.; Uehara, H.; Mamalis, C.; Luo, L.; Oakey, Z.; Ambati, B.K. Soluble vascular endothelial growth factor receptor 3 is essential for corneal alymphaticity. *Blood* **2013**, *121*, 4242–4249. [CrossRef] [PubMed]

29. Gabison, E.; Chang, J.-H.; Hernández-Quintela, E.; Javier, J.; Lu, P.C.S.; Ye, H.; Kure, T.; Kato, T.; Azar, D.T. Anti-angiogenic role of angiostatin during corneal wound healing. *Exp. Eye Res.* **2004**, *78*, 579–589. [CrossRef] [PubMed]

30. Cursiefen, C.; Masli, S.; Ng, T.F.; Dana, M.R.; Bornstein, P.; Lawler, J.; Streilein, J.W. Roles of thrombospondin-1 and -2 in regulating corneal and iris angiogenesis. *Investig. Ophthalmol. Vis. Sci.* **2004**, *45*, 1117–1124. [CrossRef]

31. Lin, H.C.; Chang, J.H.; Jain, S.; Gabison, E.E.; Kure, T.; Kato, T.; Fukai, N.; Azar, D.T. Matrilysin cleavage of corneal collagen type XVIII NC1 domain and generation of a 28-kDa fragment. *Investig. Ophthalmol. Vis. Sci.* **2001**, *42*, 2517–2524.

32. Dawson, D.W.; Volpert, O.V.; Gillis, P.; Crawford, S.E.; Xu, H.; Benedict, W.; Bouck, N.P. Pigment epithelium-derived factor: A potent inhibitor of angiogenesis. *Science* **1999**, *285*, 245–248. [CrossRef] [PubMed]

33. Makino, Y.; Cao, R.; Svensson, K.; Bertilsson, G.; Asman, M.; Tanaka, H.; Cao, Y.; Berkenstam, A.; Poellinger, L. Inhibitory PAS domain protein is a negative regulator of hypoxia-inducible gene expression. *Nature* **2001**, *414*, 550–554. [CrossRef] [PubMed]

34. Cursiefen, C.; Küchle, M.; Naumann, G.O. Angiogenesis in corneal diseases: Histopathologic evaluation of 254 human corneal buttons with neovascularization. *Cornea* **1998**, *17*, 611–613. [CrossRef] [PubMed]

35. Cursiefen, C.; Schlötzer-Schrehardt, U.; Küchle, M.; Sorokin, L.; Breiteneder-Geleff, S.; Alitalo, K.; Jackson, D. Lymphatic vessels in vascularized human corneas: Immunohistochemical investigation using LYVE-1 and podoplanin. *Investig. Ophthalmol. Vis. Sci.* **2002**, *43*, 2127–2135.

36. Bock, F.; Maruyama, K.; Regenfuss, B.; Hos, D.; Steven, P.; Heindl, L.M.; Cursiefen, C. Novel anti(lymph)angiogenic treatment strategies for corneal and ocular surface diseases. *Prog. Retin. Eye Res.* **2013**, *34*, 89–124. [CrossRef] [PubMed]

37. Niederkorn, J.Y. Immune privilege in the anterior chamber of the eye. *Crit. Rev. Immunol.* **2002**, *22*, 13–46. [CrossRef] [PubMed]

38. Kezuka, T.; Streilein, J.W. Evidence for multiple CD95-CD95 ligand interactions in anteriorchamber-associated immune deviation induced by soluble protein antigen. *Immunology* **2000**, *99*, 451–457. [CrossRef] [PubMed]

39. Meng, Q.; Yang, P.; Li, B.; Zhou, H.; Huang, X.; Zhu, L.; Ren, Y.; Kijlstra, A. CD4+PD-1+ T cells acting as regulatory cells during the induction of anterior chamber-associated immune deviation. *Investig. Ophthalmol. Vis. Sci.* **2006**, *47*, 4444–4452. [CrossRef] [PubMed]

40. Streilein, J.W.; Masli, S.; Takeuchi, M.; Kezuka, T. The eye's view of antigen presentation. *Hum. Immunol.* **2002**, *63*, 435–443. [CrossRef]

41. Skelsey, M.E.; Mayhew, E.; Niederkorn, J.Y. CD25+, interleukin-10-producing CD4+ T cells are required for suppressor cell production and immune privilege in the anterior chamber of the eye. *Immunology* **2003**, *110*, 18–29. [CrossRef] [PubMed]

42. Sonoda, K.-H.; Stein-Streilein, J. CD1d on antigen-transporting APC and splenic marginal zone B cells promotes NKT cell-dependent tolerance. *Eur. J. Immunol.* **2002**, *32*, 848–857. [CrossRef]

43. Lin, H.-H.; Faunce, D.E.; Stacey, M.; Terajewicz, A.; Nakamura, T.; Zhang-Hoover, J.; Kerley, M.; Mucenski, M.L.; Gordon, S.; Stein-Streilein, J. The macrophage F4/80 receptor is required for the induction of antigen-specific efferent regulatory T cells in peripheral tolerance. *J. Exp. Med.* **2005**, *201*, 1615–1625. [CrossRef] [PubMed]

44. Streilein, J.W.; Niederkorn, J.Y. Induction of anterior chamber-associated immune deviation requires an intact, functional spleen. *J. Exp. Med.* **1981**, *153*, 1058–1067. [CrossRef] [PubMed]

45. Cursiefen, C. Immune privilege and angiogenic privilege of the cornea. *Chem. Immunol. Allergy* **2007**, *92*, 50–57. [PubMed]

46. Niederkorn, J.Y. High-risk corneal allografts and why they lose their immune privilege. *Curr. Opin. Allergy Clin. Immunol.* **2010**, *10*, 493–497. [CrossRef] [PubMed]

47. Bachmann, B.; Taylor, R.S.; Cursiefen, C. Corneal neovascularization as a risk factor for graft failure and rejection after keratoplasty: An evidence-based meta-analysis. *Ophthalmology* **2010**, *117*, 1300–1305. [CrossRef] [PubMed]

48. Hos, D.; Cursiefen, C. Lymphatic vessels in the development of tissue and organ rejection. *Adv. Anat. Embryol. Cell Biol.* **2014**, *214*, 119–141. [PubMed]

49. Niederkorn, J.Y. The immune privilege of corneal allografts. *Transplantation* **1999**, *67*, 1503–1508. [CrossRef] [PubMed]

50. Hamrah, P.; Huq, S.O.; Liu, Y.; Zhang, Q.; Dana, M.R. Corneal immunity is mediated by heterogeneous population of antigen-presenting cells. *J. Leukoc. Biol.* **2003**, *74*, 172–178. [CrossRef] [PubMed]

51. Brissette-Storkus, C.S.; Reynolds, S.M.; Lepisto, A.J.; Hendricks, R.L. Identification of a novel macrophage population in the normal mouse corneal stroma. *Investig. Ophthalmol. Vis. Sci.* **2002**, *43*, 2264–2271.

52. Chinnery, H.R.; Humphries, T.; Clare, A.; Dixon, A.E.; Howes, K.; Moran, C.B.; Scott, D.; Zakrzewski, M.; Pearlman, E.; McMenamin, P.G. Turnover of bone marrow-derived cells in the irradiated mouse cornea. *Immunology* **2008**, *125*, 541–548. [CrossRef] [PubMed]

53. Gordon, S.; Taylor, P.R. Monocyte and macrophage heterogeneity. *Nat. Rev. Immunol.* **2005**, *5*, 953–964. [CrossRef] [PubMed]

54. Liu, J.; Xue, Y.; Dong, D.; Xiao, C.; Lin, C.; Wang, H.; Song, F.; Fu, T.; Wang, Z.; Chen, J.; et al. CCR2⁻ and CCR2+ corneal macrophages exhibit distinct characteristics and balance inflammatory responses after epithelial abrasion. *Mucosal Immunol.* **2017**. [CrossRef] [PubMed]

55. Hamrah, P.; Zhang, Q.; Liu, Y.; Dana, M.R. Novel characterization of MHC class II-negative population of resident corneal Langerhans cell-type dendritic cells. *Investig. Ophthalmol. Vis. Sci.* **2002**, *43*, 639–646.

56. Chen, W.; Hara, K.; Tian, Q.; Zhao, K.; Yoshitomi, T. Existence of small slow-cycling Langerhans cells in the limbal basal epithelium that express ABCG2. *Exp. Eye Res.* **2007**, *84*, 626–634. [CrossRef] [PubMed]

57. Hamrah, P.; Liu, Y.; Zhang, Q.; Dana, M.R. The corneal stroma is endowed with a significant number of resident dendritic cells. *Investig. Ophthalmol. Vis. Sci.* **2003**, *44*, 581–589. [CrossRef]

58. Nakamura, T.; Ishikawa, F.; Sonoda, K.-H.; Hisatomi, T.; Qiao, H.; Yamada, J.; Fukata, M.; Ishibashi, T.; Harada, M.; Kinoshita, S. Characterization and distribution of bone marrow-derived cells in mouse cornea. *Investig. Ophthalmol. Vis. Sci.* **2005**, *46*, 497–503. [CrossRef] [PubMed]

59. Chinnery, H.R.; Ruitenberg, M.J.; Plant, G.W.; Pearlman, E.; Jung, S.; McMenamin, P.G. The chemokine receptor CX3CR1 mediates homing of MHC class II-positive cells to the normal mouse corneal epithelium. *Investig. Ophthalmol. Vis. Sci.* **2007**, *48*, 1568–1574. [CrossRef] [PubMed]

60. Ueta, M.; Koga, A.; Kikuta, J.; Yamada, K.; Kojima, S.; Shinomiya, K.; Ishii, M.; Kinoshita, S. Intravital imaging of the cellular dynamics of LysM-positive cells in a murine corneal suture model. *Br. J. Ophthalmol.* **2016**, *100*, 432–435. [CrossRef] [PubMed]

61. Liu, J.; Fu, T.; Song, F.; Xue, Y.; Xia, C.; Liu, P.; Wang, H.; Zhong, J.; Li, Q.; Chen, J.; et al. Mast cells participate in corneal development in mice. *Sci. Rep.* **2015**, *5*, 17569. [CrossRef] [PubMed]

62. Mott, K.R.; Osorio, Y.; Brown, D.J.; Morishige, N.; Wahlert, A.; Jester, J.V.; Ghiasi, H. The corneas of naive mice contain both CD4+ and CD8+ T cells. *Mol. Vis.* **2007**, *13*, 1802–1812. [PubMed]

63. Imanishi, J.; Kamiyama, K.; Iguchi, I.; Kita, M.; Sotozono, C.; Kinoshita, S. Growth factors: Importance in wound healing and maintenance of transparency of the cornea. *Prog. Retin. Eye Res.* **2000**, *19*, 113–129. [CrossRef]

64. Almadi, A.J.; Jakobiec, F.A. Corneal wound healing: Cytokines and extracellular matrix proteins. *Int. Ophthalmol. Clin.* **2002**, *42*, 13–22. [CrossRef] [PubMed]

65. Yu, F.-S.X.; Yin, J.; Xu, K.; Huang, J. Growth factors and corneal epithelial wound healing. *Brain Res. Bull.* **2010**, *81*, 229–235. [CrossRef] [PubMed]

66. Ljubimov, A.V.; Saghizadeh, M. Progress in corneal wound healing. *Prog. Retin. Eye Res.* **2015**, *49*, 17–45. [CrossRef] [PubMed]

67. Zieske, J.D. Extracellular matrix and wound healing. *Curr. Opin. Ophthalmol.* **2001**, *12*, 237–241. [CrossRef] [PubMed]

68. Buck, R.C. Hemidesmosomes of normal and regenerating mouse corneal epithelium. *Virchows Arch. B Cell Pathol. Incl. Mol. Pathol.* **1982**, *41*, 1–16. [CrossRef] [PubMed]

69. Spadea, L.; Giammaria, D.; Trabucco, P. Corneal wound healing after laser vision correction. *Br. J. Ophthalmol.* **2016**, *100*, 28–33. [CrossRef] [PubMed]

70. Müller, L.J.; Marfurt, C.F.; Kruse, F.; Tervo, T.M.T. Corneal nerves: Structure, contents and function. *Exp. Eye Res.* **2003**, *76*, 521–542. [CrossRef]

71. Garcia-Hirschfeld, J.; Lopez-Briones, L.G.; Belmonte, C. Neurotrophic influences on corneal epithelial cells. *Exp. Eye Res.* **1994**, *59*, 597–605. [CrossRef] [PubMed]

72. Reid, T.W.; Murphy, C.J.; Iwahashi, C.K.; Foster, B.A.; Mannis, M.J. Stimulation of epithelial cell growth by the neuropeptide substance P. *J. Cell. Biochem.* **1993**, *52*, 476–485. [CrossRef] [PubMed]

73. Nakamura, M.; Chikama, T.; Nishida, T. Up-regulation of integrin α5 expression by combination of substance P and insulin-like growth factor-1 in rabbit corneal epithelial cells. *Biochem. Biophys. Res. Commun.* **1998**, *246*, 777–782. [CrossRef] [PubMed]

74. Mikulec, A.A.; Tanelian, D.L. CGRP increases the rate of corneal re-epithelialization in an in vitro whole mount preparation. *J. Ocul. Pharmacol. Ther.* **1996**, *12*, 417–423. [CrossRef] [PubMed]

75. Lambiase, A.; Bonini, S.; Aloe, L.; Rama, P.; Bonini, S. Anti-inflammatory and healing properties of nerve growth factor in immune corneal ulcers with stromal melting. *Arch. Ophthalmol.* **2000**, *118*, 1446–1449. [CrossRef] [PubMed]

76. Lambiase, A.; Rama, P.; Bonini, S.; Caprioglio, G.; Aloe, L. Topical treatment with nerve growth factor for corneal neurotrophic ulcers. *N. Engl. J. Med.* **1998**, *338*, 1174–1180. [CrossRef] [PubMed]

77. Pastor, J.C.; Calonge, M. Epidermal growth factor and corneal wound healing. A multicenter study. *Cornea* **1992**, *11*, 311–314. [CrossRef] [PubMed]

78. Meduri, A.; Aragona, P.; Grenga, P.L.; Roszkowska, A.M. Effect of basic fibroblast growth factor on corneal epithelial healing after photorefractive keratectomy. *J. Refract. Surg.* **2012**, *28*, 220–223. [CrossRef] [PubMed]
79. Yamada, N.; Matsuda, R.; Morishige, N.; Yanai, R.; Chikama, T.-I.; Nishida, T.; Ishimitsu, T.; Kamiya, A. Open clinical study of eye-drops containing tetrapeptides derived from substance P and insulin-like growth factor-1 for treatment of persistent corneal epithelial defects associated with neurotrophic keratopathy. *Br. J. Ophthalmol.* **2008**, *92*, 896–900. [CrossRef] [PubMed]
80. Wilson, S.E.; He, Y.G.; Weng, J.; Li, Q.; McDowall, A.W.; Vital, M.; Chwang, E.L. Epithelial injury induces keratocyte apoptosis: Hypothesized role for the interleukin-1 system in the modulation of corneal tissue organization and wound healing. *Exp. Eye Res.* **1996**, *62*, 325–327. [CrossRef] [PubMed]
81. Mohan, R.R.; Liang, Q.; Kim, W.J.; Helena, M.C.; Baerveldt, F.; Wilson, S.E. Apoptosis in the cornea: Further characterization of Fas/Fas ligand system. *Exp. Eye Res.* **1997**, *65*, 575–589. [CrossRef] [PubMed]
82. Mohan, R.R.; Mohan, R.R.; Kim, W.J.; Wilson, S.E. Modulation of TNF-α-induced apoptosis in corneal fibroblasts by transcription factor NF-κB. *Investig. Ophthalmol. Vis. Sci.* **2000**, *41*, 1327–1336.
83. Ambrósio, R.; Kara-José, N.; Wilson, S.E. Early keratocyte apoptosis after epithelial scrape injury in the human cornea. *Exp. Eye Res.* **2009**, *89*, 597–599. [CrossRef] [PubMed]
84. Wilson, S.E.; Mohan, R.R.; Mohan, R.R.; Ambrósio, R.; Hong, J.; Lee, J. The corneal wound healing response: Cytokine-mediated interaction of the epithelium, stroma, and inflammatory cells. *Prog. Retin. Eye Res.* **2001**, *20*, 625–637. [CrossRef]
85. Petroll, W.M.; Kivanany, P.B.; Hagenasr, D.; Graham, E.K. Corneal fibroblast migration patterns during intrastromal wound healing correlate with ECM structure and alignment. *Investig. Ophthalmol. Vis. Sci.* **2015**, *56*, 7352–7361. [CrossRef] [PubMed]
86. Andresen, J.L.; Ledet, T.; Ehlers, N. Keratocyte migration and peptide growth factors: The effect of PDGF, bFGF, EGF, IGF-I, aFGF and TGF-β on human keratocyte migration in a collagen gel. *Curr. Eye Res.* **1997**, *16*, 605–613. [CrossRef] [PubMed]
87. Jester, J.V.; Barry-Lane, P.A.; Petroll, W.M.; Olsen, D.R.; Cavanagh, H.D. Inhibition of corneal fibrosis by topical application of blocking antibodies to TGF β in the rabbit. *Cornea* **1997**, *16*, 177–187. [CrossRef] [PubMed]
88. Jester, J.V.; Ho-Chang, J. Modulation of cultured corneal keratocyte phenotype by growth factors/cytokines control in vitro contractility and extracellular matrix contraction. *Exp. Eye Res.* **2003**, *77*, 581–592. [CrossRef]
89. Masur, S.K.; Dewal, H.S.; Dinh, T.T.; Erenburg, I.; Petridou, S. Myofibroblasts differentiate from fibroblasts when plated at low density. *Proc. Natl. Acad. Sci. USA* **1996**, *93*, 4219–4223. [CrossRef] [PubMed]
90. Jester, J.V.; Huang, J.; Barry-Lane, P.A.; Kao, W.W.; Petroll, W.M.; Cavanagh, H.D. Transforming growth factor(β)-mediated corneal myofibroblast differentiation requires actin and fibronectin assembly. *Investig. Ophthalmol. Vis. Sci.* **1999**, *40*, 1959–1967.
91. Jester, J.V.; Huang, J.; Petroll, W.M.; Cavanagh, H.D. TGFβ induced myofibroblast differentiation of rabbit keratocytes requires synergistic TGFβ, PDGF and integrin signaling. *Exp. Eye Res.* **2002**, *75*, 645–657. [CrossRef] [PubMed]
92. Kaur, H.; Chaurasia, S.S.; de Medeiros, F.W.; Agrawal, V.; Salomao, M.Q.; Singh, N.; Ambati, B.K.; Wilson, S.E. Corneal stroma PDGF blockade and myofibroblast development. *Exp. Eye Res.* **2009**, *88*, 960–965. [CrossRef] [PubMed]
93. He, J.; Bazan, H.E.P. Epidermal growth factor synergism with TGF-β1 via PI-3 kinase activity in corneal keratocyte differentiation. *Investig. Ophthalmol. Vis. Sci.* **2008**, *49*, 2936–2945. [CrossRef] [PubMed]
94. Garana, R.M.; Petroll, W.M.; Chen, W.T.; Herman, I.M.; Barry, P.; Andrews, P.; Cavanagh, H.D.; Jester, J.V. Radial keratotomy. II. Role of the myofibroblast in corneal wound contraction. *Investig. Ophthalmol. Vis. Sci.* **1992**, *33*, 3271–3282.
95. Jester, J.V.; Petroll, W.M.; Barry, P.A.; Cavanagh, H.D. Expression of α-smooth muscle (α-SM) actin during corneal stromal wound healing. *Investig. Ophthalmol. Vis. Sci.* **1995**, *36*, 809–819.
96. Tomasek, J.J.; Gabbiani, G.; Hinz, B.; Chaponnier, C.; Brown, R.A. Myofibroblasts and mechano-regulation of connective tissue remodelling. *Nat. Rev. Mol. Cell Biol.* **2002**, *3*, 349–363. [CrossRef] [PubMed]
97. Jester, J.V.; Moller-Pedersen, T.; Huang, J.; Sax, C.M.; Kays, W.T.; Cavangh, H.D.; Petroll, W.M.; Piatigorsky, J. The cellular basis of corneal transparency: Evidence for 'corneal crystallins'. *J. Cell Sci.* **1999**, *112*, 613–622. [PubMed]

98. Chaurasia, S.S.; Kaur, H.; de Medeiros, F.W.; Smith, S.D.; Wilson, S.E. Dynamics of the expression of intermediate filaments vimentin and desmin during myofibroblast differentiation after corneal injury. *Exp. Eye Res.* **2009**, *89*, 133–139. [CrossRef] [PubMed]
99. Torricelli, A.A.M.; Wilson, S.E. Cellular and extracellular matrix modulation of corneal stromal opacity. *Exp. Eye Res.* **2014**, *129*, 151–160. [CrossRef] [PubMed]
100. Blanco-Mezquita, J.T.; Hutcheon, A.E.K.; Zieske, J.D. Role of thrombospondin-1 in repair of penetrating corneal wounds. *Investig. Ophthalmol. Vis. Sci.* **2013**, *54*, 6262–6268. [CrossRef] [PubMed]
101. Barbosa, F.L.; Chaurasia, S.S.; Cutler, A.; Asosingh, K.; Kaur, H.; de Medeiros, F.W.; Agrawal, V.; Wilson, S.E. Corneal myofibroblast generation from bone marrow-derived cells. *Exp. Eye Res.* **2010**, *91*, 92–96. [CrossRef] [PubMed]
102. Direkze, N.C.; Forbes, S.J.; Brittan, M.; Hunt, T.; Jeffery, R.; Preston, S.L.; Poulsom, R.; Hodivala-Dilke, K.; Alison, M.R.; Wright, N.A. Multiple organ engraftment by bone-marrow-derived myofibroblasts and fibroblasts in bone-marrow-transplanted mice. *Stem Cells* **2003**, *21*, 514–520. [CrossRef] [PubMed]
103. Singh, V.; Agrawal, V.; Santhiago, M.R.; Wilson, S.E. Stromal fibroblast-bone marrow-derived cell interactions: Implications for myofibroblast development in the cornea. *Exp. Eye Res.* **2012**, *98*, 1–8. [CrossRef] [PubMed]
104. Roy, O.; Leclerc, V.B.; Bourget, J.-M.; Thériault, M.; Proulx, S. Understanding the process of corneal endothelial morphological change in vitro. *Investig. Ophthalmol. Vis. Sci.* **2015**, *56*, 1228–1237. [CrossRef] [PubMed]
105. Kaimori, A.; Potter, J.; Kaimori, J.; Wang, C.; Mezey, E.; Koteish, A. Transforming growth factor-β1 induces an epithelial-to-mesenchymal transition state in mouse hepatocytes in vitro. *J. Biol. Chem.* **2007**, *282*, 22089–22101. [CrossRef] [PubMed]
106. Lee, J.G.; Ko, M.K.; Kay, E.P. Endothelial mesenchymal transformation mediated by IL-1β-induced FGF-2 in corneal endothelial cells. *Exp. Eye Res.* **2012**, *95*, 35–39. [CrossRef] [PubMed]
107. Lee, J.G.; Kay, E.P. NF-κB is the transcription factor for FGF-2 that causes endothelial mesenchymal transformation in cornea. *Investig. Ophthalmol. Vis. Sci.* **2012**, *53*, 1530–1538. [CrossRef] [PubMed]
108. Okumura, N.; Koizumi, N.; Kay, E.P.; Ueno, M.; Sakamoto, Y.; Nakamura, S.; Hamuro, J.; Kinoshita, S. The ROCK inhibitor eye drop accelerates corneal endothelium wound healing. *Investig. Ophthalmol. Vis. Sci.* **2013**, *54*, 2493–2502. [CrossRef] [PubMed]
109. Miyamoto, T.; Sumioka, T.; Saika, S. Endothelial mesenchymal transition: A therapeutic target in retrocorneal membrane. *Cornea* **2010**, *29*, S52–S56. [CrossRef] [PubMed]
110. Okumura, N.; Kinoshita, S.; Koizumi, N. Cell-based approach for treatment of corneal endothelial dysfunction. *Cornea* **2014**, *33*, S37–S41. [CrossRef] [PubMed]
111. Sumioka, T.; Ikeda, K.; Okada, Y.; Yamanaka, O.; Kitano, A.; Saika, S. Inhibitory effect of blocking TGF-β/Smad signal on injury-induced fibrosis of corneal endothelium. *Mol. Vis.* **2008**, *14*, 2272–2281. [PubMed]
112. Okumura, N.; Kay, E.P.; Nakahara, M.; Hamuro, J.; Kinoshita, S.; Koizumi, N. Inhibition of TGF-β signaling enables human corneal endothelial cell expansion in vitro for use in regenerative medicine. *PLoS ONE* **2013**, *8*, e58000. [CrossRef] [PubMed]
113. Andresen, J.L.; Ledet, T.; Hager, H.; Josephsen, K.; Ehlers, N. The influence of corneal stromal matrix proteins on the migration of human corneal fibroblasts. *Exp. Eye Res.* **2000**, *71*, 33–43. [CrossRef] [PubMed]
114. Hassell, J.R.; Cintron, C.; Kublin, C.; Newsome, D.A. Proteoglycan changes during restoration of transparency in corneal scars. *Arch. Biochem. Biophys.* **1983**, *222*, 362–369. [CrossRef]
115. Cintron, C.; Gregory, J.D.; Damle, S.P.; Kublin, C.L. Biochemical analyses of proteoglycans in rabbit corneal scars. *Investig. Ophthalmol. Vis. Sci.* **1990**, *31*, 1975–1981.
116. Stramer, B.M.; Zieske, J.D.; Jung, J.-C.; Austin, J.S.; Fini, M.E. Molecular mechanisms controlling the fibrotic repair phenotype in cornea: Implications for surgical outcomes. *Investig. Ophthalmol. Vis. Sci.* **2003**, *44*, 4237–4246. [CrossRef]
117. Stramer, B.M.; Austin, J.S.; Roberts, A.B.; Fini, M.E. Selective reduction of fibrotic markers in repairing corneas of mice deficient in Smad3. *J. Cell. Physiol.* **2005**, *203*, 226–232. [CrossRef] [PubMed]
118. Li, Z.; Rumbaut, R.E.; Burns, A.R.; Smith, C.W. Platelet response to corneal abrasion is necessary for acute inflammation and efficient re-epithelialization. *Investig. Ophthalmol. Vis. Sci.* **2006**, *47*, 4794–4802. [CrossRef] [PubMed]
119. Gupta, A.; Monroy, D.; Ji, Z.; Yoshino, K.; Huang, A.; Pflugfelder, S.C. Transforming growth factor β-1 and β-2 in human tear fluid. *Curr. Eye Res.* **1996**, *15*, 605–614. [CrossRef] [PubMed]

120. Vesaluoma, M.; Teppo, A.M.; Grönhagen-Riska, C.; Tervo, T. Platelet-derived growth factor-BB (PDGF-BB) in tear fluid: A potential modulator of corneal wound healing following photorefractive keratectomy. *Curr. Eye Res.* **1997**, *16*, 825–831. [CrossRef] [PubMed]

121. Kim, W.J.; Mohan, R.R.; Mohan, R.R.; Wilson, S.E. Effect of PDGF, IL-1α, and BMP2/4 on corneal fibroblast chemotaxis: Expression of the platelet-derived growth factor system in the cornea. *Investig. Ophthalmol. Vis. Sci.* **1999**, *40*, 1364–1372.

122. Kamiyama, K.; Iguchi, I.; Wang, X.; Imanishi, J. Effects of PDGF on the migration of rabbit corneal fibroblasts and epithelial cells. *Cornea* **1998**, *17*, 315–325. [CrossRef] [PubMed]

123. Han, K.-Y.; Tran, J.A.; Chang, J.-H.; Azar, D.T.; Zieske, J.D. Potential role of corneal epithelial cell-derived exosomes in corneal wound healing and neovascularization. *Sci. Rep.* **2017**, *7*, 40548. [CrossRef] [PubMed]

124. Spurlin, J.W.; Lwigale, P.Y. Wounded embryonic corneas exhibit nonfibrotic regeneration and complete innervation. *Investig. Ophthalmol. Vis. Sci.* **2013**, *54*, 6334–6344. [CrossRef] [PubMed]

125. Lee, J.Y.-Y.; Yang, C.-C.; Chao, S.-C.; Wong, T.-W. Histopathological differential diagnosis of keloid and hypertrophic scar. *Am. J. Dermatopathol.* **2004**, *26*, 379–384. [CrossRef] [PubMed]

126. Bourcier, T.; Baudrimont, M.; Boutboul, C.; Thomas, F.; Borderie, V.; Laroche, L. Corneal keloid: Clinical, ultrasonographic, and ultrastructural characteristics. *J. Cataract Refract. Surg.* **2004**, *30*, 921–924. [CrossRef] [PubMed]

127. Bakhtiari, P.; Agarwal, D.R.; Fernandez, A.A.; Milman, T.; Glasgow, B.; Starr, C.E.; Aldave, A.J. Corneal keloid: Report of natural history and outcome of surgical management in two cases. *Cornea* **2013**, *32*, 1621–1624. [CrossRef] [PubMed]

128. Lee, H.K.; Choi, H.J.; Kim, M.K.; Wee, W.R.; Oh, J.Y. Corneal keloid: Four case reports of clinicopathological features and surgical outcome. *BMC Ophthalmol.* **2016**, *16*, 198. [CrossRef] [PubMed]

129. Holbach, L.M.; Font, R.L.; Shivitz, I.A.; Jones, D.B. Bilateral keloid-like myofibroblastic proliferations of the cornea in children. *Ophthalmology* **1990**, *97*, 1188–1193. [CrossRef]

130. Mejía, L.F.; Acosta, C.; Santamaría, J.P. Clinical, surgical, and histopathologic characteristics of corneal keloid. *Cornea* **2001**, *20*, 421–424. [CrossRef] [PubMed]

131. Jung, J.J.; Wojno, T.H.; Grossniklaus, H.E. Giant corneal keloid: Case report and review of the literature. *Cornea* **2010**, *29*, 1455–1458. [CrossRef] [PubMed]

132. McElvanney, A.M.; Adhikary, H.P. Corneal keloid: Aetiology and management in Lowe's syndrome. *Eye* **1995**, *9*, 375–376. [CrossRef] [PubMed]

133. Esquenazi, S.; Eustis, H.S.; Bazan, H.E.; Leon, A.; He, J. Corneal keloid in Lowe syndrome. *J. Pediatr. Ophthalmol. Strabismus* **2005**, *42*, 308–310. [PubMed]

134. Rao, S.K.; Fan, D.S.P.; Pang, C.P.; Li, W.W.Y.; Ng, J.S.K.; Good, W.V.; Lam, D.S.C. Bilateral congenital corneal keloids and anterior segment mesenchymal dysgenesis in a case of Rubinstein-Taybi syndrome. *Cornea* **2002**, *21*, 126–130. [CrossRef] [PubMed]

135. Hendrix, J.D.; Greer, K.E. Rubinstein-Taybi syndrome with multiple flamboyant keloids. *Cutis* **1996**, *57*, 346–348. [PubMed]

136. Van de Kar, A.L.; Houge, G.; Shaw, A.C.; de Jong, D.; van Belzen, M.J.; Peters, D.J.M.; Hennekam, R.C.M. Keloids in Rubinstein-Taybi syndrome: A clinical study. *Br. J. Dermatol.* **2014**, *171*, 615–621. [CrossRef] [PubMed]

137. Shilpashree, P.; Jaiswal, A.K.; Kharge, P.M. Keloids: An unwanted spontaneity in rubinstein-taybi syndrome. *Indian J. Dermatol.* **2015**, *60*, 214. [PubMed]

138. Gipson, I.K.; Kiorpes, T.C. Epithelial sheet movement: Protein and glycoprotein synthesis. *Dev. Biol.* **1982**, *92*, 259–262. [CrossRef]

139. Dana, M.R.; Streilein, J.W. Loss and restoration of immune privilege in eyes with corneal neovascularization. *Investig. Ophthalmol. Vis. Sci.* **1996**, *37*, 2485–2494.

140. Mahoney, J.M.; Waterbury, L.D. Drug effects on the neovascularization response to silver nitrate cauterization of the rat cornea. *Curr. Eye Res.* **1985**, *4*, 531–535. [CrossRef] [PubMed]

141. Stepp, M.A.; Zieske, J.D.; Trinkaus-Randall, V.; Kyne, B.M.; Pal-Ghosh, S.; Tadvalkar, G.; Pajoohesh-Ganji, A. Wounding the cornea to learn how it heals. *Exp. Eye Res.* **2014**, *121*, 178–193. [CrossRef] [PubMed]

142. Pal-Ghosh, S.; Tadvalkar, G.; Jurjus, R.A.; Zieske, J.D.; Stepp, M.A. BALB/c and C57BL6 mouse strains vary in their ability to heal corneal epithelial debridement wounds. *Exp. Eye Res.* **2008**, *87*, 478–486. [CrossRef] [PubMed]

143. Tran, M.T.; Tellaetxe-Isusi, M.; Elner, V.; Strieter, R.M.; Lausch, R.N.; Oakes, J.E. Proinflammatory cytokines induce RANTES and MCP-1 synthesis in human corneal keratocytes but not in corneal epithelial cells. B-chemokine synthesis in corneal cells. *Investig. Ophthalmol. Vis. Sci.* **1996**, *37*, 987–996.

144. Hong, J.W.; Liu, J.J.; Lee, J.S.; Mohan, R.R.; Mohan, R.R.; Woods, D.J.; He, Y.G.; Wilson, S.E. Proinflammatory chemokine induction in keratocytes and inflammatory cell infiltration into the cornea. *Investig. Ophthalmol. Vis. Sci.* **2001**, *42*, 2795–2803.

145. Stapleton, W.M.; Chaurasia, S.S.; Medeiros, F.W.; Mohan, R.R.; Sinha, S.; Wilson, S.E. Topical interleukin-1 receptor antagonist inhibits inflammatory cell infiltration into the cornea. *Exp. Eye Res.* **2008**, *86*, 753–757. [CrossRef] [PubMed]

146. Ebihara, N.; Matsuda, A.; Nakamura, S.; Matsuda, H.; Murakami, A. Role of the IL-6 classic- and trans-signaling pathways in corneal sterile inflammation and wound healing. *Investig. Ophthalmol. Vis. Sci.* **2011**, *52*, 8549–8557. [CrossRef] [PubMed]

147. Sotozono, C.; He, J.; Matsumoto, Y.; Kita, M.; Imanishi, J.; Kinoshita, S. Cytokine expression in the alkali-burned cornea. *Curr. Eye Res.* **1997**, *16*, 670–676. [CrossRef] [PubMed]

148. Brien, T.P.O.; Li, Q.; Ashraf, M.F.; Matteson, D.M.; Stark, W.J.; Chan, C.C. Inflammatory response in the early stages of wound healing after excimer laser keratectomy. *Arch. Ophthalmol.* **1998**, *116*, 1470–1474. [CrossRef]

149. Li, Z.; Burns, A.R.; Smith, C.W. Two waves of neutrophil emigration in response to corneal epithelial abrasion: Distinct adhesion molecule requirements. *Investig. Ophthalmol. Vis. Sci.* **2006**, *47*, 1947–1955. [CrossRef] [PubMed]

150. Marrazzo, G.; Bellner, L.; Halilovic, A.; Volti, G.L.; Drago, F.; Dunn, M.W.; Schwartzman, M.L. The role of neutrophils in corneal wound healing in HO-2 null mice. *PLoS ONE* **2011**, *6*, e21180. [CrossRef] [PubMed]

151. Li, Z.; Burns, A.R.; Han, L.; Rumbaut, R.E.; Smith, C.W. IL-17 and VEGF are necessary for efficient corneal nerve regeneration. *Am. J. Pathol.* **2011**, *178*, 1106–1116. [CrossRef] [PubMed]

152. Pan, Z.; Fukuoka, S.; Karagianni, N.; Guaiquil, V.H.; Rosenblatt, M.I. Vascular endothelial growth factor promotes anatomical and functional recovery of injured peripheral nerves in the avascular cornea. *FASEB J.* **2013**, *27*, 2756–2767. [CrossRef] [PubMed]

153. Cursiefen, C.; Chen, L.; Borges, L.P.; Jackson, D.; Cao, J.; Radziejewski, C.; DÁmore, P.A.; Dana, M.R.; Wiegand, S.J.; Streilein, J.W. VEGF-A stimulates lymphangiogenesis and hemangiogenesis in inflammatory neovascularization via macrophage recruitment. *J. Clin. Investig.* **2004**, *113*, 1040–1050. [CrossRef] [PubMed]

154. Watari, K.; Nakao, S.; Fotovati, A.; Basaki, Y.; Hosoi, F.; Bereczky, B.; Higuchi, R.; Miyamoto, T.; Kuwano, M.; Ono, M. Role of macrophages in inflammatory lymphangiogenesis: Enhanced production of vascular endothelial growth factor C and D through NF-κB activation. *Biochem. Biophys. Res. Commun.* **2008**, *377*, 826–831. [CrossRef] [PubMed]

155. Maruyama, K.; Ii, M.; Cursiefen, C.; Jackson, D.G.; Keino, H.; Tomita, M.; van Rooijen, N.; Takenaka, H.; DÁmore, P.A.; Stein-Streilein, J.; et al. Inflammation-induced lymphangiogenesis in the cornea arises from CD11b-positive macrophages. *J. Clin. Investig.* **2005**, *115*, 2363–2372. [CrossRef] [PubMed]

156. Barnett, F.H.; Rosenfeld, M.; Wood, M.; Kiosses, W.B.; Usui, Y.; Marchetti, V.; Aguilar, E.; Friedlander, M. Macrophages form functional vascular mimicry channels in vivo. *Sci. Rep.* **2016**, *6*, 35569. [CrossRef] [PubMed]

157. Li, S.; Li, B.; Jiang, H.; Wang, Y.; Qu, M.; Duan, H.; Zhou, Q.; Shi, W. Macrophage depletion impairs corneal wound healing after autologous transplantation in mice. *PLoS ONE* **2013**, *8*, e61799. [CrossRef] [PubMed]

158. Bellner, L.; Marrazzo, G.; van Rooijen, N.; Dunn, M.W.; Abraham, N.G.; Schwartzman, M.L. Heme oxygenase-2 deletion impairs macrophage function: Implication in wound healing. *FASEB J.* **2015**, *29*, 105–115. [CrossRef] [PubMed]

159. Yamagami, S.; Hamrah, P.; Miyamoto, K.; Miyazaki, D.; Dekaris, I.; Dawson, T.; Lu, B.; Gerard, C.; Dana, M.R. CCR5 chemokine receptor mediates recruitment of MHC class II-positive Langerhans cells in the mouse corneal epithelium. *Investig. Ophthalmol. Vis. Sci.* **2005**, *46*, 1201–1207. [CrossRef] [PubMed]

160. Chen, W.; Lin, H.; Dong, N.; Sanae, T.; Liu, Z.; Yoshitomi, T. Cauterization of central cornea induces recruitment of major histocompatibility complex class II+ Langerhans cells from limbal basal epithelium. *Cornea* **2010**, *29*, 73–79. [CrossRef] [PubMed]

161. Hamrah, P.; Liu, Y.; Zhang, Q.; Dana, M.R. Alterations in corneal stromal dendritic cell phenotype and distribution in inflammation. *Arch. Ophthalmol.* **2003**, *121*, 1132–1140. [CrossRef] [PubMed]

162. Gao, Y.; Li, Z.; Hassan, N.; Mehta, P.; Burns, A.R.; Tang, X.; Smith, C.W. NK cells are necessary for recovery of corneal CD11c+ dendritic cells after epithelial abrasion injury. *J. Leukoc. Biol.* **2013**, *94*, 343–351. [CrossRef] [PubMed]

163. Gao, N.; Yin, J.; Yoon, G.S.; Mi, Q.-S.; Yu, F.-S.X. Dendritic cell-epithelium interplay is a determinant factor for corneal epithelial wound repair. *Am. J. Pathol.* **2011**, *179*, 2243–2253. [CrossRef] [PubMed]

164. Liu, Q.; Smith, C.W.; Zhang, W.; Burns, A.R.; Li, Z. NK cells modulate the inflammatory response to corneal epithelial abrasion and thereby support wound healing. *Am. J. Pathol.* **2012**, *181*, 452–462. [CrossRef] [PubMed]

165. Li, Z.; Burns, A.R.; Rumbaut, R.E.; Smith, C.W. Gamma delta T cells are necessary for platelet and neutrophil accumulation in limbal vessels and efficient epithelial repair after corneal abrasion. *Am. J. Pathol.* **2007**, *171*, 838–845. [CrossRef] [PubMed]

166. Byeseda, S.E.; Burns, A.R.; Dieffenbaugher, S.; Rumbaut, R.E.; Smith, C.W.; Li, Z. ICAM-1 is necessary for epithelial recruitment of gammadelta T cells and efficient corneal wound healing. *Am. J. Pathol.* **2009**, *175*, 571–579. [CrossRef] [PubMed]

167. Brien, R.L.O.; Taylor, M.A.; Hartley, J.; Nuhsbaum, T.; Dugan, S.; Lahmers, K.; Aydintug, M.K.; Wands, J.M.; Roark, C.L.; Born, W.K. Protective role of gammadelta T cells in spontaneous ocular inflammation. *Investig. Ophthalmol. Vis. Sci.* **2009**, *50*, 3266–3274. [CrossRef] [PubMed]

168. Spandau, U.H.M.; Toksoy, A.; Verhaart, S.; Gillitzer, R.; Kruse, F.E. High expression of chemokines Gro-α (CXCL-1), IL-8 (CXCL-8), and MCP-1 (CCL-2) in inflamed human corneas in vivo. *Arch. Ophthalmol.* **2003**, *121*, 825–831. [CrossRef] [PubMed]

169. Jin, Y.; Shen, L.; Chong, E.-M.; Hamrah, P.; Zhang, Q.; Chen, L.; Dana, M.R. The chemokine receptor CCR7 mediates corneal antigen-presenting cell trafficking. *Mol. Vis.* **2007**, *13*, 626–634. [PubMed]

170. Oshima, T.; Sonoda, K.-H.; Tsutsumi-Miyahara, C.; Qiao, H.; Hisatomi, T.; Nakao, S.; Hamano, S.; Egashira, K.; Charo, I.F.; Ishibashi, T. Analysis of corneal inflammation induced by cauterisation in CCR2 and MCP-1 knockout mice. *Br. J. Ophthalmol.* **2006**, *90*, 218–222. [CrossRef] [PubMed]

171. Ebihara, N.; Yamagami, S.; Yokoo, S.; Amano, S.; Murakami, A. Involvement of C-C chemokine ligand 2-CCR2 interaction in monocyte-lineage cell recruitment of normal human corneal stroma. *J. Immunol.* **2007**, *178*, 3288–3292. [CrossRef] [PubMed]

172. Goede, V.; Brogelli, L.; Ziche, M.; Augustin, H.G. Induction of inflammatory angiogenesis by monocyte chemoattractant protein-1. *Int. J. Cancer* **1999**, *82*, 765–770. [CrossRef]

173. Goyal, S.; Chauhan, S.K.; Zhang, Q.; Dana, R. Amelioration of murine dry eye disease by topical antagonist to chemokine receptor 2. *Arch. Ophthalmol.* **2009**, *127*, 882–887. [CrossRef] [PubMed]

174. Willenborg, S.; Lucas, T.; van Loo, G.; Knipper, J.A.; Krieg, T.; Haase, I.; Brachvogel, B.; Hammerschmidt, M.; Nagy, A.; Ferrara, N.; et al. CCR2 recruits an inflammatory macrophage subpopulation critical for angiogenesis in tissue repair. *Blood* **2012**, *120*, 613–625. [CrossRef] [PubMed]

175. Uchiyama, M.; Shimizu, A.; Masuda, Y.; Nagasaka, S.; Fukuda, Y.; Takahashi, H. An ophthalmic solution of a peroxisome proliferator-activated receptor gamma agonist prevents corneal inflammation in a rat alkali burn model. *Mol. Vis.* **2013**, *19*, 2135–2150. [PubMed]

176. Dua, H.S.; Azuara-Blanco, A. Limbal stem cells of the corneal epithelium. *Surv. Ophthalmol.* **2000**, *44*, 415–425. [CrossRef]

177. Branch, M.J.; Hashmani, K.; Dhillon, P.; Jones, D.R.E.; Dua, H.S.; Hopkinson, A. Mesenchymal stem cells in the human corneal limbal stroma. *Investig. Ophthalmol. Vis. Sci.* **2012**, *53*, 5109–5116. [CrossRef] [PubMed]

178. Funderburgh, M.L.; Du, Y.; Mann, M.M.; SundarRaj, N.; Funderburgh, J.L. PAX6 expression identifies progenitor cells for corneal keratocytes. *FASEB J.* **2005**, *19*, 1371–1373. [CrossRef] [PubMed]

179. O'callaghan, A.R.; Daniels, J.T. Concise review: Limbal epithelial stem cell therapy: Controversies and challenges. *Stem Cells* **2011**, *29*, 1923–1932. [CrossRef] [PubMed]

180. Haagdorens, M.; van Acker, S.I.; van Gerwen, V.; Ní Dhubhghaill, S.; Koppen, C.; Tassignon, M.-J.; Zakaria, N. Limbal Stem Cell Deficiency: Current Treatment Options and Emerging Therapies. *Stem Cells Int.* **2016**, *2016*, 9798374. [CrossRef] [PubMed]

181. Du, Y.; Funderburgh, M.L.; Mann, M.M.; SundarRaj, N.; Funderburgh, J.L. Multipotent stem cells in human corneal stroma. *Stem Cells* **2005**, *23*, 1266–1275. [CrossRef] [PubMed]

182. Funderburgh, J.L.; Funderburgh, M.L.; Du, Y. Stem Cells in the Limbal Stroma. *Ocul. Surf.* **2016**, *14*, 113–120. [CrossRef] [PubMed]

183. Du, Y.; Sundarraj, N.; Funderburgh, M.L.; Harvey, S.A.; Birk, D.E.; Funderburgh, J.L. Secretion and organization of a cornea-like tissue in vitro by stem cells from human corneal stroma. *Investig. Ophthalmol. Vis. Sci.* **2007**, *48*, 5038–5045. [CrossRef] [PubMed]

184. Wu, J.; Du, Y.; Watkins, S.C.; Funderburgh, J.L.; Wagner, W.R. The engineering of organized human corneal tissue through the spatial guidance of corneal stromal stem cells. *Biomaterials* **2012**, *33*, 1343–1352. [CrossRef] [PubMed]

185. Karamichos, D.; Funderburgh, M.L.; Hutcheon, A.E.K.; Zieske, J.D.; Du, Y.; Wu, J.; Funderburgh, J.L. A role for topographic cues in the organization of collagenous matrix by corneal fibroblasts and stem cells. *PLoS ONE* **2014**, *9*, e86260. [CrossRef] [PubMed]

186. Wu, J.; Rnjak-Kovacina, J.; Du, Y.; Funderburgh, M.L.; Kaplan, D.L.; Funderburgh, J.L. Corneal stromal bioequivalents secreted on patterned silk substrates. *Biomaterials* **2014**, *35*, 3744–3755. [CrossRef] [PubMed]

187. Zhang, X.; Sun, H.; Li, X.; Yuan, X.; Zhang, L.; Zhao, S. Utilization of human limbal mesenchymal cells as feeder layers for human limbal stem cells cultured on amniotic membrane. *J. Tissue Eng. Regen. Med.* **2010**, *4*, 38–44. [CrossRef] [PubMed]

188. Xie, H.-T.; Chen, S.-Y.; Li, G.-G.; Tseng, S.C.G. Isolation and expansion of human limbal stromal niche cells. *Investig. Ophthalmol. Vis. Sci.* **2012**, *53*, 279–286. [CrossRef] [PubMed]

189. Basu, S.; Hertsenberg, A.J.; Funderburgh, M.L.; Burrow, M.K.; Mann, M.M.; Du, Y.; Lathrop, K.L.; Syed-Picard, F.N.; Adams, S.M.; Birk, D.E.; et al. Human limbal biopsy-derived stromal stem cells prevent corneal scarring. *Sci. Transl. Med.* **2014**, *6*, 266ra172. [CrossRef] [PubMed]

190. Du, Y.; Carlson, E.C.; Funderburgh, M.L.; Birk, D.E.; Pearlman, E.; Guo, N.; Kao, W.W.-Y.; Funderburgh, J.L. Stem cell therapy restores transparency to defective murine corneas. *Stem Cells* **2009**, *27*, 1635–1642. [CrossRef] [PubMed]

191. Garg, P.; Krishna, P.V.; Stratis, A.K.; Gopinathan, U. The value of corneal transplantation in reducing blindness. *Eye* **2005**, *19*, 1106–1114. [CrossRef] [PubMed]

192. Congdon, N.G.; Friedman, D.S.; Lietman, T. Important causes of visual impairment in the world today. *JAMA* **2003**, *290*, 2057–2060. [CrossRef] [PubMed]

193. Whitcher, J.P.; Srinivasan, M.; Upadhyay, M.P. Corneal blindness: A global perspective. *Bull. World Health Organ.* **2001**, *79*, 214–221. [PubMed]

194. Rama, P.; Matuska, S.; Paganoni, G.; Spinelli, A.; de Luca, M.; Pellegrini, G. Limbal stem-cell therapy and long-term corneal regeneration. *N. Engl. J. Med.* **2010**, *363*, 147–155. [CrossRef] [PubMed]

195. Harkin, D.G.; Foyn, L.; Bray, L.J.; Sutherland, A.J.; Li, F.J.; Cronin, B.G. Concise reviews: Can mesenchymal stromal cells differentiate into corneal cells? A systematic review of published data. *Stem Cells* **2015**, *33*, 785–791. [CrossRef] [PubMed]

196. Nishida, K.; Yamato, M.; Hayashida, Y.; Watanabe, K.; Yamamoto, K.; Adachi, E.; Nagai, S.; Kikuchi, A.; Maeda, N.; Watanabe, H.; et al. Corneal reconstruction with tissue-engineered cell sheets composed of autologous oral mucosal epithelium. *N. Engl. J. Med.* **2004**, *351*, 1187–1196. [CrossRef] [PubMed]

197. Nakamura, T.; Takeda, K.; Inatomi, T.; Sotozono, C.; Kinoshita, S. Long-term results of autologous cultivated oral mucosal epithelial transplantation in the scar phase of severe ocular surface disorders. *Br. J. Ophthalmol.* **2011**, *95*, 942–946. [CrossRef] [PubMed]

198. Tanioka, H.; Kawasaki, S.; Yamasaki, K.; Ang, L.P.K.; Koizumi, N.; Nakamura, T.; Yokoi, N.; Komuro, A.; Inatomi, T.; Kinoshita, S. Establishment of a cultivated human conjunctival epithelium as an alternative tissue source for autologous corneal epithelial transplantation. *Investig. Ophthalmol. Vis. Sci.* **2006**, *47*, 3820–3827. [CrossRef] [PubMed]

199. Ono, K.; Yokoo, S.; Mimura, T.; Usui, T.; Miyata, K.; Araie, M.; Yamagami, S.; Amano, S. Autologous transplantation of conjunctival epithelial cells cultured on amniotic membrane in a rabbit model. *Mol. Vis.* **2007**, *13*, 1138–1143. [PubMed]

200. Ricardo, J.R.S.; Cristovam, P.C.; Filho, P.A.N.; Farias, C.C.; de Araujo, A.L.; Loureiro, R.R.; Covre, J.L.; de Barros, J.N.; Barreiro, T.P.; dos Santos, M.S.; et al. Transplantation of conjunctival epithelial cells cultivated ex vivo in patients with total limbal stem cell deficiency. *Cornea* **2013**, *32*, 221–228. [CrossRef] [PubMed]

201. Saichanma, S.; Bunyaratvej, A.; Sila-Asna, M. In vitro transdifferentiation of corneal epithelial-like cells from human skin-derived precursor cells. *Int. J. Ophthalmol.* **2012**, *5*, 158–163. [PubMed]

202. Meyer-Blazejewska, E.A.; Call, M.K.; Yamanaka, O.; Liu, H.; Schlötzer-Schrehardt, U.; Kruse, F.E.; Kao, W.W. From hair to cornea: Toward the therapeutic use of hair follicle-derived stem cells in the treatment of limbal stem cell deficiency. *Stem Cells* **2011**, *29*, 57–66. [CrossRef] [PubMed]

203. Yang, X.; Moldovan, N.I.; Zhao, Q.; Mi, S.; Zhou, Z.; Chen, D.; Gao, Z.; Tong, D.; Dou, Z. Reconstruction of damaged cornea by autologous transplantation of epidermal adult stem cells. *Mol. Vis.* **2008**, *14*, 1064–1070. [PubMed]

204. Syed-Picard, F.N.; Du, Y.; Lathrop, K.L.; Mann, M.M.; Funderburgh, M.L.; Funderburgh, J.L. Dental pulp stem cells: A new cellular resource for corneal stromal regeneration. *Stem Cells Transl. Med.* **2015**, *4*, 276–285. [CrossRef] [PubMed]

205. Hayashi, R.; Ishikawa, Y.; Ito, M.; Kageyama, T.; Takashiba, K.; Fujioka, T.; Tsujikawa, M.; Miyoshi, H.; Yamato, M.; Nakamura, Y.; et al. Generation of corneal epithelial cells from induced pluripotent stem cells derived from human dermal fibroblast and corneal limbal epithelium. *PLoS ONE* **2012**, *7*, e45435. [CrossRef] [PubMed]

206. Naylor, R.W.; McGhee, C.N.J.; Cowan, C.A.; Davidson, A.J.; Holm, T.M.; Sherwin, T. Derivation of Corneal Keratocyte-Like Cells from Human Induced Pluripotent Stem Cells. *PLoS ONE* **2016**, *11*, e0165464. [CrossRef] [PubMed]

207. Yu, D.; Chen, M.; Sun, X.; Ge, J. Differentiation of mouse induced pluripotent stem cells into corneal epithelial-like cells. *Cell Biol. Int.* **2013**, *37*, 87–94. [CrossRef] [PubMed]

208. Sareen, D.; Saghizadeh, M.; Ornelas, L.; Winkler, M.A.; Narwani, K.; Sahabian, A.; Funari, V.A.; Tang, J.; Spurka, L.; Punj, V.; et al. Differentiation of human limbal-derived induced pluripotent stem cells into limbal-like epithelium. *Stem Cells Transl. Med.* **2014**, *3*, 1002–1012. [CrossRef] [PubMed]

209. Foster, J.W.; Wahlin, K.; Adams, S.M.; Birk, D.E.; Zack, D.J.; Chakravarti, S. Cornea organoids from human induced pluripotent stem cells. *Sci. Rep.* **2017**, *7*, 41286. [CrossRef] [PubMed]

210. Chien, Y.; Liao, Y.-W.; Liu, D.-M.; Lin, H.-L.; Chen, S.-J.; Chen, H.-L.; Peng, C.-H.; Liang, C.-M.; Mou, C.-Y.; Chiou, S.-H. Corneal repair by human corneal keratocyte-reprogrammed iPSCs and amphiphatic carboxymethyl-hexanoyl chitosan hydrogel. *Biomaterials* **2012**, *33*, 8003–8016. [CrossRef] [PubMed]

211. Monteiro, B.G.; Serafim, R.C.; Melo, G.B.; Silva, M.C.P.; Lizier, N.F.; Maranduba, C.M.C.; Smith, R.L.; Kerkis, A.; Cerruti, H.; Gomes, J.A.P.; et al. Human immature dental pulp stem cells share key characteristic features with limbal stem cells. *Cell Prolif.* **2009**, *42*, 587–594. [CrossRef] [PubMed]

212. Gomes, J.A.P.; Monteiro, B.G.; Melo, G.B.; Smith, R.L.; da Silva, M.C.P.; Lizier, N.F.; Kerkis, A.; Cerruti, H.; Kerkis, I. Corneal reconstruction with tissue-engineered cell sheets composed of human immature dental pulp stem cells. *Investig. Ophthalmol. Vis. Sci.* **2010**, *51*, 1408–1414. [CrossRef] [PubMed]

213. Gu, S.; Xing, C.; Han, J.; Tso, M.O.M.; Hong, J. Differentiation of rabbit bone marrow mesenchymal stem cells into corneal epithelial cells in vivo and ex vivo. *Mol. Vis.* **2009**, *15*, 99–107. [PubMed]

214. Katikireddy, K.R.; Dana, R.; Jurkunas, U.V. Differentiation potential of limbal fibroblasts and bone marrow mesenchymal stem cells to corneal epithelial cells. *Stem Cells* **2014**, *32*, 717–729. [CrossRef] [PubMed]

215. Du, Y.; Roh, D.S.; Funderburgh, M.L.; Mann, M.M.; Marra, K.G.; Rubin, J.P.; Li, X.; Funderburgh, J.L. Adipose-derived stem cells differentiate to keratocytes in vitro. *Mol. Vis.* **2010**, *16*, 1680–2689.

216. Liu, H.; Zhang, J.; Liu, C.-Y.; Hayashi, Y.; Kao, W.W.-Y. Bone marrow mesenchymal stem cells can differentiate and assume corneal keratocyte phenotype. *J. Cell. Mol. Med.* **2012**, *16*, 1114–1124. [CrossRef] [PubMed]

217. Demirayak, B.; Yüksel, N.; Çelik, O.S.; Subaşı, C.; Duruksu, G.; Unal, Z.S.; Yıldız, D.K.; Karaöz, E. Effect of bone marrow and adipose tissue-derived mesenchymal stem cells on the natural course of corneal scarring after penetrating injury. *Exp. Eye Res.* **2016**, *151*, 227–235. [CrossRef] [PubMed]

218. Yao, L.; Li, Z.-R.; Su, W.-R.; Li, Y.-P.; Lin, M.-L.; Zhang, W.-X.; Liu, Y.; Wan, Q.; Liang, D. Role of mesenchymal stem cells on cornea wound healing induced by acute alkali burn. *PLoS ONE* **2012**, *7*, e30842. [CrossRef] [PubMed]

219. Almaliotis, D.; Koliakos, G.; Papakonstantinou, E.; Komnenou, A.; Thomas, A.; Petrakis, S.; Nakos, I.; Gounari, E.; Karampatakis, V. Mesenchymal stem cells improve healing of the cornea after alkali injury. *Graefe's Arch. Clin. Exp. Ophthalmol.* **2015**, *253*, 1121–1135. [CrossRef] [PubMed]

220. Zeppieri, M.; Salvetat, M.L.; Beltrami, A.P.; Cesselli, D.; Bergamin, N.; Russo, R.; Cavaliere, F.; Varano, G.P.; Alcalde, I.; Merayo, J.; et al. Human adipose-derived stem cells for the treatment of chemically burned rat cornea: Preliminary results. *Curr. Eye Res.* **2013**, *38*, 451–463. [CrossRef] [PubMed]

221. Lin, H.-F.; Lai, Y.-C.; Tai, C.-F.; Tsai, J.-L.; Hsu, H.-C.; Hsu, R.-F.; Lu, S.-N.; Feng, N.-H.; Chai, C.-Y.; Lee, C.-H. Effects of cultured human adipose-derived stem cells transplantation on rabbit cornea regeneration after alkaline chemical burn. *Kaohsiung J. Med. Sci.* **2013**, *29*, 14–18. [CrossRef] [PubMed]

222. Mimura, T.; Shimomura, N.; Usui, T.; Noda, Y.; Kaji, Y.; Yamgami, S.; Amano, S.; Miyata, K.; Araie, M. Magnetic attraction of iron-endocytosed corneal endothelial cells to Descemet's membrane. *Exp. Eye Res.* **2003**, *76*, 745–751. [CrossRef]

223. Patel, S.V.; Bachman, L.A.; Hann, C.R.; Bahler, C.K.; Fautsch, M.P. Human corneal endothelial cell transplantation in a human ex vivo model. *Investig. Ophthalmol. Vis. Sci.* **2009**, *50*, 2123–2131. [CrossRef] [PubMed]

224. Okumura, N.; Koizumi, N.; Ueno, M.; Sakamoto, Y.; Takahashi, H.; Tsuchiya, H.; Hamuro, J.; Kinoshita, S. ROCK inhibitor converts corneal endothelial cells into a phenotype capable of regenerating in vivo endothelial tissue. *Am. J. Pathol.* **2012**, *181*, 268–277. [CrossRef] [PubMed]

© 2017 by the authors. Licensee MDPI, Basel, Switzerland. This article is an open access article distributed under the terms and conditions of the Creative Commons Attribution (CC BY) license (http://creativecommons.org/licenses/by/4.0/).

International Journal of
Molecular Sciences

MDPI

Review

Macrophage Phenotypes Regulate Scar Formation and Chronic Wound Healing

Mark Hesketh, Katherine B. Sahin, Zoe E. West and Rachael Z. Murray *

The Institute for Health and Biomedical Innovation, School of Biomedical Sciences, Faculty of Health, Queensland University of Technology, Brisbane QLD 4059, Australia; m.hesketh@connect.qut.edu.au (M.H.); katherine.sahin@connect.qut.edu.au (K.B.S.); zoe.west@connect.qut.edu.au (Z.E.W.)
* Correspondence: rachael.murray@qut.edu.au; Tel.: +61-7-3138-6081

Received: 27 May 2017; Accepted: 16 July 2017; Published: 17 July 2017

Abstract: Macrophages and inflammation play a beneficial role during wound repair with macrophages regulating a wide range of processes, such as removal of dead cells, debris and pathogens, through to extracellular matrix deposition re-vascularisation and wound re-epithelialisation. To perform this range of functions, these cells develop distinct phenotypes over the course of wound healing. They can present with a pro-inflammatory M1 phenotype, more often found in the early stages of repair, through to anti-inflammatory M2 phenotypes that are pro-repair in the latter stages of wound healing. There is a continuum of phenotypes between these ranges with some cells sharing phenotypes of both M1 and M2 macrophages. One of the less pleasant consequences of quick closure, namely the replacement with scar tissue, is also regulated by macrophages, through their promotion of fibroblast proliferation, myofibroblast differentiation and collagen deposition. Alterations in macrophage number and phenotype disrupt this process and can dictate the level of scar formation. It is also clear that dysregulated inflammation and altered macrophage phenotypes are responsible for hindering closure of chronic wounds. The review will discuss our current knowledge of macrophage phenotype on the repair process and how alterations in the phenotypes might alter wound closure and the final repair quality.

Keywords: macrophage; monocyte; wound healing; fibrosis; chronic wound; diabetes; chronic venous disease

1. Introduction

To overcome the repeated dermal insults and injuries that occur on a daily basis, evolution has provided a swift but robust mechanism for tissue repair. The drawback of this hasty mechanism is scar formation and imbalances in this mechanism can lead to non-healing wounds. The repair process is highly complex, consisting of four stages: haemostasis, followed by inflammatory, proliferative and remodeling phases, that involve complex interactions between skin cells, immune cells and extracellular matrix (ECM) components, as well as a plethora of soluble mediators that help orchestrate the process [1,2]. At the start of the repair process, blood loss is stemmed by formation of a blood clot containing platelets, red blood cells, white blood cells and fibrin fibers. These fibers then act as a scaffold for immune cells that are attracted by soluble factors released from platelets and injured tissue, to enter the wound. Mouse wound models have shown that within hours neutrophils enter the injured skin along with a small wave of pioneer monocytes that leak into tissue through areas of microhemorrhaging [3,4]. Neutrophil numbers peak around days 1–2 [4]. These cells are the first responders and kill potential pathogens, debride the wound, and release a number of soluble mediators that attract other immune cells to the site of injury. Once they have undergone their role in the repair process, they apoptose. Monocytes begin migrating into the wound and en route they differentiate into wound associated macrophages in a process driven by factors in the extracellular milieu [2]. Wound associated macrophage numbers then increase until day 2 when their numbers remain stable until around day 5 when they begin to gradually decrease to steady state levels by day 14 [5]. During this time in the wound, macrophages remove spent neutrophils and secrete various cytokines, growth factors and other mediators [6]. Through these mediators, macrophages either directly or indirectly regulate the proliferative stage. They stimulate fibroblasts, keratinocytes and endothelial cells to differentiate, proliferate and migrate, leading to the deposition of new ECM, re-epithelialisation and neovascularisation of the wound [7]. In the remodelling phase, macrophages can release enzymes that alter the composition of the ECM and the structure of the wound.

2. Macrophages Phenotypes

Differentiated wound macrophages are not a homogeneous population of cells but exist as multiple phenotypes that can be broadly classified as M1 and M2 phenotypes (Table 1) [8–10]. These are not discrete populations in vivo but instead they represent a continuum of phenotypes from M1-M2 that evolve as the wound matures. Pro-inflammatory mediators interferon-γ (IFN-γ) and tumour necrosis factor (TNF), and damage associated pattern molecules (DAMPs) activate cells to produce 'classically activated' M1 macrophages (Figure 1) [11]. This results in a pro-inflammatory macrophage phenotype that prolifically produces pro-inflammatory cytokines, such as TNF and Interleukin-6 (IL-6), and other mediators that facilitate the initial stages of wound healing (Figure 1). These M1 cells are highly phagocytic, enabling them to phagocytose neutrophils that have apoptosed and remove any pathogens or debris in the wound. Alternatively activated M2 macrophages are typically anti-inflammatory and can be produced in vitro through the addition of IL-4 or IL-13 or through their phagocytosis of apoptosed neutrophils. M2 cells have been divided into four discrete types—M2a, M2b, M2c and M2d—in vitro based on their function and key markers (Table 1) [12]. Whether they all exist in wounds is currently unclear. There is some controversy in the literature as to how the M2 wound phenotype is derived [13]. Different wound macrophage subsets could be derived from monocytes with different phenotypes recruited at different times that differentiate into macrophages with distinct phenotypes, or monocytes could be recruited at different times in the course of wound healing where the ever-changing wound milieu affecting their polarisation, or M1 macrophages could differentiate into M2 macrophages driven by cues in the local wound environment [13]. A number of studies point to the latter suggestion that it is the same macrophages that regulate early inflammatory functions and later wound reparative functions [13,14]. This suggests that the local environment or some function of the macrophage results in their switch in phenotype in wounds.

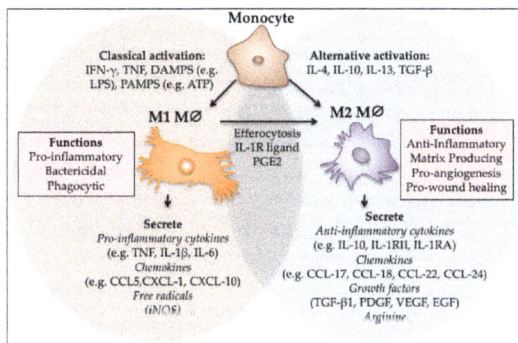

Figure 1. M1 and M2 polarisation of macrophages. Monocytes can be classically or alternatively activated to form M1 and M2 macrophages respectively. M1 macrophages can also differentiate into M2 macrophages through local cues and after efferocytosis. The M1 phenotype is pro-inflammatory, phagocytic and bactericidal, while the M2 macrophages act to switch off inflammation and regulate re-vascularisation and wound closure.

Table 1. Human M1 and M2 phenotypes.

Products	M1	M2a	M2b	M2c	M2d	Reference
Marker expression	CD14^{++}, CD16$^-$, CD68, CD86, CD80, MHC II^{high}	CD14$^+$ CD16^{++} CD68 CD163 CD206 CD3001, MHC-II^{low}	CD14$^+$ CD16^{++} CD68 CD86, MHC-II$^+$	CD14$^+$, CD16^{++}, CD68, CD150, CD163 CD206	CD14$^+$, CD16^{++}, CD68	[9,10,15–17]
Cytokines	IL-1β, IL-6, IL-12, IL-18, IL-23, TNF	IL-1ra, sIL-1R, IL-10	IL-1β, IL-6, IL-10, TNF	IL-10	IL-10, IL-12 TNF	[15,18]
Chemokines	CXCL1, CXCL10, CXCL 11, CXCL 16, CCL2-5, CCL8-11	CCL-17, 18, CCL22, CCL24	CCL1 CCL20 CXCL1, CXCL3	CXCL$^-$-13, CCL-16, CCL-18 CCR2	CXCL-10, CXCL-16, CCL-5	[19]
Signaling	STAT-1, STAT-4, SOCS-3	STAT-3	SOCS3$^+$	STAT-6, SOCS3$^+$		[18,20]
Proteases	MMP-1, 2, 3, 7, 9 & 12	MMP-1, 2, 9, 12,13, 14, TIMP-1				[18]
Growth factors/Other	ROS, RNS, NO	EGF, TGFβ, IGF1, Arg1,	COX-2	EGF, TGFβ Arg1,	VEGF, TGFβ iNOS	[10,21]

$^-$, negative; $^+$ positive (low); $^{++}$ positive (high); IL, interleukin; MMP, matrix metalloproteinase; TNF, tumour necrosis factor; MHC, major histocompatibility complex; NOS, Nitric Oxide Synthase; Arg, arginase; ROS, reactive oxygen species; RNS, reactive nitrogen species; EGF, epidermal growth factor; TGFβ, transforming growth factor β; IGF1, Insulin-like growth factor 1; VEGF, vascular endothelial growth factor.

3. Macrophages Regulate Wound Closure and Scar Formation

Historically, a number of studies addressing the role of macrophages in wound healing provided some doubt as to the necessity and role of these cells in the repair process [22,23]. Depletion of macrophages using anti-serum suggested that macrophages were necessary for the repair process [22]. However, these studies required steroids to deplete most macrophages and so this may also have affected the repair process. The advent of genetic engineering mice has since improved our understanding of the role in macrophages in repair. Mice that lack macrophages (PU.1 mice) repair skin at the same rate as control mice with greatly reduced scar formation, suggesting that macrophages regulate scar formation but not wound closure [23]. However, the PU.1 mice also lack functioning neutrophils [23]. More recently, elegant studies using diphtheria toxin-driven lysozyme M-specific ablation or CD11b specific cell lineage ablation to selectively deplete all macrophage phenotypes, show that these cells are necessary for timely wound healing and they regulate scar formation [24–26]. Depletion of macrophages using the diphtheria toxin driven (DTR) lysozyme M-specific strategy just prior to and throughout the repair process leads to impaired wound healing compared with control mice [24]. Without macrophages, wounds are not cleared of neutrophils and contain high levels of pro-inflammatory cytokines [24]. They contain less anti-inflammatory cytokines, so the ability to dampen inflammation in the wound is reduced, as is the expression of transforming growth factor-beta 1 (TGF-β1), a growth factor that has roles in regulating proliferation, migration, differentiation, and ECM production during the repair process [24]. Consequently, reduced myofibroblast differentiation leads to less wound contraction, and a temporal shift in VEGF expression to later time points resulting in a reduction in angiogenesis [24]. Similar results were seen when ablating CD11b macrophages with diphtheria toxin, except, curiously, there were no changes in wound neutrophil levels [25]. These results suggest that macrophages are necessary for efficient repair.

Wound macrophages are not a homogenous population of cells and at different stages of the repair process they can have distinct phenotypes. Using the diphtheria toxin driven lysozyme M-specific strategy, macrophages have been selectively depleted at distinct times in the repair process to tease out the roles of these macrophages over the course of wound healing [26]. While loss of macrophages during the final stages of repair had no impact on tissue maturation, the removal of macrophages in the mid repair stage resulted in wounds that were significantly delayed with signs of haemorrhaging, suggesting they play a crucial role at this point of the repair process. The mid repair macrophages, presumably M2 macrophages, were found to be required for vascular stability through their production of vascular epithelial growth factor-A (VEGF-A) and TGF-β1 [26]. Depletion of macrophages only in the early stages (days 1–5) suggests macrophages regulate the degree of scar formation and that this pool might be a good target for reducing scar formation [26]. Loss of the early pool of macrophages did significantly delay wound closure, but macrophage repopulation of the wound rescued this delay with wounds reepithelialised by day 14 similar to the control mice [26]. Attractively, these latter wounds have reduced fibrosis, indicating that this early pool of macrophages is necessary for scar formation [26]. This suggests that having less macrophages in the early stages of the repair process might be beneficial in terms of reducing scar formation. This data together suggests that macrophages regulate wound closure and scar formation, and macrophages at different stages of the repair process perform different functions. Interestingly, the distinct roles of different macrophage populations in fibrosis and on the repair process is not limited to skin repair [27]. In liver injury, macrophage depletion using a CD11b-DTR macrophage mouse model shows that during the progressive inflammatory liver injury stage a lack of macrophages reduces fibrosis [27]. However, macrophage depletion during the recovery phase leads to failure of resolution with persistence of cellular and matrix components of the fibrotic response [27]. This suggests that macrophages can also play dual roles during the repair process in other types of injury.

While macrophages found in the wound are derived predominantly from circulatory monocytes, skin does contain low numbers of resident macrophages [8]. The exact contribution of these resident macrophages during the repair process is currently unclear. They potentially play a role in the initial recruitment of immune cells to the site of injury, although the absence of skin macrophages prior to wounding does not affect the overall recruitment of neutrophils [24]. It is known that alternatively activating macrophages derived from monocytes and from tissue macrophages can produce phenotypically and functionally distinct macrophages [28]. This suggests that the small population of resident macrophages might play different roles to the recruited macrophages derived from monocytes during the wound repair process, but this has yet to be tested.

4. M2 Macrophages and Wound Closure

Wounds contain macrophages with phenotypes associated with both classical and alternative activation, the ratio of which alters as the wound matures [29]. During the early stages of inflammation, around 85% of macrophages have an M1 phenotype and avidly secrete pro-inflammatory mediators, while 15% have an M2 phenotype that secrete anti-inflammatory cytokines and growth factors [29]. This ratio switches as the wound matures so that 5–7 days post injury only 15–20% of macrophages have an M1 phenotype and the wound is primarily populated by cells with an M2 phenotype [29,30]. At 5 days post injury, M2 macrophages dominating the acute wound express CD301b, along with high levels of the anti-inflammatory cytokine IL-10 and growth factors such as platelet-derived growth factor (PDGFβ), and TGF-β1 [30]. By using mice expressing a diphtheria toxin receptor/GFP fusion protein under the endogenous CD301b promoter to deplete these CD301b expressing population of M2 cells in mice wounds 3 days post injury, it has been shown that these cells are necessary for timely re-epithelisation, revascularization and fibroblast regeneration [30]. This mirrors the results seen when both M1 and M2 phenotypes are depleted [24–26]. To further tease out the role of the M1 and M2 phenotypes, the cFMS kinase inhibitor GW2580 has been used to selectively inhibit the transition from M1 to M2 macrophages in an acute wound [31]. Blocking this transition prolongs the inflammatory phase; mice treated with this inhibitor prior to and during wound healing have higher number of neutrophils and macrophages at 14 days post injury compared to untreated mice [31]. M2 macrophages secrete factors, such as TGF-β1, that induce the proliferation of fibroblasts and their differentiation into myofibroblasts. These cells are responsible for collagen production in the wound.

Consequently, the GW2580 treated mice have significantly reduced levels of total collagen suggesting that M2 macrophages regulate scar formation.

Since M2 cells have been found to promote repair, it has been proposed that increasing M2 macrophage numbers in the wound might accelerate wound closure. Transplantation of the CD301b M2 population from a day 5 wound into a day 3 wound increases fibroblast proliferation and vascular regeneration, confirming that these CD301b positive M2 macrophages are key cells regulating the reparative phases during wound healing [30]. They do not, however, alter wound closure rates [30]. Similarly, injection of anti-inflammatory M2a or M2c macrophages, that have been stimulated in vitro with IL-4 or IL-10 respectively, into excisional wounds of mice had no effect on wound closure [32]. However, data from chronic diabetic wounds suggest that it may well be how the M2 macrophage is activated that is crucial for timely wound healing, and that IL-4 or IL-10 are not key drivers of the reparative M2 phenotype in wounds. Collectively, these results suggest that in an acute wound the extended and/or increased presence of M1 macrophages, reduced numbers of M2 macrophages and/or increased M2 activation potentially dictate repair and the level of scar formation in part through their secretion of TGF-β1. High levels of TGF-β1 lead to an expansion of fibroblast population and increased differentiation into myofibroblasts, which then secrete more collagen, strengthening the wound. However, excess collagen increases fibrosis. Collectively, these results suggest that M2 macrophages regulate re-vascularisation, fibroblast regeneration and myofibroblast differentiation and collagen production.

5. Macrophage Phenotype and Dysregulated Inflammation in Chronic Wounds

5.1. Chronic Wound Macrophages Are Predominantly M1 Pro-Inflammatory Cells

Chronic wounds, such as diabetic foot ulcers, venous leg ulcers, and pressure ulcers, do not heal in a timely manner. The pathophysiology of these chronic wounds is complex and can include diverse factors such diabetic status, venous insufficiency, arterial perfusion or persistent pressure [33]. What these wounds all have in common is an increased and prolonged inflammatory stage with a distinct wound cell composition, showing changes in both in cell number and the predominant macrophage phenotype, compared to acute wounds [34–36]. The typical shift in M1 to M2 macrophages seen in acute wounds is dysregulated in chronic wounds (Figure 2) [34–37]. At the chronic wound margin, approximately 80% of cells are pro-inflammatory M1 macrophages with these cells playing a major role in the pathogenesis and the on-going chronicity of wounds [35]. Mouse models of diabetic wounds show that reduced M2 macrophage levels lead to a reduction in growth factor levels that regulate the proliferative stage of repair, such as TGFβ1, insulin-like growth factor-1 (IGF-1) and vascular endothelial growth factor (VEGF) [37,38]. These wounds contain high levels of pro-inflammatory cytokines and mediators, such as TNF, IL-1β, IL-17 and iNOS, which contribute to the non-healing phenotype [5,35,37–39]. Patients with chronic wounds present with high levels of pro-inflammatory cytokines in their wound fluid [35,40,41]. At a cellular level, high levels of pro-inflammatory mediators, such as TNF in the wound extracellular milieu, lead to alterations in the secretion of MMP-1, MMP-3 and TIMP-1 from dermal fibroblasts [42]. This changes the fine balance between MMPs and TIMPs, leading to the detrimental excessive ECM proteolysis seen in the wound environment, which contributes to wound chronicity [42].

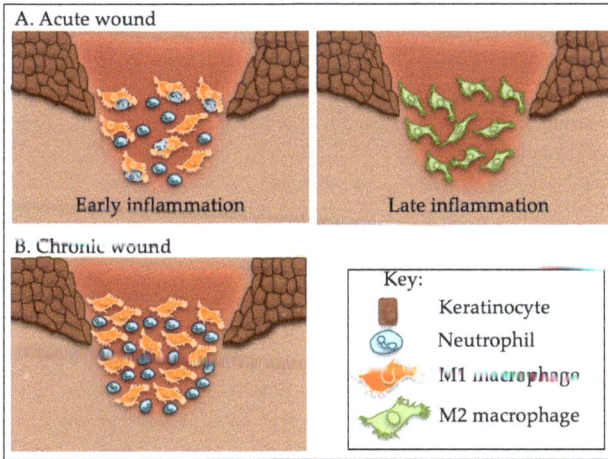

Figure 2. The M1 to M2 switch is dysregulated in chronic wounds and, unlike in acute wounds, macrophages are unable to phagocytose neutrophils. (**A**) In the early inflammatory stage of an acute wound, macrophages phagocytose spent neutrophils. In the later stages, having performed this role, macrophages switch phenotype and are predominantly M2 macrophages; (**B**) In a chronic wound, macrophages are predominantly M1 that are unable phagocytosis spent neutrophils. This leads to the recruitment of more macrophages and an increase in inflammation.

5.2. Hyperglycaemia Alters Efferocytosis and the Macrophage Phenotypic Switch in Diabetes

Alternative activation and macrophage phenotype can be driven by factors in the extracellular environment. In vitro M2 alternate activation is driven by stimulation with IL-4/IL-13, but in an acute wound extracellular IL-4 and IL-13 is lacking [29]. This, along with the fact that transplantation of in vitro IL-4 or IL-10 differentiated M2 macrophages has no effect on wound closure, suggest that there may be some other driver of differentiation and perhaps a different phenotype involved [32]. Other than factors in the local environment, macrophage phenotype is also altered after the phagocytosis of neutrophils, in a process known as efferocytosis ('to carry to the grave') [43–45]. In vitro, macrophages that engulf apoptosed neutrophils switch to an M2b phenotype that expresses low levels of IL-12 and produces high levels of IL-10 and TGF-β [46]. During the initial stages of wound healing, neutrophils enter the wound and apoptose once they have performed their roles in clearing pathogens and debris. These spent neutrophils are removed by macrophages, which in turn is proposed to lead to their switching to an anti-inflammatory M2 phenotype that dampens inflammation and so aids the resolution of wounds. In acute wounds, neutrophil efferocytosis drives the resolution of inflammation through the induction of micro-RNA-21, which silences the pro-inflammatory activity of the programmed cell death 4 (*PDCD4*) gene and favours cJun-AP1 activity leading to an increased production of IL-10, a major suppressor of inflammation [47]. Thus, the efferocytosis of neutrophils, and possibly the local environment, can determine whether macrophages differentiate and effectively resolve inflammation in a timely manner [43]. Alterations in external cues and in efferocytosis could prevent these macrophages from differentiating, leaving a pro inflammatory phenotype that amplifies inflammation and contributes to the impaired wound resolution phenotype.

Monocytes can be preprogramed prior to entering the wound environment, particularly in diseases that induce a chronic state of inflammation. In diabetes, the diabetic environment triggers changes in haematopoietic cells generating monocytes that are pre-programmed towards a pro-inflammatory phenotype. Hyperglycemia and the associated advanced glycated end products (AGEs) alter macrophages such that they present an impaired ability to efferocytose spent neutrophils [48–51]. This results in high levels of neutrophils in diabetic wounds. In mice diabetic models, the increase in apoptotic cells leads to increases in pro-inflammatory cytokines TNF and IL-6 in tissue and wound fluid, with a concomitant decrease in the anti-inflammatory cytokine IL-10 that would normally act to turn off the inflammatory response [48,49]. TNF is one of the key cytokines that can impair wound healing [52]. Increased apoptotic cell load leads to increased inflammation as seen in chronic wounds. Increased apoptosis in an acute wound model leads to an increased inflammatory response with reduced anti-inflammatory cytokines [48]. This, along with the lack of IL-4 and IL-13 in acute wounds suggests that engulfment of apoptotic cells is a major trigger pushing macrophages from an M1 pro-inflammatory phenotype towards the M2 anti-inflammatory phenotype to resolve inflammation (Figure 2).

5.3. Iron Regulates the Macrophage Phenotype in Chronic Venous Ulcers (CVU)

Like diabetes, chronic venous disease is a chronic inflammatory disease. These patients have persistently raised venous pressure (venous hypertension) in lower limb deep and superficial veins. Increased pressure and shear stress leads to changes in the endothelium and the activation of leukocytes including monocytes, macrophages and neutrophils which become "trapped" in the limbs. These activated leukocytes in turn cause injury to the endothelium, leading to a chronic inflammatory state. So, like the diabetic monocytes/macrophages, these cells also have an altered phenotype prior to entering a wound. Within a human chronic venous ulcer (CVU), approximately 80% of the cells at wound margin are macrophages, similar to what is seen in diabetic wounds [35]. These cells differ from those in other wound types in that they contain high amounts of iron from their engulfment of erythrocytes that have leaked into tissue [35]. In iron-dextran–treated mice, iron laden macrophages isolated from wound margins at day 5 post injury are dominantly pro-inflammatory M1 with an intermediate anti-inflammatory M2 phenotype (TNF-α^{hi}, IL-12^{hi}, CCR2^{hi}, Ly6C^{hi}, Dectin-1^{med}, IL-4Rα^{med}, and CD204^{med}) compared with control mice [35]. Treatment of iron loaded mice with the iron chelator desferrioxamine improved wound healing, suggesting that the removal of iron might improve these wounds [35]. Thus, like hypoglycaemia in diabetes, high levels of iron in patients with chronic venous disease leads to an alteration in the macrophage phenotype. Whether these altered macrophages have a defect in efferocytosis, as is seen in diabetes, is yet to be tested. Impaired wound healing is often accompanied by low grade inflammation which can alter cell phenotype, pushing macrophages towards a premature pro-inflammatory phenotype. The high levels of M1 macrophages seen in CVUs results an increased production of TNF which correlates with delayed healing in patients. Addition of recombinant TNF to the margins of an acute wound delays healing [35], while the TNF inhibitors to chronic wounds rescues delayed healing [35,39]. This suggests one of the key factors in wound chronicity is TNF with its increased secretion leading to a delayed phenotype and making TNF a good target to improve wound healing [35].

6. Summary

Collectively, these studies suggest that the extended and/or increased presence of M1 macrophages, reduced numbers of M2 macrophages and/or increased M2 activation can all alter the delicate balance of inflammation in the wound and so drastically affect the repair process. Any failure to switch from an M1 phenotype to an M2 phenotype leads to increased inflammation and the secretion of copious amounts of TNF, which inhibits wound closure in chronic wounds. A driver of this M2 switch in wounds is efferocytosis, a process that is dysregulated in diabetic wounds and potentially in CVU, possibly through the chronic inflammation seen in these patients. The anti-inflammatory M2 cells regulate both the repair process and regulate the final scar outcome. These cells secrete growth factors that stimulate proliferation, angiogenesis and fibroblast differentiation to myofibroblasts in the wound. The latter cells secrete collagen, which strengthens the wound but excess collagen leads to increased fibrosis. Great progress has been made in recent years in our understanding of cell types within the wound. Yet there are many more questions to answer, such as how many different phenotypes are present in the wound, what are the key phenotype regulating scar formation and can we improve efferocytosis and the M2 switch in cells from diabetic patients? While macrophage targeted therapies have not been developed, the progress made in the last decade has provided key data that will help inform future research and take us one step further toward perfect healing.

Conflicts of Interest: The authors declare no conflict of interest.

Abbreviations

AGEs	advanced glycated end products
CVU	chronic venous ulcers
DAMPs	damage associated pattern molecules
ECM	extracellular matrix
IGF-1	insulin-like growth factor-1
PDGFβ	platelet derived growth factor
TGF-1	transforming growth factor-1
VEGF	vascular endothelial growth factor
cJun-AP1	cJun activator protein 1
IFN-γ	interferon gamma
iNOS	inducible nitric oxide synthase
MMP	matrix metalloproteinase
TIMP	tissue inhibitor of matrix metalloproteinase
TNF	tumour necrosis factor

References

1. Eming, S.A.; Krieg, T.; Davidson, J.M. Inflammation in wound repair: Molecular and cellular mechanisms. *J. Investig. Dermatol.* **2007**, *127*, 514–525. [CrossRef] [PubMed]
2. Eming, S.A.; Hammerschmidt, M.; Krieg, T.; Roers, A. Interrelation of immunity and tissue repair or regeneration. *Semin. Cell Dev. Biol.* **2009**, *20*, 517–527. [CrossRef] [PubMed]
3. Rodero, M.P.; Licata, F.; Poupel, L.; Hamon, P.; Khosrotehrani, K.; Combadiere, C.; Boissonnas, A. In vivo imaging reveals a pioneer wave of monocyte recruitment into mouse skin wounds. *PLoS ONE* **2014**, *9*, e108212. [CrossRef] [PubMed]
4. Kim, M.H.; Liu, W.; Borjesson, D.L.; Curry, F.R.; Miller, L.S.; Cheung, A.L.; Liu, F.T.; Isseroff, R.R.; Simon, S.I. Dynamics of neutrophil infiltration during cutaneous wound healing and infection using fluorescence imaging. *J. Investig. Dermatol.* **2008**, *128*, 1812–1820. [CrossRef] [PubMed]
5. Rodero, M.P.; Hodgson, S.S.; Hollier, B.; Combadiere, C.; Khosrotehrani, K. Reduced Il17a expression distinguishes a Ly6c(lo)MHCII(hi) macrophage population promoting wound healing. *J. Investig. Dermatol.* **2013**, *133*, 783–792. [CrossRef] [PubMed]
6. Mahdavian Delavary, B.; van der Veer, W.M.; van Egmond, M.; Niessen, F.B.; Beelen, R.H. Macrophages in skin injury and repair. *Immunobiology* **2011**, *216*, 753–762. [CrossRef] [PubMed]
7. Rodero, M.P.; Khosrotehrani, K. Skin wound healing modulation by macrophages. *Int. J. Clin. Exp. Pathol.* **2010**, *3*, 643–653. [PubMed]
8. Brancato, S.K.; Albina, J.E. Wound macrophages as key regulators of repair: Origin, phenotype, and function. *Am. J. Pathol.* **2011**, *178*, 19–25. [CrossRef] [PubMed]
9. Murray, P.J.; Wynn, T.A. Protective and pathogenic functions of macrophage subsets. *Nat. Rev. Immunol.* **2011**, *11*, 723–737. [CrossRef] [PubMed]
10. Roszer, T. Understanding the mysterious M2 macrophage through activation markers and effector mechanisms. *Mediat. Inflamm.* **2015**, *2015*, 816460. [CrossRef] [PubMed]
11. Zhang, X.; Mosser, D.M. Macrophage activation by endogenous danger signals. *J. Pathol.* **2008**, *214*, 161–178. [CrossRef] [PubMed]
12. Novak, M.L.; Koh, T.J. Macrophage phenotypes during tissue repair. *J. Leukoc. Biol.* **2013**, *93*, 875–881. [CrossRef] [PubMed]

13. Das, A.; Sinha, M.; Datta, S.; Abas, M.; Chaffee, S.; Sen, C.K.; Roy, S. Monocyte and macrophage plasticity in tissue repair and regeneration. *Am. J. Pathol.* **2015**, *185*, 2596–2606. [CrossRef] [PubMed]
14. Crane, M.J.; Daley, J.M.; van Houtte, O.; Brancato, S.K.; Henry, W.L., Jr.; Albina, J.E. The monocyte to macrophage transition in the murine sterile wound. *PLoS ONE* **2014**, *9*, e86660. [CrossRef] [PubMed]
15. Butcher, M.J.; Galkina, E.V. Phenotypic and functional heterogeneity of macrophages and dendritic cell subsets in the healthy and atherosclerosis-prone aorta. *Front. Physiol.* **2012**, *3*, 44. [CrossRef] [PubMed]
16. Castoldi, A.; Naffah de Souza, C.; Camara, N.O.; Moraes-Vieira, P.M. The macrophage switch in obesity development. *Front. Immunol.* **2015**, *6*, 637. [CrossRef] [PubMed]
17. Duluc, D.; Delneste, Y.; Tan, F.; Moles, M.P.; Grimaud, L.; Lenoir, J.; Preisser, L.; Anegon, I.; Catala, L.; Ifrah, N.; et al. Tumor-associated leukemia inhibitory factor and IL-6 skew monocyte differentiation into tumor-associated macrophage-like cells. *Blood* **2007**, *110*, 4319–4330. [CrossRef] [PubMed]
18. Foey, A.D. Macrophage Polarisation: A collaboration of differentiation and activation signals as well as monocyte pre-programming in health and disease? *J. Clin. Cell Immunol.* **2015**, *6*, 693. [CrossRef]
19. Rees, P.A.; Greaves, N.S.; Baguneid, M.; Bayat, A. Chemokines in wound healing and as potential therapeutic targets for reducing cutaneous scarring. *Adv. Wound Care* **2015**, *4*, 687–703. [CrossRef] [PubMed]
20. Sica, A.; Mantovani, A. Macrophage plasticity and polarization: In vivo veritas. *J. Clin. Investig.* **2012**, *122*, 787–795. [CrossRef] [PubMed]
21. Weagel, E.; Smith, C.; Liu, G.P.; Robison, R.; O'Neill, K. Macrophage polarization and its role in cancer. *J. Clin. Cell Immunol.* **2015**, *6*, 338.
22. Leibovich, S.J.; Ross, R. The role of the macrophage in wound repair. A study with hydrocortisone and antimacrophage serum. *Am. J. Pathol.* **1975**, *78*, 71–100. [PubMed]
23. Martin, P.; D'Souza, D.; Martin, J.; Grose, R.; Cooper, L.; Maki, R.; McKercher, S.R. Wound healing in the PU.1 null mouse—Tissue repair is not dependent on inflammatory cells. *Curr. Biol.* **2003**, *13*, 1122–1128. [CrossRef]
24. Goren, I.; Allmann, N.; Yogev, N.; Schurmann, C.; Linke, A.; Holdener, M.; Waisman, A.; Pfeilschifter, J.; Frank, S. A transgenic mouse model of inducible macrophage depletion: Effects of diphtheria toxin-driven lysozyme M-specific cell lineage ablation on wound inflammatory, angiogenic, and contractive processes. *Am. J. Pathol.* **2009**, *175*, 132–147. [CrossRef] [PubMed]
25. Mirza, R.; DiPietro, L.A.; Koh, T.J. Selective and specific macrophage ablation is detrimental to wound healing in mice. *Am. J. Pathol.* **2009**, *175*, 2454–2462. [CrossRef] [PubMed]
26. Lucas, T.; Waisman, A.; Ranjan, R.; Roes, J.; Krieg, T.; Muller, W.; Roers, A.; Eming, S.A. Differential roles of macrophages in diverse phases of skin repair. *J. Immunol.* **2010**, *184*, 3964–3977. [CrossRef] [PubMed]
27. Duffield, J.S.; Forbes, S.J.; Constandinou, C.M.; Clay, S.; Partolina, M.; Vuthoori, S.; Wu, S.; Lang, R.; Iredale, J.P. Selective depletion of macrophages reveals distinct, opposing roles during liver injury and repair. *J. Clin. Investig.* **2005**, *115*, 56–65. [CrossRef] [PubMed]
28. Gundra, U.M.; Girgis, N.M.; Ruckerl, D.; Jenkins, S.; Ward, L.N.; Kurtz, Z.D.; Wiens, K.E.; Tang, M.S.; Basu-Roy, U.; Mansukhani, A.; et al. Alternatively activated macrophages derived from monocytes and tissue macrophages are phenotypically and functionally distinct. *Blood* **2014**, *123*, 110–122. [CrossRef] [PubMed]
29. Daley, J.M.; Brancato, S.K.; Thomay, A.A.; Reichner, J.S.; Albina, J.E. The phenotype of murine wound macrophages. *J. Leukoc. Biol.* **2010**, *87*, 59–67. [CrossRef] [PubMed]
30. Shook, B.; Xiao, E.; Kumamoto, Y.; Iwasaki, A.; Horsley, V. CD301b+ macrophages are essential for effective skin wound healing. *J. Investig. Dermatol.* **2016**, *136*, 1885–1891. [CrossRef] [PubMed]
31. Klinkert, K.; Whelan, D.; Clover, A.J.P.; Leblond, A.L.; Kumar, A.H.S.; Caplice, N.M. Selective M2 macrophage depletion leads to prolonged inflammation in surgical wounds. *Eur. Surg. Res.* **2017**, *58*, 109–120. [CrossRef] [PubMed]
32. Jetten, N.; Roumans, N.; Gijbels, M.J.; Romano, A.; Post, M.J.; de Winther, M.P.; van der Hulst, R.R.; Xanthoulea, S. Wound administration of M2-polarized macrophages does not improve murine cutaneous healing responses. *PLoS ONE* **2014**, *9*, e102994. [CrossRef] [PubMed]
33. Guo, S.; Dipietro, L.A. Factors affecting wound healing. *J. Dent. Res.* **2010**, *89*, 219–229. [CrossRef] [PubMed]
34. Loots, M.A.; Lamme, E.N.; Zeegelaar, J.; Mekkes, J.R.; Bos, J.D.; Middelkoop, E. Differences in cellular infiltrate and extracellular matrix of chronic diabetic and venous ulcers versus acute wounds. *J. Investig. Dermatol.* **1998**, *111*, 850–857. [CrossRef] [PubMed]

35. Sindrilaru, A.; Peters, T.; Wieschalka, S.; Baican, C.; Baican, A.; Peter, H.; Hainzl, A.; Schatz, S.; Qi, Y.; Schlecht, A.; et al. An unrestrained proinflammatory M1 macrophage population induced by iron impairs wound healing in humans and mice. *J. Clin. Investig.* **2011**, *121*, 985–997. [CrossRef] [PubMed]

36. Miao, M.; Niu, Y.; Xie, T.; Yuan, B.; Qing, C.; Lu, S. Diabetes-impaired wound healing and altered macrophage activation: A possible pathophysiologic correlation. *Wound Repair Regen.* **2012**, *20*, 203–213. [CrossRef] [PubMed]

37. Okizaki, S.; Ito, Y.; Hosono, K.; Oba, K.; Ohkubo, H.; Amano, H.; Shichiri, M.; Majima, M. Suppressed recruitment of alternatively activated macrophages reduces TGF-β1 and impairs wound healing in streptozotocin-induced diabetic mice. *Biomed. Pharmacother.* **2015**, *70*, 317–325. [CrossRef] [PubMed]

38. Mirza, R.; Koh, T.J. Dysregulation of monocyte/macrophage phenotype in wounds of diabetic mice. *Cytokine* **2011**, *56*, 256–264. [CrossRef] [PubMed]

39. Ashcroft, G.S.; Jeong, M.J.; Ashworth, J.J.; Hardman, M.; Jin, W.; Moutsopoulos, N.; Wild, T.; McCartney-Francis, N.; Sim, D.; McGrady, G.; et al. Tumor necrosis factor-alpha (TNF-α) is a therapeutic target for impaired cutaneous wound healing. *Wound Repair Regen.* **2012**, *20*, 38–49. [CrossRef] [PubMed]

40. Wallace, H.J.; Stacey, M.C. Levels of tumor necrosis factor-alpha (TNF-α) and soluble TNF receptors in chronic venous leg ulcers—Correlations to healing status. *J. Investig. Dermatol.* **1998**, *110*, 292–296. [CrossRef] [PubMed]

41. Patel, S.; Maheshwari, A.; Chandra, A. Biomarkers for wound healing and their evaluation. *J. Wound Care* **2016**, *25*, 46–55. [CrossRef] [PubMed]

42. Subramaniam, K.; Pech, C.M.; Stacey, M.C.; Wallace, H.J. Induction of MMP-1, MMP-3 and TIMP-1 in normal dermal fibroblasts by chronic venous leg ulcer wound fluid. *Int. Wound J.* **2008**, *5*, 79–86. [CrossRef] [PubMed]

43. Greenlee-Wacker, M.C. Clearance of apoptotic neutrophils and resolution of inflammation. *Immunol. Rev.* **2016**, *273*, 357–370. [CrossRef] [PubMed]

44. Savill, J.; Dransfield, I.; Gregory, C.; Haslett, C. A blast from the past: Clearance of apoptotic cells regulates immune responses. *Nat. Rev. Immunol.* **2002**, *2*, 965–975. [CrossRef] [PubMed]

45. Serhan, C.N.; Savill, J. Resolution of inflammation: The beginning programs the end. *Nat. Immunol.* **2005**, *6*, 1191–1197. [CrossRef] [PubMed]

46. Filardy, A.A.; Pires, D.R.; Nunes, M.P.; Takiya, C.M.; Freire-de-Lima, C.G.; Ribeiro-Gomes, F.L.; DosReis, G.A. Proinflammatory clearance of apoptotic neutrophils induces an IL-12(low)IL-10(high) regulatory phenotype in macrophages. *J. Immunol.* **2010**, *185*, 2044–2450. [CrossRef] [PubMed]

47. Das, A.; Ganesh, K.; Khanna, S.; Sen, C.K.; Roy, S. Engulfment of apoptotic cells by macrophages: A role of microRNA-21 in the resolution of wound inflammation. *J. Immunol.* **2014**, *192*, 1120–1129. [CrossRef] [PubMed]

48. Khanna, S.; Biswas, S.; Shang, Y.; Collard, E.; Azad, A.; Kauh, C.; Bhasker, V.; Gordillo, G.M.; Sen, C.K.; Roy, S. Macrophage dysfunction impairs resolution of inflammation in the wounds of diabetic mice. *PLoS ONE* **2010**, *5*, e9539. [CrossRef] [PubMed]

49. Darby, I.A.; Bisucci, T.; Hewitson, T.D.; MacLellan, D.G. Apoptosis is increased in a model of diabetes-impaired wound healing in genetically diabetic mice. *Int. J. Biochem. Cell Biol.* **1997**, *29*, 191–200. [CrossRef]

50. Liu, B.F.; Miyata, S.; Kojima, H.; Uriuhara, A.; Kusunoki, H.; Suzuki, K.; Kasuga, M. Low phagocytic activity of resident peritoneal macrophages in diabetic mice: Relevance to the formation of advanced glycation end products. *Diabetes* **1999**, *48*, 2074–2082. [CrossRef] [PubMed]

51. Guo, Y.; Lin, C.; Xu, P.; Wu, S.; Fu, X.; Xia, W.; Yao, M. AGEs Induced autophagy impairs cutaneous Wound healing via stimulating macrophage polarization to M1 in diabetes. *Sci. Rep.* **2016**, *6*, 36416. [CrossRef] [PubMed]

52. Goren, I.; Muller, E.; Schiefelbein, D.; Christen, U.; Pfeilschifter, J.; Muhl, H.; Frank, S. Systemic anti-TNFalpha treatment restores diabetes-impaired skin repair in ob/ob mice by inactivation of macrophages. *J. Investig. Dermatol.* **2007**, *127*, 2259–2267. [CrossRef] [PubMed]

© 2017 by the authors. Licensee MDPI, Basel, Switzerland. This article is an open access article distributed under the terms and conditions of the Creative Commons Attribution (CC BY) license (http://creativecommons.org/licenses/by/4.0/).

International Journal of
Molecular Sciences

MDPI

Review

Association of Extracellular Membrane Vesicles with Cutaneous Wound Healing

Uyen Thi Trang Than [1,2,3], Dominic Guanzon [1,2,3], David Leavesley [3,4] and Tony Parker [1,2,3,*]

1   Tissue Repair and Translational Physiology Program, Institute of Health and Biomedical Innovation, Queensland University of Technology, Kelvin Grove, Queensland 4059, Australia; thitranguyen.than@hdr.qut.edu.au (U.T.T.T.); dominic.guanzon@hdr.qut.edu.au (D.G.)
2   School of Biomedical Science, Faculty of Health, Queensland University of Technology, Kelvin Grove, Queensland 4059, Australia
3   Wound Management Innovation Cooperative Research Centre, 25 Donkin, West End, Queensland 4101, Australia; d.leavesley@imb.a-star.edu.sg
4   Institute of Medical Biology, Agency for Science, Technology and Research, 8A Biomedical Grove, Singapore 138648, Singapore
*   Correspondence: a.parker@qut.edu.au; Tel.: +61-7-3138-6187

Academic Editor: Allison Cowin
Received: 21 March 2017; Accepted: 27 April 2017; Published: 1 May 2017

Abstract: Extracellular vesicles (EVs) are membrane-enclosed vesicles that are released into the extracellular environment by various cell types, which can be classified as apoptotic bodies, microvesicles and exosomes. EVs have been shown to carry DNA, small RNAs, proteins and membrane lipids which are derived from the parental cells. Recently, several studies have demonstrated that EVs can regulate many biological processes, such as cancer progression, the immune response, cell proliferation, cell migration and blood vessel tube formation. This regulation is achieved through the release and transport of EVs and the transfer of their parental cell-derived molecular cargo to recipient cells. This thereby influences various physiological and sometimes pathological functions within the target cells. While intensive investigation of EVs has focused on pathological processes, the involvement of EVs in normal wound healing is less clear; however, recent preliminarily investigations have produced some initial insights. This review will provide an overview of EVs and discuss the current literature regarding the role of EVs in wound healing, especially, their influence on coagulation, cell proliferation, migration, angiogenesis, collagen production and extracellular matrix remodelling.

Keywords: wound healing; proliferation; migration; angiogenesis; extracellular membrane vesicles; microvesicles; apoptotic bodies; exosomes; endothelial cells; keratinocytes

---

1. Introduction

Membrane-enclosed extracellular vesicles (EVs) include apoptotic bodies, microvesicles and exosomes which are released by various cell types, and have been found in cell culture media, as well as body fluids such as breast milk, urine, amniotic fluids, saliva and blood [1–11]. During EV biogenesis, a number of biological molecules are encapsulated into the EV, including DNA, small RNAs, proteins and lipids which are derived from parental cells [6,11]. However, the real significance of EVs lies in their ability to deliver their contents to recipient cells, thereby altering biological and cellular processes. Furthermore, aberrant delivery of EV cargo to recipient cells has been implicated in several pathologies such as some autoimmune disorders and cancers [12–15].

Wound healing is a complicated process which involves an overlapping cascade of events characterised into four distinct phases (Figure 1). These include:

- Haemostasis, where blood loss ceases;
- The inflammatory phase, characterised by infiltration of immune cells to combat infection and remove cellular debris;
- The proliferative phase, where fibroblasts and keratinocytes at the wound margins migrate into the wound and increase in cell number to re-establish the barrier function of the skin; and
- The remodelling phase, during which reorganisation of the dermis occurs and the preliminary extracellular matrix (ECM), laid down during the earlier phases of the healing response, is remodelled to strengthen the wound area through the reduction of scar tissue [16].

Preliminary studies have shown that EVs may be involved in wound healing through the control of a number of cellular processes [17–21]. Therefore, this review will summarise the role of EVs derived from predominant cells involved in the wound healing process. An overview of EV classification, biogenesis, components, and cell-to-cell communication via EVs is also described.

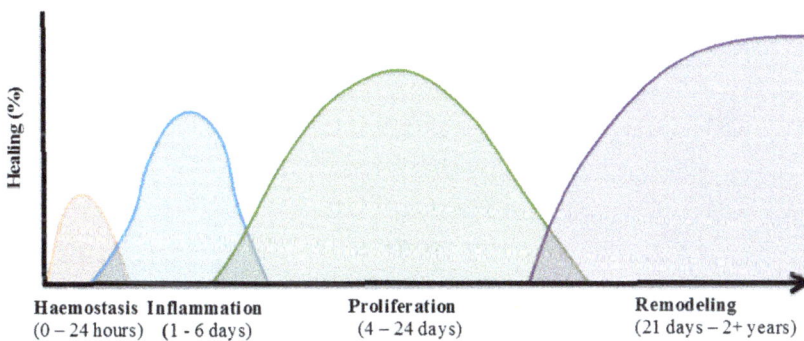

Figure 1. Wound healing process. The normal tissue repair process is comprised of continuous and overlapping phases. These four phases include: (i) Haemostasis; (ii) inflammation; (iii) proliferation; and (iv) remodelling. Each phase consists of different cellular events which requires the interplay of multiple cell populations [22].

2. Extracellular Membrane Vesicles (EVs)

EVs have a phospholipid bilayer similar to the cell membrane, with diameters ranging from 40 nm to 5 μm. In general, EVs can be classified into three subtypes: apoptotic bodies (also known as apoptotic vesicles); microvesicles (also known as shedding vesicles); and exosomes.

2.1. Apoptotic Bodies

Apoptotic bodies are the largest vesicle population, with a diameter ranging from 1 to 5 μm and have a heterogeneous morphology. Apoptotic bodies are released when cells undergo apoptosis and therefore they contain various components from their parental cells often including organelles and DNA fragments [23].

Apoptosis and necrosis are major mechanisms of cell death, which produce cell debris but are activated by diverse biological stimuli [23–25]. In contrast to apoptosis which is programed cell death, necrosis is passive and activated by mechanical damage or disease [25]. Interestingly, apoptotic bodies are released into the extracellular environment through several stages. During the early and intermediate stages, the cell membrane is contracted, the cytoplasm is condensed, and the cells become smaller in size [23]. Simultaneously, nuclear chromatin is also condensed and undergoes alteration, and the plasma membrane deteriorates such that its permeability increases in the late stage. As a result, the plasma membrane undergoes a process that is commonly known as blebbing [24]; and the cellular content is disintegrated into distinct membrane enclosed vesicles known as apoptotic bodies [24].

Therefore, apoptotic bodies contain the cytoplasm, but does not necessarily include tightly packed organelles or nuclear fragments. However, if organelles are encapsulated within apoptotic bodies, these organelles have been shown to have their integrity maintained [23]. During the formation process of apoptotic bodies, phosphatidylserine (PS) residues that are normally located on the internal surface of the plasma membrane, subsequently translocate to the external surface. This process presents extracellular signals that attract macrophages to clear the apoptotic bodies via phagocytosis (Figure 2).

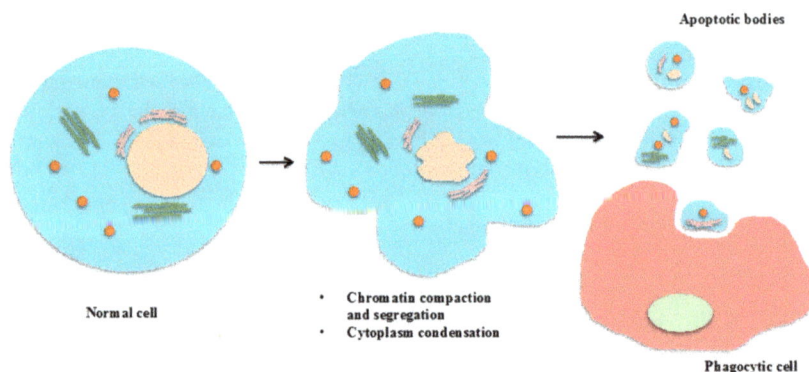

Figure 2. Formation of apoptotic bodies and clearance by phagocytosis. Formation of apoptotic bodies includes the condensation and segregation of the nucleus, and the deterioration and blebbing of the plasma membrane. The result of these processes is a separation of the cellular contents into membrane-enclosed vesicles which can be cleared by phagocytic cells.

2.2. Microvesicles

Microvesicles, also known as "ectosomes", are the second largest vesicle type between 100 and 1000 nm in diameter, which are formed by the outward budding and fission of the plasma membrane [26,27]. However, there is limited understanding about the formation and shedding mechanism of microvesicles at the cell surface, although it is hypothesised that the formation of membrane microvesicles may be a result of the dynamic interplay between phospholipid redistribution and cytoskeletal protein contraction [28,29]. This process is regulated by several enzymes such as calpain, flippase, floppase, scramblase and gelsolin [30]. Flippases transfer phospholipids from the outer leaflet to the inner leaflet while floppases transfer phospholipids from the inner leaflet to the outer leaflet [31]. The translocation of PS to the outer-membrane leaflet is a signal that induces the membrane budding/vesicle formation (Figure 3) [28]. In addition, microvesicle formation was associated with ADP-ribosylation factor 6 (ARF6), a small GTPase protein, which regulating the activation of myosin light chain kinase (MLCK) and subsequent phosphorylation of MLCK lead to a promotion of contraction of actin-based cytoskeleton [32]. Thereby, the budding process is completed through the contraction of cytoskeletal structures via actin and myosin interactions [32,33].

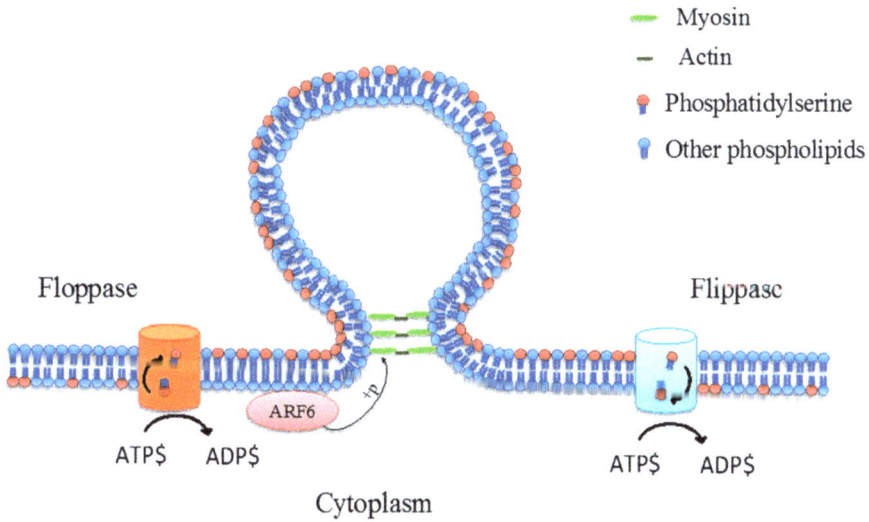

Figure 3. Phospholipid translocase activity via floppase and flippase which translocates phosphatidylserine and other phospholipids from the inner leaflet to the outer leaflet, and outer leaflet to inner leaflet, respectively, during microvesicle formation. These processes are adenosine triphosphate (ATP)-dependant [28,30,32,33]

2.3. Exosomes

Exosomes are the smallest class of EVs with diameters between 40 and 100 nm and a cup shape morphology according to previous studies using electron microscopy [34]. Although the mechanism of exosome biogenesis is not fully understood, it is commonly accepted that exosomes are formed and developed via the endocytic pathway, and are subsequently released to the extracellular environment by exocytosis [35,36]. The formation and release process begins when fluids, solutes, macromolecules, plasma components and particles are internalized by various endocytic trafficking pathways into transport vesicles [37,38]. The transport vesicles then fuse with one another or with an existing sorting endosome to form early endosomes. During the next developmental stage, early endosomes may collect proteins and other components and develop into late endosomes, and that late endosome may recycle its components back to the plasma membrane or be subjected to degradation by lysosomes [37,39]. Alternatively, that late endosomes may develop to become multi-vesicular bodies (MVBs) carrying and releasing exosomes when MVBs fuse with the cellular membrane (Figure 4) [39,40]. Thus while complex and much remains unknown about exosome biogenesis, it is clear that their formation and release are tightly regulated by multiple signalling mechanisms. For instance, some evidence suggests that Endosomal Sorting Complexes Required for Transport (ESCRT) pathway is needed for exosome biogenesis [39,41] as is Rab27a/b which is involved in endosome development and docking of MVBs to the cellular plasma membrane [42].

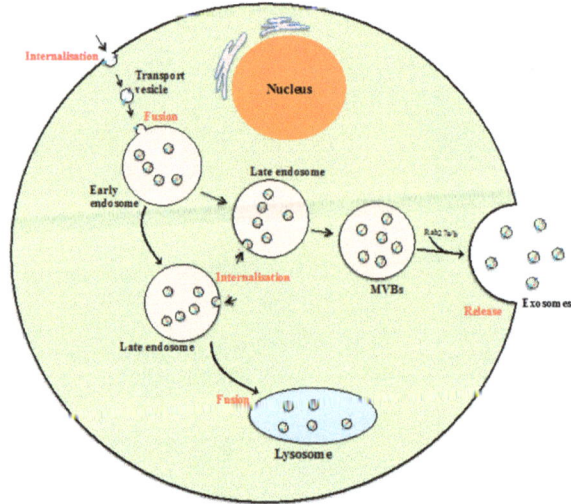

Figure 4. Exosome biogenesis. Beginning from internalization of membrane proteins and lipid complexes by endocytosis, endocytotic vesicles are delivered to early endosomes, which fuse with each other resulting in formation of late endosomes/multivesicular bodies (MVB). MVBs either release exosomes by fusion with the cellular membrane, or their contents are degraded if they fuse with lysosomes [40,42,43]. The key steps of the exosomal formation and development process are highlighted in red.

3. Cell-to-Cell Communication

Cell-to-cell communication is a pivotal mechanism that enables the differentiation of cells and the development of multicellular organisms. The mechanisms of cell-to-cell communication are very complex and involve intercellular and intracellular signals, which includes cell junctions, adhesion contacts and soluble factors [44,45]. Cells can form bridges (cytonemes) to connect and exchange surface-associated cargo to neighbouring cells based on the mechanism of adhesion [46], or tunnelling nanotubes to contact and transfer both cell surface molecules and cytoplasmic components to other cells [46,47]. Recently, EVs were found to be a new means of communication because they carry functional molecules and can horizontally transfer these to neighbouring cells [19].

Cell-to-cell communication by EVs is described as a way for cells to interact with neighbouring cells or one another over long distances, when EVs were detected in the circulation and other body fluids [44]. The binding of EVs to target cells is specific with examples including: platelet-derived EVs binding to neutrophils [48]; neutrophil-derived EVs binding to dendritic cells [49] or to monocytes and endothelial cells [50]; and leukocyte-derived EVs binding to platelets [51]. The interaction between EVs and target cells are thought to require the coordinated action of the cytoskeleton and vesicle fusion machinery [28]. Despite the limited understanding about EVs transport, different mechanisms of interaction between EVs and recipient cells include ligand–receptor interaction, internalisation and direct membrane fusion have been studied [52] (Figure 5).

Examples of ligand–receptor interaction mechanisms include studies of EVs released from dendritic cells and platelets [53,54]. Dendritic cell-derived exosomes contain MHC class II and CD9 on their membrane and were found to bind to the surface of activated T cells [53]. This binding could be an interaction between MHC II molecules on exosome membranes and T cell receptors on T cell membranes [53]. Interestingly, exosomes only bound to the surface of the plasma membrane without fusing and internalisation into T cells [53]. Similarly, platelet-derived microvesicles containing the CD41 antigen were discovered to bind to the membrane of human bone marrow CD34+ cells, which then stimulated the adhesion of these cells to the endothelium as well as directed them from peripheral blood back into the bone marrow [54].

Regarding the mechanism of direct membrane fusion, some evidences suggested that microvesicles fused directly with the plasma membrane and transferred their contents to the intracellular milieu of the recipient cells [55–58]. In the case of microvesicles enriched selectively with P-selectin glycoprotein ligand-1 (PSGL-1), the fusion may be controlled by Annexin V or antibody to PSGL-1 [55]. Additionally, it seems that the fusion of EVs to the plasma membrane is also dependant on SNARE proteins, which regulate the fusion and target specificity in intracellular vesicle trafficking [58]. When EVs fuse with and transfer membrane components such as receptors and ligands to their targets, these can increase the resistance to apoptosis in the case of macrophages receiving chemokine receptors; or induce an increased frequency of apoptosis in the case of T lymphocytes receiving the Fas ligand (a death-receptor ligand) [56,57].

Morelli et al. found evidence of the internalisation of circulating EVs into dendritic cells, phagocytes of spleen, and Kupffer cells in the liver via clathrin-dependent endocytosis [59]. This internalisation of EVs by dendritic cells requires participation of the dendritic cell cytoskeleton as well as surface molecules such as externalised PS, CD11a, CD54, CD9 and CD81 [59]. Additionally, cellular maturity also influences the internalisation of EVs by dendritic cells, since immature dendritic cells exhibit a higher internalisation capability than mature dendritic cells [59]. This internalisation of EVs into target cells could induce peripheral T-cell tolerance in the absence of danger signals [59] or stimulate cell proliferation and migration [21].

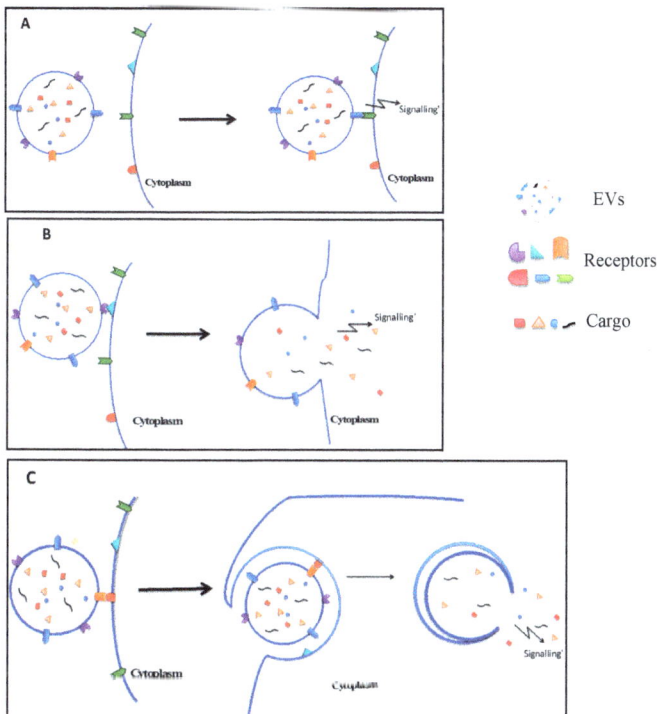

Figure 5. Interaction of EVs with target cells: (**A**) intracellular signalling due to EV membrane ligand cell surface receptor interactions [53,54]; (**B**) direct membrane fusion which induces cell function through release of EV cargo into target cells [55–58]; and (**C**) internalisation of EVs into target cells, prior to the release of their cargo into the recipient cell cytoplasm inducing functional effects [21,59].

4. Functional Components of EVs

4.1. Proteins

In general, all three populations of mammalian EVs share some common characteristics such as structure (lipid bi-layer), and tend to have common types of cargo such as protein, lipid or genetic material. However, these characteristics may be distinct depending on the manner of formation as well as the nature of their parental cells. Specific molecules that are unique to each vesicle population can be used to distinguish between them. For example, PS is only transferred to the external plasma membrane surface when cells undergo apoptosis and where microvesicles are formed and shed [28,60]. In another example, the tetraspanin family (e.g., CD9, CD63, and CD81) form a complex network of interacting molecules which play a role in trafficking of transmembrane proteins [61]. Thus CD9, CD63, and CD81 are enriched and often detected in exosomes since they are often involved in exosome biogenesis pathways [8,52,62], however these components are also detected in other vesicles [63]. Interestingly, CD24 could be considered as a marker of exosomes isolated from urine and amniotic fluids [8]. Additionally, the CD24 has been found to be over-expressed in various cancer cell lines such as ovarian, breast, non-small cell lung, prostate and pancreatic carcinomas [64]. In terms of other functional proteins, Dujarin et al. reported that Tau, a microtubule-associated protein, and matrix metalloproteinase 1 (MMP-1) protein, were enriched in microvesicles compared to exosomes [17,65]. However, no studies have reported whether or not Tau and MMP-1 proteins are enriched in apoptotic bodies.

Importantly, specific EV biomolecular cargo has previously been linked to cancer and tumour development [50,66,67]. For example, apoptotic bodies, which contained DNA encoding the oncogenes H-rasV12 and c-myc, were found to induce tumour formation and growth in severe combined immunodeficient (SCID) mice [66]. Interestingly, the authors also found that these H-rasV12 and c-myc genes were replicated and incorporated into the recipient genome in newly formed tumour cells in vivo (SCID mice model) [66]. Furthermore, microvesicles carrying MMP-9 released from breast carcinoma cells, fibrosarcoma cells and polymorphonuclear leucocytes functioned in the digestion of the extracellular matrix, which is necessary for the progress of cancer growth and inflammation [50,67].

EVs have also been found to carry components associated with angiogenesis and coagulation processes [55,68,69]. For instance, endothelial cell-derived microvesicles, which were enriched with MMPs, could stimulate the invasion and formation of capillary-like structures of human umbilical vein endothelial cells (HUVECs) [69]. Indeed, MMPs have been identified as important mediators of angiogenesis [70,71], but exactly how the EV MMPs contribute to the angiogenic process has not yet been revealed. Interestingly, the accumulation of MMPs within microvesicles was shown to be stimulated by angiogenic factors including, fibroblast growth factor (FGF)-2 and vascular endothelial growth factor (VEGF) [69]. Moreover, EVs were also associated with coagulation since microvesicles contained tissue factor (TF) derived from platelets and monocytes [55,68].

EVs also have a function in the regulation of immune responses, since exosomes have been found to contain major histocompatibility complex (MHC) class I and II molecules and may transfer these molecules to dendritic cells [5,72,73]. The transferral of MHC I molecules from melanoma cell line-derived exosomes to dendritic cells, triggered the production of Interferon γ in these cells [73]. Furthermore, MHC I molecules carried by dendritic cell derived exosomes could activate CD8+ T cell responses [5,73]. These studies indicate that exosomes can potentially act as extracellular antigen sources, which could help to develop immune interventions.

Finally, cytoplasmic proteins, such as tubulin, actin, actin-binding proteins, annexins and Rab proteins, which are involved in intracellular membrane fusion and transport, and ESCRT machinery complex proteins are also found in exosomes [39,41,74,75]. In addition, molecules responsible for signal transduction, such as protein kinases, 14-3-3 and heterotrimeric G proteins, are enveloped during exosome formation and release [39,74,76,77]. While there is evidence to support the conclusion that the molecular components of EVs may be delivered to and elicit functional responses in recipient cells, unequivocal evidence is still required [19]. Taken together, these studies indicate that EVs and their cargo are functional and could influence recipient cell behaviour.

4.2. Lipids

Interestingly, EV lipid composition, which includes cholesterol, sphingomyelin, ceramide, phospholipids and glucans can provide an additional means of identification for exosomes [78–81]. Indeed, exosomes have been found to be enriched with higher amounts of cholesterol and sphingolipids, including sphingomyelin and hexosylceramide, but contain lower amounts of accumulated phosphatidylcholine, compared to parent cell membranes [81]. Importantly, Trajkovic et al. showed that high concentrations of ceramide exist in the micro-domains where multi-vesicular endosomes are formed. This led to the hypothesis that ceramide could be useful for enlarging micro-domains and facilitating the inducement of domain budding [81]. These studies indicate that lipids can partially regulate the formation and release of vesicles [78,81].

4.3. Genetic Materials (Messenger RNAs and miRNAs)

Regarding genetic biomolecules, RNAs including messenger RNAs (mRNAs) and microRNAs (miRNAs) have also been found in EVs which have been increasingly found to elicit various functions in recipient cells [11,82–90]. In particular, large amounts of mRNAs and miRNAs were found in all three EV types released from both cell cultures and body fluids [14,82–84,87–90]. Importantly, these EV-derived genetic molecules were selectively sorted into EVs and as such could be used for diagnostics, such as cancer markers, or as potential treatment targets [90–94]. In addition to carrying the information from secreting cells, EV mRNAs and miRNAs can also function to promote biological processes, including proliferation, angiogenesis and apoptosis [86,92,95,96]. Interestingly, there is evidence showing that EV mRNAs were horizontally transferred to recipient cells and are able to be translated into proteins [3,95,96]. Additionally, the proteins translated from transferred EV mRNAs have been shown to activate PI3K/AKT signalling pathways in recipient endothelial cells [96] and accelerate morphological and functional recovery in rat liver [95]. While these studies provide some evidence for the role of EV mRNAs and miRNAs in the regulation of biological processes, it is clear that further studies are required to more fully understand the functional consequences of EV genetic cargo.

5. The Role of EVs in Cutaneous Wound Healing

Wound healing is an essential process that enables the restoration of the structure and function of damaged or injured tissues. For most tissues, the process requires multiple cell types from several distinct lineages where each responds to and generates a range of signals at different times [97]. The wound healing process involves a series of overlapping phases including: haemostasis; inflammation; migration; and tissue remodelling [16]. During these phases, a series of precise biological events occur including: rapid vascular spasms, platelet plug formation and coagulation (collectively known as haemostasis); inflammation; cell migration into the wound site; cell proliferation and differentiation to form granulation tissue; angiogenesis; and extracellular-matrix reorganisation [97]. Evidence for the involvement of EVs in wound healing has been described for coagulation, cell proliferation, cell migration and angiogenesis and summarised in Table 1 [17–19,21].

5.1. Coagulation

The blood coagulation cascade is complex and regulated by a range of factors such as calcium ions, PS and TF in addition to many others. TF is an initiator of coagulation activation that was found to be present on the plasma membrane of EVs, including microvesicles and exosomes, which were derived from various cell types such as human monocytes/macrophages [55] human platelets [98]; and EVs isolated from human saliva [99]. Indeed, monocyte-derived microvesicles enriched with TF and PSDL-1 were found to bind, fuse and transfer their content to collagen-activated platelets [55]. The result of this TF transfer from microvesicles to the platelets enabled TF to become "decrypted"/activated, and thus initiate the extrinsic coagulation cascade leading to conversion of prothrombin to thrombin and fibrin clot formation [55]. Furthermore, in an in vitro study, salivary EVs that contain TF in association with coagulation factor VII significantly reduced the clotting time of human wound blood (Figure 6) [99]. Similarly, EVs carrying TF from pericardial blood of cardiac surgery patients induced coagulation in vitro and stimulated thrombogenicity in rats [98]. Conversely, EVs from the venous blood of healthy individuals also carrying TF did not reduce clotting time [98]. This may be due to soluble inhibitors including Tissue Factor Pathway Inhibitor (TFPI) or conformational regulation of TF in plasma that impedes the binding of coagulation factor VII and inhibits its activity [55].

5.2. Cell Proliferation

Cell proliferation, a pivotal cellular process underpinning wound healing, has been shown to be regulated by EVs derived from a plethora of cell types, such as human mesenchymal stem cells (MSCs) [21,100,101], fibroblasts [18], murine embryonic stem cells [102], and human endothelial progenitor cells [103]. The regulation of cell proliferation by EVs occurred through the binding to or internalisation of EVs into recipient cells and delivery of EV content such as RNAs [104,105]. For instance, the internalisation of EVs into fibroblasts greatly enhanced the cell proliferation due to the activation of the Akt and Erk1/2 effector pathway by EVs [18,21,100,106,107]. Additionally, embryonic stem cell-derived EVs may facilitate fibroblast proliferation, possibly through the activation of the mitogen-activated protein kinase (MAPK) pathway [102]. Importantly, these aforementioned activated pathways induce the expression of a number of genes involved in cell cycle progression such as c-myc, cyclin A1 and cyclin D2 which then support cell proliferation [21,106]. Furthermore, activation of these pathways also promoted the production of growth factors, including insulin-like growth factor-1 (IGF-1), stromal-derived growth factor-1 (SDF-1) and cytokine interleukin-6 (IL-6), VEGF-α and transforming growth factor beta (TGF-β) [18,21,101,102]. Thus, it seems clear that EVs can promote cell proliferation by activation of signalling pathways not only directly involved in stimulation of cell cycle but also involved in the regulation of growth factor expression. Additionally, this growth factor expression can act in a paracrine or autocrine fashion to further stimulate cell growth responses.

5.3. Migration

Following clotting, immune cells including neutrophils and macrophages are recruited to remove necrotic tissues, debris, and bacteria from the wound site. Within hours after injury, other cells including epithelial cells and fibroblasts migrate into the wound site to perform specific tasks, such as induction of growth factor secretion, the synthesis of extracellular matrix, angiogenesis and stimulation of wound closure [97]. Preliminary studies have been conducted investigating the involvement of EVs in the migration of various cells associated with cutaneous repair, such as human and murine epithelial cells [108], bovine endothelial cells [109], human MSCs [100], human keratinocytes and fibroblasts in vitro [110].

Evidently, keratinocyte migration has been shown to be influenced by EVs released from various cell types; however, different results were observed depending on the specific biomolecules carried by the EVs [110,111]. For example, CoCl$_2$-treated tumour cell-derived exosomes which contained C4.4A (Ly6/PLAUR domain-containing protein 3), $\alpha 6\beta 4$ integrin and MMP-14 inhibited keratinocyte migration through degradation of laminin 332 resulting in delayed wound closure [111]. Conversely, keratinocyte derived exosomes with heat-shock protein (HSP) 90α cargo significantly enhanced the migration of primary human keratinocytes in a wound scratch model [110]. Additionally, the migration of endothelial cells, which are crucial for vascular repair and regeneration, has also been shown to be influenced by EVs released from cells such as human keratinocytes [108], human MSCs [112,113]; and bovine endothelial cells [109]. However, the mechanisms that facilitate EV mediated endothelial migration are unclear.

In a wound healing context, fibroblasts are known to release growth factors which induce other cells to proliferate and migrate, and also produce collagen (COL) that provides structure to the wound [16,114]. Interestingly, EVs have been shown to regulate the migration of fibroblasts towards wound sites [104,105,110]. Cheng et al. showed that TGFα stimulates HSP90α secretion from human keratinocytes via the exosome pathway [110]. Most exosomes contain some member of the heat-shock protein (HSP) family, such as HSP70 and HSP90 [39,76,115]. This protein family supports the folding of nascent peptides; prevents the aggregation of proteins; assists with the transportation of other proteins across cell membranes [116], and can induce cell motility [115,117]. Interestingly, human keratinocyte conditioned media containing exosomes with HSP90α cargo was found to stimulate the rapid migration of dermal fibroblasts [110]. HSP90α is comprised of four domains, including an N-terminal domain, a charged sequence connected to the N-terminal domain, a middle domain and a C-terminal domain [118]. It is thought that HSP90α may promote cell migration through binding interactions with the cell surface receptor LRP-1/CD91 [110,117]. Furthermore, in corroboration with cell migration enhancement, EVs released from mesenchymal stem cells and keratinocytes were also shown to promote the expression of important genes such as TGF-β, transforming growth factor beta receptor II, COL I, COL III, N-cadherin, cyclin-1, MMP-1, MMP-3 and IL-6 in fibroblasts [104,105]. These genes are involved in the Erk1/2 signalling pathway, which has roles in both cell proliferation and migration.

5.4. Angiogenesis

New blood vessel formation is critical for wound healing in order to supply nutrients and oxygen to newly formed tissues. The formation of new blood vessels requires the proliferation of endothelial cells, as well as the interaction between endothelial cells, angiogenic factors (such as VEGF and fibroblast growth factor) and surrounding ECM proteins [119]. Under chemotaxis, endothelial cells penetrate the underlying vascular basement membrane, invade ECM stroma and form tube-like structures that continue to extend, branch, and create networks [119,120]. Interestingly, all three populations of EVs (apoptotic bodies, microvesicles and exosomes) contribute to the regulation of vessel formation by enhancing the expression of pro-angiogenic factors [21,92,102,103,106,107,109,112,120,121]. For example, exosomes released from human embryonic MSCs [92,113] and human endothelial cells [103,106] enhanced angiogenesis by promoting endothelial cell proliferation and migration towards the wound site. Furthermore, a large number of blood vessels were observed in exosome-treated sites compared to control treatment [106,107,112].

Interestingly, various critical pro-angiogenic genes, including IL-6, IL-8, angiopoietin-1, E-selectin, and fibroblast growth factor 2, which activate the Erk1/2 signalling pathway, were up-regulated after platelet-rich plasma and endothelial progenitor cell-derived exosome stimulation [106,122]. Additionally, the downstream target of the Erk1/2 signalling pathway such as inhibitor of DNA binding 1, cyclooxygenase 1, VEGFA, VEGFR-2, c-Myc and cyclin-D1, were also increased after treatment with exosomes [106,122,123]. This provides further evidence that exosomes enhance endothelial cell function through activation of the Erk1/2 signalling pathway (Figure 7) [106]. In contrast, microvesicles from multiple cellular sources inhibited tube-like structure formation of microvascular endothelial cells via CD36 on the treated cell membrane, while exosome derived from the same cell source as the microvesicles did not have the same inhibitory activity [124]. Taken together, the previous studies imply that EVs could potentially be used in vascular regenerative medicine. Moreover, the EV bioactivities may depend on EV types and the physiological conditions of administered cells However, further research is required to investigate which specific EV derived biomolecules are responsible for the stimulation of angiogenesis.

5.5. Collagen Production and ECM Remodelling

Successful wound healing requires the contribution of many cellular events and biological processes in addition to coagulation, cell proliferation, cell migration and angiogenesis. Besides the involvement of EVs in the above events, EVs have also been shown to regulate ECM remodelling which is the last phase of wound healing. For example, EVs have been shown to stimulate an increase in elastin secretion, which is a structural protein of the ECM [102,112]. Furthermore, when wound sites are treated with MSC-derived exosomes, an increase of COL I and III were observed in the early stage of wound healing [105,112,125]. However, in the late stage of wound healing, exosomes may inhibit collagen expression [105]. Interestingly, EVs catalyse the crosslinking of collagen in the ECM via lysyl oxidase-like 2 (LOXL2), which is located on the exterior of the exosome membrane [126]. In addition, a study by Huleihel et al. (2016) observed that EVs present within the ECM were closely associated with the collagen network of the matrix, but these EVs could be separated from the fibre network [127]. This may indicate that EVs attend to the ECM formation and function. Furthermore, EVs have also been demonstrated to significantly increase the healing rate and reduce scar widths by interaction with Annexin A1 and formyl peptide receptors in a rat model [107,108,125]. Taken together, these studies indicate that EVs play a pivotal role in the ECM remodelling phase of wound healing.

Table 1. Summary of research investigating EVs involvement in wound healing.

Events	EV Types	Parental Cells	Target Cells	Secreted Factors/Factors Presented in EVs	Molecules/Pathways Activated	References
Coagulation	Mv, Ex	Saliva/granulocytes, EPC	Blood	TF	Trigger coagulation by initiating TF/factor VII	[99]
	Mv	Monocytes	Activated Platelet	TF, PSGL-1		[55]
	Mv	Plasma/Platelet, erythrocytes, granulocytes	Blood	TF	Promote thrombus formation	[98]
Proliferation	Ex	MSC	FB, EC		Increase expression levels of HGF, IGF1, NGF, SDF1; increase re-epithelialisation; reduce scar widths; promote collagen maturity and the creation of newly formed vessels; accelerate maturation of wound sites; activate Akt, Erk and Stat3 signalling	[21,112]
	Ex/nanoparticles *	ESC	FB		Enhance the expression levels of mRNA, EVGFα, TGFβ, collagen I, Ki-67	[102]
	Mv, Ex	MyoFB	FB, EC			[18]
	Ex	KC	FB	HSP90	Could not promote cell proliferation	[110]
	Ex	Platelet-rich plasma	EC	VEGF, bFGF, PDGFBB	Activating Pi3K/Akt and Erk signalling pathway	[107]
	Ex	EC, EPC	EC		Activating Erk1/2 signalling pathway	[103,106]
	Mv	Platelet-rich plasma after exercise	HUVEC			[128]
Migration	Ex	MSC	FB, EC		Induction the expression of HGF, IGF1, NGF, SDF1; activate Akt, Erk and Stat3 signalling	[21,112]
	Ex	KC	KC, HDMECs	HSP90	Hsp90-Ex increased cell migration without the need to bind any cofactor or ATP; CD91 is receptor of extracellular Hsp90; TGFβ could not inhibit cell migration	[110]
	Ex/Ev	EC	Murine wound	Annexin-I		[108]
	Ex/nanop-articles *	ESC	FB		Higher expression of mRNA, EVGFα, TGFβ, collagen I, Ki-67	[102]
	Ex	Platelet-rich plasma	EC	VEGF, bFGF, PDGFBB	Activating Pi3K/Akt and Erk signalling pathway	[107]
	Ex	EC, EPC	EC		Activating Erk1/2 signalling pathway	[103,106]
	Ex	hUSC	HUVEC			[121]
	Ev	Multiple cellular sources	EC	CD63		[124]

Table 1. *Cont.*

Events	EV Types	Parental Cells	Target Cells	Secreted Factors/Factors Presented in EVs	Molecules/Pathways Activated	References
Angiogenesis	Ex	MSC	FB, EC		Induction the expression of HGF, IGF1, NGF, SDF1; promote the creation and maturation of newly formed vessels, increase re-epithelialisation, reduce scar widths	[21,112]
	Ex/nanop-articles*	ESC	FB		Enhance the expression levels of mRNA, EVGFα, TGFβ, collagen I, KI-67	[102]
	Mv, Ex	MyoFB	FB, EC	VEGF, FGF2	Increase angiogenesis	[18]
	Ex	Platelet-rich plasma	EC	VEGF, bFGF, PDGFBB	Activating PI3K/Akt and Erk signalling pathway	[107]
	Ex	EC	EC		Activating Erk1/2 signalling pathway	[103,106]
	Ex	MSC	EC	EMMPRIN, VEGF, MMP9	ERK/Akt pathway	[123]
	Ex	MSC	EC	miR-125a	Direct target DLL4	[92]
	Ex	hUSC	HUVEC			[121]
	Ex	Epithelium cells	EC	VEGFR		[129]
	Ev	Multiple cellular sources	EC	CD63		[124]
	Mv	Platelet-rich plasma after exercise	HUVEC			[128]
Collagen production and ECM remodelling	Ex	MSC	FB	TF	Increase reepithelialisation, reduce scar widths, promote collagen maturity and maturation of wound sites	[112]
	Ex	MSCs, FB	FB		Increase collagen production	[105,125]
	Ex	Hypoxic EC	ECM	LOXL2	ECM remodelling	[126]

FB: Fibroblasts, KC: Keratinocytes, MSCs: Mesenchymal stem cells, EC: Endothelial cells, ESC: Embryonic stem cells, EPC: Epithelial cells, hUSC: Human urine derived stem cells, HUVUEC: Human umbilical vein endothelial cells, Mv: Microvesicle(s), Ex: Exosome(s), EMMPRIN: Extracellular matrix metalloproteinase inducer, (*) nanoparticles mimicking. Exosome were extracted from living cells.

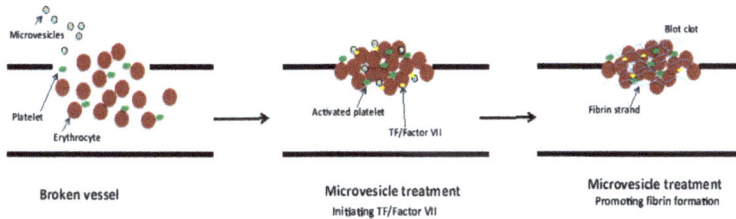

Figure 6. Promotion of coagulation by TF-barring microvesicle treatment. Rapid coagulation is triggered by the initiating TF/Factor VII and promotion of fibrin strand formation [98,99].

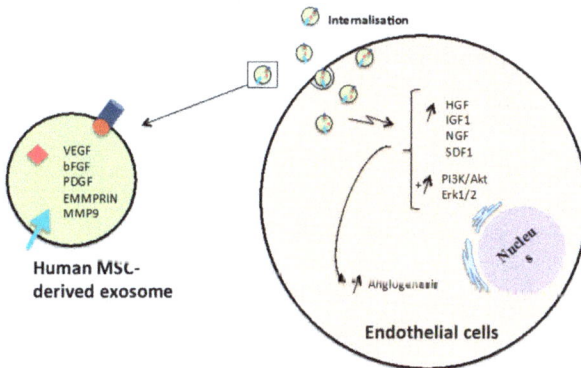

Figure 7. Promotion of angiogenesis by MSC-derived exosomes. Exosomes released from human MSCs can induce expression of genes and activate PI3K/Akt and Erk1/2 signalling pathways in endothelial cells leading to promotion tube formation and newly formed vessels.

6. Conclusions

In the past, EVs were disregarded simply as cellular debris. However, current research has demonstrated that EVs contain bioactive molecules and are able to deliver these to recipient cells via newly described cell-to-cell communication mechanisms. In wound healing, EVs were initially shown to regulate inflammation, proliferation, migration, angiogenesis, collagen production and ECM remodelling. Such regulation could be mediated through the enhancement of gene expression, suppression of gene translation, and/or activation of signalling pathways important for wound healing processes. However, the current level of knowledge regarding how the EV molecular cargo regulates the process of wound healing remains unclear. Therefore, more research is required to clarify the regulation of EVs during the wound healing process, and to translate these results to the clinic.

Acknowledgments: The authors would like to acknowledge the support of the Australian Government's Cooperative Research Centre Programme.

Author Contributions: All authors contributed significantly to the study. Uyen Thi Trang Than did the search and extracted the information. Uyen Thi Trang Than, Tony Parker, David Leavesley and Dominic Guanzon drafted the manuscript. All authors contributed to interpretation of data and critically reviewed the manuscript for important intellectual content.

Conflicts of Interest: The authors declare no conflict of interest.

Abbreviations

ARF6	ADP-ribosylation factor 6
COL	Collagen
DNA	Deoxyribonucleic acid
EC(s)	Endothelial cell(s)
ECM	Extracellular matrix
EMMPRIN	Extracellular matrix metalloproteinase inducer
EPC(s)	Epithelial cell(s)
ESC(s)	Embryonic stem cell(s)
ESCRT	Endosomal sorting complex required for transport
EV(s)	Extracellular membrane vesicle(s)/extracellular vesicle(s)
Ex	Exosome(s)
FB	Fibroblast
FGF	Fibroblast growth factor
HSP	Heat-shock protein
HUVEC(s)	Human umbilical vein endothelial cell(s)
hUSC	human urine derived stem cells
IGF	Insulin-like growth factor
IL	Interleukin
KC(s)	Keratinocyte(s)
LOXL2	Lysyl oxidase-like 2
MAPK	Mitogen-activated protein kinase
MHC	Major histocompatibility complex
miRNA(s)	microRNAs(s)
MLCK	Myosin light chain kinase
MMP	Matrix metalloproteinase
MSC(s)	Mesenchymal stem cell(s)
Mv	Microvesicle(s)
MVB(s)	Multi-vesicular bodie(s)
P	Platelet
PS	Phosphatidylserine
PDGFBB	Platelet-derived growth factor-BB
PSGL-1	P-selectin glycoprotein ligand-1

RNA	Ribonucleic acid
SCID	Severe combined immunodeficient
SDF	Stroma-derived growth factor
TF	Tissue factor
TFPI	Tissue factor pathway inhibitor
TGF	Transforming growth factor
VEGF	Vascular endothelial growth factor

References

1. Baietti, M.F.; Zhang, Z.; Mortier, E.; Melchior, A.; Degeest, G.; Geeraerts, A.; Ivarsson, Y.; Depoortere, F.; Coomans, C.; Vermeiren, E. Syndecan-syntenin-ALIX regulates the biogenesis of exosomes. *Nat. Cell Biol.* **2012**, *14*, 677–685. [CrossRef] [PubMed]
2. Meister, G.; Landthaler, M.; Peters, L.; Chen, P.Y.; Urlaub, H.; Lührmann, R.; Tuschl, T. Identification of novel argonaute associated proteins. *Curr. Biol.* **2005**, *15*, 2149–2155. [CrossRef] [PubMed]
3. Valadi, H.; Ekström, K.; Bossios, A.; Sjöstrand, M.; Lee, J.J.; Lötvall, J.O. Exosome-mediated transfer of mRNAs and microRNAs is a novel mechanism of genetic exchange between cells. *Nat. Cell Biol.* **2007**, *9*, 654–659. [CrossRef] [PubMed]
4. Vlassov, A.V.; Magdaleno, S.; Setterquist, R.; Conrad, R. Exosomes: Current knowledge of their composition, biological functions, and diagnostic and therapeutic potentials. *Biochim. Biophys. Acta* **2012**, *1820*, 940–948. [CrossRef] [PubMed]
5. Hao, S.; Bai, O.; Yuan, J.; Qureshi, M.; Xiang, J. Dendritic cell-derived exosomes stimulate stronger CD8+ CTL responses and antitumor immunity than tumor cell-derived exosomes. *Cell. Mol. Immunol.* **2006**, *3*, 205–211. [PubMed]
6. Lee, T.H.; D'Asti, E.; Magnus, N.; Al Nedawi, K.; Meehan, B.; Rak, J. Microvesicles as mediators of intercellular communication in cancer—The emerging science of cellular "debris". *Semin Immunopathol.* **2011**, *33*, 455–467. [CrossRef] [PubMed]
7. Michael, A.; Bajracharya, S.D.; Yuen, P.S.; Zhou, H.; Star, R.A.; Illei, G.G.; Alevizos, I. Exosomes from human saliva as a source of microRNA biomarkers. *Oral Dis.* **2010**, *16*, 34–38. [CrossRef] [PubMed]
8. Keller, S.; Rupp, C.; Stoeck, A.; Runz, S.; Fogel, M.; Lugert, S.; Hager, H.-D.; Abdel-Bakky, M.; Gutwein, P.; Altevogt, P. CD24 is a marker of exosomes secreted into urine and amniotic fluid. *Kidney Int.* **2007**, *72*, 1095–1102. [CrossRef] [PubMed]
9. Zonneveld, M.I.; Brisson, A.R.; van Herwijnen, M.J.; Tan, S.; van de Lest, C.H.; Redegeld, F.A.; Garssen, J.; Wauben, M.H.; Nolte-'t Hoen, E.N. Recovery of extracellular vesicles from human breast milk is influenced by sample collection and vesicle isolation procedures. *J. Extracell. Vesicles* **2014**, *3*, 24215. [CrossRef] [PubMed]
10. Rupp, A.-K.; Rupp, C.; Keller, S.; Brase, J.C.; Ehehalt, R.; Fogel, M.; Moldenhauer, G.; Marmé, F.; Sültmann, H.; Altevogt, P. Loss of epcam expression in breast cancer derived serum exosomes: Role of proteolytic cleavage. *Gynecol. Oncol.* **2011**, *122*, 437–446. [CrossRef] [PubMed]
11. Crescitelli, R.; Lässer, C.; Szabo, T.G.; Kittel, A.; Eldh, M.; Dianzani, I.; Buzás, E.I.; Lötvall, J. Distinct RNA profiles in subpopulations of extracellular vesicles: Apoptotic bodies, microvesicles and exosomes. *J. Extracell. Vesicles* **2013**, *2*. [CrossRef] [PubMed]
12. Skriner, K.; Adolph, K.; Jungblut, P.R.; Burmester, G.R. Association of citrullinated proteins with synovial exosomes. *Arthritis Rheum.* **2006**, *54*, 3809–3814. [CrossRef] [PubMed]
13. Ostenfeld, M.S.; Jeppesen, D.K.; Laurberg, J.R.; Boysen, A.T.; Bramsen, J.B.; Primdal-Bengtson, B.; Hendrix, A.; Lamy, P.; Dagnaes-Hansen, F.; Rasmussen, M.H. Cellular disposal of miR23b by RAB27-dependent exosome release is linked to acquisition of metastatic properties. *Cancer Res.* **2014**, *74*, 5758–5771. [CrossRef] [PubMed]
14. Mittelbrunn, M.; Gutiérrez-Vázquez, C.; Villarroya-Beltri, C.; González, S.; Sánchez-Cabo, F.; González, M.Á.; Bernad, A.; Sánchez-Madrid, F. Unidirectional transfer of microRNA-loaded exosomes from T cells to antigen-presenting cells. *Nat. Commun.* **2011**, *2*, 282. [CrossRef] [PubMed]
15. Yáñez-Mó, M.; Siljander, P.R.-M.; Andreu, Z.; Bedina Zavec, A.; Borràs, F.E.; Buzas, E.I.; Buzas, K.; Casal, E.; Cappello, F.; Carvalho, J. Biological properties of extracellular vesicles and their physiological functions. *J. Extracell. Vesicles* **2015**, *4*, 27066. [CrossRef] [PubMed]
16. Guo, S.A.; di Pietro, L.A. Factors affecting wound healing. *J. Dent. Res.* **2010**, *89*, 219–229. [CrossRef] [PubMed]

17. Keerthikumar, S.; Gangoda, L.; Liem, M.; Fonseka, P.; Atukorala, I.; Ozcitti, C.; Mechler, A.; Adda, C.G.; Ang, C.-S.; Mathivanan, S. Proteogenomic analysis reveals exosomes are more oncogenic than ectosomes. *Oncotarget* **2015**, *6*, 15375–15396. [CrossRef] [PubMed]

18. Moulin, V.J.; Mayrand, D.; Messier, H.; Martinez, M.C.; Lopez-Vallé, C.A.; Genest, H. Shedding of microparticles by myofibroblasts as mediator of cellular cross-talk during normal wound healing. *J. Cell. Physiol.* **2010**, *225*, 734–740. [CrossRef] [PubMed]

19. Cocucci, E.; Racchetti, G.; Meldolesi, J. Shedding microvesicles: Artefacts no more. *Trends Cell Biol.* **2009**, *19*, 43–51. [CrossRef] [PubMed]

20. Rani, S.; Ritter, T. The exosome—A naturally secreted nanoparticle and its application to wound healing. *Adv. Mater.* **2016**, *27*, 5542–5552. [CrossRef] [PubMed]

21. Shabbir, A.; Cox, A.; Rodriguez-Menocal, L.; Salgado, M.; Badiavas, E.V. Mesenchymal stem cell exosomes induce proliferation and migration of normal and chronic wound fibroblasts, and enhance angiogenesis in vitro. *Stem Cells Dev.* **2015**, *24*, 1635–1647. [CrossRef] [PubMed]

22. Velnar, T.; Bailey, T.; Smrkolj, V. The wound healing process: An overview of the cellular and molecular mechanisms. *J. Int. Med. Res.* **2009**, *37*, 1528–1542. [CrossRef] [PubMed]

23. Elmore, S. Apoptosis: A review of programmed cell death. *Toxicol. Pathol.* **2007**, *35*, 495–516. [CrossRef] [PubMed]

24. Kerr, J.F.; Wyllie, A.H.; Currie, A.R. Apoptosis: A basic biological phenomenon with wide-ranging implications in tissue kinetics. *Br. J. Cancer* **1972**, *26*, 239. [CrossRef] [PubMed]

25. Edinger, A.L.; Thompson, C.B. Death by design: Apoptosis, necrosis and autophagy. *Curr. Opin. Cell Biol.* **2004**, *16*, 663–669. [CrossRef] [PubMed]

26. Simpson, R.; Mathivanan, S. Extracellular microvesicles: The need for internationally recognised nomenclature and stringent purification criteria. *J. Proteom. Bioinform.* **2012**, *5*, 1–2. [CrossRef]

27. Kalra, H.; Simpson, R.J.; Ji, H.; Aikawa, E.; Altevogt, P.; Askenase, P.; Bond, V.C.; Borràs, F.E.; Breakefield, X.; Budnik, V. Vesiclepedia: A compendium for extracellular vesicles with continuous community annotation. *PLoS Biol.* **2012**, *10*, e1001450. [CrossRef] [PubMed]

28. Akers, J.C.; Gonda, D.; Kim, R.; Carter, B.S.; Chen, C.C. Biogenesis of extracellular vesicles (EV): Exosomes, microvesicles, retrovirus-like vesicles, and apoptotic bodies. *J. Neurooncol.* **2013**, *113*, 1–11. [CrossRef] [PubMed]

29. Maezawa, S.; Yoshimura, T.; Hong, K.; Duzgunes, N.; Papahadjopoulos, D. Mechanism of protein-induced membrane fusion: Fusion of phospholipid vesicles by clathrin associated with its membrane binding and conformational change. *Biochemistry* **1989**, *28*, 1422–1428. [CrossRef] [PubMed]

30. Piccin, A.; Murphy, W.G.; Smith, O.P. Circulating microparticles: Pathophysiology and clinical implications. *Blood Rev.* **2007**, *21*, 157–171. [CrossRef] [PubMed]

31. Devaux, P.F.; Herrmann, A.; Ohlwein, N.; Kozlov, M.M. How lipid flippases can modulate membrane structure. *Biochim. Biophys. Acta* **2008**, *1778*, 1591–1600. [CrossRef] [PubMed]

32. Muralidharan-Chari, V.; Clancy, J.; Plou, C.; Romao, M.; Chavrier, P.; Raposo, G.; D'Souza-Schorey, C. ARF6-regulated shedding of tumor cell-derived plasma membrane microvesicles. *Curr. Biol.* **2009**, *19*, 1875–1885. [CrossRef] [PubMed]

33. McConnell, R.E.; Higginbotham, J.N.; Shifrin, D.A.; Tabb, D.L.; Coffey, R.J.; Tyska, M.J. The enterocyte microvillus is a vesicle-generating organelle. *J. Cell Biol.* **2009**, *185*, 1285–1298. [CrossRef] [PubMed]

34. Théry, C.; Ostrowski, M.; Segura, E. Membrane vesicles as conveyors of immune responses. *Nat. Rev. Immunol.* **2009**, *9*, 581–593. [CrossRef] [PubMed]

35. Raposo, G.; Nijman, H.W.; Stoorvogel, W.; Liejendekker, R.; Harding, C.V.; Melief, C.J.; Geuze, H.J. B lymphocytes secrete antigen-presenting vesicles. *J. Exp. Med.* **1996**, *183*, 1161–1172. [CrossRef] [PubMed]

36. Pan, B.-T.; Teng, K.; Wu, C.; Adam, M.; Johnstone, R.M. Electron microscopic evidence for externalization of the transferrin receptor in vesicular form in sheep reticulocytes. *J. Cell Biol.* **1985**, *101*, 942–948. [CrossRef] [PubMed]

37. Harrison, P.; Gardiner, C.; Sargent, I.L. *Extracellular vesicles in health and disease*; Pan Stanford Publishing: Hoboken, NJ, USA, 2014.

38. Huotari, J.; Helenius, A. Endosome maturation. *EMBO J.* **2011**, *30*, 3481–3500. [CrossRef] [PubMed]

39. Théry, C.; Zitvogel, L.; Amigorena, S. Exosomes: Composition, biogenesis and function. *Nat. Rev. Immunol.* **2002**, *2*, 569–579. [PubMed]

40. Harding, C.; Heuser, J.; Stahl, P. Receptor-mediated endocytosis of transferrin and recycling of the transferrin receptor in rat reticulocytes. *J. Cell Biol.* **1983**, *97*, 329–339. [CrossRef] [PubMed]
41. Wang, Z.; Hill, S.; Luther, J.M.; Hachey, D.L.; Schey, K.L. Proteomic analysis of urine exosomes by multidimensional protein identification technology (MudPIT). *Proteomics* **2012**, *12*, 329–338. [CrossRef] [PubMed]
42. Ostrowski, M.; Carmo, N.B.; Krumeich, S.; Fanget, I.; Raposo, G.; Savina, A.; Moita, C.F.; Schauer, K.; Hume, A.N.; Freitas, R.P. RAB27a and RAB27b control different steps of the exosome secretion pathway. *Nat. Cell Biol.* **2010**, *12*, 19–30. [CrossRef] [PubMed]
43. Raposo, G.; Stoorvogel, W. Extracellular vesicles: Exosomes, microvesicles, and friends. *J. Cell Biol.* **2013**, *200*, 373–383. [CrossRef] [PubMed]
44. Camussi, G.; Deregibus, M.C.; Bruno, S.; Cantaluppi, V.; Biancone, L. Exosomes/microvesicles as a mechanism of cell-to-cell communication. *Kidney Int.* **2010**, *78*, 838–848. [CrossRef] [PubMed]
45. Majka, M.; Janowska-Wieczorek, A.; Ratajczak, J.; Ehrenman, K.; Pietrzkowski, Z.; Kowalska, M.A.; Gewirtz, A.M.; Emerson, S.G.; Ratajczak, M.Z. Numerous growth factors, cytokines, and chemokines are secreted by human CD34+ cells, myeloblasts, erythroblasts, and megakaryoblasts and regulate normal hematopoiesis in an autocrine/paracrine manner. *Blood* **2001**, *97*, 3075–3085. [CrossRef] [PubMed]
46. Sherer, N.M.; Mothes, W. Cytonemes and tunneling nanotubules in cell-cell communication and viral pathogenesis. *Trends Cell Biol.* **2008**, *18*, 414–420. [CrossRef] [PubMed]
47. Rustom, A.; Saffrich, R.; Markovic, I.; Walther, P.; Gerdes, H.-H. Nanotubular highways for intercellular organelle transport. *Science* **2004**, *303*, 1007–1010. [CrossRef] [PubMed]
48. Lösche, W.; Scholz, T.; Temmler, U.; Oberle, V.; Claus, R.A. Platelet-derived microvesicles transfer tissue factor to monocytes but not to neutrophils. *Platelets* **2004**, *15*, 109–115. [CrossRef] [PubMed]
49. Eken, C.; Gasser, O.; Zenhaeusern, G.; Oehri, I.; Hess, C.; Schifferli, J.A. Polymorphonuclear neutrophil-derived ectosomes interfere with the maturation of monocyte-derived dendritic cells. *J. Immunol.* **2008**, *180*, 817–824. [CrossRef] [PubMed]
50. Gasser, O.; Hess, C.; Miot, S.; Deon, C.; Sanchez, J.-C. Characterisation and properties of ectosomes released by human polymorphonuclear neutrophils. *Exp. Cell. Res.* **2003**, *285*, 243–257. [CrossRef]
51. Pluskota, E.; Woody, N.M.; Szpak, D.; Ballantyne, C.M.; Soloviev, D.A.; Simon, D.I.; Plow, E.F. Expression, activation, and function of integrin αMβ2 (Mac-1) on neutrophil-derived microparticles. *Blood* **2008**, *112*, 2327–2335. [CrossRef] [PubMed]
52. Smith, J.; Leonardi, T.; Huang, B.; Iraci, N.; Vega, B.; Pluchino, S. Extracellular vesicles and their synthetic analogues in aging and age-associated brain diseases. *Biogerontology* **2015**, *16*, 147–185. [CrossRef] [PubMed]
53. Nolte, E.N.; Buschow, S.I.; Anderton, S.M.; Stoorvogel, W.; Wauben, M.H. Activated T cells recruit exosomes secreted by dendritic cells via LFA-1. *Blood* **2009**, *113*, 1977–1981. [CrossRef] [PubMed]
54. Janowska-Wieczorek, A.; Majka, M.; Kijowski, J.; Baj-Krzyworzeka, M.; Reca, R.; Turner, A.R.; Ratajczak, J.; Emerson, S.G.; Kowalska, M.A.; Ratajczak, M.Z. Platelet-derived microparticles bind to hematopoietic stem/progenitor cells and enhance their engraftment. *Blood* **2001**, *98*, 3143–3149. [CrossRef] [PubMed]
55. Del Conde, I.; Shrimpton, C.N.; Thiagarajan, P.; López, J.A. Tissue-factor–bearing microvesicles arise from lipid rafts and fuse with activated platelets to initiate coagulation. *Blood* **2005**, *106*, 1604–1611. [CrossRef] [PubMed]
56. Kim, J.W.; Wieckowski, E.; Taylor, D.D.; Reichert, T.E.; Watkins, S.; Whiteside, T.L. Fas ligand–positive membranous vesicles isolated from sera of patients with oral cancer induce apoptosis of activated T lymphocytes. *Clin. Cancer Res.* **2005**, *11*, 1010–1020. [PubMed]
57. Mack, M.; Kleinschmidt, A.; Brühl, H.; Klier, C.; Nelson, P.J.; Cihak, J.; Plachý, J.; Stangassinger, M.; Erfle, V.; Schlöndorff, D. Transfer of the chemokine receptor CCR5 between cells by membrane-derived microparticles: A mechanism for cellular human immunodeficiency virus 1 infection. *Nat. Med.* **2000**, *6*, 769–775. [CrossRef] [PubMed]
58. Fader, C.M.; Sánchez, D.G.; Mestre, M.B.; Colombo, M.I. Ti-VAMP/VAMP7 and VAMP3/cellubrevin: Two v-SNARE proteins involved in specific steps of the autophagy/multivesicular body pathways. *Biochim. Biophys. Acta* **2009**, *1793*, 1901–1916. [CrossRef] [PubMed]
59. Morelli, A.E.; Larregina, A.T.; Shufesky, W.J.; Sullivan, M.L.; Stolz, D.B.; Papworth, G.D.; Zahorchak, A.F.; Logar, A.J.; Wang, Z.; Watkins, S.C. Endocytosis, intracellular sorting, and processing of exosomes by dendritic cells. *Blood* **2004**, *104*, 3257–3266. [CrossRef] [PubMed]

60. Bilyy, R.O.; Shkandina, T.; Tomin, A.; Muñoz, L.E.; Franz, S.; Antonyuk, V.; Kit, Y.Y.; Zirngibl, M.; Fürnrohr, B.G.; Janko, C. Macrophages discriminate glycosylation patterns of apoptotic cell-derived microparticles. *J. Biol. Chem.* **2012**, *287*, 496–503. [CrossRef] [PubMed]

61. Berditchevski, F.; Odintsova, E. Tetraspanins as regulators of protein trafficking. *Traffic* **2007**, *8*, 89–96. [CrossRef] [PubMed]

62. Escola, J.M.; Kleijmeer, M.J.; Stoorvogel, W.; Griffith, J.M.; Yoshie, O.; Geuze, H.J. Selective enrichment of tetraspan proteins on the internal vesicles of multivesicular endosomes and on exosomes secreted by human B-lymphocytes. *J. Biol. Chem.* **1998**, *273*, 20121–20127. [CrossRef] [PubMed]

63. Kowal, J.; Arras, G.; Colombo, M.; Jouve, M.; Morath, J.P.; Primdal-Bengtson, B.; Dingli, F.; Loew, D.; Tkach, M.; Théry, C. Proteomic comparison defines novel markers to characterize heterogeneous populations of extracellular vesicle subtypes. *Proc. Natl. Acad. Sci. USA* **2016**, *113*, E968–E977. [CrossRef] [PubMed]

64. Kristiansen, G.; Machado, E.; Bretz, N.; Rupp, C.; Winzer, K.-J.; König, A.-K.; Moldenhauer, G.; Marmé, F.; Costa, J.; Altevogt, P. Molecular and clinical dissection of CD24 antibody specificity by a comprehensive comparative analysis. *Lab. Investig.* **2010**, *90*, 1102–1116. [CrossRef] [PubMed]

65. Dujardin, S.; Bégard, S.; Caillierez, R.; Lachaud, C.; Delattre, L.; Carrier, S.; Loyens, A.; Galas, M.-C.; Bousset, L.; Melki, R. Ectosomes: A new mechanism for non-exosomal secretion of tau protein. *PLoS ONE* **2014**, *9*, e100760. [CrossRef] [PubMed]

66. Bergsmedh, A.; Szeles, A.; Henriksson, M.; Bratt, A.; Folkman, M.J.; Spetz, A.-L.; Holmgren, L. Horizontal transfer of oncogenes by uptake of apoptotic bodies. *Proc. Natl. Acad. Sci. USA* **2001**, *98*, 6407–6411. [CrossRef] [PubMed]

67. Dolo, V.; Ginestra, A.; Cassarà, D.; Violini, S.; Lucania, G.; Torrisi, M.R.; Nagase, H.; Canevari, S.; Pavan, A.; Vittorelli, M.L. Selective localization of matrix metalloproteinase 9, β1 integrins, and human lymphocyte antigen class I molecules on membrane vesicles shed by 8701-BC breast carcinoma cells. *Cancer Res.* **1998**, *58*, 4468–4474. [PubMed]

68. Heijnen, H.F.; Schiel, A.E.; Fijnheer, R.; Geuze, H.J.; Sixma, J.J. Activated platelets release two types of membrane vesicles: Microvesicles by surface shedding and exosomes derived from exocytosis of multivesicular bodies and α-granules. *Blood* **1999**, *94*, 3791–3799. [PubMed]

69. Taraboletti, G.; D'Ascenzo, S.; Borsotti, P.; Giavazzi, R.; Pavan, A.; Dolo, V. Shedding of the matrix metalloproteinases MMP-2, MMP-9, and MT1-MMP as membrane vesicle-associated components by endothelial cells. *Am. J. Pathol.* **2002**, *160*, 673–680. [CrossRef]

70. Giavazzi, R.; Taraboletti, G. Preclinical development of metalloproteasis inhibitors in cancer therapy. *Crit. Rev. Oncol. Hematol.* **2001**, *37*, 53–60. [CrossRef]

71. Hidalgo, M.; Eckhardt, S.G. Development of matrix metalloproteinase inhibitors in cancer therapy. *J. Natl. Cancer Inst.* **2001**, *93*, 178–193. [CrossRef] [PubMed]

72. Blanchard, N.; Lankar, D.; Faure, F.; Regnault, A.; Dumont, C.; Raposo, G.; Hivroz, C. TCR activation of human T cells induces the production of exosomes bearing the TCR/CD3/ζ complex. *J. Immunol.* **2002**, *168*, 3235–3241. [CrossRef] [PubMed]

73. Wolfers, J.; Lozier, A.; Raposo, G.; Regnault, A.; Théry, C.; Masurier, C.; Flament, C.; Pouzieux, S.; Faure, F.; Tursz, T. Tumor-derived exosomes are a source of shared tumor rejection antigens for CTL cross-priming. *Nat. Med.* **2001**, *7*, 297–303. [CrossRef] [PubMed]

74. Théry, C.; Boussac, M.; Véron, P.; Ricciardi-Castagnoli, P.; Raposo, G.; Garin, J.; Amigorena, S. Proteomic analysis of dendritic cell-derived exosomes: A secreted subcellular compartment distinct from apoptotic vesicles. *J. Immunol.* **2001**, *166*, 7309–7318. [CrossRef] [PubMed]

75. Théry, C.; Regnault, A.; Garin, J.; Wolfers, J.; Zitvogel, L.; Ricciardi-Castagnoli, P.; Raposo, G.; Amigorena, S. Molecular characterization of dendritic cell-derived exosomes. *J. Cell Biol.* **1999**, *147*, 599–610. [CrossRef] [PubMed]

76. Bard, M.P.; Hegmans, J.P.; Hemmes, A.; Luider, T.M.; Willemsen, R.; Severijnen, L.-A.A.; van Meerbeeck, J.P.; Burgers, S.A.; Hoogsteden, H.C.; Lambrecht, B.N. Proteomic analysis of exosomes isolated from human malignant pleural effusions. *Am. J. Respir. Cell Mol. Biol.* **2004**, *31*, 114–121. [CrossRef] [PubMed]

77. Wang, T.; Feng, Y.; Sun, H.; Zhang, L.; Hao, L.; Shi, C.; Wang, J.; Li, R.; Ran, X.; Su, Y. miR-21 regulates skin wound healing by targeting multiple aspects of the healing process. *Am. J. Pathol.* **2012**, *181*, 1911–1920. [CrossRef] [PubMed]

78. Batista, B.S.; Eng, W.S.; Pilobello, K.T.; Hendricks-Muñoz, K.D.; Mahal, L.K. Identification of a conserved glycan signature for microvesicles. *J. Proteome Res.* **2011**, *10*, 4624. [CrossRef] [PubMed]

79. Laulagnier, K.; Motta, C.; Hamdi, S.; Sébastien, R.; Fauvelle, F.; Pageaux, J.-F.; Kobayashi, T.; Salles, J.-P.; Perret, B.; Bonnerot, C. Mast cell-and dendritic cell-derived exosomes display a specific lipid composition and an unusual membrane organization. *Biochem. J.* **2004**, *380*, 161–171. [CrossRef] [PubMed]

80. Subra, C.; Laulagnier, K.; Perret, B.; Record, M. Exosome lipidomics unravels lipid sorting at the level of multivesicular bodies. *Biochimie* **2007**, *89*, 205–212. [CrossRef] [PubMed]

81. Trajkovic, K.; Hsu, C.; Chiantia, S.; Rajendran, L.; Wenzel, D.; Wieland, F.; Schwille, P.; Brügger, B.; Simons, M. Ceramide triggers budding of exosome vesicles into multivesicular endosomes. *Science* **2008**, *319*, 1244–1247. [CrossRef] [PubMed]

82. Lunavat, T.R.; Cheng, L.; Kim, D.-K.; Bhadury, J.; Jang, S.C.; Lässer, C.; Sharples, R.A.; López, M.D.; Nilsson, J.; Gho, Y.S. Small RNA deep sequencing discriminates subsets of extracellular vesicles released by melanoma cells–evidence of unique microRNA cargos. *RNA Biol.* **2015**, *12*, 810–823. [CrossRef] [PubMed]

83. Hunter, M.P.; Ismail, N.; Zhang, X.; Aguda, B.D.; Lee, E.J.; Yu, L.; Xiao, T.; Schafer, J.; Lee, M.-L.T.; Schmittgen, T.D. Detection of microRNA expression in human peripheral blood microvesicles. *PLoS ONE* **2008**, *3*, e3694. [CrossRef] [PubMed]

84. Ji, H.; Chen, M.; Greening, D.W.; He, W.; Rai, A.; Zhang, W.; Simpson, R.J. Deep sequencing of RNA from three different extracellular vesicle (EV) subtypes released from the human lim1863 colon cancer cell line uncovers distinct miRNA-enrichment signatures. *PLoS ONE* **2014**, *9*, e110314. [CrossRef] [PubMed]

85. Ekström, K.; Valadi, H.; Sjöstrand, M.; Malmhäll, C.; Bossios, A.; Eldh, M.; Lötvall, J. Characterization of mRNA and microRNA in human mast cell-derived exosomes and their transfer to other mast cells and blood CD34 progenitor cells. *J. Extracell. Vesicles* **2012**, *1*. [CrossRef] [PubMed]

86. Hong, B.S.; Cho, J.-H.; Kim, H.; Choi, E.-J.; Rho, S.; Kim, J.; Kim, J.H.; Choi, D.-S.; Kim, Y.-K.; Hwang, D. Colorectal cancer cell-derived microvesicles are enriched in cell cycle-related mRNAs that promote proliferation of endothelial cells. *BMC Genom.* **2009**, *10*, 556. [CrossRef] [PubMed]

87. Noerholm, M.; Balaj, L.; Limperg, T.; Salehi, A.; Zhu, L.D.; Hochberg, F.H.; Breakefield, X.O.; Carter, B.S.; Skog, J. RNA expression patterns in serum microvesicles from patients with glioblastoma multiforme and controls. *BMC Cancer* **2012**, *12*, 22. [CrossRef] [PubMed]

88. Palanisamy, V.; Sharma, S.; Deshpande, A.; Zhou, H.; Gimzewski, J.; Wong, D.T. Nanostructural and transcriptomic analyses of human saliva derived exosomes. *PLoS ONE* **2010**, *5*, e8577. [CrossRef] [PubMed]

89. Rauschenberger, L.; Staar, D.; Thom, K.; Scharf, C.; Venz, S.; Homuth, G.; Schlüter, R.; Brandenburg, L.O.; Ziegler, P.; Zimmermann, U. Exosomal particles secreted by prostate cancer cells are potent mRNA and protein vehicles for the interference of tumor and tumor environment. *Prostate* **2016**, *76*, 409–424. [CrossRef] [PubMed]

90. Yang, J.; Wei, F.; Schafer, C.; Wong, D.T. Detection of tumor cell-specific mRNA and protein in exosome-like microvesicles from blood and saliva. *PLoS ONE* **2014**, *9*, e110641. [CrossRef] [PubMed]

91. Taylor, D.D.; Gercel-Taylor, C. MicroRNA signatures of tumor-derived exosomes as diagnostic biomarkers of ovarian cancer. *Gynecol. Oncol.* **2008**, *110*, 13–21. [CrossRef] [PubMed]

92. Liang, X.; Zhang, L.; Wang, S.; Han, Q.; Zhao, R.C. Exosomes secreted by mesenchymal stem cells promote endothelial cell angiogenesis by transferring miR-125a. *J. Cell Sci.* **2016**, *129*, 2182–2189. [CrossRef] [PubMed]

93. Miller, I.V.; Raposo, G.; Welsch, U.; Prazeres da Costa, O.; Thiel, U.; Lebar, M.; Maurer, M.; Bender, H.U.; Luettichau, I.; Richter, G.H. First identification of ewing's sarcoma-derived extracellular vesicles and exploration of their biological and potential diagnostic implications. *Biol. Cell* **2013**, *105*, 289–303. [CrossRef] [PubMed]

94. Shao, H.; Chung, J.; Lee, K.; Balaj, L.; Min, C.; Carter, B.S.; Hochberg, F.H.; Breakefield, X.O.; Lee, H.; Weissleder, R. Chip-based analysis of exosomal mRNA mediating drug resistance in glioblastoma. *Nat. Commun.* **2015**, *6*. [CrossRef] [PubMed]

95. Herrera, M.; Fonsato, V.; Gatti, S.; Deregibus, M.; Sordi, A.; Cantarella, D.; Calogero, R.; Bussolati, B.; Tetta, C.; Camussi, G. Human liver stem cell-derived microvesicles accelerate hepatic regeneration in hepatectomized rats. *J. Cell. Mol. Med.* **2010**, *14*, 1605–1618. [CrossRef] [PubMed]

96. Deregibus, M.C.; Cantaluppi, V.; Calogero, R.; Iacono, M.L.; Tetta, C.; Biancone, L.; Bruno, S.; Bussolati, B.; Camussi, G. Endothelial progenitor cell–derived microvesicles activate an angiogenic program in endothelial cells by a horizontal transfer of mRNA. *Blood* **2007**, *110*, 2440–2448. [CrossRef] [PubMed]

97. Singer, A.J.; Clark, R.A. Cutaneous wound healing. *N. Engl. J. Med.* **1999**, *341*, 738–746. [PubMed]
98. Biró, É.; Sturk-Maquelin, K.N.; Vogel, G.M.; Meuleman, D.G.; Smit, M.J.; Hack, C.E.; Sturk, A.; Nieuwland, R. Human cell-derived microparticles promote thrombus formation in vivo in a tissue factor-dependent manner. *J. Thromb. Haemost.* **2003**, *1*, 2561–2568. [CrossRef] [PubMed]
99. Berckmans, R.J.; Sturk, A.; van Tienen, L.M.; Schaap, M.C.; Nieuwland, R. Cell-derived vesicles exposing coagulant tissue factor in saliva. *Blood* **2011**, *117*, 3172–3180. [CrossRef] [PubMed]
100. Zhang, B.; Wu, X.; Zhang, X.; Sun, Y.; Yan, Y.; Shi, H.; Zhu, Y.; Wu, L.; Pan, Z.; Zhu, W. Human umbilical cord mesenchymal stem cell exosomes enhance angiogenesis through the WNT4/β-catenin pathway. *Stem Cells Trans. Med.* **2015**, *4*, 513–522. [CrossRef] [PubMed]
101. Du, T.; Ju, G.; Wu, S.; Cheng, Z.; Cheng, J.; Zou, X.; Zhang, G.; Miao, S.; Liu, G.; Zhu, Y. Microvesicles derived from human Wharton's jelly mesenchymal stem cells promote human renal cancer cell growth and aggressiveness through induction of hepatocyte growth factor. *PLoS ONE* **2014**, *9*, e96836. [CrossRef] [PubMed]
102. Jeong, D.; Jo, W.; Yoon, J.; Kim, J.; Gianchandani, S.; Gho, Y.S.; Park, J. Nanovesicles engineered from ES cells for enhanced cell proliferation. *Biomaterials* **2014**, *35*, 9302–9310. [CrossRef] [PubMed]
103. Li, X.; Jiang, C.; Zhao, J. Human endothelial progenitor cells-derived exosomes accelerate cutaneous wound healing in diabetic rats by promoting endothelial function. *J. Diabetes Complicat.* **2016**, *30*, 986–992. [CrossRef] [PubMed]
104. Huang, P.; Bi, J.; Owen, G.R.; Chen, W.; Rokka, A.; Koivisto, L.; Heino, J.; Häkkinen, L.; Larjava, H. Keratinocyte microvesicles regulate the expression of multiple genes in dermal fibroblasts. *J. Investig. Dermatol.* **2015**, *135*, 3051–3059. [CrossRef] [PubMed]
105. Hu, L.; Wang, J.; Zhou, X.; Xiong, Z.; Zhao, J.; Yu, R.; Huang, F.; Zhang, H.; Chen, L. Exosomes derived from human adipose mensenchymal stem cells accelerates cutaneous wound healing via optimizing the characteristics of fibroblasts. *Sci. Rep.* **2016**, *6*, 32993. [CrossRef] [PubMed]
106. Zhang, J.; Chen, C.; Hu, B.; Niu, X.; Liu, X.; Zhang, G.; Zhang, C.; Li, Q.; Wang, Y. Exosomes derived from human endothelial progenitor cells accelerate cutaneous wound healing by promoting angiogenesis through Erk1/2 signaling. *Int. J. Biol. Sci.* **2016**, *12*, 1472. [CrossRef] [PubMed]
107. Guo, S.-C.; Tao, S.-C.; Yin, W.-J.; Qi, X.; Yuan, T.; Zhang, C.-Q. Exosomes derived from platelet-rich plasma promote the re-epithelization of chronic cutaneous wounds via activation of YAP in a diabetic rat model. *Theranostics* **2017**, *7*, 81. [CrossRef] [PubMed]
108. Leoni, G.; Neumann, P.-A.; Kamaly, N.; Quiros, M.; Nishio, H.; Jones, H.R.; Sumagin, R.; Hilgarth, R.S.; Alam, A.; Fredman, G. Annexin A1-containing extracellular vesicles and polymeric nanoparticles promote epithelial wound repair. *J. Clin. Investig.* **2015**, *125*, 1215–1227. [CrossRef] [PubMed]
109. Bhatwadekar, A.D.; Glenn, J.V.; Curtis, T.M.; Grant, M.B.; Stitt, A.W.; Gardiner, T.A. Retinal endothelial cell apoptosis stimulates recruitment of endothelial progenitor cells. *Investig. Ophthalmol. Vis. Sci.* **2009**, *50*, 4967–4973. [CrossRef] [PubMed]
110. Cheng, C.-F.; Fan, J.; Fedesco, M.; Guan, S.; Li, Y.; Bandyopadhyay, B.; Bright, A.M.; Yerushalmi, D.; Liang, M.; Chen, M. Transforming growth factor α (TGFα)-stimulated secretion of HSP90α: Using the receptor LRP-1/CD91 to promote human skin cell migration against a TGFβ-rich environment during wound healing. *Mol. Cell. Biol.* **2008**, *28*, 3344–3358. [CrossRef] [PubMed]
111. Ngora, H.; Galli, U.M.; Miyazaki, K.; Zöller, M. Membrane-bound and exosomal metastasis-associated C4. 4A promotes migration by associating with the α6β4 integrin and MT1-MMP. *Neoplasia* **2012**, *14*, 95IN91–107IN102. [CrossRef]
112. Zhang, J.; Guan, J.; Niu, X.; Hu, G.; Guo, S.; Li, Q.; Xie, Z.; Zhang, C.; Wang, Y. Exosomes released from human induced pluripotent stem cells-derived MSCs facilitate cutaneous wound healing by promoting collagen synthesis and angiogenesis. *J. Transl. Med.* **2015**, *13*, 49. [CrossRef] [PubMed]
113. Van Koppen, A.; Joles, J.A.; van Balkom, B.W.; Lim, S.K.; de Kleijn, D.; Giles, R.H.; Verhaar, M.C. Human embryonic mesenchymal stem cell-derived conditioned medium rescues kidney function in rats with established chronic kidney disease. *PLoS ONE* **2012**, *7*, e38746. [CrossRef] [PubMed]
114. Gospodarowicz, D. Biological activities of fibroblast growth factors. *Ann. N. Y. Acad. Sci.* **1991**, *638*, 1–8. [CrossRef] [PubMed]

115. Gastpar, R.; Gehrmann, M.; Bausero, M.A.; Asea, A.; Gross, C.; Schroeder, J.A.; Multhoff, G. Heat shock protein 70 surface-positive tumor exosomes stimulate migratory and cytolytic activity of natural killer cells. *Cancer Res.* **2005**, *65*, 5238–5247. [CrossRef] [PubMed]

116. Atalay, M.; Oksala, N.; Lappalainen, J.; Laaksonen, D.E.; Sen, C.K.; Roy, S. Heat shock proteins in diabetes and wound healing. *Curr. Protein Pept. Sci.* **2009**, *10*, 85–95. [CrossRef] [PubMed]

117. McCready, J.; Sims, J.D.; Chan, D.; Jay, D.G. Secretion of extracellular HSP90α via exosomes increases cancer cell motility: A role for plasminogen activation. *BMC Cancer* **2010**, *10*, 294. [CrossRef] [PubMed]

118. Pearl, L.H.; Prodromou, C. Structure and in vivo function of HSP90. *Curr. Opin. Struct. Biol.* **2000**, *10*, 46–51. [CrossRef]

119. Liekens, S.; de Clercq, E.; Neyts, J. Angiogenesis: Regulators and clinical applications. *Biochem. Pharmacol.* **2001**, *61*, 253–270. [CrossRef]

120. Hughes, C.C. Endothelial–stromal interactions in angiogenesis. *Curr. Opin. Hematol.* **2008**, *15*, 204–209. [CrossRef] [PubMed]

121. Yuan, H.; Guan, J.; Zhang, J.; Zhang, R.; Li, M. Exosomes secreted by human urine-derived stem cells accelerate skin wound healing by promoting angiogenesis in rat. *Cell Biol. Int.* **2016**. [CrossRef] [PubMed]

122. Li, X.; Chen, C.; Wei, L.; Li, Q.; Niu, X.; Xu, Y.; Wang, Y.; Zhao, J. Exosomes derived from endothelial progenitor cells attenuate vascular repair and accelerate reendothelialization by enhancing endothelial function. *Cytotherapy* **2016**, *18*, 253–262. [CrossRef] [PubMed]

123. Vrijsen, K.R.; Maring, J.A.; Chamuleau, S.A.; Verhage, V.; Mol, E.A.; Deddens, J.C.; Metz, C.H.; Lodder, K.; van Eeuwijk, E.; van Dommelen, S.M. Exosomes from cardiomyocyte progenitor cells and mesenchymal stem cells stimulate angiogenesis via emmprin. *Adv. Healthc. Mater.* **2016**, *5*, 2555–2565. [CrossRef] [PubMed]

124. Ramakrishnan, D.P.; Hajj-Ali, R.A.; Chen, Y.; Silverstein, R.L. Extracellular vesicles activate a CD36-dependent signaling pathway to inhibit microvascular endothelial cell migration and tube formationsignificance. *Arterioscler. Thromb. Vasc. Biol.* **2016**, *36*, 534–544. [CrossRef] [PubMed]

125. Nakamura, K.; Jinnin, M.; Harada, M.; Kudo, H.; Nakayama, W.; Inoue, K.; Ogata, A.; Kajihara, I.; Fukushima, S.; Ihn, H. Altered expression of CD63 and exosomes in scleroderma dermal fibroblasts. *J. Dermatol. Sci.* **2016**, *84*, 30–39. [CrossRef] [PubMed]

126. Jong, O.G.; Balkom, B.W.; Gremmels, H.; Verhaar, M.C. Exosomes from hypoxic endothelial cells have increased collagen crosslinking activity through up-regulation of lysyl oxidase-like 2. *J. Cell. Mol. Med.* **2016**, *20*, 342–350. [CrossRef] [PubMed]

127. Huleihel, L.; Hussey, G.S.; Naranjo, J.D.; Zhang, L.; Dziki, J.L.; Turner, N.J.; Stolz, D.B.; Badylak, S.F. Matrix-bound nanovesicles within ECM bioscaffolds. *Sci. Adv.* **2016**, *2*, e1600502. [CrossRef] [PubMed]

128. Wilhelm, E.N.; González-Alonso, J.; Parris, C.; Rakobowchuk, M. Exercise intensity modulates the appearance of circulating microvesicles with proangiogenic potential upon endothelial cells. *Am. J. Physiol. Heart Circ. Physiol.* **2016**, *311*, H1297–H1310. [CrossRef] [PubMed]

129. Atienzar-Aroca, S.; Flores-Bellver, M.; Serrano-Heras, G.; Martinez-Gil, N.; Barcia, J.M.; Aparicio, S.; Perez-Cremades, D.; Garcia-Verdugo, J.M.; Diaz-Llopis, M.; Romero, F.J. Oxidative stress in retinal pigment epithelium cells increases exosome secretion and promotes angiogenesis in endothelial cells. *J. Cell Mol. Med.* **2016**, *20*, 1457–1466. [CrossRef] [PubMed]

© 2017 by the authors. Licensee MDPI, Basel, Switzerland. This article is an open access article distributed under the terms and conditions of the Creative Commons Attribution (CC BY) license (http://creativecommons.org/licenses/by/4.0/).

International Journal of
Molecular Sciences

MDPI

Review

Regulation of T$_H$17 Cells and Associated Cytokines in Wound Healing, Tissue Regeneration, and Carcinogenesis

Leonie Brockmann [1], Anastasios D. Giannou [1], Nicola Gagliani [1,2,3] and Samuel Huber [1,*]

1   I. Department of Medicine, University Medical Center Hamburg-Eppendorf, 20246 Hamburg, Germany; l.brockmann@uke.de (L.B.); a.giannou@uke.de (A.D.G.); n.gagliani@uke.de (N.G.)
2   Department of General, Visceral and Thoracic Surgery, University Medical Center Hamburg-Eppendorf, 20246 Hamburg, Germany
3   Department of Medicine Solna (MedS), Karolinska Institute, 17177 Stochkolm, Sweden
*   Correspondence: s.huber@uke.de; Tel.: +49-40-7410-57273; Fax: +49-40-7410-59038

Academic Editor: Allison Cowin
Received: 30 March 2017; Accepted: 8 May 2017; Published: 11 May 2017

Abstract: Wound healing is a crucial process which protects our body against permanent damage and invasive infectious agents. Upon tissue damage, inflammation is an early event which is orchestrated by a multitude of innate and adaptive immune cell subsets including T$_H$17 cells. T$_H$17 cells and T$_H$17 cell associated cytokines can impact wound healing positively by clearing pathogens and modulating mucosal surfaces and epithelial cells. Injury of the gut mucosa can cause fast expansion of T$_H$17 cells and their induction from naïve T cells through Interleukin (IL)-6, TGF-β, and IL-1β signaling. T$_H$17 cells produce various cytokines, such as tumor necrosis factor (TNF)-α, IL-17, and IL-22, which can promote cell survival and proliferation and thus tissue regeneration in several organs including the skin, the intestine, and the liver. However, T$_H$17 cells are also potentially pathogenic if not tightly controlled. Failure of these control mechanisms can result in chronic inflammatory conditions, such as Inflammatory Bowel Disease (IBD), and can ultimately promote carcinogenesis. Therefore, there are several mechanisms which control T$_H$17 cells. One control mechanism is the regulation of T$_H$17 cells via regulatory T cells and IL-10. This mechanism is especially important in the intestine to terminate immune responses and maintain homeostasis. Furthermore, T$_H$17 cells have the potential to convert from a pro-inflammatory phenotype to an anti-inflammatory phenotype by changing their cytokine profile and acquiring IL-10 production, thereby limiting their own pathological potential. Finally, IL-22, a signature cytokine of T$_H$17 cells, can be controlled by an endogenous soluble inhibitory receptor, Interleukin 22 binding protein (IL-22BP). During tissue injury, the production of IL-22 by T$_H$17 cells is upregulated in order to promote tissue regeneration. To limit the regenerative program, which could promote carcinogenesis, IL-22BP is upregulated during the later phase of regeneration in order to terminate the effects of IL-22. This delicate balance secures the beneficial effects of IL-22 and prevents its potential pathogenicity. An important future goal is to understand the precise mechanisms underlying the regulation of T$_H$17 cells during inflammation, wound healing, and carcinogenesis in order to design targeted therapies for a variety of diseases including infections, cancer, and immune mediated inflammatory disease.

Keywords: T$_H$17 cells; cytokines; wound healing; tissue regeneration; carcinogenesis; immune regulation

---

1. Inflammation in Wound Healing and Carcinogenesis

In 1986, Dvorak published an essay with the vivid title: "Tumors: Wounds that do not heal", summarizing in this one statement the relation between wound healing and carcinogenesis [1]. Wound healing normally follows sequential but overlapping steps. The immediate reaction is hemostasis to provisionally close the wound. A fibrin clot is formed and platelets aggregate. This is followed by the inflammatory phase which is characterized by the presence of neutrophils, macrophages, and lymphocytes in the wound. These cells are attracted by chemokines, which are, for example, released by platelet cells [2]. The inflammatory phase is important for tissue regeneration due to the release of pro-inflammatory cytokines and growth factors from immune cells [3–5]. Additionally, phagocytes can ingest cell debris and invading pathogens. Therefore, the inflammatory phase is essential to prevent spreading of infections. The proliferative phase follows several days later leading to re-epithelialization, formation of new blood vessels, and fibrogenesis. During the last phase, the resolution phase, vessel regression and collagen remodeling occur [6]. Thus, occurrence of a wound has a dramatic impact on the body. Multiple cell types are necessary to secure a prompt healing process. Research elucidating this process mainly focuses on innate immunity contributing to tissue regeneration. However, adaptive immunity also plays its part during this process, even though its contribution and regulation is much less understood.

Tissue damage, especially at barrier organs such as the intestine, the lung, and the skin, is a potential gateway for invading pathogens, therefore inflammation is an essential part of wound healing. In this regard, the involvement of T cells during wound repair has been under investigation for a long time. In 1987, the hypothesis that T lymphocytes represent the most frequent leucocyte population in skin wounds was published [7]. Several studies indicate that delayed infiltration of T cells and a lower concentration of these cells at the site of the wound are associated with impaired wound healing. Furthermore, CD4$^+$ T cells seem to play a beneficial role during the process of wound healing and regeneration [8–10]. However, very little is known about the contribution of different T cell subsets. Furthermore, these mechanisms, which are designed to promote wound healing, also have the potential to promote chronic inflammation and carcinogenesis. Two important predisposing factors for colorectal carcinogenesis are chronic intestinal inflammation and tissue injury. This association is based on the fact that wound healing and carcinogenesis are driven by several common factors and signaling pathways. Upon tissue injury, factors promoting healing are stimulated and their action must be tightly controlled in order to avoid carcinogenesis [11]. A chronically inflamed and wounded tissue is associated with a long-lasting healing response which may lead to fibrosis, tissue dysfunction, and ultimately the development of cancer. Thus, carcinogenesis could be considered as a consequence of failing regulatory mechanisms allowing abnormal excessive healing [11].

2. CD4$^+$ T Helper Cells

Conventional αβ CD4$^+$ T cells are one of the main players during an adaptive immune response. Due to their great variety, CD4$^+$ T cells can orchestrate the immune response and react to the whole spectrum of immune challenges. Two major CD4$^+$ T helper cell subsets, T_H1 and T_H2, were discovered in 1989 [12]. However, the T_H1/T_H2 paradigm was challenged in 2005 by the description of T_H17 cells, which are primarily needed for the defense against extracellular bacteria and fungi [13]. Additionally, T_H17 cell associated cytokines can promote epithelial proliferation and tissue regeneration [14].

The gastrointestinal tract, like skin and lung, is constantly in contact with hundreds of different species of commensal bacteria and fungi [15]. How the commensal microbiota modulate the immune system of the host and vice versa has been an area of intensive research and is still not completely understood. Nonetheless, it is known that the induction of T_H17 cells is dependent on the presence of microbiota, especially segmented filamentous bacteria (SFB) in the terminal ileum of the small intestine in mice. Hence, under physiological conditions, T_H17 cells are mainly located in the small intestine [16]. Upon tissue damage, T_H17 cells rapidly expand at mucosal surfaces to guarantee a quick clearance of the invading microorganisms such as commensal bacteria. To this end, T_H17 cells release several chemokines and cytokines, such as IL-17A, IL-17F, and IL-22 which are the signature cytokines of T_H17 cells. One major defense mechanism of T_H17 cells is the attraction of inflammatory cells via chemotaxis to the site of infection, which is crucial for the fast clearance of pathogens [17,18].

Additionally, T$_H$17 cells contribute to the crosstalk of immune cells and epithelial cells or other innate immune cells by inducing the release of anti-microbial peptides [19–22]. The signature cytokines, IL-17A and IL-17F, both binding to IL-17RA [23], and IL-22, binding to IL-22R1, mediate this effect of T$_H$17 cells. IL-22 does not only promotes the secretion of anti-microbial peptides from the epithelium, but also exhibits tissue protective properties. An important effect of IL-22 signaling on epithelial cells is the induction of proliferation, survival, and tissue repair via induction of STAT3 [14,24,25]. Thus, T$_H$17 cells act at different front lines during the defense of the body: (a) T$_H$17 cells activate and attract other immune cells, mainly neutrophils, which can phagocyte pathogens. Therefore, tissue resident T$_H$17 cells could contribute to the inflammatory phase of wound healing. (b) T$_H$17 cells induce the release of anti-microbial peptides from non-immune cells. (c) Finally, T$_H$17 cell associated cytokines, such as IL-22, can promote wound healing and tissue regeneration, leading to a faster closing of potential entryways of microorganisms. In the next sections, we will give an overview of how T$_H$17 cell associated cytokines affect wound healing and carcinogenesis.

3. Effects of T$_H$17 Cell Associated Cytokines on Wound Healing and Carcinogenesis

3.1. Effects of IL-17 on Wound Healing

Interleukin-17 (IL-17), also known as IL-17A, is the main member of the IL-17 family, which consists of six members: IL-17A, IL-17B, IL-17C, IL-17D, IL-17E (also named IL-25), and IL-17F. This mediator is not only secreted by T$_H$17 cells alone, but also by other immune cell types such as natural killer (NK), natural killer T (NKT), lymphoid tissue inducer (LTi), and LTi-like cells. Apart from immune cells, IL-17A can also be produced by Paneth cells which are epithelial cells located in the small intestine. To conduct its biological functions, IL-17 binds to a heteromeric receptor complex consisting of IL-17RA and IL-17RC. Along with three additional variants, namely IL-17RB, IL-17RD and IL-17RE, IL-17RA and IL-17RC, they comprise the IL-17 receptor family [26]. IL-17 is a pro-inflammatory cytokine, which is commonly found at high levels in inflamed sites. Recent studies have identified a link between the IL-23 induced expansion of T$_H$17 cells followed by IL-17 secretion and the pathogenesis of IBD [27]. T$_H$17 cells are mainly responsible for IL-17 production, which is elevated in IBD. Controversially, a role of IL-17 in IBD-related mucosal healing was recently found. Song et al. showed that following epithelial barrier damage, FGF2 is highly expressed and cooperates with IL-17 in order to enhance tissue repair. Specifically, upon barrier destruction, dysregulated microbiota cause an upregulation of TGF-β which subsequently stimulates FGF2 production by regulatory T cells. Then, FGF2 together with IL-17 upregulates genes associated with epithelial healing and boosts epithelial cell proliferation [28].

Recent studies focusing on skin wound healing have concluded that IL-17 plays a role in the early inflammatory stage of the wound and could act as an inhibitor to normal wound repair. As seen in psoriasis, IL-17 has several functions in the skin. The expression of IL-17 receptor on keratinocytes, fibroblasts, and inflammatory cells reveals that IL-17 can affect numerous cell types in the skin. In skin lesions, it is macrophages and not T cells which are the main cellular source of IL-17. To elucidate the role of IL-17 in skin wound healing, Rodero et al. used *Il17*$^{-/-}$ mice and showed that absence of IL-17 enhances and accelerates skin tissue repair. Blocking IL-17 with specific antibodies during the early steps of healing also resulted in wound closure promotion [29]. On the other hand, a beneficial role of IL-17 in sensory nerve regeneration has been recently revealed. In a mouse model of corneal abrasion, an IL-17 dependent signaling pathway involving IL-17, neutrophils, platelets, and vascular endothelial growth factor (VEGF)-A was found to promote corneal repair and nerve regeneration [30].

In conclusion, IL-17 seems to have dual functions during wound healing. It may have beneficial functions in the intestine and for nerve regeneration [31]. However, in the skin, especially in psoriasis, it seems to be mainly pathogenic. Environmental factors may play a crucial role determining the part IL-17 has in wound healing, but further studies are needed to clarify this point.

Likewise, the role of IL-17 during carcinogenesis is still not completely understood. T$_H$17-related events and specifically the role of IL-17 were first thought to promote tumor growth and invasion and to enhance angiogenesis [32]. However, other studies identified a protective role in tumor immune surveillance together with an inhibition of cancer cell proliferation and metastasis [33–35]. Using spontaneous intestinal carcinogenesis models, the role of IL-17A in colorectal cancer development was further elucidated. Recent studies showed that blocked or genetically-induced absence of IL-17A resulted in significantly attenuated tumor burden in Apc$^{Min/+}$ mice infected with enterotoxigenic *Bacteroides fragilis* and in the standard Apc$^{Min/+}$ model [36,37]. In the dextran sulfate sodium/azoxymethane (DSS/AOM) carcinogenesis model, *Il17a$^{-/-}$* mice displayed reduced tumorigenesis, which could be attributed to lower levels of intestinal TNF-α, interferon (IFN)-γ, IL-6 ,and STAT3 activity [38]. Interestingly, a highly expressed T$_H$17 cell related mRNA pattern correlates with poor prognosis in human colorectal cancer [39,40]. This observation is likely associated with the fact that T$_H$17 cells are the main IL-17A producers in the tumor. However, other immune cell subsets such as CD8$^+$ CTLs (Tc17 cells), $\gamma\delta$ T cells ($\gamma\delta$T17 cells), and innate lymphoid cells (ILCs) also produce IL-17A [41–43]. Notably, upon dendritic cell (DC)-mediated stimulation, $\gamma\delta$T17 cells produce IL-17 and enhance the recruitment and expansion of myeloid-derived suppressor cells. This finding suggests that IL-17A is likely to be associated with immune silencing in colorectal cancer. Due to the well-described implication of IL-17A in colorectal cancer, targeting of IL-17A may serve as a promising therapeutic approach. Interestingly, a recent study showed that blocking of IL-17A in an adenomatous polyposis coli (APC)-mediated colon carcinogenesis model enhanced tumor sensitivity to the anticancer agent 5-fluorouracil [44]. Similarly, antiangiogenic therapies might fail due to the IL-17A-driven emergence of resistant tumor stromal cells [45].

Apart from IL-17A, another member of the IL-17 superfamily and ligand of IL-17RE, IL-17C, was shown to be expressed in intestinal epithelial cells during early stages of colon carcinogenesis. In both the Apc$^{Min/+}$ and the DSS/AOM-induced carcinogenesis models, mice lacking IL-17RE displayed a decreased tumor burden together with a lower expression of the anti-apoptotic proteins BCL-2 and BCLXL. Interestingly, IL-17C can be associated with the human condition, since it is overexpressed in human colon cancers [46]. IL-17F, which also belongs to the IL-17 family, resembles IL-17A and acts via the same receptor, and may play an opposite role in colon carcinogenesis. In contrast to IL-17A and C, IL-17F is significantly downregulated in human colorectal cancer. Notably, it reduces carcinogenesis in the DSS/AOM carcinogenesis model by inhibiting angiogenesis [47].

3.2. Effects of IL-22 on Wound Healing

Interleukin-22 (IL-22), a member of the IL-10 cytokine family, participates in the signaling between the immune system and the peripheral tissues. IL-22 can be produced by several other cell types such as T$_H$22, T cell receptor (TCR)-$\gamma\delta$, NK, NKT, ILCs, and LTi cells, with the notable exception of T$_H$17 cells. IL-22 acts via binding to the heterodimer IL-10R2/IL-22R1 complex [48]. IL-22R1 is expressed on non-hematopoietic cells, such as intestinal epithelial cells, hepatocytes, and fibroblasts in the skin. After binding of IL-22 to the receptor complex, STAT3, STAT1, and STAT5, as well as the Janus kinase (JNK) and mitogen-activated protein kinases are activated. The translocation of activated STAT dimers into the nucleus leads to the activation of several genes linked to proliferation and cell survival. IL-22 is known to play a key role in tissue regeneration and wound healing [3]. In a mouse model of acute skin injury, an upregulation of *Il22* mRNA expression was observed during the inflammatory stage and IL-22 was identified as a critical mediator for normal fibroblast function, extracellular matrix protein production, and myofibroblast differentiation during skin wound healing [49]. IL-22 was found to facilitate the crosstalk between immune cells and fibroblasts during skin wound healing. It has also been shown to promote keratinocyte proliferation and migration while acting as an inhibitor for keratinocyte differentiation [3,50–52]. IL-22 is unlikely to play a major role in the early stages of skin wound healing such as immune cell accumulation and angiogenesis. Similarly, loss of IL-22 does not affect keratinocyte function during skin wound healing. However, upon wound healing, fibroblast function was shown to be IL-22-dependent. Specifically, absence of IL-22 leads to impaired granulation, tissue formation, production of extracellular matrix components

(ECM), and wound contraction. Primary dermal fibroblasts are directly affected by IL-22, since they express IL-22R1, whose IL-22-triggered activation can lead to STAT3 phosphorylation. IL-22 stimulates ECM production by inducing ECM gene expression in fibroblasts and by promoting myofibroblast differentiation. A decreased number of myofibroblasts in the wound may lead to defective wound contraction and impaired ECM formation, as seen in *Il22*$^{-/-}$ mice. IL-22 may induce the expression of ECM genes, mainly via STAT3 activation. IL-22-mediated STAT3 phosphorylation leads to activation of the promoters of fibronectin and collagen.

Interestingly, IL-22 is also essential for intestinal healing and the maintenance of the mucosal barrier. In studies using the DSS-induced acute intestinal injury in *Il22*$^{-/-}$ mice, IL-22 was shown to enhance intestinal wound healing, specifically via STAT3 activation, which in turn regulates signaling pathways commonly associated with tissue repair and gut homeostasis. Mice lacking IL-22 exhibited an impaired and delayed recovery from DSS-caused intestinal injury [14]. Similarly, targeting of IL-22 with specific neutralizing antibodies led to impaired wound healing in wildtype mice. On the other hand, increase of intestinal IL-22 expression via a gene delivery system boosted the recovery of the injured intestine [24]. Interestingly, another study showed that IL-22 deficiency leads to altered intestinal microbiota and therefore increases the severity of the disease in a mouse model of experimental colitis. This IL-22-mediated changed microbiota can be transferred to co-housed wild type mice, which subsequently become more susceptible to experimental colitis, suggesting that IL-22 is essential for maintaining the balance between immunity and intestinal microbiota [52].

Although the beneficial role of IL-22 in wound healing is well documented, in some cases the delicate balance between protection and harm shifts in favor of the pathogenic direction leading to carcinogenesis. A tumor-promoting function of IL-22 via STAT3 stimulation has already been identified in cancers such as hepatocellular carcinoma and lung cancer [53,54]. Recent studies suggest that in the absence of close regulation, IL-22 also promotes colon carcinogenesis via STAT3 activation [55]. To this end, Huber et al. recently reported that IL-22 is involved in colitis associated colon cancer in a dual manner. On the one hand, IL-22 deficiency can delay tissue repair, thereby sustaining inflammation and leading to tumor development. On the other hand, high levels of IL-22 in *Il22bp*$^{-/-}$ mice may prolong the regenerative program and promote colon carcinogenesis [56]. Clinical data show an association between high serum IL-22 levels and resistance to chemotherapy in patients with colorectal cancer, an observation which was further confirmed in vitro [57,58]. Recently, Kryczek et al. suggested that IL-22 can enhance colon cancer stemness. Specifically, IL-22 was found to activate STAT3 in human colon cancer cells, resulting in the expression of the H3K9-specific *N*-methyltransferase DOT1L, which subsequently induced the expression of core stem cell genes such as SOX2, NANOG, and POU5F1. This pathway promoted colon carcinogenesis and the expression of the implicated genes was associated with poor patient prognosis [59]. IL-22 is known to promote and sustain the survival of normal intestinal stem cells in mice [60]. Therefore, a similar function resulting in maintaining the cancer stem cell niche is likely to be one of the main contributions of IL-22 to colorectal carcinogenesis. Other studies showed that high expression of RORγt, which is essential for IL-22 production, and IL-17A, which is commonly found in association with IL-22, indicates poor prognosis in patients with colorectal cancer [39].

3.3. Effects of TNF-α on Wound Healing

Upon inflammation-induced disruption of the intestinal mucosal barrier, intestinal epithelial cells (IECs) need to orchestrate the process of tissue restitution and regeneration. To this end, a crosstalk between epithelial and immune cells needs to take place. TNF-α, an important pro-inflammatory cytokine, is one of the implicated molecules, playing a key role during inflammation and subsequent wound healing [61]. Two forms of TNF-α, the soluble and its precursor, transmembrane one, participate in the inflammatory process. Transmembrane TNF-α acts via cell-to-cell contact, whereas soluble TNF-α performs its biological effects at distant sites from the TNF-α-secreting cells [62]. Transmembrane TNF-α acts both as a ligand by interacting with TNF-α receptors as well as a receptor that receives and transfers signals back into the same transmembrane TNF-α-expressing cells [63]. During intestinal inflammation, TNF-α is produced by immune cells such as T_H17 cells, stromal cells, and IECs and subsequently interacts with the latter via two receptors: TNF-R1 (TNFRSF1A) and TNF-R2 (TNFRSF1B). TNF-α signaling mediates the production of inflammatory molecules, regulates cell survival, proliferation, and death, and affects epithelial wound healing. Wound healing is a complicated process involving cellular migration, re-differentiation, and proliferation. TNF-α/TNF-R2 signaling mediates epithelial migration and enhances the survival and proliferation of IECs [64]. Furthermore, TNF-α promotes intestinal wound healing by protecting against epithelial apoptosis via ErbB pathway activation. TNF-α also enhances re-epithelialization and thus wound healing by promoting the FGF-7 production. TNF-α induced effects depend on concentration and duration of exposure. Specifically, low levels of TNF-α promote inflammation and stimulate the production of macrophage-derived growth factors facilitating wound healing. However, long exposure to high levels of TNF-α can have a negative impact on healing, as TNF-α can lead to reduced production of ECM components while promoting the synthesis of metalloproteinases (MMP-1, MMP-2, MMP-3, MMP-9, MMP-13, and MT1-MMP). In line with this, TNF-α levels are increased in chronic wounds. Infection can further promote TNF-α accumulation by prolonging inflammation.

A frequent consequence of IBD and especially ulcerative colitis is colon cancer. Although initially, TNF-α was considered to serve as a tumor-suppressive factor, due to its conditional pro-apoptotic function, it was recently found to promote colitis-associated tumor development by linking inflammation and cancer [65–67]. TNF-α can interact strongly with intestinal epithelial cells, which express high levels of TNFR1, leading to the activation of NF-κB-dependent oncogenic pathways. Recently, Popivanova et al. identified TNF-α as a key factor for the development of colitis-related colon cancer. Specifically, upon DSS/AOM-induced colon carcinogenesis, the expression of TNF-α and the intestinal recruitment of leukocytes expressing the main TNF receptor, namely TNF-Rp55, were boosted, resulting in the formation of several intestinal tumors. Absence of TNF-Rp55 or specific blocking of TNF-α led to reduced mucosal injury and inflammation followed by decreased tumor formation [67]. In other studies, TNF-α deficiency was associated with severe colitis and cancer along with increased blood levels of IL-6, IFN-γ, and IL-17A. Similarly, recent studies showed that TNF-α mRNA expression was increased in colorectal tumors compared to surrounding healthy intestinal tissue. Interestingly, TNF-α was overexpressed in Stage III and IV tumors, suggesting that high TNF-α expression in tumor cells can be associated with advanced stages of carcinogenesis [68]. Additionally, a genetic link between TNF-α and colorectal cancer has been identified recently. Furthermore, TNF was found to play a key role in the colon cancer promoting effect of obesity [69]. Finally, TNF may also enhance metastasis by promoting epithelial to mesenchymal transition (EMT) in colorectal cancer [70].

4. Control of T_H17 Cells

T_H17 cells were originally discovered in an autoimmune setting. In a mouse model of multiple sclerosis, experimental autoimmune encephalomyelitis (EAE), and a mouse model of arthritis, both formerly linked to an uncontrolled T_H1 response, it was discovered that not IL-12, the cytokine driving T_H1 differentiation, but IL-23 was essential for disease development [71,72]. From in vitro studies, it has been shown that IL-23 can induce the production of IL-17 from effector and memory T cells [73]. Finally, in 2005, Langrish et al. demonstrated that IL-23 induced IL-17 producing T cells displayed stronger pathogenic properties in EAE than T_H1 cells, and that these T_H17 cells have a distinct gene-expression profile [74]. Besides the described protective properties of T_H17 cells, this cell type is apparently also associated with autoimmune diseases, chronic inflammatory conditions, and carcinogenesis. Multiple sclerosis, rheumatoid arthritis, IBD, and psoriasis are amongst the diseases with a strong T_H17 cell involvement [75–78]. Additionally, as discussed above, uncontrolled IL-22, IL-17, and TNF-α level can promote carcinogenesis. Therefore, the immune system needs control mechanisms to keep T_H17 cells in check. There are several layers to control T_H17 cells, which are discussed in the next sections.

4.1. T$_H$17 Cell Differentiation

The first level of control already occurs under physiological conditions by regulating T$_H$17 cell differentiation. In the past decade, intensive studies have further elucidated the signaling pathways leading to the differentiation of T$_H$17 cells. Noteworthy, IL-23 signaling is not essential for the induction of T$_H$17 cells from naïve T cells, since naïve T cells only express very low amounts of IL-23R [79]. Nonetheless, IL-23 signaling is crucial for the terminal differentiation, expansion, and maintenance of T$_H$17 cells. IL23R-deficient T$_H$17 cells fail to maintain IL-17 expression in vivo and cannot induce EAE [80]. IL-6 signaling can induce the expression of IL-23R, a crucial step in the early priming phase of T$_H$17 cells. This leads to the activation of STAT3. Translocation of phosphorylated STAT3 dimers to the nucleus results in induction of T$_H$17-related genes such as *Rorc*, *Il17*, and also *Il23r*. The induction of *Rorc* (encoding RORγt) is indispensable for T$_H$17 cell development [81–83]. RORγt is the master transcriptional regulator of T$_H$17 cells, demonstrated by the absence of IL-17 producing T cells in RORγt-deficient mice [17].

TGF-β is another cytokine contributing to the development of T$_H$17 cells, even though its part in this process is still controversial. In low concentrations, TGF-β can inhibit T$_H$1 and T$_H$2 differentiation by inhibiting IL-2 dependent STAT5 activation and expression of T-bet and GATA3, the master regulators of T$_H$1 and T$_H$2, respectively [84]. Nonetheless, higher concentrations of TGF-β result in downregulation of IL-23R and consequently counter regulate T$_H$17 cell expansion [79]. Additionally, in 2010, it was demonstrated that T$_H$17 cells can occur in the absence of TGF-β signaling in the gut mucosa in vivo [85]. On the contrary, TGF-β signaling can induce the differentiation of inducible regulatory T cells (pTreg). TGF-β is dispensable for T$_H$17 cell differentiation but non-redundant for the induction of pTregs [85]. TGF-β signaling induces both FOXP3, the master transcription factor of Treg cells, and RORγt expression. However, in the absence of IL-6 signaling, FOXP3 abrogates the effects of RORγt [79,86,87]. Additionally, IL-2 signaling can both enhance FOXP3 expression and induce STAT5, which leads to impaired binding of STAT3 to IL-17 related genes and inhibits T$_H$17 cell differentiation [88–90]. In the absence of pro-inflammatory cytokines, such as IL-6 or IL-1β, TGF-β favors the development of regulatory T cells to maintain immune homeostasis. T$_H$17 cells and Treg cells are cell subsets with opposite functions for the immune system, however, they share common pathways for their differentiation. This close relationship demonstrates the important and delicate balance the immune system has to maintain to guarantee immune homeostasis in the presence of foreign antigens from commensal microorganisms and food and to guarantee effective protection against pathogens. Besides TGF-β, the cytokine IL-27 is known to negatively regulate T$_H$17 cell induction. IL-27 signaling inhibits the expression of RORγt and therefore suppresses T$_H$17 cell differentiation [91]. On the other hand, IL-27 can induce the differentiation of another regulatory T cell subset, type one regulatory T cells (T$_R$1). These cells are characterized by high expression levels of IL-10, they, however, lack FOXP3 expression [92,93]. In summary, the differentiation of the two major regulatory T cell subsets is inversely related with T$_H$17 cell induction, a phenomenon not known for other effector T cell subsets, such as T$_H$1 and T$_H$2.

Furthermore, IL-1β is important for the differentiation of T$_H$17 cells, which was already established in human T cells in 2007 [94,95]. Unlike TGF-β signaling, it was demonstrated by using mice that IL-1β signaling was crucial for T$_H$17 cell induction in all tissues in vivo [96]. IL-1β signaling has multiple effects on the differentiation of T$_H$17 cells. However, one essential consequence is the induction of the transcription factor IRF4, which is strictly needed for RORγt expression [97]. Interestingly, T$_H$17 cells, which differentiated in the absence of TGF-β signaling, seem to have an altered phenotype. Since TGF-β is required for the suppression of T-bet expression in T$_H$17 cells, IL-1β induced T$_H$17 cells are also T-bet positive and co-express IFN-γ, the signature cytokine of T$_H$1 cells [85]. These IFN-γ producing T$_H$17 cells are frequently linked with the occurrence of autoimmune disease such as multiple sclerosis [98].

Finally, the microbiota plays an important role in T$_H$17 cell differentiation. Under physiological conditions, T$_H$17 cells are most abundant in the lamina propria of the small intestine [17] due to the presence of intestinal microbiota. Studies demonstrated that germ free mice have dramatically reduced levels of T_H17 cells, which can be induced by colonialization with conventional microbiota [99]. SFB were identified as contributing to the expansion of T_H17 cells in the small intestine due to the induction of serum amyloid A (SAA), which can stimulate DCs to release IL-6 and IL-23 and finally promote T_H17 cell differentiation [16]. Another effect of the microbiota is the induction of IL-1β, further contributing to T_H17 cell development [100].

Last, but not least, ligands for the aryl hydrocarbon receptor (AHR) also derive from diet or are products of the intestinal microbiota. AHR is another transcription factor which plays a non-redundant role for T_H17 cell biology. It has been reported that AHR can promote T_H17 cell differentiation and is already highly expressed during the early T_H17 cell polarization [101,102]. However, contradicting studies reported an increase in T_H17 cells in AHR-deficient mice, especially in the small intestine, demonstrating that AHR is not essential for T_H17 cell development [103]. Nonetheless, AHR expression is crucial for IL-22 secretion by T_H17 cells and therefore important for some tissue regenerative functions of T_H17 cells [101,104,105].

In conclusion, T_H17 cell differentiation is strongly influenced by the cytokine environment in different tissues of the body and the presence or absence of environmental factors such as microbiota. In the last decade, immense efforts have been made to understand the regulation of T_H17 cell induction. Environmental factors, such as microbiota or diet, can directly or indirectly via DCs influence the development and phenotype of T_H17 cells. Various cytokines are involved in the differentiation of T_H17 cells and a complex transcriptional network orchestrates this process. Understanding the whole picture could facilitate the design of new therapeutic strategies targeting T_H17 cells.

4.2. Regulation of T_H17 Cells Expansion

A second mechanism is to control the expansion of T_H17 cells via regulatory T cells. Upon tissue damage and infections with extracellular bacteria or fungi, T_H17 cell immunity is strictly required. A pro-inflammatory environment favors the differentiation of T_H17 cells over regulatory T cells. Nonetheless, this immune response must be regulated to prevent the onset of chronic inflammatory conditions. Likewise, during wound healing the inflammatory phase needs to be ended. Very little is known about the direct role of T_H17 cells in the inflammatory phase. However, it was demonstrated that the absence of regulatory T cells results in decreased inflammation resolution after myocardial infarct injury and delayed wound healing in skin, further underlining the non-redundant role of CD4$^+$ T cells in regulation of wound repair and regeneration [106,107]. Regulatory T cells, both Treg cells and T_R1 cells, are key to terminate T_H17 cell associated immune responses by suppressing the expansion of T_H17 cells. Both regulatory T cell subsets can suppress T_H17 cell expansion in vivo [108,109]. A major suppressive mechanism of regulatory T cells is the release of anti-inflammatory factors, such as TGF-β and IL-10. IL-10 signaling is a key factor to dampen inflammatory responses. IL-10 deficiency leads to severe inflammatory diseases in humans [110]. T_H17 cells express IL-10 receptor and can be directly controlled via IL-10 released by regulatory T cells [111]. Another effect of IL-10 signaling is the reinforcing of regulatory T cell stability [112,113]. An environment high in IL-10 will therefore directly inhibit or terminate a T_H17 cell immune response and amplify anti-inflammatory T cell subsets.

4.3. T_H17 Cell Plasticity

A third mechanism to regulate T_H17 cell immune responses lies within the T_H17 cells themselves. T_H17 cells display a great plasticity depending on their cytokine environment. The acquisition of IFN-γ production occurs frequently during inflammation and is linked to disease progression in multiple human diseases [114–116]. T_H17 cells can also acquire the production of the T_H2 signature cytokine, IL-4 [117]. These cells are present in patients suffering from allergic asthma, and in mice it has been demonstrated that these IL-4 producing T_H17 cells have greater potential to induce asthma than conventional T_H2 cells [117,118]. However, T_H17 cells can also acquire a regulatory phenotype. T_H17 cells can start producing IL-10 themselves, and it has been demonstrated that these regulatory

T$_H$17 cells can suppress other effector T cells in vitro and display a non-inflammatory gene expression profile [119]. Finally, due to the usage of IL-17A fate-mapping mice, it was possible to prove that T$_H$17 cells can completely transdifferentiate into T$_R$1 cells and therefore contribute to the resolution of inflammation [120]. The key for new therapeutic approaches for inflammatory diseases may lie especially within this T$_H$17 cell plasticity.

4.4. Regulation of IL-22 via IL-22BP

Finally, there are ways to control the activity of T$_H$17 cell associated cytokines, such as IL-22. Apart from its protective characteristics, IL-22 is known to play a pathogenic role in autoimmune diseases, several cancers, and chronic liver damage. In order to successfully maintain the balance between protection and harm, endogenous mechanisms controlling the activity of IL-22 are required. IL-22 binding protein (IL-22BP, IL-22Ra2) is a soluble IL-22 receptor and inhibitor. It has been shown that IL-22BP binds to IL-22 and blocks its interaction with the membrane bound IL-22R1 in vitro using human and mouse cells. In vivo studies using mouse models have concluded that the effect of IL-22BP is dependent on the presence of IL-22, revealing a specific in vivo binding of IL-22BP to IL-22 with higher affinity than to the membrane-bound IL-22R1 [121–124]. IL-22BP is present in lymphatic organs, the gastrointestinal system, the lung, the skin, the liver, and in the female reproductive system. Normally, cellular sources of IL-22BP detected in lymphoid organs and the intestine are conventional dendritic cells (DCs), T cells, and eosinophils. It has been shown previously that endogenous IL-22BP is responsible for controlling IL-22-induced effects in the intestine. Recently, Pelczar et al. identified the role of IL-22BP in the development of IBD [125]. It was shown that T cell derived IL-22BP promotes IBD development. CD4$^+$ T cells from patients with IBD were found to produce high levels of IL-22BP. Interestingly, reduced IL-22BP expression was found in intestinal CD4$^+$ T cells derived from IBD patients treated with anti-TNF-α antibodies. Therefore, these studies suggest that the regulation of IL-22 by IL-22BP is crucial and may serve as a therapeutic target in diseases like IBD.

5. Concluding Remarks and Future Perspective

Adaptive immunity plays a non-redundant role not only for defense against infections, but also for wound healing, tissue regeneration, or carcinogenesis. However, adaptive immunity has both beneficial and potential pathogenic characteristics. The immune system performs a constant balancing act to maintain beneficial properties over pathogenic ones. T$_H$17 cells have been a main focus of immunological research since their discovery in 2005. As described above, T$_H$17 cells and T$_H$17 cell related cytokines can act in a beneficial fashion during wound repair and regeneration, but they can also cause chronic inflammation and carcinogenesis. Therefore, the immune system developed several control mechanisms to regulate T$_H$17 cell mediated immunity. However, failure of these control mechanisms results in chronic inflammatory diseases and cancer. Thus, new therapeutic strategies targeting T$_H$17 cells have been a main focus of clinical research in recent years. Nonetheless, there are many remaining open questions regarding the involvement and regulation of T$_H$17 cells during tissue regeneration and wound healing. It has long been known that $\alpha\beta$ CD4$^+$ T cells infiltrate the wound bed, however, recent research lacks detailed analysis of these cells during wound healing. T$_H$17 cells are especially of great interest in this matter, since they can attract neutrophils and other innate immune cells and induce anti-microbial peptides from epithelial cells. Therefore, T$_H$17 cells could help to prevent spreading infections in a wound. Additionally, T$_H$17 cell associated cytokines such as IL-22 can promote epithelial cell proliferation. However, the basic question if the inflammatory phase during wound healing is altered in the absence of T$_H$17 cells has not been fully addressed yet. A detailed characterization of T$_H$17 cell immune responses during wound healing in different tissues is needed. T$_H$17 cells are most abundant in the intestine and other barrier organs such as lung and skin, therefore it can be assumed that T$_H$17 cells are most important during wound healing in these organs. Besides the beneficial properties of T$_H$17 cells, these cells are strongly associated with chronic inflammatory conditions. Prolonged inflammation is a common hallmark for chronic wounds, which represent an increasing health threat and a therapeutic challenge. Whether or not an uncontrolled T_H17 cell immune response plays an important part during this process is unknown so far. However, if this is the case, it is crucial to understand control mechanisms of T_H17 cells that could allow the reprogramming of the immune system so that this chronic inflammatory stage can be resolved. Establishing a better understanding of this process and the underlying mechanisms could potentially facilitate the design of new therapeutic approaches for a wide variety of diseases including infections, cancer, and immune mediated inflammatory diseases.

Acknowledgments: This work was supported by the ERC (ERC Stg 337251 to SH).

Author Contributions: Leonie Brockmann and Anastasios D. Giannou wrote the manuscript, Nicola Gagliani and Samuel Huber supervised and revised the manuscript.

Conflicts of Interest: The authors declare no conflicts of interest.

References

1. Dvorak, H.F. Tumors: Wounds that do not heal. Similarities between tumor stroma generation and wound healing. *N. Engl. J. Med.* **1986**, *315*, 1650–1659. [PubMed]
2. Minutti, C.M.; Knipper, J.A.; Allen, J.E.; Zaiss, D.M. Tissue-specific contribution of macrophages to wound healing. *Semin. Cell Dev. Biol.* **2017**, *61*, 3–11. [CrossRef] [PubMed]
3. Avitabile, S.; Odorisio, T.; Madonna, S.; Eyerich, S.; Guerra, L.; Eyerich, K.; Zambruno, G.; Cavani, A.; Cianfarani, F. Interleukin-22 promotes wound repair in diabetes by improving keratinocyte pro-healing functions. *J. Investig. Dermatol.* **2015**, *135*, 2862–2870. [CrossRef] [PubMed]
4. Xiao, W.A.; Hu, Z.Y.; Li, T.W.; Li, J.J. Bone fracture healing is delayed in splenectomic rats. *Life Sci.* **2017**, *173*, 55–61. [CrossRef] [PubMed]
5. Broekman, W.; Amatngalim, G.D.; de Mooij-Eijk, Y.; Oostendorp, J.; Roelofs, H.; Taube, C.; Stolk, J.; Hiemstra, P.S. TNF-α and IL-1β-activated human mesenchymal stromal cells increase airway epithelial wound healing in vitro via activation of the epidermal growth factor receptor. *Respir. Res.* **2016**, *17*, 3. [CrossRef] [PubMed]
6. Gosain, A.; DiPietro, L.A. Aging and wound healing. *World J. Surg.* **2004**, *28*, 321–326. [CrossRef] [PubMed]
7. Fishel, R.S.; Barbul, A.; Beschorner, W.E.; Wasserkrug, H.L.; Efron, G. Lymphocyte participation in wound healing. Morphologic assessment using monoclonal antibodies. *Ann. Surg.* **1987**, *206*, 25–29. [CrossRef] [PubMed]
8. Hofmann, U.; Beyersdorf, N.; Weirather, J.; Podolskaya, A.; Bauersachs, J.; Ertl, G.; Kerkau, T.; Frantz, S. Activation of CD4+ T lymphocytes improves wound healing and survival after experimental myocardial infarction in mice. *Circulation* **2012**, *125*, 1652–1663. [CrossRef] [PubMed]
9. Park, J.E.; Barbul, A. Understanding the role of immune regulation in wound healing. *Am. J. Surg.* **2004**, *187*, 11S–16S. [CrossRef]
10. Swift, M.E.; Burns, A.L.; Gray, K.L.; DiPietro, L.A. Age-related alterations in the inflammatory response to dermal injury. *J. Investig. Dermatol.* **2001**, *117*, 1027–1035. [CrossRef] [PubMed]
11. Antonio, N.; Bonnelykke-Behrndtz, M.L.; Ward, L.C.; Collin, J.; Christensen, I.J.; Steiniche, T.; Schmidt, H.; Feng, Y.; Martin, P. The wound inflammatory response exacerbates growth of pre-neoplastic cells and progression to cancer. *EMBO J.* **2015**, *34*, 2219–2236. [CrossRef] [PubMed]
12. Mosmann, T.R.; Coffman, R.L. T_H1 and T_H2 cells: Different patterns of lymphokine secretion lead to different functional properties. *Annu. Rev. Immunol.* **1989**, *7*, 145–173. [CrossRef] [PubMed]
13. Khader, S.A.; Gaffen, S.L.; Kolls, J.K. T_H17 cells at the crossroads of innate and adaptive immunity against infectious diseases at the mucosa. *Mucosal Immunol.* **2009**, *2*, 403–411. [CrossRef] [PubMed]
14. Pickert, G.; Neufert, C.; Leppkes, M.; Zheng, Y.; Wittkopf, N.; Warntjen, M.; Lehr, H.A.; Hirth, S.; Weigmann, B.; Wirtz, S.; et al. STAT3 links IL-22 signaling in intestinal epithelial cells to mucosal wound healing. *J. Exp. Med.* **2009**, *206*, 1465–1472. [CrossRef] [PubMed]
15. Savage, D.C. Microbial ecology of the gastrointestinal tract. *Annu. Rev. Microbiol.* **1977**, *31*, 107–133. [CrossRef] [PubMed]

16. Ivanov, I.I.; Atarashi, K.; Manel, N.; Brodie, E.L.; Shima, T.; Karaoz, U.; Wei, D.; Goldfarb, K.C.; Santee, C.A.; Lynch, S.V.; et al. Induction of intestinal T_H17 cells by segmented filamentous bacteria. *Cell* **2009**, *139*, 485–498. [CrossRef] [PubMed]

17. Ivanov, I.I.; McKenzie, B.S.; Zhou, L.; Tadokoro, C.E.; Lepelley, A.; Lafaille, J.J.; Cua, D.J.; Littman, D.R. The orphan nuclear receptor rorgammat directs the differentiation program of proinflammatory IL-17$^+$ T helper cells. *Cell* **2006**, *126*, 1121–1133. [CrossRef] [PubMed]

18. Moseley, T.A.; Haudenschild, D.R.; Rose, L.; Reddi, A.H. Interleukin-17 family and IL-17 receptors. *Cytokine Growth Factor Rev.* **2003**, *14*, 155–174. [CrossRef]

19. Ouyang, W.; Kolls, J.K.; Zheng, Y. The biological functions of T helper 17 cell effector cytokines in inflammation. *Immunity* **2008**, *28*, 454–467. [CrossRef] [PubMed]

20. Liang, S.C.; Tan, X.Y.; Luxenberg, D.P.; Karim, R.; Dunussi-Joannopoulos, K.; Collins, M.; Fouser, L.A. Interleukin (IL)-22 and IL-17 are coexpressed by T_H17 cells and cooperatively enhance expression of antimicrobial peptides. *J. Exp. Med.* **2006**, *203*, 2271–2279. [CrossRef] [PubMed]

21. Archer, N.K.; Adappa, N.D.; Palmer, J.N.; Cohen, N.A.; Harro, J.M.; Lee, S.K.; Miller, L.S.; Shirtliff, M.E. Interleukin-17A (IL-17A) and IL-17F are critical for antimicrobial peptide production and clearance of staphylococcus aureus nasal colonization. *Infect. Immun.* **2016**, *84*, 3575–3583. [CrossRef] [PubMed]

22. Meller, S.; Di Domizio, J.; Voo, K.S.; Friedrich, H.C.; Chamilos, G.; Ganguly, D.; Conrad, C.; Gregorio, J.; Le Roy, D.; Roger, T.; et al. T_H17 cells promote microbial killing and innate immune sensing of DNA via interleukin 26. *Nat. Immunol.* **2015**, *16*, 970–979. [CrossRef] [PubMed]

23. Hymowitz, S.G.; Filvaroff, E.H.; Yin, J.P.; Lee, J.; Cai, L.; Risser, P.; Maruoka, M.; Mao, W.; Foster, J.; Kelley, R.F.; et al. IL-17s adopt a cystine knot fold: Structure and activity of a novel cytokine, IL-17f, and implications for receptor binding. *EMBO J.* **2001**, *20*, 5332–5341. [CrossRef] [PubMed]

24. Sugimoto, K.; Ogawa, A.; Mizoguchi, E.; Shimomura, Y.; Andoh, A.; Bhan, A.K.; Blumberg, R.S.; Xavier, R.J.; Mizoguchi, A. IL-22 ameliorates intestinal inflammation in a mouse model of ulcerative colitis. *J. Clin. Investig.* **2008**, *118*, 534–544. [CrossRef] [PubMed]

25. Zenewicz, L.A.; Yancopoulos, G.D.; Valenzuela, D.M.; Murphy, A.J.; Stevens, S.; Flavell, R.A. Innate and adaptive interleukin-22 protects mice from inflammatory bowel disease. *Immunity* **2008**, *29*, 947–957. [CrossRef] [PubMed]

26. Quesniaux, V.R.; Ryffel, B.; Di Padova, F. *IL-17, IL-22 and Their Producing Cells: Role in Inflammation and Autoimmunity*, 2nd ed.; Springer: Basel, Switzerland, 2013.

27. Catana, C.S.; Berindan Neagoe, I.; Cozma, V.; Magdas, C.; Tabaran, F.; Dumitrascu, D.L. Contribution of the IL-17/IL-23 axis to the pathogenesis of inflammatory bowel disease. *World J. Gastroenterol.* **2015**, *21*, 5823–5830. [PubMed]

28. Song, X.; Dai, D.; He, X.; Zhu, S.; Yao, Y.; Gao, H.; Wang, J.; Qu, F.; Qiu, J.; Wang, H.; et al. Growth factor FGF2 cooperates with interleukin-17 to repair intestinal epithelial damage. *Immunity* **2015**, *43*, 488–501. [CrossRef] [PubMed]

29. Rodero, M.P.; Hodgson, S.S.; Hollier, B.; Combadiere, C.; Khosrotehrani, K. Reduced IL17A expression distinguishes a Ly6c(lo)MHCII(hi) macrophage population promoting wound healing. *J. Investig. Dermatol.* **2013**, *133*, 783–792. [CrossRef] [PubMed]

30. Li, Z.; Burns, A.R.; Han, L.; Rumbaut, R.E.; Smith, C.W. IL-17 and VEGF are necessary for efficient corneal nerve regeneration. *Am. J. Pathol.* **2011**, *178*, 1106–1116. [CrossRef] [PubMed]

31. Ono, T.; Okamoto, K.; Nakashima, T.; Nitta, T.; Hori, S.; Iwakura, Y.; Takayanagi, H. IL-17-producing gammadelta T cells enhance bone regeneration. *Nat. Commun.* **2016**, *7*, 10928. [CrossRef] [PubMed]

32. Numasaki, M.; Fukushi, J.; Ono, M.; Narula, S.K.; Zavodny, P.J.; Kudo, T.; Robbins, P.D.; Tahara, H.; Lotze, M.T. Interleukin-17 promotes angiogenesis and tumor growth. *Blood* **2003**, *101*, 2620–2627. [CrossRef] [PubMed]

33. Muranski, P.; Boni, A.; Antony, P.A.; Cassard, L.; Irvine, K.R.; Kaiser, A.; Paulos, C.M.; Palmer, D.C.; Touloukian, C.E.; Ptak, K.; et al. Tumor-specific T_H17-polarized cells eradicate large established melanoma. *Blood* **2008**, *112*, 362–373. [CrossRef] [PubMed]

34. Kryczek, I.; Wei, S.; Szeliga, W.; Vatan, L.; Zou, W. Endogenous IL-17 contributes to reduced tumor growth and metastasis. *Blood* **2009**, *114*, 357–359. [CrossRef] [PubMed]

35. Martin-Orozco, N.; Muranski, P.; Chung, Y.; Yang, X.O.; Yamazaki, T.; Lu, S.; Hwu, P.; Restifo, N.P.; Overwijk, W.W.; Dong, C. T helper 17 cells promote cytotoxic T cell activation in tumor immunity. *Immunity* **2009**, *31*, 787–798. [CrossRef] [PubMed]

36. Wu, S.; Rhee, K.J.; Albesiano, E.; Rabizadeh, S.; Wu, X.; Yen, H.R.; Huso, D.L.; Brancati, F.L.; Wick, E.; McAllister, F.; et al. A human colonic commensal promotes colon tumorigenesis via activation of T helper type 17 T cell responses. *Nat. Med.* **2009**, *15*, 1016–1022. [CrossRef] [PubMed]

37. Housseau, F.; Wu, S.; Wick, E.C.; Fan, H.; Wu, X.; Llosa, N.J.; Smith, K.N.; Tam, A.; Ganguly, S.; Wanyiri, J.W.; et al. Redundant innate and adaptive sources of IL17 production drive colon tumorigenesis. *Cancer Res.* **2016**, *76*, 2115–2124. [CrossRef] [PubMed]

38. Hyun, Y.S.; Han, D.S.; Lee, A.R.; Eun, C.S.; Youn, J.; Kim, H.Y. Role of IL-17A in the development of colitis-associated cancer. *Carcinogenesis* **2012**, *33*, 931–936. [CrossRef] [PubMed]

39. Tosolini, M.; Kirilovsky, A.; Mlecnik, B.; Fredriksen, T.; Mauger, S.; Bindea, G.; Berger, A.; Bruneval, P.; Fridman, W.H.; Pages, F.; et al. Clinical impact of different classes of infiltrating T cytotoxic and helper cells (T_H1, T_H2, Treg, T_H17) in patients with colorectal cancer. *Cancer Res.* **2011**, *71*, 1263–1271. [CrossRef] [PubMed]

40. Bindea, G.; Mlecnik, B.; Tosolini, M.; Kirilovsky, A.; Waldner, M.; Obenauf, A.C.; Angell, H.; Fredriksen, T.; Lafontaine, L.; Berger, A.; et al. Spatiotemporal dynamics of intratumoral immune cells reveal the immune landscape in human cancer. *Immunity* **2013**, *39*, 782–795. [CrossRef] [PubMed]

41. Kirchberger, S.; Royston, D.J.; Boulard, O.; Thornton, E.; Franchini, F.; Szabady, R.L.; Harrison, O.; Powrie, F. Innate lymphoid cells sustain colon cancer through production of interleukin-22 in a mouse model. *J. Exp. Med.* **2013**, *210*, 917–931. [CrossRef] [PubMed]

42. Wu, P.; Wu, D.; Ni, C.; Ye, J.; Chen, W.; Hu, G.; Wang, Z.; Wang, C.; Zhang, Z.; Xia, W.; et al. GammadeltaT17 cells promote the accumulation and expansion of myeloid-derived suppressor cells in human colorectal cancer. *Immunity* **2014**, *40*, 785–800. [CrossRef] [PubMed]

43. Zhuang, Y.; Peng, L.S.; Zhao, Y.L.; Shi, Y.; Mao, X.H.; Chen, W.; Pang, K.C.; Liu, X.F.; Liu, T.; Zhang, J.Y.; et al. CD8+ T cells that produce interleukin-17 regulate myeloid-derived suppressor cells and are associated with survival time of patients with gastric cancer. *Gastroenterology* **2012**, *143*, 951–962. [CrossRef] [PubMed]

44. Wang, K.; Kim, M.K.; Di Caro, G.; Wong, J.; Shalapour, S.; Wan, J.; Zhang, W.; Zhong, Z.; Sanchez-Lopez, E.; Wu, L.W.; et al. Interleukin-17 receptor a signaling in transformed enterocytes promotes early colorectal tumorigenesis. *Immunity* **2014**, *41*, 1052–1063. [CrossRef] [PubMed]

45. Chung, A.S.; Wu, X.; Zhuang, G.; Ngu, H.; Kasman, I.; Zhang, J.; Vernes, J.M.; Jiang, Z.; Meng, Y.G.; Peale, F.V.; et al. An interleukin-17-mediated paracrine network promotes tumor resistance to anti-angiogenic therapy. *Nat. Med.* **2013**, *19*, 1114–1123. [CrossRef] [PubMed]

46. Song, X.; Gao, H.; Lin, Y.; Yao, Y.; Zhu, S.; Wang, J.; Liu, Y.; Yao, X.; Meng, G.; Shen, N.; et al. Alterations in the microbiota drive interleukin-17c production from intestinal epithelial cells to promote tumorigenesis. *Immunity* **2014**, *40*, 140–152. [CrossRef] [PubMed]

47. Tong, Z.; Yang, X.O.; Yan, H.; Liu, W.; Niu, X.; Shi, Y.; Fang, W.; Xiong, B.; Wan, Y.; Dong, C. A protective role by interleukin-17F in colon tumorigenesis. *PLoS ONE* **2012**, *7*, e34959. [CrossRef] [PubMed]

48. Khare, V.; Paul, G.; Movadat, O.; Frick, A.; Jambrich, M.; Krnjic, A.; Marian, B.; Wrba, F.; Gasche, C. IL10R2 overexpression promotes IL22/STAT3 signaling in colorectal carcinogenesis. *Cancer Immunol. Res.* **2015**, *3*, 1227–1235. [CrossRef] [PubMed]

49. McGee, H.M.; Schmidt, B.A.; Booth, C.J.; Yancopoulos, G.D.; Valenzuela, D.M.; Murphy, A.J.; Stevens, S.; Flavell, R.A.; Horsley, V. IL-22 promotes fibroblast-mediated wound repair in the skin. *J. Investig. Dermatol.* **2013**, *133*, 1321–1329. [CrossRef] [PubMed]

50. Boniface, K.; Bernard, F.X.; Garcia, M.; Gurney, A.L.; Lecron, J.C.; Morel, F. IL-22 inhibits epidermal differentiation and induces proinflammatory gene expression and migration of human keratinocytes. *J. Immunol.* **2005**, *174*, 3695–3702. [CrossRef] [PubMed]

51. Zheng, Y.; Danilenko, D.M.; Valdez, P.; Kasman, I.; Eastham-Anderson, J.; Wu, J.; Ouyang, W. Interleukin-22, a T_H17 cytokine, mediates IL-23-induced dermal inflammation and acanthosis. *Nature* **2007**, *445*, 648–651. [CrossRef] [PubMed]

52. Zenewicz, L.A.; Yin, X.; Wang, G.; Elinav, E.; Hao, L.; Zhao, L.; Flavell, R.A. IL-22 deficiency alters colonic microbiota to be transmissible and colitogenic. *J. Immunol.* **2013**, *190*, 5306–5312. [CrossRef] [PubMed]

53. Sun, X.; Zhang, J.; Wang, L.; Tian, Z. Growth inhibition of human hepatocellular carcinoma cells by blocking STAT3 activation with decoy-odn. *Cancer Lett.* **2008**, *262*, 201–213. [CrossRef] [PubMed]

54. Bi, Y.; Cao, J.; Jin, S.; Lv, L.; Qi, L.; Liu, F.; Geng, J.; Yu, Y. Interleukin-22 promotes lung cancer cell proliferation and migration via the IL-22R1/STAT3 and IL-22R1/AKT signaling pathways. *Mol. Cell. Biochem.* **2016**, *415*, 1–11. [CrossRef] [PubMed]

55. Sun, D.; Lin, Y.; Hong, J.; Chen, H.; Nagarsheth, N.; Peng, D.; Wei, S.; Huang, E.; Fang, J.; Kryczek, I.; et al. T$_H$22 cells control colon tumorigenesis through STAT3 and polycomb repression complex 2 signaling. *Oncoimmunology* **2016**, *5*, e1082704. [CrossRef] [PubMed]

56. Huber, S.; Gagliani, N.; Zenewicz, L.A.; Huber, F.J.; Bosurgi, L.; Hu, B.; Hedl, M.; Zhang, W.; O'Connor, W., Jr.; Murphy, A.J.; et al. IL-22BP is regulated by the inflammasome and modulates tumorigenesis in the intestine. *Nature* **2012**, *491*, 259–263. [CrossRef] [PubMed]

57. Wu, T.; Cui, L.; Liang, Z.; Liu, C.; Liu, Y.; Li, J. Elevated serum IL-22 levels correlate with chemoresistant condition of colorectal cancer. *Clin. Immunol.* **2013**, *147*, 38–39. [CrossRef] [PubMed]

58. Wu, T.; Wang, Z.; Liu, Y.; Mei, Z.; Wang, G.; Liang, Z.; Cui, A.; Hu, X.; Cui, L.; Yang, Y.; et al. Interleukin 22 protects colorectal cancer cells from chemotherapy by activating the STAT3 pathway and inducing autocrine expression of interleukin 8. *Clin. Immunol.* **2014**, *154*, 116–126. [CrossRef] [PubMed]

59. Kryczek, I.; Lin, Y.; Nagarsheth, N.; Peng, D.; Zhao, L.; Zhao, E.; Vatan, L.; Szeliga, W.; Dou, Y.; Owens, S.; et al. IL-22$^+$CD4$^+$ T cells promote colorectal cancer stemness via STAT3 transcription factor activation and induction of the methyltransferase DOT1L. *Immunity* **2014**, *40*, 772–784. [CrossRef] [PubMed]

60. Hanash, A.M.; Dudakov, J.A.; Hua, G.; O'Connor, M.H.; Young, L.F.; Singer, N.V.; West, M.L.; Jenq, R.R.; Holland, A.M.; Kappel, L.W.; et al. Interleukin-22 protects intestinal stem cells from immune-mediated tissue damage and regulates sensitivity to graft versus host disease. *Immunity* **2012**, *37*, 339–350. [CrossRef] [PubMed]

61. Ritsu, M.; Kawakami, K.; Kanno, E.; Tanno, H.; Ishii, K.; Imai, Y.; Maruyama, R.; Tachi, M. Critical role of tumor necrosis factor-α in the early process of wound healing in skin. *J. Dermatol. Dermatol. Surg.* **2017**, *21*, 14–19. [CrossRef]

62. Perez, C.; Albert, I.; DeFay, K.; Zachariades, N.; Gooding, L.; Kriegler, M. A nonsecretable cell surface mutant of tumor necrosis factor (TNF) kills by cell-to-cell contact. *Cell* **1990**, *63*, 251–258. [CrossRef]

63. Eissner, G.; Kolch, W.; Scheurich, P. Ligands working as receptors: Reverse signaling by members of the tnf superfamily enhance the plasticity of the immune system. *Cytokine Growth Factor Rev.* **2004**, *15*, 353–366. [CrossRef] [PubMed]

64. Mizoguchi, E.; Mizoguchi, A.; Takedatsu, H.; Cario, E.; de Jong, Y.P.; Ooi, C.J.; Xavier, R.J.; Terhorst, C.; Podolsky, D.K.; Bhan, A.K. Role of tumor necrosis factor receptor 2 (TNFR2) in colonic epithelial hyperplasia and chronic intestinal inflammation in mice. *Gastroenterology* **2002**, *122*, 134–144. [CrossRef] [PubMed]

65. Balkwill, F. Tumour necrosis factor and cancer. *Nat. Rev. Cancer* **2009**, *9*, 361–371. [CrossRef] [PubMed]

66. Wu, Y.; Deng, J.; Rychahou, P.G.; Qiu, S.; Evers, B.M.; Zhou, B.P. Stabilization of snail by NF-kappaB is required for inflammation-induced cell migration and invasion. *Cancer Cell* **2009**, *15*, 416–428. [CrossRef] [PubMed]

67. Popivanova, B.K.; Kitamura, K.; Wu, Y.; Kondo, T.; Kagaya, T.; Kaneko, S.; Oshima, M.; Fujii, C.; Mukaida, N. Blocking TNF-α in mice reduces colorectal carcinogenesis associated with chronic colitis. *J. Clin. Investig.* **2008**, *118*, 560–570. [CrossRef] [PubMed]

68. Al Obeed, O.A.; Alkhayal, K.A.; Al Sheikh, A.; Zubaidi, A.M.; Vaali-Mohammed, M.A.; Boushey, R.; McKerrow, J.H.; Abdulla, M.H. Increased expression of tumor necrosis factor-α is associated with advanced colorectal cancer stages. *World J. Gastroenterol.* **2014**, *20*, 18390–18396. [CrossRef] [PubMed]

69. Flores, M.B.; Rocha, G.Z.; Damas-Souza, D.M.; Osorio-Costa, F.; Dias, M.M.; Ropelle, E.R.; Camargo, J.A.; de Carvalho, R.B.; Carvalho, H.F.; Saad, M.J.; et al. Obesity-induced increase in tumor necrosis factor-α leads to development of colon cancer in mice. *Gastroenterology* **2012**, *143*, 741–753. [CrossRef] [PubMed]

70. Huang, L.; Wang, X.; Wen, C.; Yang, X.; Song, M.; Chen, J.; Wang, C.; Zhang, B.; Wang, L.; Iwamoto, A.; et al. Hsa-miR-19a is associated with lymph metastasis and mediates the TNF-α induced epithelial-to-mesenchymal transition in colorectal cancer. *Sci. Rep.* **2015**, *5*, 13350. [CrossRef] [PubMed]

71. Cua, D.J.; Sherlock, J.; Chen, Y.; Murphy, C.A.; Joyce, B.; Seymour, B.; Lucian, L.; To, W.; Kwan, S.; Churakova, T.; et al. Interleukin-23 rather than interleukin-12 is the critical cytokine for autoimmune inflammation of the brain. *Nature* **2003**, *421*, 744–748. [CrossRef] [PubMed]

72. Murphy, C.A.; Langrish, C.L.; Chen, Y.; Blumenschein, W.; McClanahan, T.; Kastelein, R.A.; Sedgwick, J.D.; Cua, D.J. Divergent pro- and antiinflammatory roles for IL-23 and IL-12 in joint autoimmune inflammation. *J. Exp. Med.* **2003**, *198*, 1951–1957. [CrossRef] [PubMed]

73. Aggarwal, S.; Ghilardi, N.; Xie, M.H.; de Sauvage, F.J.; Gurney, A.L. Interleukin-23 promotes a distinct CD4 T cell activation state characterized by the production of interleukin-17. *J. Biol. Chem.* **2003**, *278*, 1910–1914. [CrossRef] [PubMed]

74. Langrish, C.L.; Chen, Y.; Blumenschein, W.M.; Mattson, J.; Basham, B.; Sedgwick, J.D.; McClanahan, T.; Kastelein, R.A., Cua, D.J. IL-23 drives a pathogenic T cell population that induces autoimmune inflammation. *J. Exp. Med.* **2005**, *201*, 233–240. [CrossRef] [PubMed]

75. Fouser, L.A.; Wright, J.F.; Dunussi-Joannopoulos, K.; Collins, M. T_H17 cytokines and their emerging roles in inflammation and autoimmunity. *Immunol. Rev.* **2008**, *226*, 87–102. [CrossRef] [PubMed]

76. Harbour, S.N.; Maynard, C.L.; Zindl, C.L.; Schoeb, T.R.; Weaver, C.T. T_H17 cells give rise to T_H1 cells that are required for the pathogenesis of colitis. *Proc. Natl. Acad. Sci. USA* **2015**, *112*, 7061–7066. [CrossRef] [PubMed]

77. Jimeno, R.; Leceta, J.; Garin, M.; Ortiz, A.M.; Mellado, M.; Rodriguez-Frade, J.M.; Martinez, C.; Perez-Garcia, S.; Gomariz, R.P.; Juarranz, Y. T_H17 polarization of memory Th cells in early arthritis: The vasoactive intestinal peptide effect. *J. Leukoc. Biol.* **2015**, *98*, 257–269. [CrossRef] [PubMed]

78. Babaloo, Z.; Aliparasti, M.R.; Babaiea, F.; Almasi, S.; Baradaran, B.; Farhoudi, M. The role of T_H17 cells in patients with relapsing-remitting multiple sclerosis: Interleukin-17A and interleukin-17F serum levels. *Immunol. Lett.* **2015**, *164*, 76–80. [CrossRef] [PubMed]

79. Zhou, L.; Lopes, J.E.; Chong, M.M.; Ivanov, I.I.; Min, R.; Victora, G.D.; Shen, Y.; Du, J.; Rubtsov, Y.P.; Rudensky, A.Y.; et al. TGF-β-induced Foxp3 inhibits T_H17 cell differentiation by antagonizing rorgammat function. *Nature* **2008**, *453*, 236–240. [CrossRef] [PubMed]

80. McGeachy, M.J.; Chen, Y.; Tato, C.M.; Laurence, A.; Joyce-Shaikh, B.; Blumenschein, W.M.; McClanahan, T.K.; O'Shea, J.J.; Cua, D.J. The interleukin 23 receptor is essential for the terminal differentiation of interleukin 17-producing effector T helper cells in vivo. *Nat. Immunol.* **2009**, *10*, 314–324. [CrossRef] [PubMed]

81. Bettelli, E.; Carrier, Y.; Gao, W.; Korn, T.; Strom, T.B.; Oukka, M.; Weiner, H.L.; Kuchroo, V.K. Reciprocal developmental pathways for the generation of pathogenic effector T_H17 and regulatory T cells. *Nature* **2006**, *441*, 235–238. [CrossRef] [PubMed]

82. Mangan, P.R.; Harrington, L.E.; O'Quinn, D.B.; Helms, W.S.; Bullard, D.C.; Elson, C.O.; Hatton, R.D.; Wahl, S.M.; Schoeb, T.R.; Weaver, C.T. Transforming growth factor-β induces development of the T(H)17 lineage. *Nature* **2006**, *441*, 231–234. [CrossRef] [PubMed]

83. Veldhoen, M.; Hocking, R.J.; Atkins, C.J.; Locksley, R.M.; Stockinger, B. TGFβ in the context of an inflammatory cytokine milieu supports de novo differentiation of IL-17-producing T cells. *Immunity* **2006**, *24*, 179–189. [CrossRef] [PubMed]

84. Qin, H.; Wang, L.; Feng, T.; Elson, C.O.; Niyongere, S.A.; Lee, S.J.; Reynolds, S.L.; Weaver, C.T.; Roarty, K.; Serra, R.; et al. TGF-β promotes T_H17 cell development through inhibition of SOCS3. *J. Immunol.* **2009**, *183*, 97–105. [CrossRef] [PubMed]

85. Ghoreschi, K.; Laurence, A.; Yang, X.P.; Tato, C.M.; McGeachy, M.J.; Konkel, J.E.; Ramos, H.L.; Wei, L.; Davidson, T.S.; Bouladoux, N.; et al. Generation of pathogenic T_H17 cells in the absence of TGF-β signalling. *Nature* **2010**, *467*, 967–971. [CrossRef] [PubMed]

86. Hori, S.; Nomura, T.; Sakaguchi, S. Control of regulatory T cell development by the transcription factor Foxp3. *Science* **2003**, *299*, 1057–1061. [CrossRef] [PubMed]

87. Selvaraj, R.K.; Geiger, T.L. A kinetic and dynamic analysis of Foxp3 induced in T cells by TGF-β. *J. Immunol.* **2007**, *178*, 7667–7677. [CrossRef] [PubMed]

88. Davidson, T.S.; DiPaolo, R.J.; Andersson, J.; Shevach, E.M. Cutting edge: IL-2 is essential for TGF-β-mediated induction of Foxp3+ T regulatory cells. *J. Immunol.* **2007**, *178*, 4022–4026. [CrossRef] [PubMed]

89. Brandenburg, S.; Takahashi, T.; de la Rosa, M.; Janke, M.; Karsten, G.; Muzzulini, T.; Orinska, Z.; Bulfone-Paus, S.; Scheffold, A. IL-2 induces in vivo suppression by CD4+CD25+Foxp3+ regulatory T cells. *Eur. J. Immunol.* **2008**, *38*, 1643–1653. [CrossRef] [PubMed]

90. Laurence, A.; Tato, C.M.; Davidson, T.S.; Kanno, Y.; Chen, Z.; Yao, Z.; Blank, R.B.; Meylan, F.; Siegel, R.; Hennighausen, L.; et al. Interleukin-2 signaling via STAT5 constrains T helper 17 cell generation. *Immunity* **2007**, *26*, 371–381. [CrossRef] [PubMed]

91. Diveu, C.; McGeachy, M.J.; Boniface, K.; Stumhofer, J.S.; Sathe, M.; Joyce-Shaikh, B.; Chen, Y.; Tato, C.M.; McClanahan, T.K.; de Waal Malefyt, R.; et al. IL-27 blocks rorc expression to inhibit lineage commitment of T_H17 cells. *J. Immunol.* **2009**, *182*, 5748–5756. [CrossRef] [PubMed]

92. Awasthi, A.; Carrier, Y.; Peron, J.P.; Bettelli, E.; Kamanaka, M.; Flavell, R.A.; Kuchroo, V.K.; Oukka, M.; Weiner, H.L. A dominant function for interleukin 27 in generating interleukin 10-producing anti-inflammatory T cells. *Nat. Immunol.* **2007**, *8*, 1380–1389. [CrossRef] [PubMed]

93. Stumhofer, J.S.; Silver, J.S.; Laurence, A.; Porrett, P.M.; Harris, T.H.; Turka, L.A.; Ernst, M.; Saris, C.J.; O'Shea, J.J.; Hunter, C.A. Interleukins 27 and 6 induce STAT3-mediated T cell production of interleukin 10. *Nat. Immunol.* **2007**, *8*, 1363–1371. [CrossRef] [PubMed]

94. Acosta-Rodriguez, E.V.; Napolitani, G.; Lanzavecchia, A.; Sallusto, F. Interleukins 1β and 6 but not transforming growth factor-β are essential for the differentiation of interleukin 17-producing human T helper cells. *Nat. Immunol.* **2007**, *8*, 942–949. [CrossRef] [PubMed]

95. Wilson, N.J.; Boniface, K.; Chan, J.R.; McKenzie, B.S.; Blumenschein, W.M.; Mattson, J.D.; Basham, B.; Smith, K.; Chen, T.; Morel, F.; et al. Development, cytokine profile and function of human interleukin 17-producing helper T cells. *Nat. Immunol.* **2007**, *8*, 950–957. [CrossRef] [PubMed]

96. Hu, W.; Troutman, T.D.; Edukulla, R.; Pasare, C. Priming microenvironments dictate cytokine requirements for T helper 17 cell lineage commitment. *Immunity* **2011**, *35*, 1010–1022. [CrossRef] [PubMed]

97. Chung, Y.; Chang, S.H.; Martinez, G.J.; Yang, X.O.; Nurieva, R.; Kang, H.S.; Ma, L.; Watowich, S.S.; Jetten, A.M.; Tian, Q.; et al. Critical regulation of early T_H17 cell differentiation by interleukin-1 signaling. *Immunity* **2009**, *30*, 576–587. [CrossRef] [PubMed]

98. Kebir, H.; Ifergan, I.; Alvarez, J.I.; Bernard, M.; Poirier, J.; Arbour, N.; Duquette, P.; Prat, A. Preferential recruitment of interferon-gamma-expressing T_H17 cells in multiple sclerosis. *Ann. Neurol.* **2009**, *66*, 390–402. [CrossRef] [PubMed]

99. Ivanov, I.I.; Frutos Rde, L.; Manel, N.; Yoshinaga, K.; Rifkin, D.B.; Sartor, R.B.; Finlay, B.B.; Littman, D.R. Specific microbiota direct the differentiation of IL-17-producing T-helper cells in the mucosa of the small intestine. *Cell Host Microbe* **2008**, *4*, 337–349. [CrossRef] [PubMed]

100. Shaw, M.H.; Kamada, N.; Kim, Y.G.; Nunez, G. Microbiota-induced IL-1β, but not IL-6, is critical for the development of steady-state T_H17 cells in the intestine. *J. Exp. Med.* **2012**, *209*, 251–258. [CrossRef] [PubMed]

101. Veldhoen, M.; Hirota, K.; Westendorf, A.M.; Buer, J.; Dumoutier, L.; Renauld, J.C.; Stockinger, B. The aryl hydrocarbon receptor links T_H17-cell-mediated autoimmunity to environmental toxins. *Nature* **2008**, *453*, 106–109. [CrossRef] [PubMed]

102. Quintana, F.J.; Basso, A.S.; Iglesias, A.H.; Korn, T.; Farez, M.F.; Bettelli, E.; Caccamo, M.; Oukka, M.; Weiner, H.L. Control of Treg and T_H17 cell differentiation by the aryl hydrocarbon receptor. *Nature* **2008**, *453*, 65–71. [CrossRef] [PubMed]

103. Li, Y.; Innocentin, S.; Withers, D.R.; Roberts, N.A.; Gallagher, A.R.; Grigorieva, E.F.; Wilhelm, C.; Veldhoen, M. Exogenous stimuli maintain intraepithelial lymphocytes via aryl hydrocarbon receptor activation. *Cell* **2011**, *147*, 629–640. [CrossRef] [PubMed]

104. Yeste, A.; Mascanfroni, I.D.; Nadeau, M.; Burns, E.J.; Tukpah, A.M.; Santiago, A.; Wu, C.; Patel, B.; Kumar, D.; Quintana, F.J. IL-21 induces IL-22 production in CD4+ T cells. *Nat. Commun.* **2014**, *5*, 3753. [CrossRef] [PubMed]

105. Cochez, P.M.; Michiels, C.; Hendrickx, E.; Van Belle, A.B.; Lemaire, M.M.; Dauguet, N.; Warnier, G.; de Heusch, M.; Togbe, D.; Ryffel, B.; et al. AhR modulates the IL-22-producing cell proliferation/recruitment in imiquimod-induced psoriasis mouse model. *Eur. J. Immunol.* **2016**, *46*, 1449–1459. [CrossRef] [PubMed]

106. Weirather, J.; Hofmann, U.D.; Beyersdorf, N.; Ramos, G.C.; Vogel, B.; Frey, A.; Ertl, G.; Kerkau, T.; Frantz, S. Foxp3+ CD4+ T cells improve healing after myocardial infarction by modulating monocyte/macrophage differentiation. *Circ. Res.* **2014**, *115*, 55–67. [CrossRef] [PubMed]

107. Nosbaum, A.; Prevel, N.; Truong, H.A.; Mehta, P.; Ettinger, M.; Scharschmidt, T.C.; Ali, N.H.; Pauli, M.L.; Abbas, A.K.; Rosenblum, M.D. Cutting edge: Regulatory T cells facilitate cutaneous wound healing. *J. Immunol.* **2016**, *196*, 2010–2014. [CrossRef] [PubMed]

108. Littman, D.R.; Rudensky, A.Y. T_H17 and regulatory T cells in mediating and restraining inflammation. *Cell* **2010**, *140*, 845–858. [CrossRef] [PubMed]

109. Roncarolo, M.G.; Battaglia, M. Regulatory T-cell immunotherapy for tolerance to self antigens and alloantigens in humans. *Nat. Rev. Immunol.* **2007**, *7*, 585–598. [CrossRef] [PubMed]

110. Glocker, E.O.; Frede, N.; Perro, M.; Sebire, N.; Elawad, M.; Shah, N.; Grimbacher, B. Infant colitis—It's in the genes. *Lancet* **2010**, *376*, 1272. [CrossRef]
111. Huber, S.; Gagliani, N.; Esplugues, E.; O'Connor, W., Jr.; Huber, F.J.; Chaudhry, A.; Kamanaka, M.; Kobayashi, Y.; Booth, C.J.; Rudensky, A.Y.; et al. T$_H$17 cells express interleukin-10 receptor and are controlled by Foxp3$^-$ and Foxp3$^+$ regulatory CD4$^+$ T cells in an interleukin-10-dependent manner. *Immunity* **2011**, *34*, 554–565. [CrossRef] [PubMed]
112. Chaudhry, A.; Samstein, R.M.; Treuting, P.; Liang, Y.; Pils, M.C.; Heinrich, J.M.; Jack, R.S.; Wunderlich, F.T.; Bruning, J.C.; Muller, W.; et al. Interleukin-10 signaling in regulatory T cells is required for suppression of T$_H$17 cell-mediated inflammation. *Immunity* **2011**, *34*, 566–578. [CrossRef] [PubMed]
113. Brockmann, L.; Gagliani, N.; Steglich, B.; Giannou, A.D.; Kempski, J.; Pelczar, P.; Geffken, M.; Mfarrej, B.; Huber, F.; Herkel, J.; et al. IL-10 receptor signaling is essential for TR1 cell function in vivo. *J. Immunol.* **2017**, *198*, 1130–1141. [CrossRef] [PubMed]
114. Annunziato, F.; Cosmi, L.; Santarlasci, V.; Maggi, L.; Liotta, F.; Mazzinghi, B.; Parente, E.; Fili, L.; Ferri, S.; Frosali, F.; et al. Phenotypic and functional features of human T$_H$17 cells. *J. Exp. Med.* **2007**, *204*, 1849–1861. [CrossRef] [PubMed]
115. Annunziato, F.; Cosmi, L.; Liotta, F.; Maggi, E.; Romagnani, S. Type 17 T helper cells-origins, features and possible roles in rheumatic disease. *Nat. Rev. Rheumatol.* **2009**, *5*, 325–331. [CrossRef] [PubMed]
116. Cosmi, L.; Cimaz, R.; Maggi, L.; Santarlasci, V.; Capone, M.; Borriello, F.; Frosali, F.; Querci, V.; Simonini, G.; Barra, G.; et al. Evidence of the transient nature of the T$_H$17 phenotype of CD4$^+$CD161$^+$ T cells in the synovial fluid of patients with juvenile idiopathic arthritis. *Arthritis Rheumatol.* **2011**, *63*, 2504–2515. [CrossRef] [PubMed]
117. Cosmi, L.; Maggi, L.; Santarlasci, V.; Capone, M.; Cardilicchia, E.; Frosali, F.; Querci, V.; Angeli, R.; Matucci, A.; Fambrini, M.; et al. Identification of a novel subset of human circulating memory CD4$^+$ T cells that produce both IL-17A and IL-4. *J. Allergy Clin. Immunol.* **2010**, *125*, 222–230. [CrossRef] [PubMed]
118. Wang, Y.H.; Voo, K.S.; Liu, B.; Chen, C.Y.; Uygungil, B.; Spoede, W.; Bernstein, J.A.; Huston, D.P.; Liu, Y.J. A novel subset of CD4$^+$ T$_H$2 memory/effector cells that produce inflammatory IL-17 cytokine and promote the exacerbation of chronic allergic asthma. *J. Exp. Med.* **2010**, *207*, 2479–2491. [CrossRef] [PubMed]
119. Esplugues, E.; Huber, S.; Gagliani, N.; Hauser, A.E.; Town, T.; Wan, Y.Y.; O'Connor, W., Jr.; Rongvaux, A.; Van Rooijen, N.; Haberman, A.M.; et al. Control of T$_H$17 cells occurs in the small intestine. *Nature* **2011**, *475*, 514–518. [CrossRef] [PubMed]
120. Gagliani, N.; Amezcua Vesely, M.C.; Iseppon, A.; Brockmann, L.; Xu, H.; Palm, N.W.; de Zoete, M.R.; Licona-Limon, P.; Paiva, R.S.; Ching, T.; et al. T$_H$17 cells transdifferentiate into regulatory T cells during resolution of inflammation. *Nature* **2015**, *523*, 221–225. [CrossRef] [PubMed]
121. Dumoutier, L.; Lejeune, D.; Colau, D.; Renauld, J.C. Cloning and characterization of IL-22 binding protein, a natural antagonist of IL-10-related T cell-derived inducible factor/IL-22. *J. Immunol.* **2001**, *166*, 7090–7095. [CrossRef] [PubMed]
122. Kotenko, S.V.; Izotova, L.S.; Mirochnitchenko, O.V.; Esterova, E.; Dickensheets, H.; Donnelly, R.P.; Pestka, S. Identification, cloning, and characterization of a novel soluble receptor that binds IL-22 and neutralizes its activity. *J. Immunol.* **2001**, *166*, 7096–7103. [CrossRef] [PubMed]
123. Wei, C.C.; Ho, T.W.; Liang, W.G.; Chen, G.Y.; Chang, M.S. Cloning and characterization of mouse IL-22 binding protein. *Genes Immun.* **2003**, *4*, 204–211. [CrossRef] [PubMed]
124. Weber, G.F.; Schlautkotter, S.; Kaiser-Moore, S.; Altmayr, F.; Holzmann, B.; Weighardt, H. Inhibition of interleukin-22 attenuates bacterial load and organ failure during acute polymicrobial sepsis. *Infect. Immun.* **2007**, *75*, 1690–1697. [CrossRef] [PubMed]
125. Pelczar, P.; Witkowski, M.; Perez, L.G.; Kempski, J.; Hammel, A.G.; Brockmann, L.; Kleinschmidt, D.; Wende, S.; Haueis, C.; Bedke, T.; et al. A pathogenic role for T cell-derived IL-22BP in inflammatory bowel disease. *Science* **2016**, *354*, 358–362. [CrossRef] [PubMed]

© 2017 by the authors. Licensee MDPI, Basel, Switzerland. This article is an open access article distributed under the terms and conditions of the Creative Commons Attribution (CC BY) license (http://creativecommons.org/licenses/by/4.0/).

International Journal of
Molecular Sciences

MDPI

Review

Implications of Extracellular Matrix Production by Adipose Tissue-Derived Stem Cells for Development of Wound Healing Therapies

Kathrine Hyldig, Simone Riis, Cristian Pablo Pennisi, Vladimir Zachar and Trine Fink *

Laboratory for Stem Cell Research, Department of Health Science and Technology, Aalborg University,
9220 Aalborg, Denmark; kha12@student.aau.dk (K.H.); sriis@hst.aau.dk (S.R.); cpennisi@hst.aau.dk (C.P.P.);
vlaz@hst.aau.dk (V.Z.)
* Correspondence: trinef@hst.aau.dk; Tel.: +45-9940-7550

Academic Editor: Allison Cowin
Received: 3 April 2017; Accepted: 26 May 2017; Published: 31 May 2017

Abstract: The synthesis and deposition of extracellular matrix (ECM) plays an important role in the healing of acute and chronic wounds. Consequently, the use of ECM as treatment for chronic wounds has been of special interest—both in terms of inducing ECM production by resident cells and applying ex vivo produced ECM. For these purposes, using adipose tissue-derived stem cells (ASCs) could be of use. ASCs are recognized to promote wound healing of otherwise chronic wounds, possibly through the reduction of inflammation, induction of angiogenesis, and promotion of fibroblast and keratinocyte growth. However, little is known regarding the importance of ASC-produced ECM for wound healing. In this review, we describe the importance of ECM for wound healing, and how ECM production by ASCs may be exploited in developing new therapies for the treatment of chronic wounds.

Keywords: adipose stem cells; ASCs; extracellular matrix; wound healing

1. Introduction

Wound healing is a dynamic and well-orchestrated process with both molecular and cellular events. When for some reason the wound healing process is perturbed, the wounds may become chronic, with concomitant alterations in the microenvironment leading to prolonged inflammation, ischemia, dysfunctional extracellular matrix (ECM), and lack of re-epithelialization [1]. Traditional wound healing therapies are often not sufficient, so there is considerable interest in developing novel more efficient therapies. Among the novel strategies that are being explored, the use of adipose tissue-derived stem/stromal cells (ASCs) appears to be very promising, judging from animal studies [2] and early clinical studies [3]. The ASCs are derived from the so-called stromal vascular fraction of adipose tissue [4,5], which is a rather heterogeneous population. However, after expansion for just a few passages, the ASCs converge towards a common phenotype comprised of fewer, perhaps functionally distinct subtypes [6,7].

While it is still not clear how the ASCs mediate their effect, they have been shown to have immunomodulatory and proangiogenic properties, the ability to promote keratinocyte and fibroblast growth, as well as ability to reduce tissue scarring [8–13]. However, less is known about the putative effect of ASCs on the ECM of the chronic wounds. Consequently, in this review we will outline the role of ECM in wound healing, describe what is known regarding ASCs' effect on ECM, and speculate on how ASC-derived ECM may be exploited in novel wound healing therapies.

2. The ECM of the Skin

In human skin, the ECM contains both fibrous proteins and ground substance. The fibrous proteins comprise collagens, elastin, and fibronectin, and provide a three-dimensional scaffold upon which both individual cells and the vascular network are supported or anchored. The most abundant fibrous protein in the human skin is collagen I, with collagen III and collagen V representing only minor proportions of the total collagen [14]. During pathological conditions such as scar formation, the composition and structure of collagen fibers are altered [15]. The ground substance of the ECM contains proteoglycans and glycosaminoglycans, and surround the fibrous proteins as a jelly-like substance which provides hydration to the skin due to the strong hydrophilic characteristics.

Initially, ECM was thought to function only as structural support for the cells; however, it has become clear that the ECM plays a pivotal role in the regulation of cell behavior both under normal conditions and during wound healing [16,17]. The ECM regulates cell behavior through molecular signaling primarily mediated by integrins (a family of cell surface receptors), and it has been shown that these signals are involved in determining whether the cells proliferate, differentiate, or undergo apoptosis [18]. Among the resident skin cells that express integrins—and thus may be subjected to modulation by the ECM—are fibroblasts and keratinocytes [19]. In addition, proteins in the ECM modulate the activity of growth factors and cytokines such as platelet-derived growth factor (PDGF) and transforming growth factor-β (TGF-β), produced by activated platelets and macrophages, respectively [20,21]. Thus, the ECM functions as a reservoir by protecting the growth factors from degradation and controlling their release [22].

ECM homeostasis is partly controlled by the activity of matrix metalloproteinases (MMPs) and their counterpart, tissue inhibitors of metalloproteinases (TIMPs). The MMPs are mainly secreted by keratinocytes, fibroblasts, and endothelial cells [23], and TIMPS by—among others—mesenchymal stem cells (MSCs), keratinocytes, and fibroblasts [24,25]. Thus, the balance between MMPs and TIMPs is important for ECM remodeling, cell signaling, and cell migration [26], and it has been suggested that a high MMP/TIMP ratio could be a biomarker of non-healing wounds [27].

3. Role of ECM for Wound Healing

Acute wounds normally heal in four overlapping phases: hemostasis, inflammation, proliferation, and remodeling (Figure 1) [17,28,29]. Hemostasis occurs immediately after the injury, and is characterized by the activation and aggregation of platelets into the wounded area followed by the deposition of fibronectin and fibrin from the blood plasma. The activated platelets help initiate the inflammatory phase through the secretion of PDGF, which is important for the migration of macrophages and neutrophils to the wounded area [20], and TGF-β, which plays a major role in the transformation of monocytes to macrophages [21]. The stimulation of macrophages results in the development of polarized phenotypes termed classically activated (M1) macrophages that secrete pro-inflammatory cytokines and predominate during early wound healing and alternatively activated (M2) macrophages that are associated with a wound healing anti-inflammatory profile and which predominate in the later stages when inflammation abates and tissue undergoes remodeling [30,31].

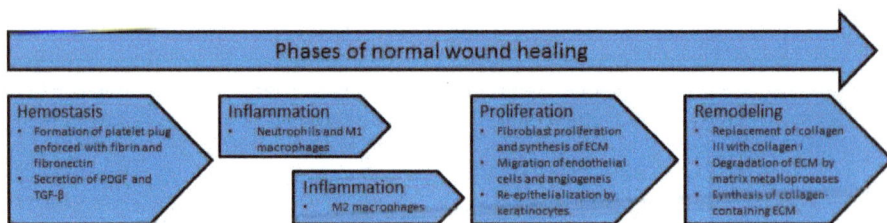

Figure 1. The phases of normal wound healing. Wound healing normally progresses through a tightly orchestrated process that is usually described as having four overlapping phases. During hemostasis, a platelet plug is formed and growth factors are secreted. The inflammatory phase has two stages. The initial stage, where neutrophils and pro-inflammatory M1 macrophages prevail, and a second stage characterized by the presence of anti-inflammatory M2 macrophages. During the proliferation phase, fibroblasts proliferate and synthesize extracellular matrix (ECM), new vessels are formed, and keratinocytes re-epithelialize the surface of the wound. In the final remodeling stage, the composition of the ECM is altered through degradation and resynthesis. PDGF: platelet-derived growth factor; TGF-β: transforming growth factor-β.

During the proliferation phase of wound healing, fibroblasts migrate to the wounded area where they proliferate and initiate ECM synthesis [32]. The temporary matrix of fibrin and fibronectin is replaced by the collagen matrix, enriched in proteoglycans, glycosaminoglycans, and glycoproteins, forming a granulation tissue. Subsequently, the abundant extracellular matrix accumulates, supporting cell migration. In response to the newly-synthesized ECM, endothelial cells migrate into the wound and initiate the process of angiogenesis to restore the circulation in the damaged area [33]. The wound environment is characterized by low oxygen supply, regulating the process of angiogenesis through hypoxia-inducible factor-1 (HIF-1) [34]. Additionally, the secreted growth factors basic fibroblast growth factor (bFGF), TGF-β, and vascular endothelial growth factor (VEGF) stimulate the angiogenic activity [35]. Concurrently, keratinocytes migrate from the basement membrane towards the wound edge and close the wound. The migration of keratinocytes is dependent on basement membrane degradation, facilitated by MMPs [36].

In the remodeling phase, fibroblasts transform into myofibroblasts and contract the wound area [37]. Remodeling of the granulation tissue is characterized by the synthesis and breakdown of collagen, regulated by the MMPs and TIMPs [38].

When the normal progression through the different phases of wound healing is perturbed as described above, the wounds may become chronic. It appears that non-healing wounds remain in a transition state between the inflammation and proliferation phases and proliferative and remodeling phases become impaired [39] (Figure 2, left panel). It is not clear what causes the prolonged inflammation; however, macrophages in chronic wounds fail to switch from the pro-inflammatory M1 to the anti-inflammatory M2 phenotype [40].

Furthermore, in mouse models of wound healing, there was a correlation between the presence of M2 macrophages, the resolution of inflammation, and wound healing, suggesting an important role of the polarization from M1 to M2 macrophages during the process of wound healing [41,42]. Interestingly, a switch in phenotype towards a more anti-inflammatory or pro-healing type has also been documented for Th1/Th2 cells and MSCs [43,44], which are possibly recruited to the site of injury from the bone marrow [45].

As wounds become chronic, the ECM homeostasis of the wound area is affected. Indeed, chronic wound fibroblasts are unresponsive to the stimulatory effect of TGF-β on collagen synthesis when compared to normal skin fibroblasts [46]. In addition, proteolytic enzymes involved in ECM degradation are dysregulated in chronic wounds, with increased expression of MMP-1, MMP-2, MMP-3, MMP-8, and MMP-9 [47,48] and decreased expression of the MMP inhibitor TIMP-2, leading to excessive proteolysis of the ECM [48]. As the balance between ECM synthesis and degradation is impaired, the ECM becomes dysfunctional in terms of supporting cell migration and proliferation as well as angiogenesis [49].

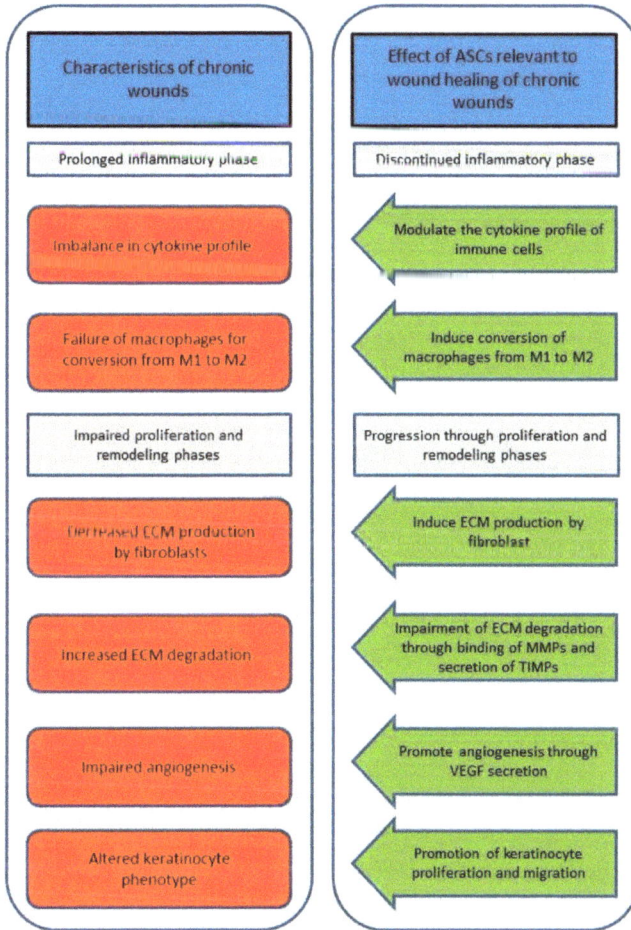

Figure 2. Characteristics of chronic wounds and the relevant regenerative effects of ASCs on these. Chronic wounds appear to unable to progress from the inflammatory phase of normal wound healing and to have impaired proliferation and remodeling phases (**left panel**). The ASCs have several regenerative characteristics that may lead to the wound progression from the inflammatory phase and through the proliferation and remodeling phases (**right panel**). ASC: adipose tissue-derived stem/stromal cells; MMP: matrix metalloproteinase; TIMP: tissue inhibitor of metalloproteinase; VEGF: vascular endothelial growth factor.

4. Using Adipose Stem Cells to Treat Chronic Wounds

The conventional treatment strategy for wound healing is based on wound bed preparation using tissue debridement, antibiotics, anti-inflammatory drugs, the restoration of moisture balance, and/or acceleration of epithelization by growth factor therapy [50,51]. Although these treatment options accelerate the wound healing process in many cases, many wounds are resistant to the current treatment options and more efficient methods are needed [49].

Recently, stem cell therapy has emerged as a novel approach for chronic wound healing. So far, most data is from studies using bone marrow-derived MSCs (BM-MSCs). However, as ASCs and BM-MSCs share numerous biological properties, much of the knowledge regarding BM-MSCs can be directly applied to the ASCs [52]. It has also become apparent the vastly higher numbers of ASCs than BM-MSCs can be obtained in a short time frame [52]. Thus, as procedures for the isolation and expansion of ASCs for clinical use have been optimized [53], ASCs are emerging as the most promising candidate for stem cell-based therapies for chronic wounds.

In the chronic wound environment, in vitro and in vivo studies suggest that the ASCs may be able to discontinue the prolonged inflammation phase and restore the progression through the phases of proliferation and remodeling (Figure 2, right panel). In terms of effects on the inflammatory processes, it is well known that ASCs may influence the functional characteristics and cytokine profile of T-, B-, and dendritic cells [54–56]. Notably, ASCs have also been shown to be able to induce a conversion of the macrophage phenotype from the pro-inflammatory M1 associated with chronic wounds to the anti-inflammatory and wound healing M2 phenotype [57,58]. During the proliferation phase, secreted factors from ASCs enhance several fibroblast characteristics, such as cell proliferation, migration and, importantly, the synthesis of collagen and other ECM proteins [59–61]. Furthermore, ASCs have been demonstrated to inhibit ECM degradation through the increased binding of MMPs and secretion of TIMPs [24]. The ability of ASCs to promote new vessel growth is also relevant to wound healing [62]. Finally, in vitro studies suggest that ASCs may promote re-epithelialization through modulation of keratinocytes in terms of promoting their proliferation and migration, but more studies are needed to confirm if this also holds true for chronic wounds [11,63].

To potentiate the wound healing effects of ASCs, the possibility of pre-conditioning the cells during in vitro expansion prior to clinical use has been suggested. In particular, the use of hypoxic culture appears interesting, as several of the wound healing properties of ASCs appear to be enhanced [64,65]. Significantly, it was recently found that hypoxic culture of ASCs altered their expression profile of several proteins related to ECM structure and function [66]. However, more data is needed to determine if the hypoxic potentiation of the regenerative properties of the ASCs in vitro can be translated into an enhanced effect in vivo.

5. ECM-Based Scaffolds for Wound Therapy

An alternative approach to using cells for wound therapy is to use acellular ECM. Acellular ECM-based scaffolds derived from natural tissues have been successfully applied in various preclinical and clinical settings for the treatment of severe wounds. These natural scaffolds appear to mediate tissue regeneration through a process known as constructive remodeling, in which the diverse ECM components orchestrate a process of scarless tissue repair [67]. There are various commercially available ECM-derived materials that are routinely used for the treatment of burns and chronic wounds, including materials obtained by the decellularization of animal tissues, such as porcine or bovine skin [68,69], or from allogeneic human skin [70]. A more detailed review of the variety of decellularized ECM scaffolds that are currently available for clinical use can be found in the literature [71]. Despite the relatively high success rates associated with these materials, some issues may still appear, such as sustained inflammatory responses and incomplete healing due to poor integrity of the native ECM molecules after decellularization [72]. In addition, xenogeneic ECM components may cause adverse host immune responses, and there is a risk of pathogen transfer [73]. To avoid these risks and the limitations associated with the supply of allogeneic human tissues, cell cultures have recently emerged as viable alternatives for the fabrication of ECM scaffolds. Depending on the cell type used for ECM synthesis, it is possible to fabricate ECM scaffolds containing specific proteins and morphogens that appear during early tissue development and which are associated with enhanced wound healing [74]. In particular, matrices derived from stem cells have shown promise as scaffolds for various tissue engineering and regenerative medicine applications, including regeneration of cartilage [75], bone [76], and neural tissue [77]. Surprisingly, despite the beneficial

Int. J. Mol. Sci. **2017**, *18*, 1167

properties of BM-MSCs or ASCs in the context of wound healing therapies, little is known regarding the use of stem cell ECM for wound healing applications. In this context, MSCs may possess a relative advantage over terminally differentiated skin fibroblasts, as they have shown an increased capacity to synthesize proteins involved in extracellular matrix, morphogenesis, and development [78,79]. The predominant upregulation of genes such as fibronectin (*FN1*) and extracellular matrix protein 2 (*ECM2*) found in MSCs suggests that the ECM derived from these cells may enhance wound healing by promoting matrix deposition and cell adhesion [78]. Accordingly, comprehensive proteomic analysis of ECM derived from MSCs has revealed an enrichment of structural proteins, including collagen I, VI, and XII, which together with an increased presence of MMPs indicates a highly dynamic matrix turnover [79]. Furthermore, MSC-derived ECM is also enriched in proteoglycans such as perlecan and hyaluronan, and glycoproteins such as fibronectin, tenascin-C, fibulin-1, and thrombospondin-1 [79]. Overall, these components of the ECM may contribute to the different phases of wound healing by supporting integrin-mediated cell adhesion and signaling, cell migration, and proliferation. In addition, decellularized stem cell ECM has demonstrated a significant angiogenic potential, which has been evidenced through the activation of endothelial cells [80]. An additional advantage of using stem cell cultures is the possibility of microenvironmental preconditioning of the cells during the fabrication process to tailor specific biological or biophysical functionalities in the scaffold that may promote wound healing [81]. In ASCs, in vitro ECM production and assembly has been shown to be controlled by mechanical and topographical cues from the microenvironment [82,83].

Decellularized ECM-scaffolds may be also used as platforms for cell delivery. It has been hypothesized that ASCs might have a better survival rate and reduce scar formation when administered in combination with ECM-components [84]. Such a co-delivery could be implemented either using a patch of ECM seeded with ASCs [84] or delivering the ASCs in a fibrin spray glue.

In summary, although fabrication of ECM scaffolds using ASC cultures or co-delivery of ASCs and ECM represent novel concepts that may offer several comparative advantages for wound healing applications, the knowledge in this field is still scarce, and more efforts are needed to further develop these approaches into a clinical reality.

Acknowledgments: This work has been supported by a grant by the Obelske Family Foundation to Trine Fink.

Author Contributions: All authors have contributed to the concept, draft and final revision of the paper.

Conflicts of Interest: The authors declare no conflict of interest. The founding sponsors had no role in the writing of the manuscript, and in the decision to publish this review.

Abbreviations

ASC	Adipose stem cell/adipose stromal cell
bFGF	Basic fibroblast growth factor
BM-MSC	Bone marrow-derived mesenchymal stem cell
ECM	Extracellular matrix
HIF-1	Hypoxia-inducible factor 1
MSC	Mesenchymal stem cell
MMP	Matrix metalloprotease
PDGF	Platelet-derived growth factor
TGF-β	Transforming growth factor-beta
TIMP	Tissue inhibitor of metalloproteases
VEGF	Vascular endothelial growth factor

References

1. Eming, S.A.; Martin, P.; Tomic-Canic, M. Wound repair and regeneration: Mechanisms, signaling, and translation. *Sci. Transl. Med.* **2014**, *6*, 265sr6. [CrossRef] [PubMed]

2. Huang, S.-P.; Huang, C.-H.; Shyu, J.-F.; Lee, H.-S.; Chen, S.-G.; Chan, J.Y.-H.; Huang, S.-M. Promotion of wound healing using adipose-derived stem cells in radiation ulcer of a rat model. *J. Biomed. Sci.* **2013**, *20*, 51. [CrossRef] [PubMed]

3. Cerqueira, M.T.; Pirraco, R.P.; Marques, A.P. Stem Cells in Skin Wound Healing: Are We There Yet? *Adv. Wound Care* **2016**, *5*, 164–175. [CrossRef] [PubMed]

4. Zuk, P.A.; Zhu, M.; Mizuno, H.; Huang, J.; Futrell, J.W.; Katz, A.J.; Benhaim, P.; Lorenz, H.P.; Hedrick, M.H. Multilineage cells from human adipose tissue: Implications for cell-based therapies. *Tissue Eng.* **2001**, *7*, 211–228. [CrossRef] [PubMed]

5. Zachar, V.; Rasmussen, J.G.; Fink, T. Isolation and growth of adipose tissue-derived stem cells. *Methods Mol. Biol.* **2011**, *698*, 37–49. [PubMed]

6. Riis, S.; Nielsen, F.M.; Pennisi, C.P.; Zachar, V.; Fink, T. Comparative Analysis of Media and Supplements on Initiation and Expansion of Adipose-Derived Stem Cells. *Stem Cells Transl. Med.* **2016**, *5*, 314–324. [CrossRef] [PubMed]

7. Nielsen, F.M.; Riis, S.E.; Andersen, J.I.; Lesage, R.; Fink, T.; Pennisi, C.P.; Zachar, V. Discrete adipose-derived stem cell subpopulations may display differential functionality after in vitro expansion despite convergence to a common phenotype distribution. *Stem Cell Res. Ther.* **2016**, *7*, 177. [CrossRef] [PubMed]

8. Mattar, P.; Bieback, K. Comparing the Immunomodulatory Properties of Bone Marrow, Adipose Tissue, and Birth-Associated Tissue Mesenchymal Stromal Cells. *Front. Immunol.* **2015**, *6*, 560. [CrossRef] [PubMed]

9. Rasmussen, J.G.; Frøbert, O.; Holst-Hansen, C.; Kastrup, J.; Baandrup, U.; Zachar, V.; Fink, T.; Simonsen, U. Comparison of human adipose-derived stem cells and bone marrow-derived stem cells in a myocardial infarction model. *Cell Transplant.* **2014**, *23*, 195–206. [CrossRef] [PubMed]

10. Rasmussen, J.G.; Riis, S.E.; Frøbert, O.; Yang, S.; Kastrup, J.; Zachar, V.; Simonsen, U.; Fink, T. Activation of Protease-Activated Receptor 2 Induces VEGF Independently of HIF-1. *PLoS ONE* **2012**, *7*, e46087. [CrossRef] [PubMed]

11. Riis, S.; Newman, R.; Ipek, H.; Andersen, J.I.; Kuninger, D.; Boucher, S.; Vemuri, M.C.; Pennisi, C.P.P.; Zachar, V.; Fink, T. Hypoxia enhances the wound-healing potential of adipose-derived stem cells in a novel human primary keratinocyte-based scratch assay. *Int. J. Mol. Med.* **2017**, *39*, 587–594. [CrossRef] [PubMed]

12. Shafy, A.; Fink, T.; Zachar, V.; Lila, N.; Carpentier, A.; Chachques, J.C. Development of cardiac support bioprostheses for ventricular restoration and myocardial regeneration. *Eur. J. Cardiothorac. Surg.* **2013**, *43*, 1211–1219. [CrossRef] [PubMed]

13. Lee, E.Y.; Xia, Y.; Kim, W.-S.; Lila, N.; Carpentier, A.; Chachques, J.C. Hypoxia-enhanced wound-healing function of adipose-derived stem cells: Increase in stem cell proliferation and up-regulation of VEGF and bFGF. *Wound Repair Regen.* **2009**, *17*, 540–547. [CrossRef] [PubMed]

14. Smith, L.T.; Holbrook, K.A.; Madri, J.A. Collagen types I, III, and V in human embryonic and fetal skin. *Am. J. Anat.* **1986**, *175*, 507–521. [CrossRef] [PubMed]

15. Verhaegen, P.D.H.M.; van Zuijlen, P.P.M.; Pennings, N.M.; van Marle, J.; Niessen, F.B.; van der Horst, C.M.A.M.; Middelkoop, E. Differences in collagen architecture between keloid, hypertrophic scar, normotrophic scar, and normal skin: An objective histopathological analysis. *Wound Repair Regen.* **2009**, *17*, 649–656. [CrossRef] [PubMed]

16. Eckes, B.; Nischt, R.; Krieg, T. Cell-matrix interactions in dermal repair and scarring. *Fibrogenesis Tissue Repair* **2010**, *3*, 4. [CrossRef] [PubMed]

17. Hodde, J.P.; Johnson, C.E. Extracellular matrix as a strategy for treating chronic wounds. *Am. J. Clin. Dermatol.* **2007**, *8*, 61–66. [CrossRef] [PubMed]

18. Giancotti, F.G.; Ruoslahti, E. Integrin Signaling. *Science* **1999**, *285*, 1028–1033. [CrossRef] [PubMed]

19. Koivisto, L.; Heino, J.; Häkkinen, L.; Larjava, H. Integrins in Wound Healing. *Adv. Wound Care* **2014**, *3*, 762–783. [CrossRef] [PubMed]

20. Lynch, S.E.; Nixon, J.C.; Colvin, R.B.; Antoniades, H.N. Role of platelet-derived growth factor in wound healing: Synergistic effects with other growth factors. *Proc. Natl. Acad. Sci. USA* **1987**, *84*, 7696–7700. [CrossRef] [PubMed]

21. Barrientos, S.; Stojadinovic, O.; Golinko, M.S.; Brem, H.; Tomic-Canic, M. PERSPECTIVE ARTICLE: Growth factors and cytokines in wound healing. *Wound Repair Regen.* **2008**, *16*, 585–601. [CrossRef] [PubMed]

22. Schultz, G.S.; Wysocki, A. Interactions between extracellular matrix and growth factors in wound healing. *Wound Repair Regen.* **2009**, *17*, 153–162. [CrossRef] [PubMed]

23. Martins, V.L.; Caley, M.; O'Toole, E.A. Matrix metalloproteinases and epidermal wound repair. *Cell Tissue Res.* **2013**, *351*, 255–268. [CrossRef] [PubMed]

24. Lozito, T.P.; Jackson, W.M.; Nesti, L.J.; Tuan, R.S. Human mesenchymal stem cells generate a distinct pericellular zone of MMP activities via binding of MMPs and secretion of high levels of TIMPs. *Matrix Biol.* **2014**, *34*, 132–143. [CrossRef] [PubMed]

25. Tandara, A.A.; Mustoe, T.A. MMP- and TIMP-secretion by human cutaneous keratinocytes and fibroblasts—Impact of coculture and hydration. *J. Plast. Reconstr. Aesthetic. Surg.* **2011**, *64*, 108–116. [CrossRef] [PubMed]

26. Nagase, H.; Visse, R.; Murphy, G. Structure and function of matrix metalloproteinases and TIMPs. *Cardiovasc. Res.* **2006**, *69*, 562–573. [CrossRef] [PubMed]

27. Patel, S.; Maheshwari, A.; Chandra, A. Biomarkers for wound healing and their evaluation. *J. Wound Care* **2016**, *25*, 46–55. [CrossRef] [PubMed]

28. Olczyk, P.; Mencner, Ł.; Komosinska-Vassev, K. The role of the extracellular matrix components in cutaneous wound healing. *BioMed Res. Int.* **2014**, *2014*, 747584. [CrossRef] [PubMed]

29. Hassan, W.U.; Greiser, U.; Wang, W. Role of adipose derived stem cells in wound healing. *Wound Repair Regen.* **2014**, *22*, 313–325. [CrossRef] [PubMed]

30. Martinez, F.O.; Sica, A.; Mantovani, A.; Locati, M. Macrophage activation and polarization. *Front. Biosci.* **2008**, *13*, 453–461. [CrossRef] [PubMed]

31. Ferrante, C.J.; Leibovich, S.J. Regulation of Macrophage Polarization and Wound Healing. *Adv. Wound Care* **2012**, *1*, 10–16. [CrossRef] [PubMed]

32. Pierce, G.F.; Mustoe, T.A.; Lingelbach, J.; Masakowski, V.R.; Griffin, G.L.; Senior, R.M.; Deuel, T.F. Platelet-derived growth factor and transforming growth factor-beta enhance tissue repair activities by unique mechanisms. *J. Cell Biol.* **1989**, *109*, 429–440. [CrossRef] [PubMed]

33. Vorotnikova, E.; McIntosh, D.; Dewilde, A.; Zhang, J.; Reing, J.E.; Zhang, L.; Cordero, K.; Bedelbaeva, K.; Gourevitch, D.; Heber-Katz, E.; et al. Extracellular matrix-derived products modulate endothelial and progenitor cell migration and proliferation in vitro and stimulate regenerative healing in vivo. *Matrix Biol.* **2010**, *29*, 690–700. [CrossRef] [PubMed]

34. Liu, Y.; Cox, S.R.; Morita, T.; Kourembanas, S. Hypoxia regulates vascular endothelial growth factor gene expression in endothelial cells. Identification of a 5′ enhancer. *Circ. Res.* **1995**, *77*, 638–643. [CrossRef] [PubMed]

35. Ucuzian, A.A.; Gassman, A.A.; East, A.T.; Greisler, H.P. Molecular mediators of angiogenesis. *J. Burn Care Res.* **2010**, *31*, 158–175. [CrossRef] [PubMed]

36. Caley, M.P.; Martins, V.L.C.; O'Toole, E.A. Metalloproteinases and Wound Healing. *Adv. Wound Care* **2015**, *4*, 225–234. [CrossRef] [PubMed]

37. Wu, M.; Ben Amar, M. Growth and remodelling for profound circular wounds in skin. *Biomech. Model. Mechanobiol.* **2015**, *14*, 357–370. [CrossRef] [PubMed]

38. Saarialho-Kere, U. Patterns of matrix metalloproteinase and TIMP expression in chronic ulcers. *Arch. Dermatol. Res.* **1998**, *290*, S47–S54. [CrossRef] [PubMed]

39. Loots, M.A.M.; Lamme, E.N.; Zeegelaar, J.; Mekkes, J.R.; Bos, J.D.; Middelkoop, E. Differences in Cellular Infiltrate and Extracellular Matrix of Chronic Diabetic and Venous Ulcers Versus Acute Wounds. *J. Investig. Dermatol.* **1998**, *111*, 850–857. [CrossRef] [PubMed]

40. Sindrilaru, A.; Peters, T.; Wieschalka, S.; Baican, C.; Baican, A.; Peter, H.; Hainzl, A.; Schatz, S.; Qi, Y.; Schlecht, A.; et al. An unrestrained proinflammatory M1 macrophage population induced by iron impairs wound healing in humans and mice. *J. Clin. Investig.* **2011**, *121*, 985–997. [CrossRef] [PubMed]

41. Mirza, R.; Koh, T.J. Dysregulation of monocyte/macrophage phenotype in wounds of diabetic mice. *Cytokine* **2011**, *56*, 256–264. [CrossRef] [PubMed]

42. Lucas, T.; Waisman, A.; Ranjan, R.; Roes, J.; Krieg, T.; Müller, W.; Roers, A.; Eming, S.A. Differential Roles of Macrophages in Diverse Phases of Skin Repair. *J. Immunol.* **2010**, *184*, 3964–3977. [CrossRef] [PubMed]

43. Park, J.E.; Barbul, A. Understanding the role of immune regulation in wound healing. *Am. J. Surg.* **2004**, *187*, S11–S16. [CrossRef]

44. Waterman, R.S.; Tomchuck, S.L.; Henkle, S.L.; Betancourt, A.M. A New Mesenchymal Stem Cell (MSC) Paradigm: Polarization into a Pro-Inflammatory MSC1 or an Immunosuppressive MSC2 Phenotype. *PLoS ONE* **2010**, *5*, e10088. [CrossRef] [PubMed]

45. Seppanen, E.; Roy, E.; Ellis, R.; Bou-Gharios, G.; Fisk, N.M.; Khosrotehrani, K. Distant Mesenchymal Progenitors Contribute to Skin Wound Healing and Produce Collagen: Evidence from a Murine Fetal Microchimerism Model. *PLoS ONE* **2013**, *8*, e62662. [CrossRef] [PubMed]

46. Hasan, A.; Murata, H.; Falabella, A.; Ochoa, S.; Zhou, L.; Badiavas, E.; Falanga, V. Dermal fibroblasts from venous ulcers are unresponsive to the action of transforming growth factor-beta 1. *J. Dermatol. Sci.* **1997**, *16*, 59–66. [CrossRef]

47. Lobmann, R.; Ambrosch, A.; Schultz, G.; Waldmann, K.; Schiweck, S.; Lehnert, H. Expression of matrix-metalloproteinases and their inhibitors in the wounds of diabetic and non-diabetic patients. *Diabetologia* **2002**, *45*, 1011–1016. [CrossRef] [PubMed]

48. Subramaniam, K.; Pech, C.M.; Stacey, M.C.; Wallace, H.J. Induction of MMP-1, MMP-3 and TIMP-1 in normal dermal fibroblasts by chronic venous leg ulcer wound fluid. *Int. Wound J.* **2008**, *5*, 79–86. [CrossRef] [PubMed]

49. Demidova-Rice, T.N.; Hamblin, M.R.; Herman, I.M. Acute and impaired wound healing: Pathophysiology and current methods for drug delivery, part 1: Normal and chronic wounds: Biology, causes, and approaches to care. *Adv. Skin Wound Care* **2012**, *25*, 304–314. [CrossRef] [PubMed]

50. Schultz, G.S.; Sibbald, R.G.; Falanga, V.; Ayello, E.A.; Dowsett, C.; Harding, K.; Romanelli, M.; Stacey, M.C.; Teot, L.; Vanscheidt, W. Wound bed preparation: A systematic approach to wound management. *Wound Repair Regen.* **2003**, *11* (Suppl. S1), S1–S28. [CrossRef] [PubMed]

51. Ayello, E.A.; Dowsett, C.; Schultz, G.S.; Sibbald, R.G.; Falanga, V.; Harding, K.; Romanelli, M.; Stacey, M.; Teot, L.; Vanscheidt, W. TIME heals all wounds. *Nursing* **2004**, *34*, 36–42. [CrossRef] [PubMed]

52. Strioga, M.; Viswanathan, S.; Darinskas, A.; Slaby, O.; Michalek, J. Same or Not the Same? Comparison of Adipose Tissue-Derived versus Bone Marrow-Derived Mesenchymal Stem and Stromal Cells. *Stem Cells Dev.* **2012**, *21*, 2724–2752. [CrossRef] [PubMed]

53. Riis, S.; Zachar, V.; Boucher, S.; Vemuri, M.C.; Pennisi, C.P.; Fink, T. Critical steps in the isolation and expansion of adipose-derived stem cells for translational therapy. *Expert Rev. Mol. Med.* **2015**, *17*, e11. [CrossRef] [PubMed]

54. Baharlou, R.; Ahmadi-Vasmehjani, A.; Faraji, F.; Atashzar, M.R.; Khoubyari, M.; Ahi, S.; Erfanian, S.; Navabi, S.-S. Human adipose tissue-derived mesenchymal stem cells in rheumatoid arthritis: Regulatory effects on peripheral blood mononuclear cells activation. *Int. Immunopharmacol.* **2017**, *47*, 59–69. [CrossRef] [PubMed]

55. Anderson, P.; Gonzalez-Rey, E.; O'Valle, F.; Martin, F.; Oliver, F.J.; Delgado, M. Allogeneic Adipose-Derived Mesenchymal Stromal Cells Ameliorate Experimental Autoimmune Encephalomyelitis by Regulating Self-Reactive T Cell Responses and Dendritic Cell Function. *Stem Cells Int.* **2017**, *2017*, 1–15. [CrossRef] [PubMed]

56. Franquesa, M.; Mensah, F.K.; Huizinga, R.; Strini, T.; Boon, L.; Lombardo, E.; DelaRosa, O.; Laman, J.D.; Grinyó, J.M.; Weimar, W.; Betjes, M.G.H.; Baan, C.C.; Hoogduijn, M.J. Human Adipose Tissue-Derived Mesenchymal Stem Cells Abrogate Plasmablast Formation and Induce Regulatory B Cells Independently of T Helper Cells. *Stem Cells* **2015**, *33*, 880–891. [CrossRef] [PubMed]

57. Manning, C.N.; Martel, C.; Sakiyama-Elbert, S.E.; Silva, M.J.; Shah, S.; Gelberman, R.H.; Thomopoulos, S. Adipose-derived mesenchymal stromal cells modulate tendon fibroblast responses to macrophage-induced inflammation in vitro. *Stem Cell Res. Ther.* **2015**, *6*, 74. [CrossRef] [PubMed]

58. Lo Sicco, C.; Reverberi, D.; Balbi, C.; Ulivi, V.; Principi, E.; Pascucci, L.; Becherini, P.; Bosco, M.C.; Varesio, L.; Franzin, C.; Pozzobon, M.; Cancedda, R.; Tasso, R. Mesenchymal Stem Cell-Derived Extracellular Vesicles as Mediators of Anti-Inflammatory Effects: Endorsement of Macrophage Polarization. *Stem Cells Transl. Med.* **2017**, *6*, 1018–1028. [CrossRef] [PubMed]

59. Zhao, J.; Hu, L.; Liu, J.; Gong, N.; Chen, L. The effects of cytokines in adipose stem cell-conditioned medium on the migration and proliferation of skin fibroblasts in vitro. *Biomed. Res. Int.* **2013**, *2013*, 578479. [CrossRef] [PubMed]

60. Hu, L.; Wang, J.; Zhou, X.; Xiong, Z.; Zhao, J.; Yu, R.; Huang, F.; Zhang, H.; Chen, L. Exosomes derived from human adipose mensenchymal stem cells accelerates cutaneous wound healing via optimizing the characteristics of fibroblasts. *Sci. Rep.* **2016**, *6*, 32993. [CrossRef] [PubMed]

61. Na, Y.K.; Ban, J.-J.; Lee, M.; Im, W.; Kim, M. Wound healing potential of adipose tissue stem cell extract. *Biochem. Biophys. Res. Commun.* **2017**, *485*, 30–34. [CrossRef] [PubMed]

62. Rasmussen, J.G.; Frøbert, O.; Pilgaard, L.; Kastrup, J.; Simonsen, U.; Zachar, V.; Fink, T. Prolonged hypoxic culture and trypsinization increase the pro-angiogenic potential of human adipose tissue-derived stem cells. *Cytotherapy* **2011**, *13*, 318–328. [CrossRef] [PubMed]

63. Lee, S.H.; Jin, S.Y.; Song, J.S.; Seo, K.K.; Cho, K.H. Paracrine Effects of Adipose-Derived Stem Cells on Keratinocytes and Dermal Fibroblasts. *Ann. Dermatol.* **2012**, *24*, 136. [CrossRef] [PubMed]

64. Zachar, V.; Duroux, M.; Emmersen, J.; Rasmussen, J.G.; Pennisi, C.P.; Yang, S.; Fink, T. Hypoxia and adipose-derived stem cell-based tissue regeneration and engineering. *Expert Opin. Biol. Ther.* **2011**, *11*, 775–786. [CrossRef] [PubMed]

65. Choi, J.R.; Yong, K.W.; Wan Safwani, W.K.Z. Effect of hypoxia on human adipose-derived mesenchymal stem cells and its potential clinical applications. *Cell. Mol. Life Sci.* **2017**. [CrossRef] [PubMed]

66. Riis, S.; Stensballe, A.; Emmersen, J.; Pennisi, C.P.; Birkelund, S.; Zachar, V.; Fink, T. Mass spectrometry analysis of adipose-derived stem cells reveals a significant effect of hypoxia on pathways regulating extracellular matrix. *Stem Cell Res. Ther.* **2016**, *7*, 52. [CrossRef] [PubMed]

67. Brown, B.N.; Badylak, S.F. Extracellular matrix as an inductive scaffold for functional tissue reconstruction. *Transl. Res.* **2014**, *163*, 268–285. [CrossRef] [PubMed]

68. Feng, X.; Shen, R.; Tan, J.; Chen, X.; Pan, Y.; Ruan, S.; Zhang, F.; Lin, Z.; Zeng, Y.; Wang, X.; Lin, Y.; Wu, Q. The study of inhibiting systematic inflammatory response syndrome by applying xenogenic (porcine) acellular dermal matrix on second-degree burns. *Burns* **2007**, *33*, 477–479. [CrossRef] [PubMed]

69. Brigido, S.A.; Boc, S.F.; Lopez, R.C. Effective management of major lower extremity wounds using an acellular regenerative tissue matrix: A pilot study. *Orthopedics* **2004**, *27*, s145–s149. [PubMed]

70. Yonehiro, L.; Burleson, G.; Sauer, V. Use of a new acellular dermal matrix for treatment of nonhealing wounds in the lower extremities of patients with diabetes. *Wounds Compend. Clin. Res. Pract.* **2013**, *25*, 340–344.

71. Parmaksiz, M.; Dogan, A.; Odabas, S.; Elçin, A.E.; Elçin, Y.M. Clinical applications of decellularized extracellular matrices for tissue engineering and regenerative medicine. *Biomed. Mater.* **2016**, *11*, 22003. [CrossRef] [PubMed]

72. Sun, W.Q.; Xu, H.; Sandor, M.; Lombardi, J. Process-induced extracellular matrix alterations affect the mechanisms of soft tissue repair and regeneration. *J. Tissue Eng.* **2013**, *4*, 204173141350530. [CrossRef] [PubMed]

73. Scobie, L.; Padler-Karavani, V.; Le Bas-Bernardet, S.; Crossan, C.; Blaha, J.; Matouskova, M.; Hector, R.D.; Cozzi, E.; Vanhove, B.; Charreau, B.; et al. Long-Term IgG Response to Porcine Neu5Gc Antigens without Transmission of PERV in Burn Patients Treated with Porcine Skin Xenografts. *J. Immunol.* **2013**, *191*, 2907–2915. [CrossRef] [PubMed]

74. Fitzpatrick, L.E.; McDevitt, T.C. Cell-derived matrices for tissue engineering and regenerative medicine applications. *Biomater. Sci.* **2015**, *3*, 12–24. [CrossRef] [PubMed]

75. Lu, H.; Hoshiba, T.; Kawazoe, N.; Koda, I.; Song, M.; Chen, G. Cultured cell-derived extracellular matrix scaffolds for tissue engineering. *Biomaterials* **2011**, *32*, 9658–9666. [CrossRef] [PubMed]

76. Zhang, Z.; Luo, X.; Xu, H.; Wang, L.; Jin, X.; Chen, R.; Ren, X.; Lu, Y.; Fu, M.; Huand, Y. Bone marrow stromal cell-derived extracellular matrix promotes osteogenesis of adipose-derived stem cells. *Cell Biol. Int.* **2015**, *39*, 291–299. [CrossRef] [PubMed]

77. Aizman, I.; Tate, C.C.; McGrogan, M.; Case, C.C. Extracellular matrix produced by bone marrow stromal cells and by their derivative, SB623 cells, supports neural cell growth. *J. Neurosci. Res.* **2009**, *87*, 3198–3206. [CrossRef] [PubMed]

78. Wagner, W.; Wein, F.; Seckinger, A.; Frankhauser, M.; Wirkner, U.; Krause, U.; Blake, J.; Schwager, C.; Eckstein, V.; Ansorge, W.; et al. Comparative characteristics of mesenchymal stem cells from human bone marrow, adipose tissue, and umbilical cord blood. *Exp. Hematol.* **2005**, *33*, 1402–1416. [CrossRef] [PubMed]

79. Ragelle, H.; Naba, A.; Larson, B.L.; Zhou, F.; Prijić, M.; Whittaker, C.A.; Del Rosario, A.; Langer, R.; Hynes, R.O. Comprehensive proteomic characterization of stem cell-derived extracellular matrices. *Biomaterials* **2017**, *128*, 147–159. [CrossRef] [PubMed]

80. Burns, J.S.; Kristiansen, M.; Kristensen, L.P.; Larsen, K.H.; Nielsen, M.O.; Christiansen, H.; Nehlin, J.; Andsern, J.S.; Kassem, M. Decellularized Matrix from Tumorigenic Human Mesenchymal Stem Cells Promotes Neovascularization with Galectin-1 Dependent Endothelial Interaction. *PLoS ONE* **2011**, *6*, e21888. [CrossRef] [PubMed]

81. Grenier, G.; Rémy-Zolghadri, M.; Larouche, D.; Gauvin, R.; Baker, K.; Bergeron, F.; Dupuis, D.; Langelier, E.; Rancourt, D.; Auger, F.A.; et al. Tissue reorganization in response to mechanical load increases functionality. *Tissue Eng.* **2005**, *11*, 90–100. [CrossRef] [PubMed]
82. Colazzo, F.; Sarathchandra, P.; Smolenski, R.T.; Chester, A.H.; Tseng, Y.T.; Czernuszka, J.T.; Yacoub, M.H.; Taylor, P.M. Extracellular matrix production by adipose-derived stem cells: Implications for heart valve tissue engineering. *Biomaterials* **2011**, *32*, 119–127. [CrossRef] [PubMed]
83. Foldberg, S.; Petersen, M.; Fojan, P.; Gurevich, L.; Fink, T.; Pennisi, C.P.; Zachar, V. Patterned poly(lactic acid) films support growth and spontaneous multilineage gene expression of adipose-derived stem cells. *Colloids Surf. B Biointerfaces* **2012**, *93*, 92–99. [CrossRef] [PubMed]
84. Lam, M.T.; Nauta, A.; Meyer, N.P.; Wu, J.C.; Longaker, M.T. Effective Delivery of Stem Cells Using an Extracellular Matrix Patch Results in Increased Cell Survival and Proliferation and Reduced Scarring in Skin Wound Healing. *Tissue Eng. Part A* **2013**, *19*, 738–747. [CrossRef] [PubMed]

© 2017 by the authors. Licensee MDPI, Basel, Switzerland. This article is an open access article distributed under the terms and conditions of the Creative Commons Attribution (CC BY) license (http://creativecommons.org/licenses/by/4.0/).

MDPI AG

St. Alban-Anlage 66

4052 Basel, Switzerland

Tel. +41 61 683 77 34

Fax +41 61 302 89 18

http://www.mdpi.com

International Journal of Molecular Sciences Editorial Office

E-mail: ijms@mdpi.com

http://www.mdpi.com/journal/ijms

www.ingramcontent.com/pod-product-compliance
Lightning Source LLC
Chambersburg PA
CBHW051718210326
41597CB00032B/5519